Springer Series in Operations Research and Financial Engineering

The Springer Series in Operations Research and Financial Engineering publishes monographs and textbooks on important topics in theory and practice of Operations Research, Management Science, and Financial Engineering. The Series is distinguished by high standards in content and exposition, and special attention to timely or emerging practice in industry, business, and government. Subject areas include:

Linear, integer and non-linear programming including applications; dynamic programming and stochastic control; interior point methods; multi-objective optimization; Supply chain management, including inventory control, logistics, planning and scheduling; Game theory Risk management and risk analysis, including actuarial science and insurance mathematics; Queuing models, point processes, extreme value theory, and heavy-tailed phenomena; Networked systems, including telecommunication, transportation, and many others; Quantitative finance: portfolio modeling, options, and derivative securities; Revenue management and quantitative marketing Innovative statistical applications such as detection and inference in very large and/or high dimensional data streams; Computational economics.

Darinka Dentcheva • Andrzej Ruszczyński

Risk-Averse Optimization and Control

Theory and Methods

Springer

Darinka Dentcheva
Department of Mathematical Sciences
Stevens Institute of Technology
Hoboken, NJ, USA

Andrzej Ruszczyński
Department of Management Science
and Information Systems
Rutgers University
Piscataway, NJ, USA

ISSN 1431-8598 ISSN 2197-1773 (electronic)
Springer Series in Operations Research and Financial Engineering
ISBN 978-3-031-57987-5 ISBN 978-3-031-57988-2 (eBook)
https://doi.org/10.1007/978-3-031-57988-2

Mathematics Subject Classification: 90C40, 90-08, 49J52, 91G70, 49K27, 91B16, 90C15

This Springer imprint is published by the registered company Springer Nature Switzerland AG
The registered company address is: Gewerbestrasse 11, 6330 Cham, Switzerland

Preface

This book is the result of our work during the last 25 years, dedicated to the modeling and optimization of risk in various contexts. Much of its content is based on our research during this period, and some results or entire sections are original. Our intention was to present a unified and cohesive theory—a very challenging task given the rapid and vigorous development of the field. As a result, the outcome is certainly far from perfect.

With the idea of reaching a broad audience, we have presented some of the results at a level of generality which is sufficient to cover major applications, but not necessarily in the most general versions. We hope that the book will be useful for both advanced graduate students and researchers in the areas such as optimization under uncertainty, statistics, data science, and other fields where risk and stochastic optimization are relevant.

We would like to express our gratitude for the support of the National Science Foundation, Air Force Office of Scientific Research, and the Office of Naval Research through research grants for various topics covered in this book.

Hoboken, NJ, USA
Piscataway, NJ, USA
January 2024

Darinka Dentcheva
Andrzej Ruszczyński

Introduction

Optimization is an established area of applied mathematics that plays a significant role in engineering, economics, statistics, business, and many other areas. In the simplest possible formulation, an optimization problem is to find an element $\hat{x}$ in some subset $\mathcal{X}$ of a space $\mathfrak{X}$, such that the value of a function $f : \mathfrak{X} \to \mathbb{R}$ at the point $\hat{x}$ is the smallest (or the largest) among all possible values of $f(x)$ at points $x \in \mathcal{X}$. For different classes of spaces $\mathfrak{X}$, sets $\mathcal{X}$, and functions $f(\cdot)$, the theory provides us with the characterizations of the optimal solutions and numerical methods for their identification.

The fundamental assumption behind the theory and methods of optimization is the existence of the function $f(\cdot)$, called the objective function, that allows us to compare the decisions and choose the best one. Such functions are evident when the optimization problem has a clear mathematical or physical nature, such as finding the shortest path, the minimum mass, the lowest energy, etc. However, in many applications, the determination of such a function is not straightforward. This is especially challenging when the outcomes of decisions are uncertain. Such situations arise in most decision problems in which anticipation of future events is essential.

Several ways of modeling uncertain outcomes exist, such as stochastic models, set-valued operators, scenario representations, etc. In this book, we focus exclusively on stochastic models of uncertainty. We assume, in the simplest case, that a decision $x \in \mathfrak{X}$ results in a random outcome $Z(x)$ which is an element of a space $\mathcal{Z}$ of random variables or random vectors, such as the space of random variables having a finite expected value, or the space of random variables with a finite variance, or higher moments. In such situations, it is difficult to directly compare decisions. For example, in an investment portfolio problem, the decision x may be a finite-dimensional vector representing the investments in various assets, and $Z(x)$ may represent the random profit or return rate in a fixed period following the investment. In this situation, when making the decision, we cannot directly compare $Z(x)$ and $Z(y)$ for two different values (x

and y) of the decision vector, because this would require the anticipation of the future returns. Likewise, if x represents the resources deployed in various oil exploration projects, and $Z(x)$ is the vector of values of the projects a year from now, it is extremely difficult to rank different investment strategies involving many locations with various prospects, and many currencies. Similar difficulties arise in public health decisions, logistics, marketing campaigns, and many other applications.

A branch of the optimization theory, known as stochastic optimization or stochastic programming, studies problems involving random outcomes, such as the examples mentioned above. A large portion of the theory and methods developed in this area are concerned with the optimization of the expected values of the random outcomes. However, in many problems of practical relevance, the use of the expected value as the major measure guiding our decisions may not be appropriate and would not be accepted by the practitioners in the corresponding fields. For example, in the portfolio problem mentioned above, the expected value criterion would suggest concentrating all investments on the one or few assets that have the best prospects. In public health policy problems, the use of the expected value may lead to neglecting the needs of a particular group of patients, or to the promotion of actions involving a significant probability of highly undesirable effects. In oil exploration, all funds would go to one project. In these and other applications, an additional aspect must be considered: risk.

The concept of risk arises in most decision problems under uncertainty. In the simplest form, it is reflected by the first question of a person offered an investment opportunity: "How much can be lost?" The Latin proverb "Venture not all in one ship"[†] already contained a recipe for risk control. Today, we can go much further than that in the description, analysis, and optimization of risk. Specific application areas developed their own methods to formalize and treat risk. For example, finance uses the concept of the value at risk, a refined version of the amount that can be lost, involving a low probability level. In the portfolio theory, the variance of the outcome is used as the second measure, in addition to the expected value, of the quality of the decision. This has further been extended to other mean–risk models and led to the introduction of the theory of measures of risk. In engineering, the reliability theory provides tools to control the risk of construction. In economics, the theory of utility and stochastic dominance are recognized methods to model risk-averse preferences.

In this book, we aim to provide a unified view on the ways to model, analyze, and optimize risk.

We start from the utility theory in Chap. 1 with the goal of presenting a concise and unified view on the classical expected utility and dual utility models. We also introduce the concept of risk aversion in this context.

[†] *Uni navi ne committas omnia.*

In Chap. 2, we present the modern theory of measures of risk, provide their main theoretical properties, and discuss many examples. In this chapter, we also present the theory of systemic risk, which is essential for the operation of distributed complex systems. Our treatment is based on the understanding that we address the risk associated with vector-valued outcomes. An essential part of this chapter is the theory of risk forms, which are risk measures associated with a variable probability measure.

Chapter 3 develops the theoretical foundation of optimization problems involving measures of risk. We provide a thorough analysis of the compositions of measures of risk with various classes of stochastic operators, involving non-convex and non-differentiable operators. This is followed by the derivation of the optimality conditions and duality theory. Finally, we present new stochastic subgradient methods with special attention to non-convex and non-smooth problems arising in modern applications.

Chapter 4 discusses dynamic risk models and dynamic risk optimization. When our decisions are made in successive periods and random outcomes are observed in these periods, the question of evaluating the risk of a sequence of outcomes becomes relevant. Subtle issues of time consistency of such an evaluation led to the theory of dynamic measures of risk. We provide a modern view on this theory, use it to construct risk-averse dynamic optimization problems, and develop optimality conditions for these problems. Our techniques lead to new results even in the expected-value case. Finally, we present specialized decomposition methods for dynamic risk optimization.

In Chap. 5, we present another approach to the risk control in decision problems, which is based on stochastic orders. For a major part of it, we focus on the notion of stochastic dominance of general order, presenting its properties that are essential to risk-averse optimization. The main portion of the chapter contains an analysis of optimization problems using stochastic dominance relations as constraints. We present optimality conditions and duality theory in various forms for convex and non-convex forms of several problem formulations. These results elucidate deeper relations of stochastic dominance to utility functions, distortions (or rank-dependent utility), and risk measures.

Chapter 6 addresses the comparison of random vectors and finite sequences which can be integrated into risk-averse decision problems. Vector-valued outcomes arise in the context of systemic risk, multi-objective optimization under uncertainty, or in any problem involving high-dimensional risks. The question of constraints of distributions in a sequential decision problem is an important one as already mentioned as the decision maker needs to mind the time consistency of the model. This chapter introduces the notion of a time-consistent stochastic order for sequences and provides a theoretical foundation for constructing time-consistent multistage stochastic programming problems with sequential stochastic dominance constraints.

In Chap. 7, we present numerical methods for solving optimization problems with stochastic order constraints with a focus on the second-order stochastic dominance relation in both univariate and multivariate settings. We start with reformulations of the static problems that allow the application of available commercial solvers for optimization problems. This approach is suitable for relatively small problems, in which the probability space consists of a small number of simple events. The main focus of the chapter is on methods for solving the respective optimization problems when dealing with general probability spaces. In this case, we construct sequential approximations of the stochastic dominance constraints by inequalities, which we call event cuts. Several approximations are presented for the problems formulated in Chaps. 5 and 6 and their convergence is analyzed. Methods for evaluation of the quality of the approximation and the validity of the dominance relation at the iterates are presented as well.

Finally, Chap. 8 considers risk models for Markov dynamical systems and risk-averse control of such systems. This is a very special class of dynamic models where a deep theory and dedicated methods can be developed, which fully exploit the Markovian structure of the system. The key concept is that of a Markov risk measure, which allows us to reduce the dynamic risk evaluation into a sequence of static evaluations with transition risk mappings. This leads to specialized dynamic programming equations in the finite-horizon, infinite-horizon, absorbing, and partially observable cases.

Most of the book contains the results of our own research conducted over a period of 25 years, and several parts of the book are entirely new. We discuss the relation to earlier research and point to relevant reading in the Bibliographical Remarks at the end of the book.

Contents

Chapter 1
Elements of the Utility Theory

The utility theory is concerned with analyzing the existence and the forms of numerical representations of preference relations. We start our presentation from abstract preference relations in a space $\mathcal{X}$, which we call the *prospect space*. Our considerations in this section are rather abstract, but in the applications to risk-averse decision-making that we discuss in this book, three special cases of $\mathcal{X}$ are particularly important.

The first case is the space $\mathcal{P}(\mathcal{S})$ of probability measures on a Polish space $\mathcal{S}$ equipped with the σ-algebra $\mathcal{B}$ of Borel sets (frequently, $\mathcal{S} = \mathbb{R}^n$). We will also consider its subspaces of measures satisfying additional integrability conditions: the existence of finite moments or order $p \in [1, \infty]$. In this context, we are interested in the following question: can we derive a functional $U : \mathcal{P}(\mathcal{S}) \to \mathbb{R}$ such that we prefer $\mu \in \mathcal{P}(\mathcal{S})$ over $\nu \in \mathcal{P}(\mathcal{S})$ if and only if $U(\mu) < U(\nu)$. Here, and later in this chapter, we adopt the convention that smaller values of $U(\cdot)$ are associated with preferred prospects, and thus $U(\mu)$ can be interpreted as the "fair price" of μ.

The second case of $\mathcal{X}$ is the space of random vectors. For a probability space $(\Omega, \mathcal{F}, P)$, we consider the space $\mathcal{L}_0(\Omega, \mathcal{F}, P; \mathbb{R}^n)$ of random vectors in $\mathbb{R}^n$, or its subspaces $\mathcal{L}_p(\Omega, \mathcal{F}, P; \mathbb{R}^n)$ of random vectors having finite moments of order $p \in [1, \infty]$. Again, can we construct $U(\cdot)$ such that Z is "better than" V (both in $\mathcal{X}$) if and only if $U(Z) < U(V)$?

The third case is the space $\mathcal{Q}_{\text{b}}$ of the quantile functions of scalar and bounded random variables. While one might reduce it to the first case with $\mathcal{S} = \mathbb{R}$, it is convenient to consider it separately, because of future applications to measures of risk.

D. Dentcheva, A. Ruszczyński, *Risk-Averse Optimization and Control*, Springer Series in Operations Research and Financial Engineering,
https://doi.org/10.1007/978-3-031-57988-2_1

1.1 Preference Relations

We assume that a preference relation in the prospect space $\mathcal{X}$ is defined by a certain *total preorder*, that is, a binary relation $\trianglelefteq$ on $\mathcal{X}$ satisfying the following two conditions:

Transitivity: For all $z, v, w \in \mathcal{X}$, if $z \trianglelefteq v$ and $v \trianglelefteq w$, then $z \trianglelefteq w$; and
Completeness: For all $z, v \in \mathcal{X}$, either $z \trianglelefteq v$, or $v \trianglelefteq z$, or both are true.

It follows from the completeness condition that the preorder is *reflexive*, that is, $z \trianglelefteq z$ for all $z \in \mathcal{X}$. The corresponding *indifference relation* $\sim$ is defined in a usual way: $z \sim v$, if $z \trianglelefteq v$ and $v \trianglelefteq z$. It is an equivalence relation. We say that z is *strictly preferred* over v and write it $z \triangleleft v$, if $z \trianglelefteq v$, and $v \ntrianglelefteq z$.

In decision problems, it is convenient to represent prospects by numbers, which summarize their usefulness.

Definition 1.1. A functional $U : \mathcal{X} \to \mathbb{R}$ is a *numerical representation* of the preference relation $\trianglelefteq$ on $\mathcal{X}$, if

$$z \triangleleft v \iff U(z) < U(v).$$

In our presentation, continuous representations play a fundamental role. To speak about continuity, we need to assume that the space of prospects $\mathcal{X}$ is a topological space.

The special cases of $\mathcal{X}$ mentioned at the beginning of this chapter can be equipped with appropriate topologies. The space $\mathcal{P}(\mathcal{S})$ of probability measures on $(\mathcal{S}, \mathcal{B})$ can be equipped with the topology of weak convergence. In $\mathcal{L}_0(\Omega, \mathcal{F}, P; \mathbb{R}^n)$ we may consider the topology of almost sure convergence, or the topology of convergence in probability. The spaces $\mathcal{L}_p(\Omega, \mathcal{F}, P; \mathbb{R}^n)$, for $p \in [1, \infty]$, as Banach spaces, may be equipped with the strong (norm) topology or the weak topology. The space $\mathcal{Q}_{\rm b}$ can be considered as a subspace of the space of bounded functions and equipped with the topology of uniform convergence.

To guarantee the existence of a continuous numerical representation, we need to make additional assumptions about the preference relation and the space itself. The first one is a fundamental condition, which we use many times in our considerations.

Definition 1.2. The preference relation $\trianglelefteq$ on $\mathcal{X}$ is *continuous*, if for every $z \in \mathcal{X}$ the sets $\{v \in \mathcal{X} : v \trianglelefteq z\}$ and $\{v \in \mathcal{X} : z \trianglelefteq v\}$ are closed.

A classical result in topology states that a continuous total preorder on a separable and connected topological space has a continuous numerical representation.

Furthermore, if $U : \mathcal{X} \to \mathbb{R}$ is a numerical representation of a preorder $\trianglelefteq$, then for any strictly increasing function $\varphi : \mathbb{R} \to \mathbb{R}$ the functional $\varphi(U(\cdot))$ is also a numerical representation of the preorder $\trianglelefteq$. If $\varphi(\cdot)$ and $U(\cdot)$ are continuous, so is their composition. Conversely, for any two numerical representations $U(\cdot)$ and $\widetilde{U}(\cdot)$ of $\trianglelefteq$, a strictly increasing function $\varphi : U(\mathcal{X}) \to \mathbb{R}$ exists, such that $\widetilde{U}(\cdot) = \varphi(U(\cdot))$.

For specific prospect spaces and under additional conditions, we can say more about the structure of a numerical representation of a preorder $\trianglelefteq$. This is our main objective in this chapter.

1.2 Expected Utility Theory

The expected utility theory is concerned with comparing probability measures.

1.2.1 The Prospect Space of Probability Measures

Given a Polish space $\mathcal{S}$, equipped with its σ-algebra $\mathcal{B}$ of Borel sets, we consider the prospect space $\mathcal{X} = \mathcal{P}(\mathcal{S})$ of probability measures on $\mathcal{S}$. For most applications discussed in this book, the finite-dimensional space $\mathcal{S} = \mathbb{R}^d$ is sufficient, but we present the theory in more general settings because they does not require much additional effort and broaden the scope of applications.

We assume that the preference relation $\trianglelefteq$ satisfies two additional conditions:

Independence Axiom: For all μ, ν, and λ in $\mathcal{P}(\mathcal{S})$ one has

$$\mu \lhd \nu \implies \alpha\mu + (1-\alpha)\lambda \lhd \alpha\nu + (1-\alpha)\lambda, \quad \forall\, \alpha \in (0,1);$$

Archimedean Axiom: For all μ, ν, and λ in $\mathcal{P}(\mathcal{S})$, satisfying the relations

$$\mu \lhd \nu \lhd \lambda,$$

there exist $\alpha, \beta \in (0, 1)$ such that

$$\alpha\mu + (1-\alpha)\lambda \lhd \nu \lhd \beta\mu + (1-\beta)\lambda.$$

We can derive the following properties of a preorder satisfying these axioms.

Lemma 1.3. *Suppose a total preorder $\trianglelefteq$ on $\mathcal{P}(\mathcal{S})$ satisfies the independence axiom. Then for every $\mu \in \mathcal{P}(\mathcal{S})$ the indifference set $\{\nu \in \mathcal{P}(\mathcal{S}) : \nu \sim \mu\}$ is convex.*

Proof. Let $\nu \sim \mu$ and $\lambda \sim \mu$. Suppose $(1-\alpha)\nu + \alpha\lambda \lhd \nu$ for some $\alpha \in (0, 1)$. Then also $(1-\alpha)\nu + \alpha\lambda \lhd \lambda$. Using the independence axiom with these two relations, we obtain a contradiction as follows:

$$\begin{aligned}(1-\alpha)\nu+\alpha\lambda &= (1-\alpha)\big[(1-\alpha)\nu+\alpha\lambda\big]+\alpha\big[(1-\alpha)\nu+\alpha\lambda\big]\\ &\lhd (1-\alpha)\nu+\alpha\big[(1-\alpha)\nu+\alpha\lambda\big]\lhd (1-\alpha)\nu+\alpha\lambda.\end{aligned}$$

The case when $\nu \lhd (1-\alpha)\nu+\alpha\lambda$ is excluded similarly. We conclude that $(1-\alpha)\nu+\alpha\lambda \sim \mu$, for all $\alpha \in (0,1)$. □

Lemma 1.4. *Suppose a total preorder $\unlhd$ on $\mathcal{P}(\mathcal{S})$ satisfies the independence and Archimedean axioms. Then for all $\mu,\nu \in \mathcal{P}(\mathcal{S})$, satisfying the relation $\mu \lhd \nu$, and for all $\lambda \in \mathcal{P}(\mathcal{S})$, there exists $\bar{\alpha} > 0$ such that*

$$(1-\alpha)\mu+\alpha\lambda \lhd \nu \quad \textit{and} \quad \mu \lhd (1-\alpha)\nu+\alpha\lambda, \qquad \forall\, \alpha \in [0,\bar{\alpha}]. \tag{1.1}$$

Proof. We focus on the left relation in (1.1) and consider three cases.
Case 1: $\nu \lhd \lambda$. The left relation in (1.1) is true for some $\bar{\alpha} \in (0,1)$, owing to the Archimedean axiom. If $\alpha \in (0,\bar{\alpha})$ then for $\beta = \alpha/\bar{\alpha} \in (0,1)$ the independence axiom yields

$$(1-\alpha)\mu+\alpha\lambda = (1-\beta)\mu+\beta\big[(1-\bar{\alpha})\mu+\bar{\alpha}\lambda\big] \lhd (1-\beta)\mu+\beta\nu \lhd \nu.$$

Case 2: $\lambda \lhd \nu$. Applying the independence axiom twice, we obtain

$$(1-\alpha)\mu+\alpha\lambda \lhd (1-\alpha)\nu+\alpha\lambda \lhd \nu, \qquad \forall\, \alpha \in (0,1).$$

Case 3: $\lambda \sim \nu$. By virtue of Lemma 1.3, $(1-\alpha)\nu+\alpha\lambda \sim \nu$ for all $\alpha \in (0,1)$, and the left relation in (1.1) follows from the independence axiom.

The right relation in (1.1) is proved analogously. □

1.2.2 Affine Numerical Representation

The set $\mathcal{P}(\mathcal{S})$ is a convex subset of the vector space $\mathcal{M}(\mathcal{S})$ of signed regular finite measures on $\mathcal{S}$. It is also convenient for our derivations to consider the linear subspace $\mathcal{M}_0(\mathcal{S})$ of measures $\mu \in \mathcal{M}(\mathcal{S})$ such that $\mu(\mathcal{S}) = 0$.

Theorem 1.5. *Suppose the total preorder $\unlhd$ on $\mathcal{P}(\mathcal{S})$ satisfies the independence and Archimedean axioms. Then an affine numerical representation of $\unlhd$ on $\mathcal{P}(\mathcal{S})$ exists.*

Proof. In the space $\mathcal{M}_0(\mathcal{S})$, define the set

$$C_0 = \{\mu-\nu : \mu \in \mathcal{P}(\mathcal{S}),\ \nu \in \mathcal{P}(\mathcal{S}),\ \mu \lhd \nu\}.$$

Consider two arbitrary points ϑ and $\varkappa$ in C_0, that is,

$$\vartheta = \mu - \nu, \qquad \mu, \nu \in \mathcal{P}(\mathcal{S}), \qquad \mu \lhd \nu,$$
$$\varkappa = \lambda - \sigma, \qquad \lambda, \sigma \in \mathcal{P}(\mathcal{S}), \qquad \lambda \lhd \sigma.$$

For every $\alpha \in (0, 1)$, using the independence axiom twice, we obtain

$$\alpha\mu + (1-\alpha)\lambda \lhd \alpha\nu + (1-\alpha)\lambda \lhd \alpha\nu + (1-\alpha)\sigma.$$

Then,

$$\alpha\vartheta + (1-\alpha)\varkappa = \big[\alpha\mu + (1-\alpha)\lambda\big] - \big[\alpha\nu + (1-\alpha)\sigma\big]$$

is an element of C_0, which proves that C_0 is convex.

Define $C = \{\gamma\vartheta : \vartheta \in C_0,\ \gamma > 0\}$. It is evident that C is a convex cone, that is, for all $\vartheta, \varkappa \in C$, and all $\alpha > 0$ and $\beta > 0$ we have $\alpha\vartheta + \beta\varkappa \in C$. Moreover, $C \subset \mathcal{M}_0$.

We shall prove that the algebraic interior of C in $\mathcal{M}_0$, denoted $\operatorname{cor}(C)$, is nonempty and that in fact $C = \operatorname{cor}(C)$. Consider any $\vartheta \in C$, an arbitrary nonzero measure $\lambda \in \mathcal{M}_0$, and the ray

$$z(\tau) = \vartheta + \tau\lambda, \quad \tau > 0.$$

Our objective is to show that $z(\tau) \in C$ for a sufficiently small $\tau > 0$. Let $\lambda = \lambda^+ - \lambda^-$ be the Jordan decomposition of λ. With no loss of generality, we may assume that the direction λ is normalized so that $|\lambda| = \lambda^+(\mathcal{S}) + \lambda^-(\mathcal{S}) = 2$. As $\lambda \in \mathcal{M}_0$, we have then $\lambda^+(\mathcal{S}) = \lambda^-(\mathcal{S}) = 1$. Let $\gamma > 0$ be such that the point $\vartheta_0 = \gamma\vartheta \in C_0$. Since C is a cone, $z(\tau) \in C$ if and only if $\gamma z(\tau) \in C$. Setting $t = \gamma\tau$, we reformulate our question as follows:

Does $\vartheta_0 + t\lambda$ belong to C for sufficiently small $t > 0$?

Since $\vartheta_0 \in C_0$, we can represent it as a difference $\vartheta_0 = \mu - \nu$, with $\mu, \nu \in \mathcal{P}(\mathcal{S})$, and $\mu \lhd \nu$. Then

$$\vartheta_0 + t\lambda = \big[(1-t)\mu + t\lambda^+\big] - \big[(1-t)\nu + t\lambda^-\big] + t\vartheta_0. \tag{1.2}$$

Both expressions in brackets are probability measures for $t \in [0, 1]$. By the independence axiom,

$$\mu \lhd \frac{1}{2}\mu + \frac{1}{2}\nu \lhd \nu.$$

According to Lemma 1.4, there exists $t_0 > 0$, such that for all $t \in [0, t_0]$ we also have

$$(1-t)\mu + t\lambda^+ \lhd \frac{1}{2}\mu + \frac{1}{2}\nu \lhd (1-t)\nu + t\lambda^-.$$

This proves that

$$\big[(1-t)\mu + t\lambda^+\big] - \big[(1-t)\nu + t\lambda^-\big] \in C_0, \quad \forall t \in [0, t_0].$$

For these values of t, the right hand side of (1.2) is a sum of two elements of C. As the set C is a convex cone, this sum is an element of C as well. Consequently, $\vartheta + \tau\lambda \in C$ for all $\tau \in [0, t_0/\alpha]$.

Summing up, C is convex, $C = \text{cor}(C)$, and $\mathbb{0} \notin C$, where $\mathbb{0}$ denotes the zero element in $\mathcal{M}_0(\mathcal{S})$. By virtue of the separation theorem for a convex set with a nonempty algebraic interior, the point $\mathbb{0}$ and the set C can be separated strictly: a linear functional U_0 on $\mathcal{M}_0(\mathcal{S})$ exists, such that

$$U_0(\vartheta) < 0, \quad \forall\, \vartheta \in C. \tag{1.3}$$

We can extend the linear functional U_0 to the whole space $\mathcal{M}(\mathcal{S})$ by choosing a measure $\lambda \in \mathcal{P}(\mathcal{S})$ and setting

$$U(\mu) = U_0\big(\mu - \mu(\mathcal{S})\lambda\big), \quad \mu \in \mathcal{M}(\mathcal{S}).$$

It is linear and coincides with U_0 on $\mathcal{M}_0(\mathcal{S})$. The relation (1.3) is equivalent to the following statement: for all $\mu, \nu \in \mathcal{P}(\mathcal{S})$ such that $\mu \lhd \nu$, we have

$$U_0(\mu - \nu) = U(\mu - \nu) = U(\mu) - U(\nu) < 0.$$

It follows that U restricted to $\mathcal{P}(\mathcal{S})$ is the postulated numerical representation of the preorder $\trianglelefteq$. It is affine on $\mathcal{P}(\mathcal{S})$. □

1.2.3 Integral Representation. Utility Functions

To prove the main result of this section, we assume that the space $\mathcal{M}(\mathcal{S})$ is equipped with the topology of weak convergence of measures. This makes it a complete topological vector space. Recall that the topology of weak convergence is metrizable and $\mathcal{M}(\mathcal{S})$ is a Polish space itself.

Theorem 1.6. *Suppose the total preorder $\trianglelefteq$ on $\mathcal{P}(\mathcal{S})$ is continuous and satisfies the independence axiom. Then a continuous and bounded function $u : \mathcal{S} \to \mathbb{R}$ exists, such that the functional*

$$U(\mu) = \int_{\mathcal{S}} u(z)\, \mu(\mathrm{d}z) \tag{1.4}$$

is a numerical representation of $\trianglelefteq$ on $\mathcal{P}(\mathcal{S})$.

Proof. The continuity of the preorder $\trianglelefteq$ implies the Archimedean axiom. Indeed, the sets $\{\pi \in \mathcal{P}(\mathcal{S}) : \pi \lhd \nu\}$ and $\{\pi \in \mathcal{P}(\mathcal{S}) : \mu \lhd \pi\}$ are open, and the mapping $\alpha \mapsto \alpha\pi + (1-\alpha)\lambda$, $\alpha \in [0, 1]$, is continuous for any $\lambda \in \mathcal{P}(\mathcal{S})$.

By virtue of Theorem 1.5, an affine numerical representation $U : \mathcal{P}(\mathcal{S}) \to \mathbb{R}$ of $\trianglelefteq$ exists.

We shall prove that the functional $U(\cdot)$ is continuous on $\mathcal{P}(\mathcal{S})$, that is, for every α the sets

$$A = \{\mu \in \mathcal{P} : U(\mu) \le \alpha\} \qquad \text{and} \qquad B = \{\mu \in \mathcal{P} : U(\mu) \ge \alpha\}$$

are closed. Since $\mathcal{P}(\mathcal{S})$ is convex and $U(\cdot)$ is affine, the set $U(\mathcal{P})$ is convex. Therefore, for every α one of three cases may occur:

(i) $U(\mu) < \alpha$ for all $\mu \in \mathcal{P}(\mathcal{S})$;
(ii) $U(\mu) > \alpha$ for all $\mu \in \mathcal{P}(\mathcal{S})$;
(iii) $\alpha \in U\big(\mathcal{P}(\mathcal{S})\big)$.

In cases (i) and (ii) there is nothing to prove. In case (iii), let $\nu \in \mathcal{P}(\mathcal{S})$ be such that $U(\nu) = \alpha$. Since $U(\cdot)$ is a numerical representation of the preorder, we have

$$A = \{\mu \in \mathcal{P} : \mu \trianglelefteq \nu\} \qquad \text{and} \qquad B = \{\mu \in \mathcal{P} : \nu \trianglelefteq \mu\}.$$

Both sets are closed due to the continuity of the preorder $\trianglelefteq$. Thus, $U(\cdot)$ in continuous on $\mathcal{P}(\mathcal{S})$.

For $z \in \mathcal{S}$, let δ_z denote the Dirac measure supported at z. Define a function $u : \mathcal{S} \to \mathbb{R}$ as follows:

$$u(z) = U(\delta_z), \quad z \in \mathcal{S}.$$

The function $u(\cdot)$ is continuous, because $z_n \to z$ implies $\delta_{z_n} \xrightarrow{w} \delta_z$, and thus $u(z_n) \to u(z)$. Suppose $u(\cdot)$ is not bounded from above. Then we can find a sequence $\{z_n\}$ such that $u(z_n) > n$ for all n. Consider the sequence of probability measures

$$\mu_n = \Big(1 - \frac{1}{\sqrt{n}}\Big)\delta_{z_1} + \frac{1}{\sqrt{n}}\delta_{z_n}.$$

Then $\mu_n \xrightarrow{w} \delta_{z_1}$, but $U(\mu_n) \to +\infty$, as $n \to \infty$. This contradicts the continuity of $U(\cdot)$. The unboundedness from below is excluded in a similar way, and thus $u(\cdot)$ is bounded.

We now verify the representation (1.4). For a discrete distribution

$$\mu = \sum_{n=1}^{N} p_n \delta_{z_n}, \quad \sum_{n=1}^{N} p_n = 1, \quad p_n \ge 0, \tag{1.5}$$

the affinity of $U(\cdot)$ yields

$$U(\mu) = \sum_{n=1}^{N} p_n u(z_n),$$

which is exactly (1.4). Furthermore, the set of discrete distributions of form (1.5) is dense in $\mathcal{P}(\mathcal{S})$, in the topology of weak convergence. By the continuity of $U(\cdot)$, the formula (1.4) is true at all $\mu \in \mathcal{P}$. □

The formula (1.4) is referred to as the *expected utility representation*, and $u(\cdot)$ is called the *utility function*.

The utility function in Theorem 1.6 is bounded. If we restrict the prospect space to include only measures satisfying additional integrability conditions, representations with unbounded utility functions may occur.

Let $\psi : \mathcal{S} \to [1, \infty)$ be a continuous function (called the *gauge function* in this context), and let $C_b^{\psi}(\mathcal{S})$ be the set of functions $f : \mathcal{S} \to \mathbb{R}$, such that $f/\psi \in C_b(\mathcal{S})$, where $C_b(\mathcal{S})$ is the space of continuous and bounded functions on $\mathcal{S}$. We can define the space $\mathcal{M}^{\psi}(\mathcal{S})$ of regular signed measures μ, such that

$$\left| \int_{\mathcal{S}} f(z)\, \mu(\mathrm{d}z) \right| < \infty, \quad \forall\, f \in C_b^{\psi}(\mathcal{S}).$$

Similar to the topology of weak convergence, we can define the topology of ψ-weak convergence on $\mathcal{M}^{\psi}(\mathcal{S})$: $\{\mu_n\}_{n\in\mathbb{N}} \xrightarrow{\psi} \mu$ if and only if $\int_{\mathcal{S}} f(z)\, \mu_n(\mathrm{d}z) \to \int_{\mathcal{S}} f(z)\, \mu(\mathrm{d}z)$ for all $f \in C_b^{\psi}(\mathcal{S})$. All continuity statements will be now made for this topology. We use the symbol $\mathcal{P}^{\psi}(\mathcal{S})$ to denote the set of probability measures in $\mathcal{M}^{\psi}(\mathcal{S})$.

Theorem 1.7. *Suppose the total preorder $\trianglelefteq$ on $\mathcal{P}^{\psi}(\mathcal{S})$ is continuous and satisfies the independence axiom. Then a function $u \in C_b^{\psi}(\mathcal{S})$ exists such that the functional*

$$U(\mu) = \int_{\mathcal{S}} u(z)\, \mu(\mathrm{d}z) \tag{1.6}$$

is a numerical representation of $\trianglelefteq$ on $\mathcal{P}^{\psi}(\mathcal{S})$.

Proof. The proof is almost identical to the proof of Theorem 1.6, except that we use the continuity with respect to the ψ-weak convergence. To verify that $u \in C_b^{\psi}(\mathcal{S})$, we argue by contradiction again. Suppose u/ψ is not bounded from above. Then we can find a sequence $\{z_n\}$ such that $u(z_n) > n\psi(z_n)$ for all n. Consider the sequence of probability measures

$$\mu_n = \Big(1 - \frac{1}{\psi(z_n)\sqrt{n}}\Big)\delta_{z_1} + \frac{1}{\psi(z_n)\sqrt{n}}\delta_{z_n}.$$

Then μ_n converges ψ-weakly to δ_{z_1}, but $U(\mu_n) \to +\infty$, as $n \to \infty$. This contradicts the continuity of $U(\cdot)$. The unboundedness of u/ψ from below is excluded in a similar way and thus $u \in C_b^{\psi}(\mathcal{S})$. □

1.2.4 Monotonicity and Risk Aversion

Suppose a partial order relation $\leq$ on $\mathcal{S}$ is defined (our notation is motivated by the applications in which $\mathcal{S}$ is a finite-dimensional vector space). In this case, it makes sense to define the monotonicity of a preference relation $\trianglelefteq$.

Monotonicity Axiom: For all $z, v \in \mathcal{S}$ one has

$$z \leq v \implies \delta_z \trianglelefteq \delta_v.$$

Theorem 1.8. *Suppose the total preorder $\trianglelefteq$ on $\mathcal{P}(\mathcal{S})$ is monotonic, continuous, and satisfies the independence axiom. Then a nondecreasing, continuous and bounded function $u : \mathcal{S} \to \mathbb{R}$ exists, such that the functional* (1.4) *is a numerical representation of $\trianglelefteq$ on $\mathcal{P}(\mathcal{S})$.*

Proof. In view of Theorem 1.6, it is sufficient to verify that the function $u(\cdot)$ in (1.4) is monotonic with respect to $\leq$. To this end, we consider $z, v \in \mathcal{S}$ such that $z \leq v$. By the monotonicity of the order, $u(z) = U(\delta_z) \leq U(\delta_v) = u(v)$. □

We now focus on the case with the gauge function $\psi_p(z) = 1 + \|z\|^p$, where $p \geq 1$. Then for every $\mu \in \mathcal{P}^{\psi_p}(\mathcal{S})$ and for every σ-subalgebra $\mathcal{G}$ of $\mathcal{B}$ the conditional expectation $\mathcal{E}_{\mu|\mathcal{G}} : \mathcal{S} \to \mathcal{S}$ is well-defined, as a $\mathcal{G}$-measurable function satisfying the equation

$$\int_G \mathcal{E}_{\mu|\mathcal{G}}(z)\, \mu(\mathrm{d}z) = \int_G z\, \mu(\mathrm{d}z), \quad G \in \mathcal{G}. \tag{1.7}$$

The conditional expectation $\mathcal{E}_{\mu|\mathcal{G}}$ induces a probability measure on $(\mathcal{S}, \mathcal{B})$ as follows

$$\mu_{\mathcal{G}}(A) = \mu\{\mathcal{E}_{\mu|\mathcal{G}}^{-1}(A)\}, \quad A \in \mathcal{B}.$$

Definition 1.9. A preference relation $\trianglelefteq$ on $\mathcal{P}^{\psi_p}(\mathcal{S})$ is *risk-averse*, if $\mu_{\mathcal{G}} \trianglelefteq \mu$, for every $\mu \in \mathcal{P}^{\psi_p}(\mathcal{S})$ and every σ-subalgebra $\mathcal{G}$ of $\mathcal{B}$.

By choosing $\mathcal{G} = \{\mathcal{S}, \emptyset\}$, we observe that Definition 1.9 implies that $\delta_{\mathcal{E}_\mu} \trianglelefteq \mu$, where $\mathcal{E}_\mu = \int_{\mathcal{S}} z\, \mu(\mathrm{d}z)$ is the expected value of the measure μ.

Theorem 1.10. *Suppose a total preorder $\trianglelefteq$ on $\mathcal{P}^{\psi_p}(\mathcal{S})$ is continuous, risk-averse, and satisfies the independence axiom. Then a convex function $u \in C_{\mathrm{b}}^{\psi_p}(\mathcal{S})$ exists such that the functional* (1.6) *is a numerical representation of $\trianglelefteq$ on $\mathcal{P}^{\psi}(\mathcal{S})$.*

Proof. In view of Theorem 1.7, we only need to prove the convexity of $u(\cdot)$. By the risk aversion, for every $\mu \in \mathcal{P}^{\psi_p}(\mathcal{S})$ we have

$$u\left(\int_{\mathcal{S}} z\ \mu(\mathrm{d}z)\right) \leq \int_{\mathcal{S}} u(z)\ \mu(\mathrm{d}z).$$

This is Jensen's inequality, which is equivalent to the convexity of $u(\cdot)$. □

Remark 1.11. It is clear from the proof that the convexity of $u(\cdot)$ could have been obtained by simply assuming that $\delta_{\mathcal{E}_\mu} \trianglelefteq \mu$. The convexity of $u(\cdot)$ would imply risk aversion in the sense of Definition 1.9, by virtue of Jensen's inequality for conditional expectations. Therefore, Definition 1.9 and the requirement that $\delta_{\mathcal{E}_\mu} \trianglelefteq \mu$ are equivalent within the framework of the expected utility theory. Nonetheless, we prefer to leave Definition 1.9 in its full form, because we shall use the concept of risk aversion in connection with other axioms, where such equivalence cannot be derived. □

1.3 Dual Utility Theory

The dual utility theory is formulated in a more restrictive setting: for the probability distributions on the real line.

1.3.1 The Prospect Space of Quantile Functions

With every probability measure $\mu \in \mathcal{P}(\mathbb{R})$ we associate the distribution function: $F_\mu(\eta) \triangleq \mu\big((-\infty, \eta]\big)$. It is nondecreasing and right-continuous. We can, therefore, define its inverse

$$F_\mu^{-1}(p) \triangleq \inf\{\eta \in \mathbb{R} : F_\mu(\eta) \geq p\}, \quad p \in (0, 1).$$

By definition, $F_\mu^{-1}(p)$ is the smallest p-quantile of μ. We call $F_\mu^{-1}(\cdot)$ the *quantile function* associated with the probability measure μ. Every quantile function is nondecreasing and left-continuous on the open interval $(0, 1)$. On the other hand, every nondecreasing and left-continuous function $\Phi(\cdot)$ on $(0, 1)$ uniquely defines the following distribution function:

$$F_\mu(\eta) = \varphi^{-1}(\eta) \triangleq \sup\{p \in (0, 1) : \Phi(p) \leq \eta\}, \quad \eta \in \mathbb{R}.$$

It is nondecreasing, right-continuous, and thus corresponds to a certain probability measure $\mu \in \mathcal{P}(\mathbb{R})$.

The set $\mathcal{Q}$ of all nondecreasing and left-continuous functions on the interval $(0, 1)$ *will be our prospect space.* It is evident that $\mathcal{Q}$ is a convex cone in the vector space $\mathcal{L}_0(0, 1)$ of all Lebesgue measurable functions on the interval $(0, 1)$.

We assume that the preference relation $\trianglelefteq$ on $\mathcal{Q}$ is a total preorder and satisfies the following two conditions.

Dual Independence Axiom: For all Φ, Ψ, and Υ in $\mathcal{Q}$ one has

$$\Phi \lhd \Psi \implies \alpha\Phi + (1-\alpha)\Upsilon \lhd \alpha\Psi + (1-\alpha)\Upsilon, \quad \forall\, \alpha \in (0, 1);$$

Dual Archimedean Axiom: For all Φ, Ψ, and Υ in $\mathcal{Q}$, satisfying the relations

$$\Phi \lhd \Psi \lhd \Upsilon,$$

there exist $\alpha, \beta \in (0, 1)$ such that

$$\alpha\Phi + (1-\alpha)\Upsilon \lhd \Psi \lhd \beta\Phi + (1-\beta)\Upsilon.$$

Similar to Lemmas 1.3 and 1.4, we can easily derive the following properties of a preorder satisfying the dual axioms.

Lemma 1.12. *Suppose a total preorder $\trianglelefteq$ on $\mathcal{Q}$ satisfies the dual independence axiom. Then for every $\Phi \in \mathcal{Q}$ the indifference set $\{\Psi \in \mathcal{Q} : \Psi \sim \Phi\}$ is convex.*

Lemma 1.13. *Suppose a total preorder $\trianglelefteq$ on $\mathcal{Q}$ satisfies the dual independence and Archimedean axioms. Then for all $\Phi, \Psi \in \mathcal{Q}$, satisfying the relation $\Phi \lhd \Psi$, and for all $\Upsilon \in \mathcal{Q}$, there exists $\bar{\alpha} > 0$ such that*

$$(1-\alpha)\Phi + \alpha\Upsilon \lhd \Psi \quad \textit{and} \quad \Phi \lhd (1-\alpha)\Psi + \alpha\Upsilon, \quad \forall\, \alpha \in [0, \bar{\alpha}]. \tag{1.8}$$

The proofs of both these facts are virtually the same as the proofs of Lemmas 1.3 and 1.4.

1.3.2 Affine Numerical Representation

It is convenient for our derivations to consider the linear span of $\mathcal{Q}$ defined as follows:

$$\text{lin}(\mathcal{Q}) = \left\{ \sum_{i=1}^{k} \alpha_i \Phi_i : \alpha_i \in \mathbb{R},\ \Phi_i \in \mathcal{Q},\ i = 1, \ldots, k,\ k \in \mathbb{N} \right\} = \mathcal{Q} - \mathcal{Q},$$

where $\mathcal{Q} - \mathcal{Q}$ is the Minkowski sum of the sets $\mathcal{Q}$ and $-\mathcal{Q}$. The last equality follows from the fact that $\mathcal{Q}$ is a convex cone.

Theorem 1.14. *If a total preorder* $\trianglelefteq$ *on* $\mathcal{Q}$ *satisfies the dual independence and Archimedean axioms, then a linear functional on* $\mathrm{lin}(\mathcal{Q})$ *exists, whose restriction to* $\mathcal{Q}$ *is a numerical representation of* $\trianglelefteq$.

Proof. Define in the space $\mathrm{lin}(\mathcal{Q})$ the set

$$C = \{\Phi - \Psi : \Phi \in \mathcal{Q},\ \Psi \in \mathcal{Q},\ \Phi \lhd \Psi\}.$$

Exactly as in the proof of Theorem 1.5 on page 4, we can prove that C is convex. We shall prove that it is a cone. Suppose $\Phi \lhd \Psi$ and let $\alpha > 0$. If $\alpha \in (0,1)$, then the independence axiom implies that[†]

$$\alpha\Phi = \alpha\Phi + (1-\alpha)\mathbb{0} \lhd \alpha\Psi + (1-\alpha)\mathbb{0} = \alpha\Psi.$$

Consider $\alpha > 1$, and suppose $\alpha\Psi \trianglelefteq \alpha\Phi$. If $\alpha\Psi \lhd \alpha\Phi$, then, owing to the independence axiom, we obtain a contradiction: $\Psi = \frac{1}{\alpha}(\alpha\Psi) \lhd \frac{1}{\alpha}(\alpha\Phi) = \Phi$. Consider the case when $\alpha\Psi \sim \alpha\Phi$. By virtue of Lemma 1.12 and the independence axiom, for any $\beta \in (0, 1/\alpha)$ we obtain a contradiction in the following way:

$$\begin{aligned}\alpha\Psi &\sim \beta(\alpha\Phi) + (1-\beta)(\alpha\Psi) = (\beta\alpha)\Phi + (1-\beta\alpha)\Big[\frac{(1-\beta)\alpha}{1-\beta\alpha}\Psi\Big]\\ &\lhd (\beta\alpha)\Psi + (1-\beta)(\alpha\Psi) = \alpha\Psi.\end{aligned}$$

Therefore, $\alpha\Phi \lhd \alpha\Psi$ for all $\alpha > 0$. We conclude that for every $\alpha > 0$ the element $\alpha(\Phi - \Psi) \in C$. Consequently, C is a convex cone.

We shall prove that the algebraic interior of C, denoted $\mathrm{cor}(C)$, is nonempty, and that, in fact, $C = \mathrm{cor}(C)$. Consider any $\Gamma \in C$, a function $\Upsilon \in \mathrm{lin}(\mathcal{Q})$, and the ray

$$Z(t) = \Gamma + t\Upsilon, \quad t > 0.$$

Our objective is to show that $Z(t) \in C$ for a sufficiently small $t > 0$. By the definition of $\mathrm{lin}(\mathcal{Q})$, we can represent $\Upsilon = \Upsilon^+ - \Upsilon^-$, with $\Upsilon^+, \Upsilon^- \in \mathcal{Q}$.

Since $\Gamma \in C$, we can represent it as a difference $\Gamma = \Phi - \Psi$, with $\Phi, \Psi \in \mathcal{Q}$, and $\Phi \lhd \Psi$. Then

$$\Gamma + t\Upsilon = \big[(1-t)\Phi + t\Upsilon^+\big] - \big[(1-t)\Psi + t\Upsilon^-\big] + t\Gamma. \tag{1.9}$$

Both expressions in brackets are elements of $\mathcal{Q}$. By the dual independence axiom,

$$\Phi \lhd \frac{1}{2}\Phi + \frac{1}{2}\Psi \lhd \Psi.$$

† We use the symbol $\mathbb{0}$ to denote the quantile function identically equal to 0.

According to Lemma 1.13, there exists $t_0 > 0$ such that for all $t \in [0, t_0]$ we also have

$$(1-t)\Phi + t\Upsilon^+ \lhd \frac{1}{2}\Phi + \frac{1}{2}\Psi \lhd (1-t)\Psi + t\Upsilon^-.$$

This proves that

$$\big[(1-t)\Phi + t\Upsilon^+\big] - \big[(1-t)\Psi + t\Upsilon^-\big] \in C,$$

provided that $t \in [0, t_0]$. Thus relation (1.9) implies that for every $t \in [0, t_0]$ the point $\Gamma + t\Upsilon$ is a sum of two elements of C. Since the set C is a convex cone, this point is also an element of C.

As C is convex, $C = \text{cor}(C)$, and $\mathbb{0} \notin C$, the point $\mathbb{0}$ and the set C can be separated strictly: a linear functional U on $\text{lin}(\mathcal{Q})$ exists, such that

$$U(\Gamma) < 0, \quad \forall\, \Gamma \in C.$$

Thus,

$$U(\Phi) - U(\Psi) < 0, \quad \text{whenever} \quad \Phi \lhd \Psi,$$

as required. □

1.3.3 Integral Representation with Rank Dependent Utility Functions

In order to derive an integral representation of the numerical representation $U(\cdot)$ of the preorder $\unlhd$, we need much stronger conditions, than those of Theorem 1.14. Two issues are important in this respect:

- Continuity of $U(\cdot)$ on an appropriate complete topological vector space containing the set $\mathcal{Q}$ of quantile functions; and
- Integral representation of a continuous linear functional on this space.

To address the first issue, we recall the construction of the space of bounded functions from classical functional analysis. Consider the algebra Σ of all sets obtained by finite unions and intersections of intervals of the form $(a, b]$ in $(0, 1]$, where $0 < a < b \le 1$. With $\mathcal{S} = (0, 1]$ and with Σ, we can define simple functions on $(0, 1]$ as follows:

$$f(p) = \sum_{i=1}^{n} \alpha_i \mathbb{1}_{A_i}(p), \quad p \in \mathcal{S}, \tag{1.10}$$

where $\alpha_i \in \mathbb{R}$ for $i = 1, \dots, n$, and A_i, $i = 1, \dots, n$, are disjoint elements of the algebra Σ. In the formula above, $\mathbb{1}_A(\cdot)$ denotes the characteristic function of a set A. Next, we define the space $B\big((0,1],\Sigma\big)$ of all bounded functions on $(0,1]$ that can be obtained as uniform limits of sequences of simple functions. The space $B\big((0,1],\Sigma\big)$ equipped with the supremum norm:

$$\|\Phi\| = \sup_{0<p\leq 1} \Phi(p),$$

is a Banach space.

From now on, we shall consider only compactly supported distributions on $\mathbb{R}$. The corresponding quantile functions form the prospect space $\mathcal{Q}_b$ of all bounded, nondecreasing, and left-continuous functions on $(0,1]$.[†] The set $\mathcal{Q}_b$ is contained in $B\big((0,1],\Sigma\big)$. Indeed, every monotonic function may have only countably many jumps and their sizes are summable due to the boundedness of the function. Owing to the left-continuity, it can be represented as a uniform limit of simple functions.

We make the following assumption:

Monotonicity Axiom: For all $\Phi, \Psi \in \mathcal{Q}_b$ one has

$$\Phi \leq \Psi \implies \Phi \trianglelefteq \Psi,$$

where the inequality between the functions is understood pointwise.

Theorem 1.15. *If a total preorder $\trianglelefteq$ on $\mathcal{Q}_b$ is continuous and satisfies the dual independence and monotonicity axioms, then a linear continuous functional on $B\big((0,1],\Sigma\big)$ exists, whose restriction to $\mathcal{Q}_b$ is a numerical representation of $\trianglelefteq$.*

Proof. Since the continuity axiom implies the Archimedean axiom, Theorem 1.14 implies the existence of a linear functional $U : \text{lin}(\mathcal{Q}_b) \to \mathbb{R}$ whose restriction to $\mathcal{Q}_b$ is a numerical representation of $\trianglelefteq$. The continuity axiom implies the continuity of the functional $U(\cdot)$ on $\mathcal{Q}_b$. We shall extend $U(\cdot)$ to a continuous functional on the entire space $B\big((0,1],\Sigma\big)$.

Every simple function can be expressed as

$$\begin{aligned}\Phi &= \sum_{i=1}^{n} \alpha_i \mathbb{1}_{(p_i,p_{i+1}]}\\ &= \sum_{i:\alpha_i>0} \alpha_i \big(\mathbb{1}_{(p_i,1]} - \mathbb{1}_{(p_{i+1},1]}\big) + \sum_{i:\alpha_i<0} |\alpha_i| \big(\mathbb{1}_{(p_{i+1},1]} - \mathbb{1}_{(p_i,1]}\big).\end{aligned}$$

with $0 = p_1 < p_2 < \cdots < p_{n+1} = 1$. Rearranging terms, we observe that every simple function is a difference of two simple functions in $\mathcal{Q}_b$ and is thus

[†] Bounded nondecreasing functions on $(0,1)$ can be extended to $(0,1]$ by assigning their left limits as values at 1.

an element of $\text{lin}(\mathcal{Q}_b)$. Consequently, the linear functional $U(\cdot)$ is well-defined on the space of simple functions.

Since the preorder $\trianglelefteq$ satisfies the monotonicity axiom, the linear functional $U(\cdot)$ is monotonic on $\mathcal{Q}_b$. We shall prove that it is also monotonic on the set of simple functions in $B\big((0,1],\Sigma\big)$. Let Φ and Ψ be two simple functions, and let $\Phi \le \Psi$. Then $\Phi = \Phi_1 - \Phi_2$, $\Psi = \Psi_1 - \Psi_2$, where $\Phi_1, \Phi_2, \Psi_1, \Psi_2 \in \mathcal{Q}_b$, and

$$\Phi_1 + \Psi_2 \le \Phi_2 + \Psi_1.$$

As both sides are elements of $\mathcal{Q}_b$ and $U(\cdot)$ is nondecreasing in $\mathcal{Q}_b$ and linear, regrouping the terms we obtain

$$U(\Phi) - U(\Psi) = U(\Phi_1 - \Phi_2 - \Psi_1 + \Psi_2) = U(\Phi_1 + \Psi_2) - U(\Phi_2 + \Psi_1) \le 0.$$

This proves the monotonicity of $U(\cdot)$ on the linear subspace of simple functions.

For any function $\Gamma \in B\big((0,1],\Sigma\big)$, we construct two sequences of simple functions: $\{\Phi_n\}$ and $\{\Psi_n\}$ such that $\Phi_n \le \Gamma \le \Psi_n$, for $n = 1, 2, \ldots$, and

$$\Gamma = \lim_{n\to\infty} \Phi_n = \lim_{n\to\infty} \Psi_n.$$

The sequence $\{U(\Phi_n)\}$ is bounded from above by $U(\Psi_k)$ for any k, due to the monotonicity of $U(\cdot)$. Similar, the sequence $\{U(\Psi_n)\}$ is bounded from below by $U(\Phi_k)$ for any k. Moreover,

$$\begin{aligned} 0 \le U(\Psi_n) - U(\Phi_n) &= U(\Psi_n - \Phi_n) \\ &\le U\big(\|\Psi_n - \Phi_n\| \mathbb{1}_{(0,1]}\big) = U(\mathbb{1}_{(0,1]})\|\Psi_n - \Phi_n\| \to 0. \end{aligned}$$

Therefore, both sequences $\{U(\Phi_n)\}$ and $\{U(\Psi_n)\}$ have the same limit and we can define

$$U(\Gamma) = \lim_{n\to\infty} U(\Phi_n) = \lim_{n\to\infty} U(\Psi_n).$$

We may use any sequence of simple functions $\Gamma_n \to \Gamma$ to calculate $U(\Gamma)$. Indeed, setting $\Phi_n = \Gamma_n - \|\Gamma_n - \Gamma\|$ and $\Psi_n = \Gamma_n + \|\Gamma_n - \Gamma\|$, we obtain $\Phi_n \le \Gamma \le \Psi_n$ and $\Phi_n \le \Gamma_n \le \Psi_n$. Consequently, $U(\Phi_n) \le U(\Gamma_n) \le U(\Psi_n)$ and

$$\lim_{n\to\infty} U(\Gamma_n) = U(\Gamma).$$

The functional $U : B\big((0,1],\Sigma\big) \to \mathbb{R}$ defined in this way is linear on the set of simple functions, which is a subspace of $\text{lin}\,\mathcal{Q}$. Consider two elements Φ and Ψ of $B\big((0,1],\Sigma\big)$, and two sequences $\{\Phi_n\}$ and $\{\Psi_n\}$ of simple functions such that $\Phi_n \to \Phi$ and $\Psi_n \to \Psi$. For any $a, b \in \mathbb{R}$, we obtain

$$
\begin{aligned}
U(a\Phi + b\Psi) &= \lim_{n\to\infty} U(a\Phi_n + b\Psi_n) = \lim_{n\to\infty} \big[aU(\Phi_n) + bU(\Psi_n)\big] \\
&= a \lim_{n\to\infty} U(\Phi_n) + b \lim_{n\to\infty} U(\Psi_n) = aU(\Phi) + bU(\Psi).
\end{aligned}
$$

This proves the linearity of $U(\cdot)$ on the whole space $B\big((0,1],\Sigma\big)$.

To verify monotonicity, consider two elements $\Phi \le \Psi$ in $B\big((0,1],\Sigma\big)$, and two sequences $\{\Phi_n\}$ and $\{\Psi_n\}$ of simple functions such that $\Phi_n \to \Phi$, $\Psi_n \to \Psi$, and $\Phi_n \le \Phi \le \Psi \le \Psi_n$. As $U(\cdot)$ is monotonic on the space of simple functions, we obtain $U(\Phi_n) \le U(\Psi_n)$, and thus $U(\Phi) \le U(\Psi)$.

To prove continuity, consider any element $\Phi \in B\big((0,1],\Sigma\big)$. Owing to linearity and monotonicity of $U(\cdot)$, we obtain

$$
U(\Phi) \le U\big(\|\Phi\|\mathbb{1}_{(0,1]}\big) = U(\mathbb{1}_{(0,1]})\|\Phi\|.
$$

Consequently, $U(\cdot)$ is continuous. □

We can now prove the main result of this section. It involves integration with respect to finitely additive measures. Recall that the integral of a simple function of form (1.10) with respect to a finitely additive measure μ on Σ is defined as follows:

$$
\int_{\mathcal{S}} f(x)\, \mathrm{d}^{\Sigma}\mu \triangleq \sum_{i=1}^{n} \alpha_i \mu(A_i).
$$

The integral of a function $f \in B(\mathcal{S},\Sigma)$ is defined as the limit of the integrals of simple functions f_n converging to f in the supremum norm:

$$
\int_{\mathcal{S}} f(x)\, \mathrm{d}^{\Sigma}\mu \triangleq \lim_{n\to\infty} \int_{\mathcal{S}} f_n(x)\, \mathrm{d}^{\Sigma}\mu, \quad \text{where} \quad \|f_n - f\| \to 0.
$$

The value of the integral does not depend on the particular sequence $\{f_n\}$ converging to f.

Theorem 1.16. *Suppose the total preorder $\trianglelefteq$ on $\mathcal{Q}_{\mathrm{b}}$ is continuous and satisfies the dual independence and monotonicity axioms. Then a nonnegative, bounded, finitely additive measure μ on Σ exists, such that the functional*

$$
U(\Phi) = \int_0^1 \Phi(p)\, \mathrm{d}^{\Sigma}\mu, \quad \Phi \in \mathcal{Q}_{\mathrm{b}}, \tag{1.11}
$$

is a numerical representation of $\trianglelefteq$.

Proof. The functional $U : B\big((0,1],\Sigma\big) \to \mathbb{R}$ constructed in the proof of Theorem 1.15 is linear and continuous. By the representation theorem of continuous linear functionals on $B\big((0,1],\Sigma\big)$, it has the form (1.11) of an integral with respect to a bounded and finitely additive measure μ. Since $U(\cdot)$ is nondecreasing, μ is nonnegative. □

Under additional conditions, we can write the integral (1.11) in the form of a Stieltjes integral. We define a nondecreasing and bounded function $w : [0,1] \to \mathbb{R}_+$ as follows:

$$w(p) = \mu\big((0,p]\big), \quad p \in (0,1]; \quad w(0) = 0. \tag{1.12}$$

If the jump points of $\Phi(\cdot)$ and $w(\cdot)$ do not coincide, we can rewrite (1.11) as follows:

$$U(\Phi) = \int_0^1 \Phi(p)\,\mathrm{d}w(p). \tag{1.13}$$

In general, however, to validate the integral representation (1.13), we need a weaker topology on the prospect space. We use the $\mathcal{L}_1$-topology on the space $\mathcal{Q}_\mathrm{b}$ of quantile functions, defined by the distance function

$$\mathrm{dist}(\Phi, \Psi) = \int_0^1 |\Phi(p) - \Psi(p)|\,\mathrm{d}p.$$

Theorem 1.17. *Suppose the total preorder $\trianglelefteq$ on $\mathcal{Q}_\mathrm{b}$ is continuous in the $\mathcal{L}_1$-topology and satisfies the dual independence and monotonicity axioms. Then a bounded, nondecreasing, and continuous function $w : [0,1] \to \mathbb{R}_+$ exists, such that the functional* (1.13) *is a numerical representation of $\trianglelefteq$ on $\mathcal{Q}_\mathrm{b}$.*

Proof. The assumptions of Theorem 1.16 are satisfied, and thus a finitely additive measure μ exists such that formula (1.11) holds. Define $w(\cdot)$ by (1.12). As μ is nonnegative, $w(\cdot)$ is nondecreasing.

Consider a sequence of simple functions $\mathbb{1}_{(p_n,1]}$, with $p_n \to p \in (0,1)$, as $n \to \infty$. They are elements of $\mathcal{Q}_\mathrm{b}$ and converge in the $\mathcal{L}_1$-topology to $\mathbb{1}_{(p,1]}$. The continuity of the preorder $\trianglelefteq$ in this topology implies that the numerical representation (1.11) is continuous. We obtain

$$U\big(\mathbb{1}_{(p_n,1]}\big) = \mu\big((p_n,1]\big) = w(1) - w(p_n) \to U\big(\mathbb{1}_{(p,1]}\big) = w(1) - w(p).$$

Thus $w(\cdot)$ is continuous in $(0,1)$. If $p = 1$, then $\mathbb{1}_{(p_n,1]} \to \mathbb{0}$, and we obtain in the same way $w(p_n) \to w(1)$.

Because $w(\cdot)$ is continuous, then μ is a regular, bounded, countably additive, and atomless measure. Consequently, the integral representation (1.11) can be written in the form of the Stieltjes integral (1.13). □

The function $w(\cdot)$ appearing in the integral representation (1.13) is called the *rank-dependent utility function* or *dual utility function*.

1.3.4 Choquet Integral Representation of Dual Utility

We presented the dual utility theory in the prospect space of the quantile functions, which is most natural for it. It is interesting, though, to derive an equivalent representation in the prospect space of distribution functions. Every bounded quantile function $\Phi \in \mathcal{Q}_b$ corresponds to a distribution function $F : \mathbb{R} \to [0, 1]$ of a measure with bounded support,

$$\Phi(p) = F^{-1}(p) \triangleq \inf\{\eta \in \mathbb{R} : F(\eta) \ge p\},$$
$$F(z) = \Phi^{-1}(z) \triangleq \begin{cases} \sup\{p \in [0, 1] : \Phi(p) \le z\} & \text{if } z \ge \Phi(0), \\ 0 & \text{otherwise.} \end{cases}$$

The following theorem employs a form of integration by parts for the integral (1.11).

Theorem 1.18. *Suppose the total preorder $\trianglelefteq$ on $\mathcal{Q}_b$ is continuous with respect to uniform convergence and satisfies the dual independence and monotonicity axioms. Then a nondecreasing function $w : [0, 1] \to [0, 1]$ exists, satisfying $w(0) = 0$, $w(1) = 1$, and such that the functional*

$$I(F) = U(F^{-1}) = -\int_{-\infty}^{0} w\big(F(z)\big)\,\mathrm{d}z + \int_{0}^{\infty} \big[1 - w\big(F(z)\big)\big]\,\mathrm{d}z \qquad (1.14)$$

is a numerical representation of $\trianglelefteq$.

Proof. According to Theorem 1.16, the functional (1.11) with some nonnegative, bounded, finitely additive measure μ on Σ, is a numerical representation of $\trianglelefteq$. Without loss of generality, we may assume that $\mu\big((0, 1]\big) = 1$ and define the function $w(\cdot)$ by (1.12).

First, we check the formula (1.14) for a stepwise function $\Phi(\cdot)$, given as follows:

$$\Phi(x) = \sum_{i=1}^{n} z_i \mathbb{1}_{(p_i, p_{i+1}]}(x), \quad x \in (0, 1], \qquad (1.15)$$

where $z_1 \le z_2 \le \cdots \le z_n$ and $0 = p_1 < p_2 < \cdots < p_n < p_{n+1} = 1$.

The integral (1.11) takes on the form:

$$\int_0^1 \Phi(x)\,\mathrm{d}^{\Sigma}\mu = \sum_{i=1}^{n} z_i \mu\big((p_i, p_{i+1}]\big) = \sum_{i=1}^{n} z_i \big[w(p_{i+1}) - w(p_i)\big].$$

With no loss of generality, we may assume that $z_k = 0$ for some k. Then

$$\int_0^1 \Phi(x)\, \mathrm{d}^\Sigma\mu = \sum_{i=1}^{k-1} z_i\big[w(p_{i+1}) - w(p_i)\big] + \sum_{i=k+1}^{n} z_i\big[w(p_{i+1}) - w(p_i)\big]$$
$$= \sum_{i=1}^{k-1} w(p_{i+1})(z_i - z_{i+1}) + \sum_{i=k}^{n-1} \big[1 - w(p_{i+1})\big](z_{i+1} - z_i).$$

Substituting $p_{i+1} = F(z_i)$, $i = 1, \dots, n-1$, we obtain

$$\int_0^1 \Phi(x)\, \mathrm{d}^\Sigma\mu = -\sum_{i=1}^{k-1} w\big(F(z_i)\big)(z_{i+1} - z_i) + \sum_{i=k}^{n-1} \big[1 - w\big(F(z_i)\big)\big](z_{i+1} - z_i)$$
$$= -\int_{-\infty}^0 w\big(F(z)\big)\, \mathrm{d}z + \int_0^\infty \big[1 - w\big(F(z)\big)\big]\, \mathrm{d}z.$$

This proves the formula (1.14) for simple functions.

To prove it for a general function $\Phi \in \mathcal{Q}_\mathrm{b}$, we consider two sequences of simple functions $\{\Phi_n\}$ and $\{\Psi_n\}$, such that $\Phi_n \le \Phi \le \Psi_n$, $n = 1, 2, \dots$, and

$$\|\Psi_n - \Phi_n\|_\infty \to 0 \quad \text{as} \quad n \to \infty.$$

Since $\Phi_n^{-1} \ge \Phi^{-1} \ge \Psi_n^{-1}$ and $w(\cdot)$ is nondecreasing, we have

$$U(\Phi_n) = -\int_{-\infty}^0 w\big(\Phi_n^{-1}\big)\, \mathrm{d}z + \int_0^\infty \big[1 - w\big(\Phi_n^{-1}\big)\big]\, \mathrm{d}z$$
$$\le -\int_{-\infty}^0 w\big(\Phi^{-1}\big)\, \mathrm{d}z + \int_0^\infty \big[1 - w\big(\Phi^{-1}\big)\big]\, \mathrm{d}z$$
$$\le -\int_{-\infty}^0 w\big(\Psi_n^{-1}\big)\, \mathrm{d}z + \int_0^\infty \big[1 - w\big(\Psi_n^{-1}\big)\big]\, \mathrm{d}z = U(\Psi_n).$$

The first and last equations follow from the formula (1.14) for simple functions. Since $U(\cdot)$ is continuous, the leftmost and the rightmost members converge to $U(\Phi)$, as $n \to \infty$, and thus the middle member is equal to $U(\Phi)$ as well. □

Formula (1.14) is a special case of the Choquet integral of the function $F^{-1}(\cdot)$. In our case, we did not invoke the theory of capacities, because the prospect space contains only monotonic functions, and all derivations could be carried out directly.

1.3.5 Risk Aversion

Now, we introduce the concept of a risk-averse preference relation among quantile functions. For every quantile function $\Phi \in \mathcal{Q}_\mathrm{b}$ and every interval $(a, b] \in \Sigma$, we can define a new quantile function $\Phi_{(a,b]}$ as follows

$$\Phi_{(a,b]}(p) = \begin{cases} \dfrac{1}{b-a}\displaystyle\int_a^b \Phi(s)\,\mathrm{d}s, & \text{if } p \in (a,b], \\ \Phi(p), & \text{otherwise.} \end{cases} \tag{1.16}$$

Formula (1.16) can interpreted as a quantile version of the conditional expectation.

The following definition is a slightly modified version of Definition 1.9.

Definition 1.19. A preference relation $\trianglelefteq$ on $\mathcal{Q}_\mathfrak{b}$ is *risk-averse*, if $\Phi_{(a,b]} \trianglelefteq \Phi$, for every $\Phi \in \mathcal{Q}_\mathfrak{b}$ and all $(a,b] \in \Sigma$.

In particular, if $(a,b] = (0,1]$ we infer that the constant function equal to the expected value of Φ is preferred over Φ.

Theorem 1.20. *Suppose a total preorder $\trianglelefteq$ on $\mathcal{Q}_\mathfrak{b}$ is monotonic, continuous, and satisfies the dual independence axiom. Then it is risk averse if and only if it has the numerical representations* (1.13) *and* (1.14) *with a nondecreasing and convex function $w : [0,1] \to [0,1]$.*

Proof. In view of Theorem 1.18, we only need to prove that $w(\cdot)$ is convex. Consider any $0 < p_1 < p_2 < p_3 \le 1$. Define a four-point distribution with mass p_1 at -3, mass $p_2 - p_1$ at -2, mass $p_3 - p_2$ at -1, and mass $1 - p_3$ at 0. The corresponding quantile function Φ has, according to (1.14), the utility $U(\Phi) = -\big[w(p_1) + w(p_2) + w(p_3)\big]$.

Choose $a = p_1$ and $b = p_3$ in (1.16). The resulting quantile function $\Phi_{(p_1,p_3]}$ has the form

$$\Phi_{(p_1,p_3]}(p) = \begin{cases} -3 & \text{if } 0 < p \le p_1, \\ \frac{2p_1 - p_2 - p_3}{p_3 - p_1} & \text{if } p_1 < p \le p_3, \\ 0 & \text{if} p_3 < p \le 1. \end{cases}$$

It has the utility

$$U(\Phi_{(p_1,p_3]}) = -\left[w(p_1)\frac{-p_1 - p_2 + 2p_3}{p_3 - p_1} + w(p_3)\frac{-2p_1 + p_2 + p_3}{p_3 - p_1} \right].$$

By risk aversion, $U(\Phi_{(p_1,p_3]}) \le U(\Phi)$. After elementary manipulations, we obtain the inequality

$$w(p_1)\frac{p_3 - p_2}{p_3 - p_1} + w(p_3)\frac{p_2 - p_1}{p_3 - p_1} \ge w(p_2).$$

Let $\alpha \in (0,1)$ and let $p_2 = \alpha p_1 + (1-\alpha)p_3$. Then the last inequality reads:

$$\alpha w(p_1) + (1-\alpha) w(p_2) \ge w\big(\alpha p_1 + (1-\alpha)p_3\big).$$

This is equivalent to the convexity of $w(\cdot)$ on $(0,1]$.

To prove the converse implication, suppose $w(\cdot)$ is convex. Then, for any quantile function $\Phi(\cdot)$ we can transform the formula (1.13) as follows:

$$U(\Phi) = \int_0^a \Phi(p)\,\mathrm{d}w(p) + \int_a^b \Phi(p)\,\mathrm{d}w(p) + \int_b^1 \Phi(p)\,\mathrm{d}w(p). \qquad (1.17)$$

Consider the middle integral:

$$\int_a^b \Phi(p)\,\mathrm{d}w(p) = \int_a^b \Phi(p)\lambda(p)\,\mathrm{d}p = \int_a^b \Phi(p)\Big[\lambda(a) + \int_a^p \mathrm{d}\lambda(r)\Big]\,\mathrm{d}p,$$

with $\lambda(p) \in \partial w(p)$. By the monotonicity and convexity of $w(\cdot)$, the function $\lambda(\cdot)$ is nonnegative and nondecreasing. Changing the order of integration, we obtain:

$$\int_a^b \Phi(p)\,\mathrm{d}w(p) = \lambda(a)\int_a^b \Phi(p)\,\mathrm{d}p + \int_a^b\int_r^b \Phi(p)\,\mathrm{d}p\,\mathrm{d}\lambda(r).$$

If we replace $\Phi(\cdot)$ by $\Phi_{(a,b]}(\cdot)$ in (1.17), only the middle integral will change. Applying the last displayed formula to $\Phi_{(a,b]}(\cdot)$ we obtain

$$\begin{aligned}\int_a^b \Phi_{(a,b]}(p)\,\mathrm{d}w(p) &= \lambda(a)\int_a^b \Phi_{(a,b]}(p)\,\mathrm{d}p + \int_a^b\int_r^b \Phi_{(a,b]}(p)\,\mathrm{d}p\,\mathrm{d}\lambda(r)\\ &= \lambda(a)\int_a^b \Phi(p)\,\mathrm{d}p + \int_a^b \frac{b-r}{b-a}\int_a^b \Phi(s)\,\mathrm{d}s\,\mathrm{d}\lambda(r).\end{aligned}$$

The only difference between the two formulas is the last term. Due to the monotonicity of $\Phi(\cdot)$,

$$\frac{b-r}{b-a}\int_a^b \Phi(s)\,\mathrm{d}s \le \int_r^b \Phi(p)\,\mathrm{d}p.$$

Since $\mathrm{d}\lambda \ge 0$ and the remaining components in the expressions for $U(\Phi_{(a,b]})$ and $U(\Phi)$ are identical, we conclude that $U(\Phi_{(a,b]}) \le U(\Phi)$. □

Remark 1.21. If we assumed only that the quantile function of the expected value is preferred, that is $\Phi_{(0,1]} \trianglelefteq \Phi$, then we would not be able to infer the convexity of the function $w(\cdot)$.

In fact, the relation $\Phi_{(0,1]} \trianglelefteq \Phi$ for all Φ is equivalent to the inequality $w(p) \le p$ for all $p \in (0,1]$. Indeed, consider a two-point distribution, with mass p at 0 and mass $1-p$ at 1. Then $U(\Phi) = 1-w(p)$ and $U\big(\Phi_{(0,1]}\big) = 1-p$. Thus $w(p) \le p$. To prove the converse implication, we use the inequality $w\big(F(z)\big) \le F(z)$ in (1.14) to obtain

$$U(\Phi) \ge -\int_{-\infty}^0 F(z)\,\mathrm{d}z + \int_0^\infty \big[1-F(z)\big]\,\mathrm{d}z = U\big(\Phi_{(0.1]}\big).$$

This is in contrast to the expected utility case when the preference of the expected value was sufficient to derive preference of all conditional expectations (*cf.* Remark 1.11). □

We may also remark that the properties of monotonicity and risk aversion are both implied by the monotonicity of the preorder with respect to the *increasing convex order*. We discuss this issue in Chapter 5.

1.4 Utility Theory for Random Vectors

We now reformulate our results of sections 1.1 and 1.3 in terms of random variables or vectors.

1.4.1 Expected Utility Theory for Random Vectors

The expected utility theory can be formulated in terms of random vectors if the operation of forming a convex combination is replaced by a lottery.

1.4.1.1 Lotteries

Suppose $(\Omega, \mathcal{F}, P)$ is a probability space and the prospect space $\mathcal{Z}$ is the space of random vectors $Z : \Omega \to \mathcal{S}$, where $\mathcal{S}$ is a Polish space equipped with its Borel σ-algebra $\mathcal{B}$. The distribution of a random vector $Z \in \mathcal{Z}$ is the probability measure μ_Z on $\mathcal{B}$ defined as $\mu_Z = P \circ Z^{-1}$. We say that Z and W *have the same law* and write $Z \overset{\mathcal{D}}{\sim} W$, if $\mu_Z = \mu_W$.

Definition 1.22. The preference relation $\trianglelefteq$ on $\mathcal{Z}$ is *law-invariant* if

$$Z \overset{\mathcal{D}}{\sim} W \implies Z \sim W.$$

Every preference relation $\preceq$ on $\mathcal{P}(\mathcal{S})$, the space of probability measures on $\mathcal{S}$, defines a law-invariant preference relation $\trianglelefteq$ on $\mathcal{Z}$ as follows:

$$Z \trianglelefteq W \iff \mu_Z \preceq \mu_W.$$

The converse statement is true, if we additionally require that every probability measure μ on $\mathcal{S}$ is a distribution of some $Z \in \mathcal{Z}$. This can be guaranteed if $(\Omega, \mathcal{F}, P)$ is a standard atomless probability space. In this case, we can consider an operation on random variables in $\mathcal{Z}$ corresponding to the operation of taking a convex combination of measures on $\mathcal{S}$.

For three elements Z, V, and W in $\mathcal{Z}$ we say that W is a *lottery* of Z and V with probabilities $\alpha \in (0,1)$ and $(1-\alpha)$, if an event $A \in \mathcal{F}$ of probability α exists, such that the conditional distribution of W, given A, is the same as the (unconditional) distribution of Z; while the conditional distribution of W, given $\bar{A} = \Omega \setminus A$, is the same as the unconditional distribution of V. If A is independent of Z and V, we can construct a lottery as follows:

$$W = \begin{cases} Z & \text{if } A, \\ V & \text{if not } A. \end{cases}$$

In this case, the probability measure μ_W induced by W on $\mathcal{S}$ is the corresponding convex combination of the probability measures μ_Z and μ_V of Z and V, respectively:

$$\mu_W = \alpha\mu_Z + (1-\alpha)\mu_V.$$

We write the lottery symbolically as

$$W = \alpha Z \oplus (1-\alpha)V.$$

It should be stressed that only the distribution μ_W of the lottery is defined uniquely, not the random variable W itself. However, if the preference relation $\trianglelefteq$ on $\mathcal{Z}$ is law-invariant, it makes sense to compare lotteries.

For law-invariant preferences on the space of random vectors with values in $\mathcal{S}$, we introduce axioms corresponding to the axioms of the expected utility theory for distributions.

Independence Axiom for Random Vectors: For all $Z, V, W \in \mathcal{Z}$ one has

$$Z \lhd V \implies \alpha Z \oplus (1-\alpha)W \lhd \alpha V \oplus (1-\alpha)W, \quad \forall \alpha \in (0,1),$$

Archimedean Axiom for Random Vectors: If $Z \lhd V \lhd W$, then $\alpha, \beta \in (0,1)$ exist such that

$$\alpha Z \oplus (1-\alpha)W \lhd V \lhd \beta Z \oplus (1-\beta)W.$$

These conditions allow us to rephrase the results of Section 1.2.2 in the language of random vectors.

Theorem 1.23. *Suppose the total preorder $\trianglelefteq$ on $\mathcal{Z}$ satisfies the independence and Archimedean axioms for random vectors. Then a numerical representation $U : \mathcal{Z} \to \mathbb{R}$ of $\trianglelefteq$ exists, which satisfies for all $Z, V \in \mathcal{Z}$ and all $\alpha \in [0,1]$ the equation*

$$U(\alpha Z \oplus (1-\alpha)V) = \alpha U(Z) + (1-\alpha)U(V).$$

Proof. The assertion follows directly from Theorem 1.5. □

1.4.1.2 Integral Numerical Representation

In order to invoke the integral representation from § 1.2.3, we need to introduce an appropriate topology on the space $\mathcal{Z}$ and assume continuity of the preorder $\trianglelefteq$ in this topology. We shall use the topology of convergence in distribution. Recall that a sequence of random vectors $Z_n : \Omega \to \mathcal{S}$ converges in distribution to a random vector $Z : \Omega \to \mathcal{S}$, if the sequence of probability measures $\{\mu_{Z_n}\}$ converges weakly to the measure μ_Z.

We can now recall Theorem 1.6 to obtain an integral representation of the utility functional.

Theorem 1.24. *Suppose the total preorder $\trianglelefteq$ on $\mathcal{Z}$ is law-invariant, continuous, and satisfies the independence axiom for random vectors. Then a continuous and bounded function $u : \mathcal{S} \to \mathbb{R}$ exists, such that the functional*

$$U(Z) = \mathbb{E}\big[u(Z)\big] = \int_{\Omega} u(Z(\omega))\, P(\mathrm{d}\omega) \tag{1.18}$$

is a numerical representation of $\trianglelefteq$ on $\mathcal{Z}$.

It should be stressed, however, that the assumption of continuity with respect to convergence in distribution is rather strong. For example, if we assume only that for every $Z \in \mathcal{Z}$ the sets

$$\{V \in \mathcal{Z} : V \trianglelefteq Z\} \quad \text{and} \quad \{V \in \mathcal{Z} : Z \trianglelefteq V\}$$

are closed in the space $\mathcal{L}_1(\Omega, \mathcal{F}, P; \mathcal{S})$, the existence of a utility function is not guaranteed.

Let $\psi : \mathcal{S} \to [1, \infty)$ be a continuous *gauge function*, and let $C_{\mathrm{b}}^{\psi}(\mathcal{S})$ be the set of functions $f : \mathcal{S} \to \mathbb{R}$, such that $f/\psi \in C_{\mathrm{b}}(\mathcal{S})$. Similar to the topology of convergence in distribution, we can define the topology of ψ-convergence in distribution, which corresponds to the ψ-weak convergence of the corresponding probability measures on $\mathcal{S}$.

Theorem 1.25. *Suppose the total preorder $\trianglelefteq$ on $\mathcal{Z}$ is law-invariant, continuous in the topology of ψ-convergence in distribution, and satisfies the independence axiom for random vectors. Then a function $u \in C_{\mathrm{b}}(\mathcal{S})$ exists, such that the functional* (1.18) *is a numerical representation of $\trianglelefteq$ on $\mathcal{Z}$.*

Again, the assumption of continuity with respect to ψ-convergence in distribution is rather unusual and strong.

1.4.1.3 Monotonicity and Risk Aversion

We briefly translate the concepts and results of § 1.2.4 to the setting with random variables.

Suppose $\mathcal{S}$ is a separable Banach lattice with the partial order relation $\le$.

Monotonicity Axiom for Random Vectors: For all $z, v \in \mathcal{S}$ one has

$$z \le v \implies z\mathbb{1} \trianglelefteq v\mathbb{1}.$$

Theorem 1.26. *Suppose the total preorder $\trianglelefteq$ on $\mathcal{Z}$ is monotonic, continuous, and satisfies the independence axiom. Then a nondecreasing, continuous and bounded function $u : \mathcal{S} \to \mathbb{R}$ exists, such that the functional* (1.18) *is a numerical representation of $\trianglelefteq$ on $\mathcal{P}(\mathcal{S})$.*

Proof. It is sufficient to verify that the function $u(\cdot)$ in (1.18) is monotonic with respect to $\le$. To this end, we consider $z, v \in \mathcal{S}$ such that $z \le v$. By monotonicity of the order, $u(z) = U(z\mathbb{1}) \le U(v\mathbb{1}) = u(v)$. □

We now consider the case with $\mathcal{Z}$ containing only integrable random vectors, $\mathcal{Z} = \mathcal{L}_p(\Omega, \mathcal{F}, P; \mathcal{S})$, with $p \in [1, \infty]$. This corresponds to a gauge function $\psi_p(z) = 1 + \|z\|^p$ in Theorem 1.25. Then for every $Z \in \mathcal{Z}$ and for every σ-subalgebra $\mathcal{G}$ of $\mathcal{B}$ the conditional expectation $\mathbb{E}[Z|\mathcal{G}]$ is well-defined.

Definition 1.27. A preference relation $\trianglelefteq$ on $\mathcal{Z}$ is *risk-averse*, if $\mathbb{E}[Z|\mathcal{G}] \trianglelefteq Z$, for every $Z \in \mathcal{Z}$ and every σ-subalgebra $\mathcal{G}$ of $\mathcal{F}$.

The following theorem is an immediate consequence of Theorem 1.10.

Theorem 1.28. *Suppose a total preorder $\trianglelefteq$ on $\mathcal{Z}$ is continuous, risk-averse, and satisfies the independence axiom. Then a convex function $u \in C_{\mathrm{b}}^{\psi_p}(\mathcal{S})$ exists such that the functional* (1.18) *is a numerical representation of $\trianglelefteq$ on $\mathcal{Z}$.*

1.4.2 Dual Utility Theory for Random Variables

For the dual utility theory, forming convex combinations of the quantile functions is related to forming convex combinations of comonotonic random variables.

1.4.2.1 Comonotonic Random Variables

Consider the prospect space $\mathcal{Z}$ of real-valued random variables defined on a probability space $(\Omega, \mathcal{F}, P)$. The axioms formulated in § 1.3.1 for quantile functions can be equivalently formulated for comonotonic random variables.

Definition 1.29. Real random variables $Z_i, i = 1, \ldots, n$, are *comonotonic*, if

$$\big(Z_i(\omega) - Z_i(\omega')\big)\big(Z_j(\omega) - Z_j(\omega')\big) \geq 0$$

for all $\omega, \omega' \in \Omega$ and all $i, j = 1, \ldots, n$.

The following axioms were formulated originally when the theory of dual utility was introduced.

Dual Independence Axiom for Random Variables: For all comonotonic random variables $Z, V, W \in \mathcal{Z}$ one has

$$Z \lhd V \implies \alpha Z + (1-\alpha)W \lhd \alpha V + (1-\alpha)W, \quad \forall \alpha \in (0,1),$$

Dual Archimedean Axiom for Random Variables: For all comonotonic random variables $Z, V, W \in \mathcal{Z}$, satisfying the relations

$$Z \lhd V \lhd W,$$

there exist $\alpha, \beta \in (0,1)$ such that

$$\alpha Z + (1-\alpha)W \lhd V \lhd \beta Z + (1-\beta)W.$$

Monotonicity Axiom for Random Variables: For all $Z, V \in \mathcal{Z}$ one has

$$Z \leq V \implies Z \unlhd V.$$

It is clear that for comonotonic random variables, the first two axioms are equivalent to the axioms discussed in § 1.3.1. The monotonicity axiom implies the quantile version, due to the monotonicity of the quantile function.

Theorem 1.30. *Suppose $\mathcal{Z}$ is the set of random variables on a standard and atomless probability space $(\Omega, \mathcal{F}, P)$. If the total preorder $\unlhd$ on $\mathcal{Z}$ is law-invariant and satisfies the dual independence, Archimedean, and monotonicity axioms for random variables, then a numerical representation $U : \mathcal{Z} \to \mathbb{R}$ of $\unlhd$ exists, which satisfies for all comonotonic $Z, V \in \mathcal{Z}$ and all $\alpha, \beta \in \mathbb{R}_+$ the equation*

$$U(\alpha Z + \beta V) = \alpha U(Z) + \beta U(V). \tag{1.19}$$

Moreover,

$$U(c\mathbb{1}) = c, \quad \forall c \in \mathbb{R}. \tag{1.20}$$

Proof. Let Y be a uniform random variable on $(\Omega, \mathcal{F}, P)$. The preference relation $\unlhd$ on $\mathcal{Z}$ induces a preference relation $\preceq$ on $\mathcal{Q}$ by the formula

$$\Phi \preceq \Psi \iff \Phi(Y) \unlhd \Psi(Y).$$

The preference relation $\preceq$ does not depend on the particular choice of Y, because $\trianglelefteq$ is law-invariant.

For comonotonic random variables Z and V, and for $\alpha \in (0, 1)$, we have

$$F^{-1}_{\alpha Z+(1-\alpha)V} = \alpha F^{-1}_Z + (1-\alpha)F^{-1}_V.$$

Thus the dual independence and Archimedean axioms for the relation $\trianglelefteq$ among random variables imply these axioms for the relation $\preceq$ on $\mathcal{Q}$. By virtue of Theorem 1.14, a linear functional $\mathcal{U} : \text{lin}(\mathcal{Q}) \to \mathbb{R}$ exists, whose restriction to $\mathcal{Q}$ is a numerical representation of the preorder $\preceq$.

Then $U(Z) = \mathcal{U}(F^{-1}_Z)$, $Z \in \mathcal{Z}$, is a numerical representation of $\trianglelefteq$. For comonotonic $Z, V \in \mathcal{Z}$ and $\alpha, \beta \geq 0$, the linearity of $\mathcal{U}$ yields

$$\begin{aligned} U(\alpha Z + \beta V) &= \mathcal{U}(F^{-1}_{\alpha Z+\beta V}) = \mathcal{U}(\alpha F^{-1}_Z + \beta F^{-1}_V) \\ &= \alpha \mathcal{U}(F^{-1}_Z) + \beta \mathcal{U}(F^{-1}_V) = \alpha U(Z) + \beta U(V), \end{aligned}$$

which proves (1.19).

By monotonicity, $U(\mathbb{1}) = \mathcal{U}(F^{-1}_{\mathbb{1}}) > 0$. Hence we can normalize $U(\cdot)$ to have $U(\mathbb{1}) = 1$. For $c > 0$ the equation (1.20) follows from (1.19). Then

$$U(-c\mathbb{1}) = \mathcal{U}(F^{-1}_{-c\mathbb{1}}) = \mathcal{U}(-F^{-1}_{c\mathbb{1}}) = -\mathcal{U}(F^{-1}_{c\mathbb{1}}) = -U(c\mathbb{1}) = -c,$$

owing to the linearity of $\mathcal{U}(\cdot)$. □

1.4.2.2 Integral Numerical Representation

In our further considerations, we assume that $\mathcal{Z}$ is the space of bounded random variables equipped with the the norm topology of the space $\mathcal{L}_1(\Omega, \mathcal{F}, P)$.

Theorem 1.31. *Suppose $\mathcal{Z}$ is the set of bounded random variables on a standard and atomless probability space $(\Omega, \mathcal{F}, P)$. If the total preorder $\trianglelefteq$ on $\mathcal{Z}$ is law-invariant, continuous, and satisfies the dual independence and monotonicity axioms for random variables, then a bounded, nondecreasing, and continuous function $w : [0, 1] \to \mathbb{R}_+$ exists, such that the functional*

$$U(Z) = \int_0^1 F^{-1}_Z(p)\, dw(p), \quad Z \in \mathcal{Z}, \tag{1.21}$$

is a numerical representation of $\trianglelefteq$.

Proof. Recall that the preorder $\trianglelefteq$ induces a preorder $\preceq$ on the space $\mathcal{Q}_b$ of bounded quantile functions. The preorder $\preceq$ is defined in the proof of Theorem 1.30. It satisfies the monotonicity condition on $\mathcal{Q}_b$, because for a uniform random variable Y we have the chain of relations:

$$\Phi \leq \Psi \iff \Phi(Y) \leq \Psi(Y) \implies \Phi(Y) \trianglelefteq \Psi(Y) \iff \Phi \preceq \Psi. \quad (1.22)$$

The dual independence axiom for $\preceq$ follows from the dual independence axiom for $\trianglelefteq$ with comonotonic random variables. In order to use Theorem 1.17, we only need to verify the continuity condition for $\preceq$.

Consider a convergent sequence of functions $\{\Phi_n\}$ and a function Ψ in $\mathcal{Q}_\text{b}$, such that $\Phi_n \preceq \Psi$, $n = 1, 2 \ldots$, and let Φ be the $\mathcal{L}_1$-limit of $\{\Phi_n\}$, that is,

$$\lim_{n\to\infty} \int_0^1 |\Phi_n(p) - \Phi(p)| \, \mathrm{d}p = 0.$$

For a uniform random variable Y, we define $Z_n = \Phi_n(Y)$, $Z = \Phi(Y)$, and $V = \Psi(Y)$. By (1.22), $Z_n \trianglelefteq V$. Substituting the definitions of Z_n and Z and changing variables we obtain

$$\begin{aligned} \|Z_n - Z\|_1 &= \int_\Omega |Z_n(\omega) - Z(\omega)| \; P(\mathrm{d}\omega) = \int_\Omega |\Phi_n(Y(\omega)) - \Phi(Y(\omega))| \; P(\mathrm{d}\omega) \\ &= \int_0^1 |\Phi_n(p) - \Phi(p)| \, \mathrm{d}p \to 0, \quad \text{as} \quad n \to \infty. \end{aligned}$$

By the continuity of $\trianglelefteq$ in $\mathcal{Z}$, we conclude that $Z \trianglelefteq V$. By (1.22), $\Phi \preceq \Psi$. In a similar way, we consider the case when $\Psi \preceq \Phi_n$, $n = 1, 2 \ldots$ and we prove that $\Psi \preceq \Phi$. Consequently, the preorder $\preceq$ is continuous in $\mathcal{Q}_\text{b}$.

According to Theorem 1.17, a numerical representation $\mathcal{U}(\cdot) : \mathcal{Q}_\text{b} \to \mathbb{R}$ of $\preceq$ exists, which has the integral representation

$$\mathcal{U}(\Phi) = \int_0^1 \Phi(p) \, \mathrm{d}w(p), \quad \Phi \in \mathcal{Q}_\text{b},$$

for some continuous nondecreasing function $w : (0, 1] \to \mathbb{R}_+$. Setting $U(Z) = \mathcal{U}\big(F_Z^{-1}\big)$, we obtain (1.21). □

Another possibility is to consider the topology of uniform convergence, induced by the norm

$$\|Z\|_\infty = \sup_{\omega\in\Omega} |Z(\omega)|.$$

This means that we identify $\mathcal{Z}$ with the Banach space $B(\Omega, \mathcal{F})$ of bounded functions defined on Ω, which can be obtained as uniform limits of simple functions. We assume that the preorder $\trianglelefteq$ is continuous in this space.

Theorem 1.32. *Suppose $\mathcal{Z} = B(\Omega, \mathcal{F})$ and the probability space $(\Omega, \mathcal{F}, P)$ is standard and atomless. If the total preorder $\trianglelefteq$ on $\mathcal{Z}$ is law-invariant, continuous, and satisfies the dual independence and monotonicity axioms for random variables, then a nondecreasing function $w : [0, 1] \to [0, 1]$ exists, such that the functional*

$$U(Z) = -\int_{-\infty}^{0} w\big(F_Z(\eta)\big)\,\mathrm{d}\eta + \int_{0}^{\infty} \big[1 - w\big(F_Z(\eta)\big)\big]\,\mathrm{d}\eta \tag{1.23}$$

is a numerical representation of $\trianglelefteq$.

Proof. Recall that the preorder $\trianglelefteq$ induces a preorder $\preceq$ on $\mathcal{Q}_b$ defined in the proof of Theorem 1.30. It satisfies the monotonicity condition on $\mathcal{Q}_b$, as in (1.22). In order to use Theorem 1.18, we need to verify the continuity condition for $\preceq$.

Consider a uniformly convergent sequence of functions $\{\Phi_n\}$ and a function Ψ in $\mathcal{Q}_b$, such that $\Phi_n \preceq \Psi$, $n = 1, 2\ldots$, and let Φ be the uniform limit of $\{\Phi_n\}$, that is,

$$\lim_{n\to\infty} \sup_{0\le p\le 1} |\Phi_n(p) - \Phi(p)| = 0.$$

For a uniform random variable Y, we define $Z_n = \Phi_n(Y)$, $Z = \Phi(Y)$, and $V = \Psi(Y)$. By (1.22), $Z_n \trianglelefteq V$. Substituting the definitions of Z_n and Z and changing variables we obtain

$$\begin{aligned}\|Z_n - Z\|_\infty &= \sup_{\omega\in\Omega} |\Phi_n(Y(\omega)) - \Phi(Y(\omega))| \\ &= \sup_{0\le p\le 1} |\Phi_n(p) - \Phi(p)| \to 0, \quad \text{as} \quad n\to\infty.\end{aligned}$$

By the continuity of $\trianglelefteq$ in $\mathcal{Z}$, we conclude that $Z \trianglelefteq V$. By (1.22), $\Phi \preceq \Psi$. In a similar way, we consider the case when $\Psi \preceq \Phi_n$, $n = 1, 2\ldots$ and we prove that $\Psi \preceq \Phi$. Consequently, the preorder $\preceq$ is continuous in $\mathcal{Q}_b$.

According to Theorem 1.18, a numerical representation $\mathcal{U}(\cdot) : \mathcal{Q}_b \to \mathbb{R}$ of $\preceq$ exists, which has the integral representation (1.14) for some continuous nondecreasing function $w : (0, 1] \to \mathbb{R}_+$. Setting $U(Z) = \mathcal{U}\big(F_Z^{-1}\big)$, we obtain (1.23). □

Proof. By Theorem 1.30, a numerical representation $U : \mathcal{Z} \to \mathbb{R}$ of $\trianglelefteq$ exists, which satisfies the comonotonic additivity condition (1.19). With no loss of generality, we may re-scale $U(\cdot)$ so that $U(0) = 0$ and $U(\mathbb{1}) = 1$. Define a set function $\lambda : \mathcal{F} \to [0, 1]$ as follows:

$$\lambda(A) = U(\mathbb{1}_A), \quad A \in \mathcal{F}. \tag{1.24}$$

We shall prove that

$$U(Z) = -\int_{-\infty}^{0} \big[1 - \lambda(Z \ge \eta)\big]\,d\eta + \int_{0}^{\infty} \lambda(Z \ge \eta)\,d\eta. \tag{1.25}$$

First, consider nonnegative simple functions of the form:

$$Z(\omega) = \sum_{i=1}^{n} z_i \mathbb{1}_{A_i}(\omega), \quad \omega \in \Omega,$$

where $A_1, \dots, A_n$ are pairwise disjoint. With no loss of generality we may assume that $0 = z_1 < z_2 < \dots < z_n$, and that $A_1 = \{\omega : Z(\omega) = 0\}$. Defining the sets

$$B_i = \bigcup_{j=i}^{n} A_j = \{z \in \mathbb{R} : z \ge z_i\}, \quad i = 1, \dots, n,$$

we can equivalently write the simple function Z as follows:

$$Z = \sum_{i=2}^{n} (z_i - z_{i-1}) \mathbb{1}_{B_i}.$$

As $B_i \supseteq B_{i+1}$ and $z_i \le z_{i+1}$ for $i = 1, \dots, n-1$, all components of the sum above are comonotonic. It follows from the comonotonic additivity of $U(\cdot)$ (see (1.19)) that

$$U(Z) = \sum_{i=2}^{n} (z_i - z_{i-1}) U\big(\mathbb{1}_{B_i}\big) = \sum_{i=2}^{n} (z_i - z_{i-1}) \lambda(B_i),$$

which proves formula (1.25) for nonnegative step functions.

To prove (1.25) for a general nonnegative bounded random variable Z, we consider two sequences of simple functions $\{Z_n\}$ and $\{V_n\}$, such that $0 \le Z_n \le Z \le V_n$, $n = 1, 2, \dots$, and

$$\|V_n - Z_n\|_\infty \to 0 \quad \text{as} \quad n \to \infty.$$

As $\lambda(Z_n \ge \eta) \le \lambda(Z \ge \eta) \le \lambda(V_n \ge \eta)$ for all $\eta \in \mathbb{R}$, we obtain the following chain of relations:

$$\begin{aligned} U(Z_n) = \int_0^\infty \lambda(Z_n \ge \eta)\, d\eta &\le \int_0^\infty \lambda(Z \ge \eta)\, d\eta \\ &\le \int_0^\infty \lambda(V_n \ge \eta)\, d\eta = U(V_n). \end{aligned}$$

The first and the last equation follow from the formula (1.25) for nonnegative simple functions. Since $U(\cdot)$ is continuous, the leftmost and the rightmost members converge to $U(Z)$, as $n \to \infty$, and thus the middle member must be equal to $U(Z)$. The formula (1.25) is true for nonnegative bounded random variables.

For a general bounded random variable Z, suppose $c = \inf_{\omega \in \Omega} Z(\omega) < 0$. By (1.19)–(1.20),

$$U(Z) = U(c\mathbb{1}) + U(Z - c\mathbb{1}) = c + U(Z - c\mathbb{1}).$$

The variable $Z - c\mathbb{1}$ is nonnegative, and thus the representation (1.25) may be applied. We obtain

$$
\begin{aligned}
U(Z) &= c + \int_0^\infty \lambda(Z - c\mathbb{1} \geq \eta)\, d\eta = c + \int_0^\infty \lambda(Z \geq \eta + c)\, d\eta \\
&= c + \int_c^\infty \lambda(Z \geq t)\, dt \\
&= -\int_c^0 \big[1 - \lambda(Z \geq t)\big]\, dt + \int_0^\infty \lambda(Z \geq t)\, dt.
\end{aligned}
$$

As $\lambda(Z \geq t) = 1$ for all $t \leq c$, (1.25) is true for all bounded random variables.

Since the preorder $\trianglelefteq$ is law-invariant, we must have $U(\mathbb{1}_A) = U(\mathbb{1}_B)$, whenever $P(A) = P(B)$. It follows that there must exist a function $w : [0,1] \to [0,1]$ such that $\lambda(A) = U(\mathbb{1}_A) = w\big(P(A)\big)$ in (1.24). As $U(\cdot)$ is nondecreasing, $w(\cdot)$ is nondecreasing as well. Formula (1.23) follows from (1.25). □

Formula (1.23) is a special case of the Choquet integral of the variable Z. Clearly, if the assumptions of Theorem 1.31 are satisfied, so are the assumptions of Theorem 1.32. In this case, the representation (1.23) follows (by integration by parts and change of variables) from (1.21), provided that the function $w(\cdot)$ in (1.21) is normalized so that $w(0) = 0$ and $w(1) = 1$.

1.4.2.3 Risk Aversion

Let us discuss the meaning of the concept of risk aversion introduced in Definition 1.19 in the space $\mathcal{Z}$ of bounded real random variables on a standard atomless probability space $(\Omega, \mathcal{F}, P)$. With no loss of generality, we may assume that $\Omega = [0,1]$, $\mathcal{F}$ is the Borel σ-algebra, and P is the Lebesgue measure on $[0,1]$. A law-invariant preference relation $\preceq$ on $\mathcal{Z}$ is equivalent to a preference relation $\trianglelefteq$ on $\mathcal{Q}_b$:

$$
Z \preceq V \iff \Phi_Z \trianglelefteq \Phi_V. \tag{1.26}
$$

First, we present the relation of the concept of risk aversion to conditional expectations. Consider a random variable $Z \in \mathcal{Z}$ with the quantile function Φ, which itself can be viewed as an element of $\mathcal{Z}$. For $(a,b] \subset [0,1]$, we define the sub-algebra $\mathcal{F}'$ generated by $(a,b]$ and the sets $B \in \mathcal{F}$ such that $B \cap (a,b] = \emptyset$. Then the random variable $\mathbb{E}[\Phi|\mathcal{F}']$ has $\Phi_{(a,b]}$ as its quantile function and our risk aversion assumption can be rephrased as follows:

$$
\mathbb{E}[\Phi|\mathcal{F}'] \preceq Z.
$$

In the case of $(a,b] = (0,1]$, we obtain $\mathbb{E}[Z] \preceq Z$, which is the concept of risk aversion discussed in Remark 1.21.

Now, we show that the convexity of a preference relation on $\mathcal{Z}$ implies the property of risk aversion.

Theorem 1.33. *If a preference relation $\preceq$ on $\mathcal{Z}$ is law-invariant, convex, and closed, then the preference $\trianglelefteq$ on $\mathcal{Q}_b$ defined by* (1.26) *is risk-averse in the sense of Definition* 1.19.

Proof. Consider a random variable Z, its quantile function $\Phi = F_Z^{-1}$, and the function $\Phi_{(a,b]}$ defined in (1.16). Observe that Φ is a rearrangement of Z. We consider both Φ and $\Phi_{(a,b]}$ as elements of $\mathcal{Z}$. For any $q \in [0,1)$ we have

$$\int_q^1 \Phi_{(a,b]}(p)\,dp \le \int_q^1 \Phi(p)\,dp,$$

and

$$\int_0^1 \Phi_{(a,b]}(p)\,dp = \int_0^1 \Phi(p)\,dp.$$

Therefore, $\Phi_{(a,b]}$ is stochastically smaller than Φ with respect to the *convex order*, which we write[†]

$$\Phi_{(a,b]} \preceq_{(c)} \Phi.$$

The set of all functions $\Psi \in \mathcal{Z}$ such that $\Psi \preceq_{(c)} \Phi$ is convex and weakly compact in $\mathcal{L}_1(0,1)$, and all its extreme points are exactly those functions Z' that have the same quantile function Φ as Z.

Applying the Krein-Milman theorem, we obtain

$$\Phi_{(a,b]} \in \overline{\text{conv}}\big\{Z' : Z' \overset{\mathcal{D}}{\sim} Z\big\}.$$

Recall that the closure of a convex set is the same in all compatible topologies. Since the preference relation $\preceq$ is law-invariant, convex, and closed, we conclude that $\Phi_{(a,b]} \preceq Z$. This implies $\Phi_{(a,b]} \trianglelefteq \Phi$. □

1.4.3 Certainty Equivalents

If the prospect space can be identified with some vector space $\mathcal{Z}$, we can select an element $\mathbb{1} \in \mathcal{Z}$ and use the preorder $\trianglelefteq$ compare other elements of $\mathcal{Z}$ to $\alpha\mathbb{1}$, where $\alpha \in \mathbb{R}$. For this construction to make sense, we have to assume that $\mathbb{1} \not\sim \mathbb{0}$.

In the most important case, which we are considering here, if $\mathcal{Z}$ is the space of real random variables defined on some probability space $(\Omega, \mathcal{F}, P)$, then we can define $\mathbb{1}$ to be the sure outcome of 1, that is, the variable identically equal to 1. This explains the term "certainty equivalent" that we use in this section.

[†] We discuss stochastic orders in Chapter 5.

A particular form of a numerical representation of $\trianglelefteq$ is to construct equivalent outcomes on the line $\{\alpha\mathbb{1} : \alpha \in \mathbb{R}\}$.

Definition 1.34. A numerical representation $\varrho : \mathcal{Z} \to \mathbb{R}$ of a total preorder $\trianglelefteq$ on $\mathcal{Z}$ is called a *certainty equivalent*, if $Z \sim \varrho(Z)\mathbb{1}$ for all $Z \in \mathcal{Z}$.

Example 1.35. Consider the expected utility functional defined in (1.18) with $\mathcal{S} = \mathbb{R}$ and with a continuous and strictly increasing function $u : \mathbb{R} \to \mathbb{R}$. Then the functional

$$\varrho(Z) = u^{-1}\big(U(Z)\big) = u^{-1}\big(\mathbb{E}\big[u(Z)\big]\big)$$

is a certainty equivalent. Indeed, $U\big(\varrho(Z)\mathbb{1}\big) = \mathbb{E}\big[u(Z)\big] = U(Z)$. Applying $u^{-1}(\cdot)$ to both sides, we conclude that $\varrho\big(\varrho(Z)\mathbb{1}\big) = \varrho(Z)$. □

Example 1.36. Let $\mathcal{Z}$ be the space of bounded real random variables, and let $U(\cdot)$ the dual utility functional (1.21), with $w(1) > w(0)$. Then

$$\varrho(Z) = \frac{1}{w(1) - w(0)} U(Z) = \frac{1}{w(1) - w(0)} \int_0^1 F_Z^{-1}(p)\, dw(p), \quad Z \in \mathcal{Z},$$

is a certainty equivalence functional. Indeed, $F^{-1}_{\varrho(Z)\mathbb{1}}(p) = \varrho(Z)$ for all $p \in (0, 1)$, and thus

$$\varrho\big(\varrho(Z)\mathbb{1}\big) = \frac{1}{w(1) - w(0)} \int_0^1 F^{-1}_{\varrho(Z)\mathbb{1}}(p)\, dw(p) = \varrho(Z),$$

as postulated. □

Example 1.37. Suppose $u : \mathbb{R} \to \mathbb{R}$ is nondecreasing and convex, with $u(0) = 0$ and $1 \in \partial u(0)$. We assume that $|u(t)| \le C(1 + |t|^p)$ for all $t \in \mathbb{R}$, with some $C > 0$ and $p \ge 1$, and we consider the prospect space $\mathcal{Z} = \mathcal{L}_p(\Omega, \mathcal{F}, P)$. The *optimized certainty equivalent* is defined as follows

$$\varrho(Z) = \inf_{t\in\mathbb{R}} \Big\{t + \mathbb{E}\big[u(Z - t)\big]\Big\}.$$

If $Z = \beta\mathbb{1}$, then $\varrho(Z) = \inf_{t\in\mathbb{R}}\{t + u(\beta - t)\} = \beta$, because 0 is a subgradient of the function in braces at $t = \beta$. Consequently, $\varrho(\cdot)$ is a certainty equivalent. □

To establish the existence of a certainly equivalent in the general case, we make one additional assumption.

Definition 1.38. The preference relation $\trianglelefteq$ on $\mathcal{Z}$ is *directed*, if it satisfies the following conditions:

(i) If $\alpha < \beta$, then $\alpha\mathbb{1} \lhd \beta\mathbb{1}$;

(ii) For every $Z \in \mathcal{Z}$ numbers α and β exist such that $\alpha\mathbb{1} \unlhd Z \unlhd \beta\mathbb{1}$;

Theorem 1.39. *If a total preorder $\unlhd$ on a vector space $\mathcal{Z}$ is directed, then it has a certainty equivalent $\varrho : \mathcal{Z} \to \mathbb{R}$, which must have the following form:*

$$\varrho(Z) = \inf\{\beta \in \mathbb{R} : Z \unlhd \beta\mathbb{1}\}, \quad Z \in \mathcal{Z}. \tag{1.27}$$

In particular, for all $c \in \mathbb{R}$,

$$\varrho(c\mathbb{1}) = c. \tag{1.28}$$

Proof. Due to the conditions (i) and (ii) of Definition 1.38, the functional (1.27) is well defined for all $Z \in \mathcal{Z}$. Define $\alpha_{\max}(Z) = \sup\{\alpha \in \mathbb{R} : \alpha\mathbb{1} \unlhd Z\}$. By transitivity, $\alpha_{\max}(Z) \le \varrho(Z)$. Suppose $\alpha_{\max}(Z) < \varrho(Z)$. Then we can choose $c \in \mathbb{R}$ such that $\alpha_{\max} < c < \varrho(Z)$. It follows from the definition of $\alpha_{\max}(Z)$ and $\varrho(Z)$ that $Z \lhd c\mathbb{1} \lhd Z$, a contradiction. Therefore, we also have

$$\varrho(Z) = \sup\{\alpha \in \mathbb{R} : \alpha\mathbb{1} \unlhd Z\}, \quad Z \in \mathcal{Z}.$$

This combined with (1.27) yields the relation

$$Z \sim \varrho(Z)\mathbb{1}, \tag{1.29}$$

which means that $\varrho(\cdot)$ is a certainty equivalent.

Suppose $Z \unlhd V$. Then $V \unlhd \beta\mathbb{1}$ implies $Z \unlhd \beta\mathbb{1}$ and thus $\varrho(Z) \le \varrho(V)$. On the other hand, if $Z \lhd V$ then (1.29) implies that

$$\varrho(Z)\mathbb{1} \lhd \varrho(V)\mathbb{1}. \tag{1.30}$$

Thus $\varrho(Z) < \varrho(V)$, because otherwise condition (i) of Definition 1.38 would be violated.

To prove that $\varrho(\cdot)$ is the only possible certainty equivalent, suppose $U : \mathcal{Z} \to \mathbb{R}$ is a certainty equivalent. Then $\varrho(Z)\mathbb{1} \sim Z \sim U(Z)\mathbb{1}$. By virtue of condition (i) of Definition 1.38, $\varrho(Z) = U(Z)$; a strict inequality would lead to a contradiction.

Finally, (1.28) follows directly from (1.27) and condition (i) of Definition 1.38. □

Although a numerical representation of a preference relation is not unique, the certainty equivalent (if it exists) is uniquely defined, once the "sure outcome" $\mathbb{1}$ is fixed.

Chapter 2
Measures of Risk

When the prospect space $\mathcal{Z}$ is a vector space of real random variables defined on some probability space $(\Omega, \mathcal{F}, P)$, we may specify our preference relation $\trianglelefteq$ by directly defining its numerical representation. In this chapter, and in large portions of this book, we shall be concerned with a special class of numerical representations of preferences that we call convex or coherent measures of risk.

2.1 Foundations

We start with the formal definition of a measure of risk and several examples.

2.1.1 Axiomatic Definition of Measures of Risk

Suppose $\varrho : \mathcal{Z} \to \overline{\mathbb{R}} \triangleq \mathbb{R} \cup \{+\infty\}$ is a functional representing our preferences, with the convention that smaller values are preferred. We assume that $\varrho(Z) < +\infty$ for at least one $Z \in \mathcal{Z}$ (it is proper). In the axioms below, the symbol $\mathbb{1}$ represents the sure outcome of 1, that is, the variable identically equal to 1, and all inequalities between random variables are understood in the almost sure sense. We consider the following axioms:

Convexity: $\varrho(\alpha Z + (1-\alpha)V) \le \alpha\varrho(Z) + (1-\alpha)\varrho(V),\ \forall \alpha \in [0, 1]$;
Monotonicity: $Z \le V \implies \varrho(Z) \le \varrho(V)$;
Translation equivariance: $\varrho(Z + c\mathbb{1}) = \varrho(Z) + c,\ \forall c \in \mathbb{R}$;
Positive homogeneity: $\varrho(\gamma Z) = \gamma\varrho(Z),\ \forall \gamma \ge 0$.

D. Dentcheva, A. Ruszczyński, *Risk-Averse Optimization and Control*, Springer Series in Operations Research and Financial Engineering,
https://doi.org/10.1007/978-3-031-57988-2_2

Definition 2.1. A *convex measure of risk* is a proper functional $\varrho : \mathcal{Z} \to \overline{\mathbb{R}}$ satisfying the axioms of convexity, monotonicity, and translation equivariance. If it also satisfies the axiom of positive homogeneity, it is called a *coherent measure of risk.*

Under the conditions of convexity and positive homogeneity, the translation property is equivalent to

Certainty Equivalence: $\varrho(c\mathbb{1}) = c, \ \forall c \in \mathbb{R}.$

The term "certainty equivalence" follows from the direct relation to the certainty equivalent of § 1.4.3.

Lemma 2.2. *If a functional $\varrho : \mathcal{Z} \to \mathbb{R}$ has the properties of convexity, positive homogeneity, and certainty equivalence, then it has the property of translation equivariance. Conversely, if $\varrho(\cdot)$ has the properties of positive homogeneity and translation equivariance, then it has the property of certainty equivalence.*

Proof. The convexity and positive homogeneity of $\varrho(\cdot)$ imply the subadditivity. Using the certainty equivalence, we obtain for all $c \in \mathbb{R}$ the inequalities

$$\varrho(Z + c\mathbb{1}) \le \varrho(Z) + \varrho(c\mathbb{1}) = \varrho(Z) + c,$$

and

$$\varrho(Z) \le \varrho(Z + c\mathbb{1}) + \varrho(-c\mathbb{1}) = \varrho(Z + c\mathbb{1}) - c.$$

Therefore, the translation equivariance is true. Conversely, the translation equivariance and positive homogeneity yield

$$\varrho(c\mathbb{1}) = \varrho(\mathbb{0} + c\mathbb{1}) = \varrho(\mathbb{0}) + c = c.$$

□

We further discuss the relations to certainty equivalents in § 2.1.2.

Remark 2.3. If larger values of $Z \in \mathcal{Z}$ are preferred, the axioms of monotonicity and translation equivariance are modified accordingly:

Reverse Monotonicity: $Z \ge V \implies \varrho(Z) \le \varrho(V);$
Reverse Translation Equivariance: $\varrho(Z + c\mathbb{1}) = \varrho(Z) - c, \ \forall c \in \mathbb{R}.$

The other axioms remain unchanged. □

The axioms can be easily verified in specific cases. In the examples below, we focus on the preferences to smaller outcomes. In the case of preferences to larger outcomes, the change of the sign of Z provides the specific form of the corresponding measure of risk.

We verify the axioms and illustrate how the proper choice of the space $\mathcal{Z}$ guarantees finite values of a measure of risk in $\mathcal{Z}$.

Example 2.4. Suppose $\mathcal{Z} = \mathcal{L}_p(\Omega, \mathcal{F}, P)$, with $p \in [1, \infty]$, and let $\varrho : \mathcal{Z} \to \mathbb{R}$ be the mean–semideviation risk measure:

$$\varrho(Z) \triangleq \mathbb{E}[Z] + \varkappa \big\| \big(Z - \mathbb{E}[Z]\big)_+ \big\|_p, \quad \varkappa \in [0, 1].$$

We shall show that $\varrho(\cdot)$ is a coherent measure of risk.

The convexity of $\varrho(\cdot)$ follows from the convexity of the function $t \mapsto (t)_+$ and the linearity of the operator $Z \mapsto Z - \mathbb{E}[Z]\mathbb{1}$:

$$\big(\alpha Z + (1-\alpha)V - \mathbb{E}[\alpha Z + (1-\alpha)V]\big)_+ \le \alpha\big(Z - \mathbb{E}[Z]\big)_+ + (1-\alpha)\big(V - \mathbb{E}[V]\big)_+.$$

Since $\mathcal{L}_p(\Omega, \mathcal{F}, P)$ is a Banach lattice and both sides of the above inequality are nonnegative,

$$\begin{aligned}
&\big\| \big(\alpha Z + (1-\alpha)V - \mathbb{E}[\alpha Z + (1-\alpha)V]\big)_+ \big\|_p \\
&\quad \le \big\| \alpha\big(Z - \mathbb{E}[Z]\big)_+ + (1-\alpha)\big(V - \mathbb{E}[V]\big)_+ \big\|_p \\
&\quad \le \alpha \big\| \big(Z - \mathbb{E}[Z]\big)_+ \big\|_p + (1-\alpha) \big\| \big(V - \mathbb{E}[V]\big)_+ \big\|_p.
\end{aligned}$$

In the last step, we also used the convexity of the norm.

To prove the monotonicity, observe that the function $t \mapsto (t)_+$ is convex and positively homogeneous, and thus

$$\big(Z - \mathbb{E}[Z]\big)_+ \le \big(Z - V\big)_+ + \big(V - \mathbb{E}[V]\big)_+ + \big(\mathbb{E}[V] - \mathbb{E}[Z]\big)_+.$$

If $Z \le V$, then $\mathbb{E}[Z] \le \mathbb{E}[V]$, and the last inequality yields

$$\big(Z - \mathbb{E}[Z]\big)_+ \le \big(V - \mathbb{E}[V]\big)_+ + \mathbb{E}[V] - \mathbb{E}[Z].$$

As $\mathcal{L}_p(\Omega, \mathcal{F}, P)$ is a Banach lattice,

$$\begin{aligned}
\big\| \big(Z - \mathbb{E}[Z]\big)_+ \big\|_p &\le \big\| \big(V - \mathbb{E}[V]\big)_+ + \mathbb{E}[V] - \mathbb{E}[Z] \big\|_p \\
&\le \big\| \big(V - \mathbb{E}[V]\big)_+ \big\|_p + \mathbb{E}[V] - \mathbb{E}[Z].
\end{aligned} \tag{2.1}$$

This proves the monotonicity of $\varrho(\cdot)$ for $\varkappa = 1$. For $\varkappa \in [0, 1)$ the monotonicity follows from combining (2.1) with the inequality $\mathbb{E}[Z] \le \mathbb{E}[V]$.

The translation property can be verified directly:

$$\begin{aligned}
\varrho(Z + c\mathbb{1}) &= \mathbb{E}[Z + c\mathbb{1}] + \varkappa \big\| \big(Z + c\mathbb{1} - \mathbb{E}[Z + c\mathbb{1}]\big)_+ \big\|_p \\
&= \mathbb{E}[Z] + c + \varkappa \big\| \big(Z - \mathbb{E}[Z]\big)_+ \big\|_p = \varrho(Z) + c.
\end{aligned}$$

Finally, the positive homogeneity follows from the linearity of the mapping $Z \mapsto Z - \mathbb{E}[Z]\mathbb{1}$, the positive homogeneity of the function $t \mapsto (t)_+$, and the positive homogeneity of the norm. For $\gamma \ge 0$ we obtain

$$\begin{aligned}\varrho(\gamma Z) &= \mathbb{E}[\gamma Z] + \varkappa\big\|(\gamma Z - \mathbb{E}[\gamma Z])_+\big\|_p = \gamma\mathbb{E}[Z] + \varkappa\big\|(\gamma(Z - \mathbb{E}[Z]))_+\big\|_p \\ &= \gamma\mathbb{E}[Z] + \varkappa\gamma\big\|(Z - \mathbb{E}[Z])_+\big\|_p = \gamma\varrho(Z).\end{aligned}$$

Because of the identity

$$\mathbb{E}\big[(Z - \mathbb{E}[Z])_+\big] = \mathbb{E}\big[(Z - \mathbb{E}[Z])_-\big] = \frac{1}{2}\mathbb{E}\big[|Z - \mathbb{E}[Z]|\big],$$

the mean–semideviation risk measure for $p = 1$ can be equivalently written as

$$\varrho(Z) \triangleq \mathbb{E}[Z] + \frac{\varkappa}{2}\big\|Z - \mathbb{E}[Z]\big\|_1, \quad \varkappa \in [0, 1].$$

□

Example 2.5. Suppose $\mathcal{Z} = \mathcal{L}_p(\Omega, \mathcal{F}, P)$, with $p \in [1, \infty]$, and let $\varrho : \mathcal{Z} \to \mathbb{R}$ be the dual risk measure:

$$\varrho(Z) \triangleq \min_{t\in\mathbb{R}} \left\{ t + \frac{1}{\alpha}\big\|(Z - t)_+\big\|_p \right\}, \quad \alpha \in (0, 1). \tag{2.2}$$

We shall verify that it is a coherent measure of risk.

To prove the convexity, we observe that the function

$$(t, Z) \mapsto t + \frac{1}{\alpha}\big\|(Z - t)_+\big\|_p$$

is convex, and thus its minimum with respect to t is convex as well.

If $Z \le V$, then

$$t + \frac{1}{\alpha}\big\|(Z - t)_+\big\|_p \le t + \frac{1}{\alpha}\big\|(V - t)_+\big\|_p$$

for all $t \in \mathbb{R}$. Therefore,

$$\min_{t\in\mathbb{R}} \left\{ t + \frac{1}{\alpha}\big\|(Z - t)_+\big\|_p \right\} \le \min_{t\in\mathbb{R}} \left\{ t + \frac{1}{\alpha}\big\|(V - t)_+\big\|_p \right\},$$

which proves the monotonicity of $\varrho(\cdot)$.

The translation property follows from the substitution $\tau = t - c$ in the chain of relations:

$$\begin{aligned}\varrho(Z + c\mathbb{1}) &= \min_{t\in\mathbb{R}} \left\{ t + \frac{1}{\alpha}\big\|(Z + c - t)_+\big\|_p \right\} \\ &= \min_{\tau\in\mathbb{R}} \left\{ \tau + c + \frac{1}{\alpha}\big\|(Z - \tau)_+\big\|_p \right\} = \varrho(Z) + c.\end{aligned}$$

To verify the positive homogeneity for $\gamma > 0$, we substitute $t = \gamma\tau$ in the equation:

$$\varrho(\gamma Z) = \min_{t\in\mathbb{R}} \left\{ t + \frac{1}{\alpha} \big\| (\gamma Z - t)_+ \big\|_p \right\} = \min_{\tau\in\mathbb{R}} \left\{ \gamma\tau + \frac{\gamma}{\alpha} \big\| (Z - \tau)_+ \big\|_p \right\}$$
$$= \gamma \min_{\tau\in\mathbb{R}} \left\{ \tau + \frac{1}{\alpha} \big\| (Z - \tau)_+ \big\|_p \right\} = \gamma\varrho(Z).$$

If $\gamma = 0$, then $\varrho(\gamma Z) = \varrho(\mathbb{0}) = 0$. □

Example 2.6. Let $\mathcal{Z}$ be the space of random variables Z for such that $\mathbb{E}\left[e^{|Z|/r}\right] < \infty$ for all $r > 0$.[†] On the space $\mathcal{Z}$ we define the *entropic measure of risk* as follows

$$\varrho(Z) \triangleq \frac{1}{\kappa} \ln\left(\mathbb{E}\left[e^{\kappa Z}\right] \right),$$

where $\kappa > 0$. It is a special case of the certainty equivalent functional discussed in Example 1.35, with $u(z) = e^{\kappa z}$. We shall show that $\varrho(\cdot)$ is a convex measure of risk.

Let us verify convexity. Suppose $Z, V \in \mathcal{Z}$ and $\lambda \in (0, 1)$. We have

$$\varrho(\lambda Z + (1-\lambda)V) = \frac{1}{\kappa} \ln\left(\mathbb{E}\left[e^{\lambda\kappa Z + (1-\lambda)\kappa V}\right] \right)$$
$$= \frac{1}{\kappa} \ln\left(\mathbb{E}\left[e^{\lambda\kappa Z} e^{(1-\lambda)\kappa V}\right] \right) = \frac{1}{\kappa} \ln\left(\mathbb{E}\left[(e^{\kappa Z})^{\lambda} (e^{\kappa V})^{1-\lambda}\right] \right).$$

By Hölder's inequality,

$$\mathbb{E}\left[(e^{\kappa Z})^{\lambda} (e^{\kappa V})^{1-\lambda}\right] \le \left(\mathbb{E}\left[e^{\kappa Z}\right] \right)^{\lambda} \left(\mathbb{E}\left[e^{\kappa V}\right] \right)^{1-\lambda}.$$

Combining the last two relations we conclude that

$$\varrho(\lambda Z + (1-\lambda)V) \le \frac{1}{\kappa} \ln\left(\left(\mathbb{E}\left[e^{\kappa Z}\right] \right)^{\lambda} \left(\mathbb{E}\left[e^{\kappa V}\right] \right)^{1-\lambda} \right)$$
$$= \frac{\lambda}{\kappa} \ln\left(\mathbb{E}\left[e^{\kappa Z}\right] \right) + \frac{1-\lambda}{\kappa} \ln\left(\mathbb{E}\left[e^{\kappa V}\right] \right)$$
$$= \lambda\varrho(Z) + (1-\lambda)\varrho(V).$$

The monotonicity follows directly from the fact that the functions $z \mapsto e^{\kappa z}$ and $t \mapsto \ln(t)$ are increasing.

To verify the translation property, we note that

$$\varrho(Z + c\mathbb{1}) = \frac{1}{\kappa} \ln\left(\mathbb{E}\left[e^{\kappa(Z+c)}\right] \right) = \frac{1}{\kappa} \ln\left(\mathbb{E}\left[e^{\kappa Z}\right] e^{\kappa c} \right) = \frac{1}{\kappa} \ln\left(\mathbb{E}\left[e^{\kappa Z}\right] \right) + c.$$

[†] $\mathcal{Z}$ is a Banach space with the norm $\|Z\| = \inf\{r > 0 : \mathbb{E}\left[e^{|Z|/r} - 1\right] \le 1\}$ (the Orlicz heart).

However, the positive homogeneity property is not satisfied, unless Z is constant. This is rectified by the next example by the procedure of homogenization. □

Example 2.7. Let $\mathcal{Z}$ be the space of random variables Z such that $\mathbb{E}\left[\mathrm{e}^{|Z|/r}\right]$ for all $r > 0$, as in Example 2.6†. On the space $\mathcal{Z}$, we define the *entropic value at risk* at level $\alpha \in (0, 1]$ as follows:

$$\varrho(Z) \triangleq \inf_{\kappa>0} \left\{ \frac{1}{\kappa} \ln \left(\frac{1}{\alpha} \mathbb{E}\left[\mathrm{e}^{\kappa Z}\right] \right) \right\}. \tag{2.3}$$

We verify that it is a coherent measure of risk.

Let us verify convexity. To this end, substitute $c = 1/\kappa$ and consider the function:

$$\varphi(c, Z) = c \ln \left(\mathbb{E}\left[\mathrm{e}^{Z/c}\right] \right).$$

We shall show that it is jointly convex in (c, Z). Let $(c_1, Z_1), (c_2, Z_2) \in \mathbb{R}_+ \times \mathcal{Z}$, and let $\lambda \in (0, 1)$. By using the convexity of the entropic risk measure verified in Example 2.6, we obtain

$$\begin{aligned} &\ln \left(\mathbb{E}\left[\mathrm{e}^{(\lambda Z_1 + (1-\lambda) Z_2)/(\lambda c_1 + (1-\lambda) c_2)}\right] \right) \\ &\leq \frac{\lambda c_1}{\lambda c_1 + (1-\lambda) c_2} \ln \left(\mathbb{E}\left[\mathrm{e}^{Z_1/c_1}\right] \right) + \frac{(1-\lambda) c_2}{\lambda c_1 + (1-\lambda) c_2} \ln \left(\mathbb{E}\left[\mathrm{e}^{Z_2/c_2}\right] \right). \end{aligned}$$

Multiplication by $\lambda c_1 + (1 - \lambda) c_2$ proves the joint convexity of the function $\varphi(\cdot, \cdot)$. Since $\varrho(Z) = \inf_{c>0} \big\{ -c \ln(\alpha) + \varphi(c, Z) \big\}$, it is convex.

The monotonicity and translation properties are verified in the same way as for the entropic risk measure in Example 2.6. It remains to prove the positive homogeneity. Obviously, $\varrho(0) = 0$. For $\gamma > 0$, setting $u = \kappa\gamma$, we obtain

$$\varrho(\gamma Z) = \inf_{\kappa>0} \left\{ \frac{1}{\kappa} \ln \left(\frac{1}{\alpha} \mathbb{E}\left[\mathrm{e}^{\kappa\gamma Z}\right] \right) \right\} = \inf_{u>0} \left\{ \frac{\gamma}{u} \ln \left(\frac{1}{\alpha} \mathbb{E}\left[\mathrm{e}^{u Z}\right] \right) \right\} = \gamma \varrho(Z).$$

□

The classical certainty equivalent of Example 1.35 is convex only under very restrictive conditions on $u(\cdot)$, and satisfies the translation property only if $u(\cdot)$ is linear or exponential, as in Example 2.6. However, the optimized certainty equivalent defined in Example 1.37 is convex or coherent under very general conditions.

Example 2.8. Suppose $u : \mathbb{R} \to \mathbb{R}$ is nondecreasing and convex, and $1 \in \partial u(0)$. We assume that $|u(t)| \leq C(1 + |t|^p)$ for all $t \in \mathbb{R}$, with some $C > 0$ and

† We could also consider here the larger Banach space $\mathcal{Z}^{**}$ of random variables Z such that $\mathbb{E}\left[\mathrm{e}^{|Z|/r}\right]$ for some $r > 0$ (the Orlicz space), which is bidual to $\mathcal{Z}$.

$p \geq 1$, and we consider the prospect space $\mathcal{Z} = \mathcal{L}_p(\Omega, \mathcal{F}, P)$. The optimized certainty equivalent,

$$\varrho(Z) = \inf_{t \in \mathbb{R}} \Big\{ t + \mathbb{E}\big[u(Z - t)\big] \Big\},$$

is a convex measure of risk. Indeed, the function in braces in the formula above is jointly convex in (Z, t), and thus $\varrho(\cdot)$ is convex. It is monotonic, by the fact that $u(\cdot)$ is nondecreasing. The translation property is evident:

$$\varrho(Z + c\mathbb{1}) = \inf_{t \in \mathbb{R}} \Big\{ t + \mathbb{E}\big[u(Z + c - t)\big] \Big\} = \inf_{\eta \in \mathbb{R}} \Big\{ \eta + c + \mathbb{E}\big[u(Z - \eta)\big] \Big\}.$$

The functional $\varrho(\cdot)$ is not positively homogeneous unless $u(\cdot)$ is positively homogeneous. □

2.1.2 Relation to Certainty Equivalents

Convex and coherent measures of risk can be formally derived from assumptions on a preference relation $\unlhd$. In all conditions below Z, V, and W denote arbitrary elements of the space $\mathcal{Z}$:

(A1) $Z \unlhd W$ and $V \unlhd W \implies \alpha Z + (1 - \alpha) V \unlhd W, \ \forall \alpha \in [0, 1]$;
(A2) $Z \leq V \implies Z \unlhd V$;
(A3) $Z \unlhd V \implies Z + c\mathbb{1} \unlhd V + c\mathbb{1}, \ \forall c \in \mathbb{R}$;
(A4) $Z \unlhd V \implies \gamma Z \unlhd \gamma V, \ \forall \gamma \geq 0$.

These conditions directly correspond to the axioms of convex and coherent measures of risk. Condition (A1) is the *convexity condition*: for every $W \in \mathcal{Z}$ the set $\{Z \in \mathcal{Z} : Z \unlhd W\}$ is convex. Condition (A2) is the monotonicity condition, (A3) is the translation property, and (A4) is the *positive homogeneity condition*.

Definition 2.9. A total preorder $\unlhd$ on $\mathcal{Z}$ is *convex*, if it satisfies conditions (A1)–(A3); it is *coherent*, if it satisfies conditions (A1)–(A4).

The relation to convex and coherent measures of risk is straightforward.

Theorem 2.10. *Suppose a total preorder $\unlhd$ on $\mathcal{Z}$ is convex and directed. Then its certainty equivalent* (1.27) *exists and is a convex and finite-valued measure of risk. If the preorder additionally satisfies* (A4)*, then its certainty equivalent is a finite-valued coherent measure of risk.*

Conversely, if a functional $\varrho : \mathcal{Z} \to \mathbb{R}$ is a convex and finite-valued measure of risk, then the relation $\unlhd$ defined on $\mathcal{Z}$ by

$$Z \unlhd V \iff \varrho(Z) \leq \varrho(V)$$

is a convex and directed total preorder. If $\varrho(\cdot)$ is a finite-valued coherent measure of risk, the preorder $\trianglelefteq$ is coherent.

Proof. By virtue of Theorem 1.39, a certainty equivalent $\varrho(\cdot)$ of the relation $\trianglelefteq$ exists and has the form (1.27).

Suppose β_Z is such that $Z \trianglelefteq \beta_Z \mathbb{1}$. Condition (A3) implies that $Z - \beta_Z \mathbb{1} \trianglelefteq \mathbb{0}$. Similar, if $V \trianglelefteq \beta_V \mathbb{1}$, then $V - \beta_V \mathbb{1} \trianglelefteq \mathbb{0}$. By condition (A1),

$$\alpha Z + (1-\alpha)V - \big(\alpha\beta_Z + (1-\alpha)\beta_V\big)\mathbb{1} \trianglelefteq \mathbb{0}, \quad \forall\, \alpha \in [0,1].$$

Using condition (A3) again, we conclude that

$$\alpha Z + (1-\alpha)V \trianglelefteq \big(\alpha\beta_Z + (1-\alpha)\beta_V\big)\mathbb{1}.$$

By (1.27),

$$\varrho(\alpha Z + (1-\alpha)V) \le \alpha\beta_Z + (1-\alpha)\beta_V,$$

whenever $Z \trianglelefteq \beta_Z \mathbb{1}$ and $V \trianglelefteq \beta_V \mathbb{1}$. Taking the infimum of the right-hand side with respect to β_Z and β_V, we obtain the convexity of $\varrho(\cdot)$.

Suppose $Z \le V$. Then, for every $\beta \in \mathbb{R}$ such that $V \trianglelefteq \beta\mathbb{1}$, condition (A2) yields $Z \trianglelefteq \beta\mathbb{1}$. It follows that

$$\varrho(Z) = \inf\{\beta : Z \trianglelefteq \beta\mathbb{1}\} \le \inf\{\beta : V \trianglelefteq \beta\mathbb{1}\} = \varrho(V),$$

which is the monotonicity of $\varrho(\cdot)$.

If $Z \trianglelefteq \beta\mathbb{1}$, condition (A3) implies that $Z + c\mathbb{1} \trianglelefteq (\beta + c)\mathbb{1}$ for all $c \in \mathbb{R}$. This implies that $\varrho(Z + c\mathbb{1}) \le \beta + c$. Taking the infimum over β we conclude that

$$\varrho(Z + c\mathbb{1}) \le \varrho(Z) + c. \tag{2.4}$$

By the convexity of $\varrho(\cdot)$,

$$\varrho(Z) \le \frac{1}{2}\big[\varrho(Z + c\mathbb{1}) + \varrho(Z - c\mathbb{1})\big] \le \varrho(Z),$$

and the inequality (2.4) must be an equation.

To prove positive homogeneity, consider $\gamma > 0$. Then it follows from (A4) that

$$Z \trianglelefteq \beta\mathbb{1} \iff \gamma Z \trianglelefteq \gamma\beta\mathbb{1}.$$

Therefore, substituting $t = \gamma\beta$ we obtain

$$\begin{aligned}\varrho(Z) &= \inf\{\beta : Z \trianglelefteq \beta\mathbb{1}\} = \inf\{\beta : \gamma Z \trianglelefteq \gamma\beta\mathbb{1}\}\\ &= \frac{1}{\gamma}\inf\{t : \gamma Z \trianglelefteq t\mathbb{1}\} = \frac{1}{\gamma}\varrho(\gamma Z).\end{aligned}$$

In particular, for $Z = \mathbb{0}$ and for every $\gamma > 0$ we get $\varrho(0) = \varrho(\gamma \mathbb{0}) = \gamma \varrho(\mathbb{0})$, which implies that $\varrho(\mathbb{0}) = 0$.

Conversely, if $\varrho : \mathcal{Z} \to \mathbb{R}$ satisfies the conditions of convexity, monotonicity, translation equivariance, and positive homogeneity, the properties of the preorder $\trianglelefteq$ follow immediately from the corresponding properties of $\varrho(\cdot)$. By the translation equivariance property, $\varrho(\alpha \mathbb{1})$ attains all real values, when $\alpha \in \mathbb{R}$. Thus the preorder $\trianglelefteq$ is directed. □

Remark 2.11. If the preorder $\trianglelefteq$ is directed and satisfies the condition

$$\textbf{(A1')} \quad Z \trianglelefteq \mathbb{0} \text{ and } V \trianglelefteq \mathbb{0} \implies \alpha Z + (1-\alpha) V \trianglelefteq \mathbb{0}, \ \forall \alpha \in [0, 1],$$

as well as condition (A3), then it also satisfies condition (A1). Indeed, it follows from (A1') and (A3) that (A1) holds for all $W = \beta \mathbb{1}$, where $\beta \in \mathbb{R}$. But then it holds true for all W, because the weak preference to W is equivalent to weak preference to $\varrho(W)\mathbb{1}$. □

2.2 Quantile-Based Measures of Risk

We now introduce basic measures of risk that can be derived from quantile functions of random variables. Later, in §2.5, we provide a more general theory of such measures.

2.2.1 *Value at Risk*

We continue with our assumption that a prospect Z represents random "cost," that is, smaller values of Z are preferred. The *Value at Risk* at level $\alpha \in (0, 1)$ of a random prospect $Z \in \mathcal{Z}$ is defined as the $(1-\alpha)$-quantile of the distribution of Z:

$$\mathrm{VaR}_\alpha^+(Z) \triangleq \inf\{\eta : F_Z(\eta) \geq 1 - \alpha\} = F_Z^{-1}(1-\alpha).$$

The Value at Risk has the monotonicity property:

$$Z \leq V \implies \mathrm{VaR}_\alpha^+(Z) \leq \mathrm{VaR}_\alpha^+(V), \quad \forall \alpha \in (0, 1).$$

It also has the translation property:

$$\mathrm{VaR}_\alpha^+(Z + c\mathbb{1}) = \mathrm{VaR}_\alpha^+(Z) + c, \quad \forall c \in \mathbb{R}, \ \forall \alpha \in (0, 1).$$

It is positively homegeneous:

$$\mathrm{VaR}_\alpha^+(\gamma Z) = \gamma \, \mathrm{VaR}_\alpha^+(Z), \quad \forall \gamma \geq 0, \ \forall \alpha \in (0, 1).$$

However, it is not convex, as illustrated in the following example.

Example 2.12. Suppose Z is a Bernoulli variable,

$$Z = \begin{cases} 0 & \text{with probability } 1-p, \\ 1 & \text{with probability } p, \end{cases}$$

with $p \in (0,1)$, and V is independent of Z and has the same distribution as Z. For $p < \alpha < 1$ we have $\mathrm{VaR}_\alpha^+(Z) = \mathrm{VaR}_\alpha^+(V) = 0$. However, if $p < \alpha < 1-(1-p)^2$, we have

$$\mathrm{VaR}_\alpha^+\big(\lambda Z + (1-\lambda)V\big) > 0 = \lambda\,\mathrm{VaR}_\alpha^+(Z) + (1-\lambda)\,\mathrm{VaR}_\alpha^+(V),$$

for all $\lambda \in (0,1)$, which contradicts convexity. □

Consequently, no matter what the value of $\alpha \in (0,1)$ is, the functional $\mathrm{VaR}_\alpha^+(\cdot)$ is neither a coherent nor a convex measure of risk.

Remark 2.13. If random prospects $Z \in \mathcal{Z}$ represent "profits," that is, their larger values are preferred, the Value at Risk is defined as

$$\mathrm{VaR}_\alpha^-(Z) \triangleq -\sup\{\eta : F_Z(\eta) \le \alpha\}.$$

It is clear that $\mathrm{VaR}_\alpha^-(Z) = \mathrm{VaR}_\alpha^+(-Z)$, and thus all considerations regarding the functional $\mathrm{VaR}_\alpha^+(\cdot)$ can be translated to case of the functional $\mathrm{VaR}_\alpha^-(\cdot)$. □

2.2.2 *Average Value at Risk*

In our further considerations we assume the prospect space $\mathcal{Z} = \mathcal{L}_p(\Omega, \mathcal{F}, P)$, where $p \in [1, \infty]$. For integrable random variables we define the *Average Value at Risk* at level $\alpha \in (0,1]$ as follows:

$$\mathrm{AVaR}_\alpha^+(Z) \triangleq \frac{1}{\alpha}\int_0^\alpha \mathrm{VaR}_\beta^+(Z)\,\mathrm{d}\beta. \tag{2.5}$$

Remark 2.14. If $P[Z \ge \mathrm{VaR}_\alpha^+(Z)] = \alpha$, and we can change variables in (2.5) to obtain the equation

$$\mathrm{AVaR}_\alpha^+(Z) = \frac{1}{\alpha}\int_{\mathrm{VaR}_\alpha^+(Z)}^\infty z\,\mathrm{d}F_Z(z) = \mathbb{E}\big[Z \,|\, Z \ge \mathrm{VaR}_\alpha^+(Z)\big].$$

This is the reason why the Average Value at Risk is also called the Conditional Value at Risk. However, the formula above is valid only under the assumption that $P[Z \ge \mathrm{VaR}_\alpha^+(Z)] = \alpha$. This can be guaranteed for random variables with continuous distributions. □

In order to analyze the properties of $\mathrm{AVaR}_\alpha^+(Z)$, it is convenient to consider it as a function of *both* $Z \in \mathcal{Z}$ and $\alpha \in (0,1]$. Define the functions $L : \mathcal{Z} \times [0,1] \to \mathbb{R}$ and $S : \mathcal{Z} \times \mathbb{R} \to \mathbb{R}$ as follows:

$$L(Z,\alpha) \triangleq \int_0^\alpha F_Z^{-1}(\beta)\, \mathrm{d}\beta,$$
$$S(Z,\eta) \triangleq \mathbb{E}\big[(\eta - Z)_+\big].$$

The function $L(Z,\cdot)$ is called the *absolute Lorenz function* associated with the random variable Z. We can formally extend it to the whole real line, by setting $L(Z,\alpha) = +\infty$, for $\alpha \notin [0,1]$.

The Fenchel conjugate functions are defined in the standard way:

$$L^*(Z,\eta) \triangleq \sup_{\alpha\in[0,1]} \big\{\alpha\eta - L(Z,\alpha)\big\}, \quad \eta \in \mathbb{R}, \tag{2.6}$$
$$S^*(Z,\alpha) \triangleq \sup_{\eta\in\mathbb{R}} \big\{\alpha\eta - S(Z,\eta)\big\}, \quad \alpha \in \mathbb{R}. \tag{2.7}$$

The following relation between these functions is at the root of many properties of quantile risk measures.

Theorem 2.15. *For any $Z \in \mathcal{L}_1(\Omega,\mathcal{F},P)$ the following conjugacy relations hold:*

$$L^*(Z,\cdot) = S(Z,\cdot) \quad \text{and} \quad S^*(Z,\cdot) = L(Z,\cdot).$$

Proof. Computing left and right derivatives of $L(Z,\cdot)$ at any $\alpha \in (0,1)$ we obtain its subdifferential:

$$\partial L(Z,\alpha) = \big\{\eta \in \mathbb{R} : P[Z \le \eta] \ge \alpha \text{ and } P[Z \ge \eta] \ge 1-\alpha\big\}. \tag{2.8}$$

As the derivatives are nondecreasing functions of α, the function $L(Z,\cdot)$ is convex. It is also lower semicontinuous.

By the Fenchel duality, the following two statements are equivalent:

(i) $\eta \in \partial L(Z,\alpha)$;

(ii) $L^*(Z,\eta) + L(Z,\alpha) = \alpha\eta$.

For any $\eta \in \mathbb{R}$, by setting $\alpha = P[Z \le \eta]$ we satisfy condition (i), owing to (2.8). Then, by condition (ii), we have

$$\begin{aligned} L^*(Z,\eta) &= \eta F_Z(\eta) - \int_0^{F_Z(\eta)} F_Z^{-1}(\beta)\, \mathrm{d}\beta \\ &= \eta F_Z(\eta) - \int_{-\infty}^{\eta} t\, \mathrm{d}F_Z(t) = \int_{-\infty}^{\infty} (\eta - t)_+\, \mathrm{d}F_Z(t) = S(Z,\eta). \end{aligned}$$

The second relation follows from the Fenchel–Moreau theorem. ◻

We can now derive a useful extremal representation of $\mathrm{AVaR}_\alpha^+(Z)$.

Theorem 2.16. *For every* $\alpha \in (0,1]$ *we have*

$$\mathrm{AVaR}_\alpha^+(Z) = \inf_{\eta\in\mathbb{R}} \left\{ \eta + \frac{1}{\alpha}\mathbb{E}\big[(Z-\eta)_+\big] \right\}. \tag{2.9}$$

For $\alpha \in (0,1)$ *the infimum in* (2.9) *is achieved at* $\hat{\eta} = \mathrm{VaR}_\alpha(Z)$.

Proof. It is elementary to verify that

$$L(Z,1) = \mathrm{AVaR}_1^+(Z) = \int_0^1 F_Z^{-1}(\beta)\, \mathrm{d}\beta = \mathbb{E}[Z]. \tag{2.10}$$

Therefore,

$$\mathrm{AVaR}_\alpha^+(Z) = \frac{1}{\alpha}\int_{1-\alpha}^1 F_Z^{-1}(\beta)\, \mathrm{d}\beta = \frac{1}{\alpha}\big(\mathbb{E}[Z] - L(Z,1-\alpha)\big). \tag{2.11}$$

Owing to Theorem 2.15, we have

$$L(Z,\alpha) = \sup_{\eta\in\mathbb{R}} \left\{ \alpha\eta - \mathbb{E}\big[(\eta - Z)_+\big] \right\}. \tag{2.12}$$

Substituting this equation to (2.11), we obtain

$$\begin{aligned}
\mathrm{AVaR}_\alpha^+(Z) &= \frac{1}{\alpha}\mathbb{E}[Z] - \frac{1}{\alpha}\sup_{\eta\in\mathbb{R}} \left\{ (1-\alpha)\eta - \mathbb{E}\big[(\eta - Z)_+\big] \right\} \\
&= \frac{1}{\alpha}\inf_{\eta\in\mathbb{R}} \left\{ -(1-\alpha)\eta + \mathbb{E}[Z] + \mathbb{E}\big[(\eta - Z)_+\big] \right\}.
\end{aligned}$$

Observe that $Z + (\eta - Z)_+ = \eta + (Z-\eta)_+$. Substituting this identity to the last displayed equation, we get

$$\mathrm{AVaR}_\alpha^+(Z) = \frac{1}{\alpha}\inf_{\eta\in\mathbb{R}} \left\{ \alpha\eta + \mathbb{E}\big[(Z-\eta)_+\big] \right\},$$

as required. For $\alpha \in (0,1)$ the infimum is achieved, because the function in braces is continuous with respect to η, and goes to $+\infty$, when η diverges to $+\infty$ or $-\infty$.

By Fenchel duality, the supremum in (2.12), and thus the infimum in the last displayed equation, can be attained only at $\eta \in \partial L(Z,\alpha)$, which proves our last statement, in view of (2.8). □

It follows from Theorem 2.16 that the Average Value at Risk is a special case of the risk measures discussed in Example 2.5, for $p = 1$. It is, therefore, a coherent measure of risk. It is also a special case of the optimized certainty equivalent functional discussed in Examples 1.37 and 2.8, with $u(t) = \max(0, t/\alpha)$.

By construction, the function $\alpha \mapsto \text{AVaR}^+_\alpha(Z)$ is nonincreasing. Thus, due to (2.10),

$$\delta^+_\alpha(Z) \triangleq \text{AVaR}^+_\alpha(Z) - \mathbb{E}[Z] \geq 0, \quad \alpha \in (0, 1].$$

From (2.9), after elementary manipulations, we obtain the equation:

$$\delta^+_\alpha(Z) = \inf_{\eta \in \mathbb{R}} \mathbb{E}\Big[\max\Big(\eta - Z, \frac{1-\alpha}{\alpha}(Z - \eta)\Big)\Big].$$

For $\alpha \in (0, 1)$, the minimizer in the formula above, as well as in (2.9), exists and is equal to the $(1 - \alpha)$-quantile of Z. Because of that, the function $\delta^+_\alpha(Z)$ is called the *weighted mean deviation from quantile*.

We may observe its relation to the first-order semideviation discussed in Example 2.4.

Corollary 2.17. *For every integrable random variable Z we have*

$$\max_{\alpha \in [0,1]} \inf_{\eta \in \mathbb{R}} \mathbb{E}\Big[\max\big(\alpha(\eta - Z), (1-\alpha)(Z - \eta)\big)\Big] = \mathbb{E}\big[(Z - \mathbb{E}[Z])_+\big].$$

Proof. From equation (2.11) and Theorem 2.15 we obtain:

$$\begin{aligned} &\max_{\alpha \in [0,1]} \inf_{\eta \in \mathbb{R}} \mathbb{E}\Big[\max\big(\alpha(\eta - Z), (1-\alpha)(Z - \eta)\big)\Big] = \max_{\alpha \in [0,1]} \alpha \delta^+_\alpha(Z) \\ &= \max_{\alpha \in [0,1]} \big\{(1-\alpha)\mathbb{E}[Z] - L(Z, 1-\alpha)\big\} = L^*\big(Z, \mathbb{E}[Z]\big) \\ &= S\big(Z, \mathbb{E}[Z]\big) = \mathbb{E}\big[(Z - \mathbb{E}[Z])_+\big]. \end{aligned}$$

□

An interesting relation exists between the absolute Lorenz function and the *Gini mean difference*, defined as follows:

$$\Gamma(Z) = \frac{1}{2}\mathbb{E}\big[|Z' - Z''|\big], \tag{2.13}$$

with Z' and Z'' representing independent random variables with the same distribution as Z.

Lemma 2.18. *For any integrable random variable Z, we have*

$$\int_0^1 L(Z, \alpha)\, d\alpha = \frac{1}{2}\mathbb{E}[Z] - \frac{1}{2}\Gamma(Z).$$

Proof. We change the order of integration and the variables in the integrals:

$$\begin{aligned}\int_0^1 L(Z,\alpha)\,\mathrm{d}\alpha &= \int_0^1\int_0^\alpha F_Z^{-1}(\beta)\,\mathrm{d}\beta\,\mathrm{d}\alpha = \int_0^1\int_\beta^1 F_Z^{-1}(\beta)\,\mathrm{d}\alpha\,\mathrm{d}\beta\\ &= \int_0^1 (1-\beta)F_Z^{-1}(\beta)\,\mathrm{d}\beta = \int_{-\infty}^{\infty}(1-F_Z(z))\,z\,\mathrm{d}F_Z(z)\\ &= \mathbb{E}[Z] - \int_{-\infty}^{\infty} zF_Z(z)\,\mathrm{d}F_Z(z) = \mathbb{E}[Z] - \frac{1}{2}\int_{-\infty}^{\infty} z\,\mathrm{d}F_Z^2(z)\\ &= \mathbb{E}[Z] - \frac{1}{2}\mathbb{E}\big[\max(Z',Z'')\big].\end{aligned}$$

In the last step, we used the fact that $F_Z^2(\cdot)$ is the distribution function of the maximum of the two copies. Since $\max(Z',Z'') = \frac{1}{2}(Z'+Z'') + \frac{1}{2}|Z'-Z''|$, the last calculation can be continued as follows:

$$\int_0^1 L(Z,\alpha)\,\mathrm{d}\alpha = \frac{1}{2}\mathbb{E}[Z] - \frac{1}{4}\mathbb{E}\big[|Z'-Z''|\big],$$

which is the postulated identity. □

Corollary 2.19. *The mean–Gini risk measure*

$$\varrho(Z) \triangleq \mathbb{E}[Z] + \varkappa\Gamma[Z], \quad \varkappa \in [0,1],$$

is coherent on the space $\mathcal{L}_1(\Omega,\mathcal{F},P)$.

Proof. From (2.11) we obtain

$$\mathbb{E}[Z] - L(Z,1-\alpha) = \alpha\,\mathrm{AVaR}_\alpha^+(Z).$$

Integrating over α from 0 to 1, and using Lemma 2.18, we get

$$\frac{1}{2}\mathbb{E}[Z] + \frac{1}{2}\Gamma(Z) = \int_0^1 \alpha\,\mathrm{AVaR}_\alpha^+(Z)\,\mathrm{d}\alpha.$$

This means that with $\varkappa = 1$

$$\varrho(Z) = 2\int_0^1 \alpha\,\mathrm{AVaR}_\alpha^+(Z)\,\mathrm{d}\alpha.$$

The coherence of $\varrho(\cdot)$ follows now from the coherence of $\mathrm{AVaR}_\alpha^+(\cdot)$ and the fact that $2\int_0^1 \alpha\,\mathrm{d}\alpha = 1$. The last equation also shows that $\varrho(\cdot)$ is a *spectral* risk measure, in the sense of § 2.4.2. □

Remark 2.20. If $Z \in \mathcal{L}_1(\Omega,\mathcal{F},P)$ represents "profits," that is, larger values of Z are preferred, we define the Average Value at Risk as follows:

$$\mathrm{AVaR}_\alpha^-(Z) \triangleq \frac{1}{\alpha}\int_0^\alpha \mathrm{VaR}_\beta^-(Z)\,\mathrm{d}\beta = -\frac{1}{\alpha}\int_0^\alpha F_Z^{-1}(\beta)\,\mathrm{d}\beta.$$

As the β-quantile of Z is unique, except for a countable set of β's, we could use the left quantile $F_Z^{-1}(\beta)$ in the right side of this equation. We observe that

$$\mathrm{AVaR}_\alpha^-(Z) = -\frac{1}{\alpha}L(Z,\alpha).$$

Therefore, applying Theorem 2.15, we obtain

$$\mathrm{AVaR}_\alpha^-(Z) = -\frac{1}{\alpha}S^*(Z,\alpha) = \inf_{\eta\in\mathbb{R}}\left\{-\eta + \frac{1}{\alpha}\mathbb{E}\big[(\eta - Z)_+\big]\right\}. \tag{2.14}$$

This relation can also be obtained by substituting $-Z$ for Z in (2.9).

The corresponding weighted mean deviation from the quantile is defined as follows:

$$\delta_\alpha^-(Z) \triangleq \mathrm{AVaR}_\alpha^-(Z) + \mathbb{E}[Z] = \inf_{\eta\in\mathbb{R}}\mathbb{E}\Big[\max\Big(\frac{1-\alpha}{\alpha}(\eta - Z), Z - \eta\Big)\Big].$$

□

2.2.3 Inverse Average Value at Risk

Suppose we know the Average Value at Risk,

$$z = \mathrm{AVaR}_\alpha^+(Z), \tag{2.15}$$

and we want to find the value of α for which this equation is true. We will verify whether equation (2.15) has a solution, whether it is unique, and whether it is possible to efficiently find this solution.

Theorem 2.21. *If* $\mathbb{E}[Z] < z < \operatorname{ess\,sup} Z$ *then equation* (2.15) *has a unique solution, which is given by the following formula:*

$$\alpha_z(Z) = \min_{c\ge 0}\mathbb{E}\big[\big(c(Z - z) + 1\big)_+\big]. \tag{2.16}$$

Proof. For simplicity, suppose $z = 0$ in equation (2.15). From the extremal representation (2.9) we obtain the equation

$$0 = \inf_{\eta\in\mathbb{R}}\big\{\alpha\eta + \mathbb{E}\big[(Z - \eta)_+\big]\big\}. \tag{2.17}$$

The minimizer $\hat{\eta}$ in the problem above, as the $(1-\alpha)$-quantile of Z, must be negative, because it is smaller than $\mathrm{AVaR}_\alpha^+(Z)$, which is equal to 0. It follows from (2.17) that for all $\eta < 0$ we have

$$\mathbb{E}\big[(Z-\eta)_+\big] \geq -\alpha\eta,$$

with equality achieved at the solution $\hat{\eta}$ of problem (2.17). Dividing both sides of the last inequality by $|\eta|$ and changing the variable to $c = 1/|\eta|$, we conclude that for all $c > 0$

$$\mathbb{E}\big[(cZ+1)_+\big] \geq \alpha,$$

with equality achieved at $c = 1/|\hat{\eta}|$. Therefore,

$$\alpha_0(Z) = \min_{c>0} \mathbb{E}\big[(cZ+1)_+\big].$$

Including the point $c = 0$ into the feasible set does not change the optimal value, because $\alpha_0(Z) < 1$. This proves the theorem for $z = 0$. For any other z, the result follows from the identity $\alpha_z(Z) = \alpha_0(Z - z)$. □

The function $\alpha_z(Z)$ is called the *buffered probability of exceedance*. Roughly speaking, it gives us the probability level for which the average of the right α-tail of the distribution of Z is equal to the prescribed value z. It is not a coherent or convex measure of risk, but its intimate relation to the Average Value at Risk and convenient extremal representation are useful in risk modeling and optimization.

We may remark here that the function

$$\psi(z) = \frac{1}{\alpha_z(Z)} = \sup_{c\geq 0} \frac{1}{\mathbb{E}\big[\big(c(Z-z)+1\big)_+\big]}$$

is convex, as the supremum of a family of convex functions.

2.2.4 Distortion Measures

An important class of risk measures can be derived from dual utility models of the form (1.21), which we recall for convenience:

$$\varrho(Z) = \int_0^1 F_Z^{-1}(p)\, dw(p), \tag{2.18}$$

where $w : [0,1] \to [0,1]$ is a nondecreasing continuous function with $w(0) = 0$ and $w(1) = 1$. From Example 1.36 we know that $\varrho(\cdot)$ is a certainty equivalent. It is positively homogeneous, because the functionals $Z \mapsto F_Z^{-1}(p)$ are positively homogeneous for all $p \in (0,1)$. We shall analyze when $\varrho(\cdot)$ satisfies the axioms of a coherent measure of risk. Although in § 1.4.2 we considered only bounded random variables, formula (2.18) may be formally considered for a larger space $\mathcal{Z}$.

Theorem 2.22. *Suppose a function* $w : [0, 1] \to [0, 1]$ *is nondecreasing, convex, has values* $w(0) = 0$ *and* $w(1) = 1$, *and satisfies for some* $r \in (0, 1)$ *the condition*

$$w'(\alpha) \le \frac{C}{(1-\alpha)^r}, \quad \alpha \in (0, 1). \tag{2.19}$$

Then the functional (2.18) *defines a coherent measure of risk on the space* $\mathcal{Z} = \mathcal{L}_p(\Omega, \mathcal{F}, P)$, *with* $p > \frac{1}{1-r}$ *and*

$$\varrho(Z) = \int_0^1 \text{AVaR}_\alpha^+(Z)\, \lambda(\mathrm{d}\alpha), \quad Z \in \mathcal{Z}, \tag{2.20}$$

where λ *is a probability measure on* $(0, 1]$.

Proof. We can represent $w(\cdot)$ as the integral

$$w(\alpha) = \int_0^\alpha s(t)\, \mathrm{d}t, \quad \alpha \in [0, 1],$$

where $s : (0, 1) \to \mathbb{R}_+$ is nondecreasing and $s(t) = w'(t)$ almost everywhere. Thus, we can rewrite (2.18) as follows:

$$\varrho(Z) = \int_0^1 F_Z^{-1}(\alpha)\, s(\alpha)\, \mathrm{d}\alpha = \int_0^1 F_Z^{-1}(\alpha)\left[s(0) + \int_0^\alpha \mathrm{d}s(t)\right] \mathrm{d}\alpha. \tag{2.21}$$

To verify the finiteness of ϱ, it is sufficient to consider nonnegative Z, because $s(\cdot)$ is nondecreasing. For these Z, using condition (2.19), we obtain

$$\begin{aligned}\int_0^1 F_Z^{-1}(\alpha)\, s(\alpha)\, \mathrm{d}\alpha &= \int_0^\infty z\, s(F(z))\, \mathrm{d}F(z) \le C \int_0^\infty z \frac{1}{(1-F(z))^r}\, \mathrm{d}F(z) \\ &= -\frac{C}{1-r}\left[z(1-F(z))^{1-r}\right]_0^\infty + \frac{C}{1-r}\int_0^\infty (1-F(z))^{1-r}\, \mathrm{d}z.\end{aligned}$$

Using Markov inequality in the space $\mathcal{Z} = \mathcal{L}_p(\Omega, \mathcal{F}, P)$:

$$1 - F(z) = P[Z > z] \le \frac{\mathbb{E}[|Z|^p]}{z^p},$$

and recalling that $p(1-r) > 1$, we verify that the first term is well-defined and equal to zero. Moreover, for a sufficiently large $M > 0$,

$$\int_M^\infty (1-F(z))^{1-r}\, \mathrm{d}z \le \mathbb{E}[|Z|^p] \int_M^\infty z^{p(1-r)}\, \mathrm{d}z < \infty.$$

$$\int_0^1 F_Z^{-1}(p)\,s(p)\,\mathrm{d}p \le C \int_0^\infty z \frac{1}{(1-F(z))^q}\,\mathrm{d}F(z).$$

This proves that $\varrho(\cdot)$ is finite-valued on $\mathcal{L}_p(\Omega, \mathcal{F}, P)$.

Changing the order of integration in (2.21) we obtain

$$\begin{aligned}\varrho(Z) &= s(0)\int_0^1 F_Z^{-1}(p)\,\mathrm{d}p + \int_0^1\int_t^1 F_Z^{-1}(p)\,\mathrm{d}p\,\mathrm{d}s(t)\\ &= s(0)\,\mathrm{AVaR}_1^+(Z) + \int_0^1 (1-t)\,\mathrm{AVaR}_{1-t}^+(Z)\,\mathrm{d}s(t).\end{aligned}$$

This is formula (2.20) with the measure $\lambda(\mathrm{d}\alpha) = -\alpha\,\mathrm{d}s(1-\alpha)$ for $\alpha \in (0,1)$, and $\lambda(\{1\}) = s(0)$.

It remains to verify that λ is a probability measure. Indeed, $\lambda \ge 0$, because $s(\cdot)$ is nondecreasing. Integrating by parts, we obtain the chain of relations

$$\begin{aligned}\int_{(0,1)} \lambda(\mathrm{d}\alpha) &= -\int_{(0,1)} \alpha\,\mathrm{d}s(1-\alpha) = -\int_{(0,1)}\int_{(0,\alpha]} \mathrm{d}u\,\mathrm{d}s(1-\alpha)\\ &= -\int_{(0,1)}\int_{[u,1)} \mathrm{d}s(1-\alpha)\,\mathrm{d}u = \int_{(0,1)} \big[s(1-u) - s(0)\big]\,\mathrm{d}u\\ &= w(1) - w(0) - s(0) = 1 - \lambda(\{1\}).\end{aligned}$$

As $\mathrm{AVaR}_\alpha^+(\cdot)$ is a coherent measure of risk for every $\alpha \in (0,1)$, the mixture (2.20) is a coherent measure of risk as well. The axioms of convexity, monotonicity, translation equivariance, and positive homogeneity on page 35 can be verified for $\varrho(\cdot)$ directly from the corresponding axioms for $\mathrm{AVaR}_\alpha^+(\cdot)$. The fact that λ is a probability measure is relevant for the translation equivariance. □

Risk measures of the form (2.20) are called *spectral*. In § 2.4.2, we provide a broader view of such structures.

Owing to Theorem 2.22, coherent measures of risk can be defined by specifying the function $w(\cdot)$ in the formula (2.18). By virtue of Theorem 1.32, an equivalent expression for the functional $\varrho(\cdot)$ is the integral:

$$\varrho(Z) = -\int_{-\infty}^0 w\big(F_Z(\eta)\big)\,\mathrm{d}\eta + \int_0^\infty \big[1 - w\big(F_Z(\eta)\big)\big]\,\mathrm{d}\eta. \tag{2.22}$$

The representation (2.22) has been derived in § 1.4.2 in a much more general setting, but for continuous nondecreasing functions $w(\cdot)$ it follows from (2.18) via change of variables and integration by parts:

$$\begin{aligned}\varrho(Z) &= \int_0^1 F_Z^{-1}(p)\,\mathrm{d}w(p) = \int_{-\infty}^{\infty} \eta\,\mathrm{d}w\big(F_Z(\eta)\big)\\ &= \int\limits_{(-\infty,0]} \eta\,\mathrm{d}w\big(F_Z(\eta)\big) - \int\limits_{(0,\infty)} \eta\,\mathrm{d}\big[1-w\big(F_Z(\eta)\big)\big]\\ &= -\int_{-\infty}^{0} w\big(F_Z(\eta)\big)\,\mathrm{d}\eta + \int_0^{\infty}\big[1-w\big(F_Z(\eta)\big)\big]\,\mathrm{d}\eta.\end{aligned}$$

The following examples are known in insurance mathematics.

Example 2.23. Define the functional

$$\varrho(Z) = -\int_{-\infty}^{0}\big[F_Z(\eta)\big]^r\,\mathrm{d}\eta + \int_0^{\infty}\big[1-F_Z(\eta)\big]^r\,\mathrm{d}\eta, \quad 0<r<1,$$

It corresponds to (2.22) with $w(\alpha) = 1-(1-\alpha)^r$ and is a coherent measure of risk on $\mathcal{Z} = \mathcal{L}_p(\Omega,\mathcal{F},P)$, for any $p > \frac{1}{1-r}$. The corresponding representation (2.20) can be calculated by setting $s(\alpha) = w'(\alpha) = r(1-\alpha)^{r-1}$ and $\lambda(\mathrm{d}\alpha) = -\alpha\,\mathrm{d}s(1-\alpha)$. We obtain the following representation of the risk measure:

$$\varrho(Z) = r(1-r)\int_0^1 \mathrm{AVaR}_\alpha^+(Z)\frac{1}{\alpha^{1-r}}\,\mathrm{d}\alpha, \quad Z\in\mathcal{Z}.$$

□

Example 2.24. Define the functional

$$\varrho(Z) = -\int_{-\infty}^{0}\Big(1-\Phi\big[\Phi^{-1}\big(\bar F_Z(\eta)\big)+c\big]\Big)\,\mathrm{d}\eta + \int_0^{\infty}\Phi\big[\Phi^{-1}\big(\bar F_Z(\eta)\big)+c\big]\,\mathrm{d}\eta,$$

where $\bar F_Z(\eta) = P[Z>\eta] = 1-F_Z(\eta)$, $\Phi(\cdot)$ is the standard normal distribution function, and $c>0$. It corresponds to (2.22) with $w(\alpha) = 1-\Phi\big[\Phi^{-1}(1-\alpha)+c\big]$ for $\alpha\in(0,1)$, $w(0)=0$, and $w(1)=1$. It is, therefore, a coherent measure of risk. The functional $\varrho(\cdot)$ is known as the *Wang transform.*

The corresponding representation (2.20) can be calculated by setting

$$s(\alpha) = w'(\alpha) = \frac{\varphi\big[\Phi^{-1}(1-\alpha)+c\big]}{\varphi\big[\Phi^{-1}(1-\alpha)\big]},$$

where $\varphi(\alpha)$ is the standard normal probability density function. Then we obtain

$$\begin{aligned}
\lambda(\mathrm{d}\alpha) &= -\alpha\ \mathrm{d}s(1-\alpha) \\
&= -\alpha \frac{\varphi'\big[\Phi^{-1}(\alpha)+c\big] - \varphi\big[\Phi^{-1}(\alpha)+c\big]\frac{\varphi'\big[\Phi^{-1}(\alpha)\big]}{\varphi\big[\Phi^{-1}(\alpha)\big]}}{\big(\varphi\big[\Phi^{-1}(\alpha)\big]\big)^2}\,\mathrm{d}\alpha \\
&= -\alpha \frac{-\big[\Phi^{-1}(\alpha)+c\big]\varphi\big[\Phi^{-1}(\alpha)+c\big] + \Phi^{-1}(\alpha)\varphi\big[\Phi^{-1}(\alpha)+c\big]}{\big(\varphi\big[\Phi^{-1}(\alpha)\big]\big)^2}\,\mathrm{d}\alpha \\
&= \frac{\alpha c \varphi\big[\Phi^{-1}(\alpha)+c\big]}{\big(\varphi\big[\Phi^{-1}(\alpha)\big]\big)^2}\,\mathrm{d}\alpha.
\end{aligned}$$

We omit further manipulations after substituting of $\varphi(x) = \frac{1}{\sqrt{2\pi}}\mathrm{e}^{-x^2/2}$ into this formula. We conclude that

$$\varrho(Z) = c\int_0^1 \mathrm{AVaR}_\alpha^+(Z)\frac{\alpha\varphi\big[\Phi^{-1}(\alpha)+c\big]}{\big(\varphi\big[\Phi^{-1}(\alpha)\big]\big)^2}\,\mathrm{d}\alpha.$$

□

2.3 Dual Representation

We return to the general setting of convex or coherent risk measures $\varrho : \mathcal{Z} \to \overline{\mathbb{R}}$, satisfying the conditions of convexity, monotonicity, translation equivariance, and positive homogeneity (for coherent measures), as specified on page 35. We allow the risk measures to take the value $+\infty$ at some $Z \in \mathcal{Z}$, but we require that they are *proper*, that is, $\varrho(Z) > -\infty$ for all $z \in \mathcal{Z}$, and $\varrho(Z) < +\infty$ for at least one $Z \in \mathcal{Z}$.

2.3.1 Conjugate Duality for Risk Measures

We focus on the case when $\mathcal{Z} = \mathcal{L}_p(\Omega, \mathcal{F}, P)$, $p \in [1, \infty)$. For each such space, its dual space $\mathcal{Z}^*$ (the space of continuous linear functionals on $\mathcal{Z}$), is isomorphic to $\mathcal{L}_q(\Omega, \mathcal{F}, P)$, where $q \in (1, \infty]$ is such that $1/p + 1/q = 1$. For each $\zeta \in \mathcal{Z}^*$ the corresponding continuous linear functional on $\mathcal{Z}$ is given by the bilinear form

$$\langle Z, \zeta\rangle = \int_\Omega Z(\omega)\,\zeta(\omega)\ P(\mathrm{d}\omega). \tag{2.23}$$

Conversely, if we fix $Z \in \mathcal{Z}$, the bilinear form (2.23) defines a continuous linear functional on $\mathcal{Z}^*$ (equipped with its norm topology). For $p \in (1, \infty)$, all continuous linear functionals on $\mathcal{Z}^*$ must have this form, while for $p = 1$, we equip $\mathcal{Z}^* = \mathcal{L}_\infty(\Omega, \mathcal{F}, P)$ with its weak* topology, to have the representation (2.23) always valid. If the topologies on $\mathcal{Z}$ and $\mathcal{Z}^*$ are such that all continuous linear functionals on the spaces are given by the bilinear form (2.23), and every linear functional of form (2.23) is continuous, then we call $\mathcal{Z}$ and $\mathcal{Z}^*$ *paired spaces*.

The *Fenchel conjugate function* $\varrho^* : \mathcal{Z}^* \to \overline{\mathbb{R}}$ of a risk measure ϱ is defined as

$$\varrho^*(\zeta) = \sup_{Z \in \mathcal{Z}} \{\langle Z, \zeta \rangle - \varrho(Z)\}, \quad \zeta \in \mathcal{Z}^*. \tag{2.24}$$

It is convex, proper, and lower semicontinuous. Furthermore, we may define the conjugate function of ϱ^* (the *biconjugate function* of ϱ) as follows:

$$\varrho^{**}(Z) = \sup_{\zeta \in \mathcal{Z}^*} \{\langle Z, \zeta \rangle - \varrho^*(\zeta)\}, \quad Z \in \mathcal{Z}. \tag{2.25}$$

By virtue of the Fenchel–Moreau theorem, if ϱ is convex, proper, and lower semicontinuous, then $\varrho^{**} \equiv \varrho$. Consequently, $\varrho(\cdot)$ has the following representation

$$\varrho(Z) = \sup_{\zeta \in \mathcal{Z}^*} \{\langle Z, \zeta \rangle - \varrho^*(\zeta)\}, \quad \forall\, Z \in \mathcal{Z}. \tag{2.26}$$

We can write the representation (2.26) in the following equivalent form

$$\varrho(Z) = \sup_{\zeta \in \mathcal{A}} \{\langle Z, \zeta \rangle - \varrho^*(\zeta)\}, \quad \forall\, Z \in \mathcal{Z}, \tag{2.27}$$

where $\mathcal{A} = \mathrm{dom}(\varrho^*)$ is the domain of the conjugate function ϱ^*.

The following duality theorem for convex risk measures is a direct consequence of the Fenchel–Moreau theorem.

Theorem 2.25. *If $\varrho : \mathcal{Z} \to \overline{\mathbb{R}}$ is convex, proper, and lower semicontinuous, then the representation* (2.27) *holds with $\mathcal{A} = \mathrm{dom}(\varrho^*)$. Moreover,*

(i) *The monotonicity condition is satisfied if and only if every $\zeta \in \mathcal{A}$ is non-negative;*

(ii) *The translation equivariance condition is satisfied if and only if*

$$\int_\Omega \zeta(\omega)\, P(d\omega) = 1, \quad \forall\, \zeta \in \mathcal{A};$$

(iii) *The positive homogeneity condition is satisfied if and only if*

$$\varrho(Z) = \sup_{\zeta \in \mathcal{A}} \langle Z, \zeta \rangle, \quad \forall\, Z \in \mathcal{Z}. \tag{2.28}$$

Proof. As discussed before the formulation of the theorem, the representation (2.27) is valid by virtue of the Fenchel–Moreau theorem.

(i) Suppose $\varrho(\cdot)$ satisfies the monotonicity condition, but $\zeta \in \mathcal{A}$ exists such that the event $B = \{\zeta < 0\}$ has positive probability. Then $\langle \mathbb{1}_B, \zeta\rangle < 0$. For any $Z_0 \in \text{dom}(\varrho)$, we consider the ray $Z_t = Z_0 - t\mathbb{1}_B$. By construction, $Z_0 \geq Z_t$ for all $t \geq 0$, and the monotonicity condition implies that $\varrho(Z_0) \geq \varrho(Z_t)$ for all $t \geq 0$. Consequently,

$$\varrho^*(\zeta) \geq \sup_{t\geq 0} \big\{\langle Z_t, \zeta\rangle - \varrho(Z_t)\big\} \geq \sup_{t\geq 0} \big\{\langle Z_0, \zeta\rangle - t\langle \mathbb{1}_B, \zeta\rangle - \varrho(Z_0)\big\} = +\infty,$$

which contradicts the assumption that $\zeta \in \mathcal{A}$. Therefore, $P[\zeta < 0] = 0$ for all $\zeta \in \mathcal{A}$.

Conversely, suppose every element of $\mathcal{A}$ is nonnegative. Then for every $\zeta \in \mathcal{A}$ and all $Z \leq V$ we obtain $\langle Z, \zeta\rangle \leq \langle V, \zeta\rangle$. Using this inequality in (2.27), we conclude that $\varrho(Z) \leq \varrho(V)$.

(ii) If the condition of translation equivariance is satisfied, then for every $Z \in \text{dom}(\varrho)$ and every $\zeta \in \mathcal{Z}^*$ we have

$$\varrho^*(\zeta) \geq \sup_{c\in\mathbb{R}} \big\{\langle Z + c\mathbb{1}, \zeta\rangle - \varrho(Z + c\mathbb{1})\big\} = \sup_{c\in\mathbb{R}} \big\{c\langle \mathbb{1}, \zeta\rangle + \langle Z, \zeta\rangle - \varrho(Z) - c\big\}.$$

It follows that $\varrho^*(\zeta) = +\infty$ if $\langle \mathbb{1}, \zeta\rangle = \int_\Omega \zeta(\omega)\, P(d\omega) \neq 1$. Therefore $\langle \mathbb{1}, \zeta\rangle = 1$ for all $\zeta \in \mathcal{A}$. Conversely, if $\langle \mathbb{1}, \zeta\rangle = 1$ for all $\zeta \in \mathcal{A}$, then $\langle Z + c\mathbb{1}, \zeta\rangle = \langle Z, \zeta\rangle + c$. Using this equality in (2.27) for $Z + c\mathbb{1}$, we obtain translation equivariance.

(iii) If (2.28) is true, then $\varrho(\cdot)$ is positively homogeneous. Conversely, if $\varrho(\cdot)$ is positively homogeneous, then its conjugate function satisfies for every $\gamma > 0$ the chain of relations:

$$\begin{aligned}\varrho^*(\zeta) &= \sup_{\gamma Z\in\mathcal{Z}} \big\{\langle \gamma Z, \zeta\rangle - \varrho(\gamma Z)\big\}\\ &= \gamma \sup_{Z\in\mathcal{Z}} \big\{\langle Z, \zeta\rangle - \varrho(Z)\big\} = \gamma\varrho^*(\zeta).\end{aligned}$$

Therefore $\varrho^*(\zeta)$ is either 0 or $+\infty$; it is the indicator function of a convex closed subset of $\mathcal{Z}^*$. Consequently, the representation (2.28) follows from (2.27). □

It follows from Theorem 2.25 that if $\varrho(\cdot)$ is a lower semicontinuous convex risk measure, then the representation (2.27) is true with $\mathcal{A}$ being a subset of the set of probability density functions:

$$\mathcal{P} = \left\{\zeta \in \mathcal{Z}^* : \int_\Omega \zeta(\omega)\ P(d\omega) = 1,\ \zeta \geq 0\right\}. \tag{2.29}$$

If $\varrho(\cdot)$ is additionally positively homogeneous, then its conjugate function $\varrho^*(\cdot)$ is the indicator function of a convex set $\mathcal{A} \subset \mathcal{Z}^*$, which is equal to the subdifferential $\partial\varrho(0)$. Furthermore, $\varrho(0) = 0$ and the definition of $\partial\varrho(0)$ implies that

$$\mathcal{A} = \{\zeta \in \mathcal{P} : \langle Z, \zeta \rangle \le \varrho(Z), \ \ \forall\, Z \in \mathcal{Z}\}. \tag{2.30}$$

The set $\mathcal{A}$ is convex and weakly* closed. If $\varrho(\cdot)$ is continuous at 0, the set $\mathcal{A}$ is also bounded and thus it is weakly* compact. Consequently, the supremum in (2.28) is achieved at a certain $\zeta \in \partial\varrho(Z)$. We call $\mathcal{A}$ the *dual set* of the coherent risk measure $\varrho(\cdot)$.

For a coherent risk measure $\varrho(\cdot)$, we can view $\langle Z, \zeta \rangle$ as the expected value $\mathbb{E}_Q[Z]$ taken with respect to the probability measure $Q(\mathrm{d}\omega) = \zeta(\omega)P(\mathrm{d}\omega)$, defined by the density ζ. Consequently, the representation (2.28) can be written in the form

$$\varrho(Z) = \sup_{\frac{\mathrm{d}Q}{\mathrm{d}P} \in \mathcal{A}} \mathbb{E}_Q[Z], \quad \forall\, Z \in \mathcal{Z}. \tag{2.31}$$

The dual representations of some measures of risk on finite probability spaces can be derived by linear programming duality. We start from such cases.

Example 2.26. Consider the Average Value at Risk measure (2.5) on a finite probability space $\Omega = \{\omega_1, \ldots, \omega_N\}$ with $P[\omega_k] = p_k$, $k = 1, \ldots, N$. For a random variable Z on this space, we write $z_k = Z(\omega_k)$, $k = 1, \ldots, N$. According to Theorem 2.16, $\mathrm{AVaR}_\alpha^+(Z)$ can be represented as the optimal value of the linear programming problem:

$$\begin{aligned} \min_{\eta, v} \quad & \eta + \frac{1}{\alpha} \sum_{k=1}^{N} p_k v_k \\ \text{subject to} \quad & \eta + v_k \ge z_k, \quad k = 1, \ldots, N, \\ & v_k \ge 0, \quad k = 1, \ldots, N. \end{aligned}$$

Its dual problem has the form

$$\begin{aligned} \max \quad & \sum_{k=1}^{N} \mu_k z_k \\ \text{subject to} \quad & \sum_{k=1}^{N} \mu_k = 1, \\ & 0 \le \mu_k \le \frac{p_k}{\alpha}, \quad k = 1, \ldots, n. \end{aligned}$$

It follows from the duality theory in linear programming that the optimal values of both problems coincide, and thus formula (2.31) is true with the set

$$\mathcal{A} = \left\{ \zeta \in \mathcal{P} : \zeta \le \frac{1}{\alpha} \right\}.$$

In the general case, for every $\zeta \in \mathcal{A}$ and all $Z \in \mathcal{L}_1(\Omega, \mathcal{F}, P)$, we have

$$\frac{1}{\alpha}\mathbb{E}\big[(Z-\eta)_+\big] \ge \mathbb{E}\big[\zeta(Z-\eta)_+\big] \ge \mathbb{E}\big[\zeta(Z-\eta)\big] = \mathbb{E}\big[\zeta Z\big] - \eta.$$

Thus, from Theorem 2.16 we obtain

$$\operatorname{AVaR}_\alpha^+(Z) = \min_\eta \left\{ \eta + \frac{1}{\alpha}\mathbb{E}\big[(Z-\eta)_+\big] \right\} \ge \mathbb{E}[\zeta Z],$$

which proves that $\mathcal{A} \subseteq \partial \operatorname{AVaR}_\alpha^+(0)$. Suppose these sets are not equal: a point $\bar{\zeta} \in \partial \operatorname{AVaR}_\alpha^+(0) \setminus \mathcal{A}$ exists. As $\mathcal{A}$ is convex and closed, the point $\bar{\zeta}$ can be strictly separated: an element $\bar{Z}$ exists, such that

$$\operatorname{AVaR}_\alpha^+(\bar{Z}) \ge \langle \bar{\zeta}, \bar{Z} \rangle \ge \langle \zeta, \bar{Z} \rangle + \varepsilon$$

for all $\zeta \in \mathcal{A}$, where $\varepsilon > 0$. But we can point a particular $\hat{\zeta} \in \mathcal{A}$ such that $\operatorname{AVaR}_\alpha^+(\bar{Z}) = \langle \hat{\zeta}, \bar{Z} \rangle$, which will lead to a contradiction. For this purpose, we use formula (2.5). If $P\big[\bar{Z}(\omega) \ge \operatorname{VaR}_\alpha^+(\bar{Z})\big] = \alpha$, we set

$$\hat{\zeta}(\omega) = \begin{cases} 1/\alpha & \text{if } \bar{Z}(\omega) \ge \operatorname{VaR}_\alpha^+(\bar{Z}), \\ 0 & \text{otherwise.} \end{cases}$$

When an atom at $\operatorname{VaR}_\alpha^+(\bar{Z})$ exists and $P\big[\bar{Z}(\omega) \ge \operatorname{VaR}_\alpha^+(\bar{Z})\big] > \alpha$, a minor modification of this construction is needed. The value of $\hat{\zeta}$ at the event $\big[\bar{Z} = \operatorname{VaR}_\alpha^+(\bar{Z})\big]$ must be decreased to make sure that $\mathbb{E}[\hat{\zeta}] = 1$. In any case, we see that $\mathbb{E}[\hat{\zeta}\bar{Z}]$ satisfies formula (2.5).

In Example 2.29, we provide another derivation of $\partial \operatorname{AVaR}_\alpha^+(0)$. □

Example 2.27. Consider the mean–semideviation risk measure of Example 2.4 with $p = 1$ on a finite probability space $\Omega = \{\omega_1, \dots, \omega_N\}$ with $P[\omega_k] = p_k$, $k = 1, \dots, N$. For a random variable Z on this space, we write $z_k = Z(\omega_k)$, $k = 1, \dots, N$. The value of the risk measure is equal to the optimal value of the linear programming problem

$$\begin{aligned} \min_v \quad & \sum_{k=1}^N p_k z_k + \varkappa \sum_{k=1}^N p_k v_k \\ \text{subject to} \quad & v_k \ge z_k - \sum_{j=1}^N p_j z_j, \quad k = 1, \dots, N, \\ & v_k \ge 0, \quad k = 1, \dots, N. \end{aligned}$$

Assigning multipliers $\lambda_k \geq 0$ to the constraints, we obtain the dual problem

$$\begin{aligned} \max_{\lambda} \quad & \sum_{k=1}^{N} p_k z_k + \sum_{k=1}^{N} \lambda_k \Big(z_k - \sum_{j=1}^{N} p_j z_j \Big), \\ \text{subject to} \quad & 0 \leq \lambda_k \leq \varkappa p_k, \quad k = 1, \dots, N. \end{aligned}$$

Its solution is evident:

$$\lambda_k = \begin{cases} \varkappa p_k & \text{if } z_k \geq \sum_{j=1}^{N} p_j z_j, \\ 0 & \text{otherwise,} \end{cases} \qquad k = 1, \dots, N.$$

Let $\mu_k = p_k + \lambda_k - p_k \sum_{j=1}^{N} \lambda_j$, for $k = 1, \dots, N$. Then $\sum_{k=1}^{N} \mu_k = 1$ and also $\mu_k \geq 0$, $k = 1, \dots, N$. Denoting further $\delta_k = \frac{\lambda_k}{p_k}$, we see that the formula (2.31) is true with the set

$$\mathcal{A} = \Big\{ \zeta \in \mathbb{R}^n : \zeta_k = 1 + \delta_k - \sum_{j=1}^{N} p_j \delta_j,\ 0 \leq \delta_j \leq \varkappa,\ j = 1, \dots, N \Big\}. \tag{2.32}$$

□

In general, the subdifferential calculus needs to be employed to derive the dual representation.

Example 2.28. Consider the mean–semideviation risk measure of Example 2.4 defined on the space $\mathcal{Z} = \mathcal{L}_p(\Omega, \mathcal{F}, P)$ with $p \in [1, \infty)$:

$$\varrho(Z) \triangleq \mathbb{E}[Z] + \varkappa \big\| \big(Z - \mathbb{E}[Z]\big)_+ \big\|_p, \quad \varkappa \in [0, 1].$$

By the rules of the subdifferential calculus,

$$\mathcal{A} = \partial \varrho(0) = \partial \mathbb{E}[0] + \varkappa B^* \partial f(0), \tag{2.33}$$

where

$$f(X) = \|(X)_+\|_p, \tag{2.34}$$

and $B : \mathcal{Z} \to \mathcal{Z}$ is the linear bounded operator given by

$$B(Z) = Z - \mathbb{E}[Z].$$

The expected value is linear and $\partial \mathbb{E}[Z] = \{\mathbb{1}\}$. We claim that

$$\partial f(0) = \big\{ \xi \in \mathcal{L}_q(\Omega, \mathcal{F}, P) : \|\xi\|_q \leq 1,\ \xi \geq 0 \big\}, \tag{2.35}$$

where $1/p + 1/q = 1$. On the one hand, $f(X) \leq \|X\|_p$ and thus

$$\partial f(0) \subseteq \partial \|0\|_p = \{ \xi \in \mathcal{L}_q(\Omega, \mathcal{F}, P) : \|\xi\|_q \leq 1 \}.$$

Moreover, $f(X) = 0$ for all $X \le 0$, and thus $\partial f(0)$ may contain only nonnegative elements. Therefore, the inclusion "$\subseteq$" must be true in the relation (2.35). On the other hand, for any element ξ of the set at right hand side of (2.35) and for all $X \in \mathcal{Z}$, we obtain

$$\langle X, \xi\rangle \le \langle (X)_+, \xi\rangle \le \|(X)_+\|_p.$$

This proves the inclusion "$\supseteq$" in (2.35).

Next, we derive the conjugate operator B^*. For all $\xi \in \mathcal{Z}^*$ and $Z \in \mathcal{Z}$ we have

$$\langle B^*(\xi), Z\rangle = \langle \xi, B(Z)\rangle = \langle \xi, Z - \mathbb{E}[Z]\rangle = \langle \xi - \mathbb{E}[\xi], Z\rangle,$$

and thus $B^*(\xi) = \xi - \mathbb{E}[\xi]$. Summing up, we obtain the subdifferential of ϱ at 0:

$$\begin{aligned}\partial\varrho(0) &= \mathbb{1} + \varkappa B^*\big\{\xi \in \mathcal{L}_q(\Omega, \mathcal{F}, P) : \|\xi\|_q \le 1,\ \xi \ge 0\big\}\\ &= \big\{\mathbb{1} + \zeta - \mathbb{E}[\zeta] : \zeta \in \mathcal{L}_q(\Omega, \mathcal{F}, P),\ \|\zeta\|_q \le \varkappa,\ \zeta \ge 0\big\}.\end{aligned}$$

It is easy to verify that every element $\xi \in \mathcal{A}$ is indeed a probability density function. In the special case of $p = 1$ and a discrete probability space, our formula coincides with the result of the derivations in the previous example. □

Example 2.29. Consider the dual risk measure of Example 2.5 at the space $\mathcal{Z} = \mathcal{L}_p(\Omega, \mathcal{F}, P)$, with $p \in [1, \infty)$:

$$\varrho(Z) \triangleq \min_{t\in\mathbb{R}} \left\{ t + \frac{1}{\alpha}\big\|(Z - t)_+\big\|_p \right\}, \quad \alpha \in (0, 1).$$

Observe that

$$\varrho(Z) = \min_{t\in\mathbb{R}} \left\{ t + \frac{1}{\alpha} f(Z - t) \right\},$$

where $f(\cdot)$ is defined in (2.34). Using the subdifferential (2.35), we can equivalently express the risk measure as follows:

$$\begin{aligned}\varrho(Z) &= \min_{t\in\mathbb{R}} \left\{ t + \frac{1}{\alpha} \max_{\xi\in\partial f(0)} \langle Z - t, \xi\rangle \right\}\\ &= \max_{\xi\in\partial f(0)} \inf_{t\in\mathbb{R}} \left\{ t + \frac{1}{\alpha}\langle Z - t, \xi\rangle \right\}.\end{aligned}$$

The exchange of the "min" and "max" operations is possible, because the function in braces is bilinear in (t, ξ), and the set $\partial f(0)$ is compact. Now, the inner minimization with respect to t yields $-\infty$, unless $\mathbb{E}[\xi] = \alpha$. Consequently,

$$\varrho(Z) = \max_{\substack{\xi \in \partial f(0) \\ \mathbb{E}[\xi]=\alpha}} \frac{1}{\alpha}\langle Z, \xi\rangle.$$

Setting $\zeta = \xi/\alpha$ we identify the support set:

$$\partial\varrho(0) = \left\{\zeta \in \mathcal{L}_q(\Omega, \mathcal{F}, P) : \zeta \geq 0,\ \mathbb{E}[\zeta] = 1,\ \|\zeta\|_q \leq \frac{1}{\alpha}\right\}.$$

Again, every element of $\mathcal{A}$ is a probability density function. In the special case of $p = 1$ we obtain the dual representation of the Average Value at Risk:

$$\partial\,\mathrm{AVaR}_\alpha^+(0) = \left\{\zeta \in \mathcal{L}_\infty(\Omega, \mathcal{F}, P) : \mathbb{E}[\zeta] = 1,\ 0 \leq \zeta \leq \frac{1}{\alpha}\right\},$$

which reproduces the result of Example 2.26. □

Example 2.30. Consider the entropic risk measure of Example 2.6:

$$\varrho(Z) \triangleq \frac{1}{\kappa} \ln\Big(\mathbb{E}\big[e^{\kappa Z}\big]\Big),$$

on the space $\mathcal{Z}$ of random variables Z such that $\mathbb{E}\big[e^{|Z|/r}\big] < \infty$ for all $r > 0$. Its dual functional has the form

$$\varrho^*(\zeta) = \sup_{Z\in\mathcal{Z}} \Big\{\langle Z, \zeta\rangle - \frac{1}{\kappa}\ln\Big(\mathbb{E}\big[e^{\kappa Z}\big]\Big)\Big\}.$$

The optimality conditions for this problem yield the equation

$$\frac{e^{\kappa Z}}{\mathbb{E}\,[e^{\kappa Z}]} = \zeta.$$

This equation has a solution if and only if $\zeta > 0$ a.s. and $\mathbb{E}[\zeta] = 1$. In this case, we see that $Z = \frac{1}{\kappa}\ln(c\zeta)$ with $c = \mathbb{E}\big[e^{\kappa Z}\big]$. Substitution into the formula for the dual functional yields

$$\varrho^*(\zeta) = \frac{1}{\kappa}\mathbb{E}\big[\zeta\ln(c\zeta)\big] - \frac{1}{\kappa}\ln(c) = \frac{1}{\kappa}\mathbb{E}\big[\zeta\ln(c\zeta) - \zeta\ln(c)\big] = \frac{1}{\kappa}\mathbb{E}\big[\zeta\ln(\zeta)\big].$$

We conclude that

$$\varrho^*(\zeta) = \begin{cases} \frac{1}{\kappa}\mathbb{E}\big[\zeta\ln(\zeta)\big] & \text{if } \zeta > 0 \text{ a.s. and } \mathbb{E}[\zeta] = 1, \\ +\infty & \text{otherwise.} \end{cases}$$

For two probability measures: P and Q with the Radon–Nikodym derivative $\frac{dQ}{dP} = \zeta$, the quantity $\mathbb{E}\big[\zeta\ln(\zeta)\big]$ is called the *Kullback–Leibler divergence* or *relative entropy* of Q with respect to P. □

Example 2.31. Consider the entropic value at risk of Example 2.7:

$$\varrho(Z) \triangleq \inf_{\kappa>0} \left\{ \frac{1}{\kappa} \ln \left(\frac{1}{\alpha} \mathbb{E}\left[e^{\kappa Z} \right] \right) \right\},$$

on the space $\mathcal{Z}$ of random variables Z such that $\mathbb{E}\left[e^{|Z|/r}\right] < \infty$ for all $r > 0$. Recall that $\alpha \in (0, 1)$. Its conjugate function has the form

$$\begin{aligned} \varrho^*(\zeta) &= \sup_{Z \in \mathcal{Z}} \left\{ \langle Z, \zeta \rangle - \inf_{\kappa>0} \left\{ \frac{1}{\kappa} \ln \left(\frac{1}{\alpha} \mathbb{E}\left[e^{\kappa Z} \right] \right) \right\} \right\} \\ &= \sup_{\kappa>0} \sup_{Z \in \mathcal{Z}} \left\{ \langle Z, \zeta \rangle - \frac{1}{\kappa} \ln \left(\mathbb{E}\left[e^{\kappa Z} \right] \right) + \frac{\ln(\alpha)}{\kappa} \right\}. \end{aligned}$$

We notice that the inner "sup" is the conjugate function of the entropic risk measure calculated in Example 2.30. For the dual function to be finite it is necessary (but not sufficient) that $\zeta > 0$ and $\mathbb{E}[\zeta] = 1$. Under these conditions, we obtain

$$\varrho^*(\zeta) = \sup_{\kappa>0} \left\{ \frac{1}{\kappa} \mathbb{E}\left[\zeta \ln(\zeta) \right] + \frac{\ln(\alpha)}{\kappa} \right\}.$$

We conclude that the supremum is $+\infty$, unless $\mathbb{E}\left[\zeta \ln(\zeta)\right] + \ln(\alpha) \le 0$, in which case the supremum is 0. Therefore, the conjugate function is the indicator function of the set:

$$\partial \varrho(0) = \left\{ \zeta \in \mathcal{Z}^* : \zeta > 0 \text{ a.s.},\ \mathbb{E}[\zeta] = 1,\ \mathbb{E}\left[\zeta \ln(\zeta) \right] \le -\ln(\alpha) \right\}.$$

These are the densities of probability measures whose relative entropies with respect to P do not exceed $-\ln(\alpha)$. □

2.3.2 Subdifferentiation

To resolve the lower semicontinuity issue, we make use of the lattice structure of the space $\mathcal{L}_p(\Omega, \mathcal{F}, P)$, $p \in [1, \infty]$. This will allow to prove the continuity and the subdifferentiability of a risk measure.

Recall that $\mathcal{Z}$ is a *lattice* (with respect to the cone of nonnegative functions) if for any $Z, V \in \mathcal{Z}$ the element $Z \vee V$, defined as

$$\left[Z \vee V \right](\omega) = \max \left\{ Z(\omega), V(\omega) \right\}, \quad \omega \in \Omega,$$

belongs to $\mathcal{Z}$. Moreover, for every $Z \in \mathcal{Z}$ we can define $|Z| \in \mathcal{Z}$ in a natural way, that is, $|Z|(\omega) = |Z(\omega)|$, $\omega \in \Omega$. The Banach space $\mathcal{Z} = \mathcal{L}_p(\Omega, \mathcal{F}, P)$, with $p \in [1, \infty]$, is a *Banach lattice*, because $|Z| \le |V|$ implies $\|Z\| \le \|V\|$.

Theorem 2.32. *If $\mathcal{Z}$ is a Banach lattice and $\varrho : \mathcal{Z} \to \overline{\mathbb{R}}$ is proper and satisfies the conditions of convexity and monotonicity, then $\varrho(\cdot)$ is continuous and subdifferentiable on the interior of its domain.*

Proof. Let $\bar{Z} \in \text{int}(\text{dom}\, \varrho)$. A linear functional $\ell : \mathcal{Z} \to \mathbb{R}$ is called an algebraic subgradient of ϱ at $\bar{Z}$ if

$$\varrho(Z) \geq \varrho(\bar{Z}) + \ell(Z - \bar{Z}), \quad \forall\, Z \in \mathcal{Z}. \tag{2.36}$$

The algebraic subgradient is not required to be a continuous functional, but if $\ell \in \mathcal{Z}^*$, then it is a subgradient of ϱ at $\bar{Z}$. In fact, ϱ, as a convex functional, always possesses an algebraic subgradient at any point $\bar{Z} \in \text{int}(\text{dom}\ \varrho)$. Indeed, consider the directional derivative function $\theta(\cdot) = \varrho'(\bar{Z}; \cdot)$, where

$$\varrho'(\bar{Z}, V) \triangleq \lim_{t \downarrow 0} \frac{1}{t}\big[\varrho(\bar{Z} + tV) - \varrho(\bar{Z})\big], \quad V \in \mathcal{Z}.$$

The function $\theta(\cdot)$ is positively homogeneous. By the convexity of ϱ, it is convex and satisfies for all $W \in \mathcal{Z}$ the inequality $\varrho(W) \geq \varrho(\bar{Z}) + \theta(W - \bar{Z})$. Since $\bar{Z} \in \text{int}(\text{dom}\ \varrho)$, then $\theta(\cdot)$ is finite valued. By virtue of the Hahn–Banach theorem, a linear functional $\ell : \mathcal{Z} \to \mathbb{R}$ exists, such that $\theta(\cdot) \geq \ell(\cdot)$. It follows that ℓ satisfies (2.36).

The monotonicity of $\varrho(\cdot)$ implies that ℓ is positive in the sense that $\ell(V) \geq 0$ for all $V \geq 0$. Indeed, were $\ell(V) < 0$ for some $V \geq 0$, then (2.36) would imply that $\varrho(\bar{Z} - V) > \varrho(\bar{Z})$, which would contradict the monotonicity axiom. Since ℓ is positive, it is continuous, by virtue of the Namioka–Klee theorem[†]. Therefore, $\ell \in \partial\varrho(\bar{Z})$. It follows then from (2.36) that ϱ is lower semicontinuous at $\bar{Z}$. Since $\bar{Z}$ was an arbitrary point of $\text{int}(\text{dom}\, \varrho)$, we conclude that $\varrho(\cdot)$ is lower semicontinuous on the interior of its domain. This, combined with the fact that $\mathcal{Z}$ is a Banach space, implies the continuity of $\varrho(\cdot)$ on the interior of its domain. □

Remark 2.33. We should stress that the subdifferentiability in Theorem 2.32 means that continuous subgradients exist, where continuity is understood in the strong (norm) topology. If $\mathcal{Z} = \mathcal{L}_\infty(\Omega, \mathcal{F}, P)$, then the subgradients, as elements of its dual space, may contain singular parts: bounded finitely additive measures. This is a highly undesirable technical complication. Later, in Theorem 2.49, we demonstrate that these singular elements can be almost entirely eliminated, if the probability space is atomless and the risk measure is law-invariant (as defined in the next section). □

Finally, we establish the Lipschitz continuity of finite-valued coherent measures of risk.

[†] See also the generalization in Theorem 4.15.

Theorem 2.34. *If $\mathcal{Z}$ is a Banach lattice and $\varrho : \mathcal{Z} \to \mathbb{R}$ is coherent, then $\varrho(\cdot)$ is Lipschitz continuous on $\mathcal{Z}$.*

Proof. According to Theorem 2.32, $\varrho(\cdot)$ is subdifferentiable everywhere. For any two elements $Z, V \in \mathcal{Z}$, the following relations are true

$$\max_{\zeta \in \partial\varrho(Z)} \langle \zeta, V - Z \rangle \leq \varrho(V) - \varrho(Z) \leq \max_{\xi \in \partial\varrho(V)} \langle \xi, V - Z \rangle$$

Since $\varrho(\cdot)$ is coherent, at any point $Z \in \mathcal{Z}$, we have $\partial\varrho(Z) \subseteq \partial\varrho(0)$. Therefore,

$$\left|\varrho(V) - \varrho(Z)\right| \leq \sup\left\{\|\zeta\| : \zeta \in \partial\varrho(0)\right\} \|V - Z\|.$$

Recall that the set $\partial\varrho(0)$ is bounded (see the discussion following (2.30)). □

2.4 Law-Invariant Risk Measures

In applications, we usually deal with measures of risk that can be evaluated or estimated on the basis of the data observed. Such measures depend only on the distribution of the random variable. We follow the literature and call them *law-invariant*.

2.4.1 Basic Properties

As in the previous sections, we assume that $\mathcal{Z} = \mathcal{L}_p(\Omega, \mathcal{F}, P)$, $p \in [1, \infty]$.

Definition 2.35. A risk measure $\varrho : \mathcal{Z} \to \overline{\mathbb{R}}$ is *law-invariant* with respect to the reference probability measure P, if for all $Z, V \in \mathcal{Z}$ the following implication holds[†]

$$\{Z \overset{\mathcal{D}}{\sim} V\} \implies \{\varrho(Z) = \varrho(V)\}.$$

The concept of law invariance can be used in any probability space, but interesting results are obtained in two settings.

The first one is a finite probability space $\Omega = \{\omega_1, \dots, \omega_n\}$ with the σ-algebra $\mathcal{F}$ of all its subsets, and with the probability measure P assigning value $\frac{1}{n}$ to all elementary events ω_i, $i = 1, \dots, n$.

The second setting is a *standard atomless probability space*: a space isomorphic to the interval [0, 1] with its Borel σ-algebra, and with the Lebesgue measure μ.

[†] The relation $Z \overset{\mathcal{D}}{\sim} V$ means that Z has the same distribution as V.

Theorem 2.36. *Let $\mathcal{Z} = \mathcal{L}_p(\Omega, \mathcal{F}, P)$, $p \in [1, \infty)$, where $(\Omega, \mathcal{F}, P)$ is either finite with equally likely elementary events, or standard and atomless. Then a proper lower semicontinuous convex risk measure $\varrho : \mathcal{Z} \to \overline{\mathbb{R}}$ is law-invariant if and only if its conjugate function ϱ^* is law-invariant.*

Proof. If $\varrho(\cdot)$ is law-invariant, then

$$\begin{aligned}
\varrho^*(\zeta) &= \sup_{Z \in \mathcal{Z}} \left\{ \int \zeta(\omega) Z(\omega)\ P(d\omega) - \varrho(Z) \right\} \\
&= \sup_{Z \in \mathcal{Z}} \sup_{Z' \overset{\mathcal{D}}{\sim} Z} \left\{ \int \zeta(\omega) Z'(\omega)\ P(d\omega) - \varrho(Z') \right\} \\
&= \sup_{Z \in \mathcal{Z}} \left\{ \sup_{Z' \overset{\mathcal{D}}{\sim} Z} \int \zeta(\omega) Z'(\omega)\ P(d\omega) - \varrho(Z) \right\}.
\end{aligned} \tag{2.37}$$

In both classes of probability spaces considered, we can find a version $Z' \overset{\mathcal{D}}{\sim} Z$ which is comonotonic with ζ. Therefore, by the Hardy–Littlewood–Polya inequality, the inner supremum is

$$\sup_{Z' \overset{\mathcal{D}}{\sim} Z} \int \zeta(\omega) Z'(\omega)\ P(d\omega) = \int_0^1 F_\zeta^{-1}(u) F_Z^{-1}(u)\ du.$$

The substitution to (2.37) yields

$$\varrho^*(\zeta) = \sup_{Z \in \mathcal{Z}} \left\{ \int_0^1 F_\zeta^{-1}(u) F_Z^{-1}(u)\ du - \varrho(Z) \right\}.$$

As the right hand side depends on ζ only via $F_\zeta^{-1}(\cdot)$, it is identical for all $\zeta' \overset{\mathcal{D}}{\sim} \zeta$, which proves the law invariance of $\varrho^*(\cdot)$. The reverse implication is proved in the same way, with the roles of $\varrho(\cdot)$ and $\varrho^*(\cdot)$ reversed, as in the conjugate duality relation (2.27). □

Corollary 2.37. *In the special case of a coherent lower semicontinuous law-invariant measure of risk, we conclude that the set $\mathcal{A} = \partial\varrho(0)$ is law-invariant in the following sense: if $\zeta \in \mathcal{A}$ and $\zeta' \overset{\mathcal{D}}{\sim} \zeta$ then $\zeta' \in \mathcal{A}$ as well.*

2.4.2 Kusuoka Representation

Law-invariant measures of risk can be reconstructed as combinations of average values at risk with different risk levels. We first present this derivation for the case of a finite probability space with equally likely elementary events.

Theorem 2.38. *Let $\Omega = \{\omega_1, \dots, \omega_n\}$, with $P[\omega_i] = \frac{1}{n}$, $i = 1, \dots, n$. A coherent risk measure $\varrho : \mathcal{Z} \to \mathbb{R}$ is law-invariant if and only if a set Λ of probability vectors in $\mathbb{R}^n$ exists, such that*

$$\varrho(Z) = \sup_{\lambda \in \Lambda} \sum_{i=1}^{n} \lambda_i \operatorname{AVaR}^{+}_{i/n}(Z), \quad \forall\, Z \in \mathcal{Z}. \tag{2.38}$$

Proof. We use the dual representation (2.28):

$$\varrho(Z) = \max_{\zeta \in \mathcal{A}} \frac{1}{n} \sum_{i=1}^{n} z_i \zeta_i,$$

where we write $z_i = Z(\omega_i)$ and $\zeta_i = \zeta(\omega_i)$, $i = 1, \dots, n$. Denote by $z_{[1]} \le z_{[2]} \le \dots \le z_{[n]}$ the ordered realizations of Z. Due to the law invariance of ϱ,

$$\varrho(Z) = \max_{\zeta \in \mathcal{A}} \frac{1}{n} \sum_{i=1}^{n} z_{[i]} \zeta_i. \tag{2.39}$$

Since the set $\mathcal{A}$ is law-invariant, any permutation of ζ maps it to an element of $\mathcal{A}$. In particular, ζ' with coordinates $\zeta'_i = \zeta_{[i]}$ is an element of $\mathcal{A}$. By virtue of the Hardy–Littlewood–Polya inequality,

$$\sum_{i=1}^{n} z_{[i]} \zeta_i \le \sum_{i=1}^{n} z_{[i]} \zeta_{[i]},$$

and the maximization problem (2.39) can be restricted to ordered versions of ζ:

$$\varrho(Z) = \max_{\zeta \in \mathcal{A}} \frac{1}{n} \sum_{i=1}^{n} z_{[i]} \zeta_{[i]}.$$

The sum above can be rearranged as follows (with the convention that $\zeta_{[0]} = 0$):

$$\begin{aligned}
\frac{1}{n} \sum_{i=1}^{n} z_{[i]} \zeta_{[i]} &= \frac{1}{n} \sum_{i=1}^{n} \big(\zeta_{[i]} - \zeta_{[i-1]}\big) \sum_{j=i}^{n} z_{[j]} \\
&= \sum_{i=1}^{n} \frac{n-i+1}{n} \big(\zeta_{[i]} - \zeta_{[i-1]}\big) \frac{1}{n-i+1} \sum_{j=i}^{n} z_{[j]}.
\end{aligned}$$

Defining

$$\lambda_i = \frac{n-i+1}{n} \big(\zeta_{[i]} - \zeta_{[i-1]}\big), \quad i = 1, \dots, n,$$

we recognize the representation (2.38). The vector λ is nonnegative, because $\zeta_{[i]} \geq \zeta_{[i-1]}$ by construction. It remains to verify its sum:

$$\sum_{i=1}^{n} \lambda_i = \sum_{i=1}^{n} \frac{n-i+1}{n} \big(\zeta_{[i]} - \zeta_{[i-1]}\big) = \sum_{i=1}^{n} \frac{1}{n} \zeta_{[i]} = 1.$$

Therefore, for each ζ the corresponding λ is a probability vector. □

In the case of a standard and atomless probability space the result and its derivation are similar.

Theorem 2.39. *Let $\mathcal{Z} = \mathcal{L}_p(\Omega, \mathcal{F}, P)$, $p \in [1, \infty)$, where $(\Omega, \mathcal{F}, P)$ is standard and atomless. Then a proper lower semicontinuous coherent risk measure $\varrho : \mathcal{Z} \to \overline{\mathbb{R}}$ is law-invariant if and only if a set Λ of probability measures on the interval $(0, 1]$ exists, such that*

$$\varrho(Z) = \sup_{\lambda \in \Lambda} \int_0^1 \mathrm{AVaR}_\alpha^+(Z)\, \lambda(d\alpha), \quad \forall\, Z \in \mathcal{Z}. \tag{2.40}$$

Proof. We use the dual representation (2.28):

$$\varrho(Z) = \sup_{\zeta \in \mathcal{A}} \int Z(\omega)\, \zeta(\omega)\; P(d\omega), \quad Z \in \mathcal{Z}.$$

A uniform random variable U on a standard atomless Ω exists. As the variable $Z' = F_Z^{-1}(U)$ has the same distribution as Z and ϱ is law-invariant,

$$\varrho(Z) = \varrho(Z') = \sup_{\zeta \in \mathcal{A}} \int F_Z^{-1}(U(\omega))\, \zeta(\omega)\; P(d\omega).$$

Due to Corollary 2.37, the variable $\zeta' = F_\zeta^{-1}(U)$ is also an element of $\mathcal{A}$. By virtue of the Hardy–Littlewood–Polya inequality, the value of the integral above is the highest when both random variables, $F_Z^{-1}(U)$ and ζ, are comonotonic. Therefore,

$$\varrho(Z) = \sup_{\zeta \in \mathcal{A}} \int F_Z^{-1}(U(\omega))\, F_\zeta^{-1}(U(\omega))\; P(d\omega) = \sup_{\zeta \in \mathcal{A}} \int_{(0,1)} F_Z^{-1}(\eta)\, F_\zeta^{-1}(\eta)\, d\eta.$$

For $\zeta \in \mathcal{A}$ the quantile function $F_\zeta^{-1}(\eta)$, being nondecreasing and left-continuous, can be used to define a nonnegative measure on $(0, 1)$ as follows: $\mu(d\alpha) = -\alpha d F_\zeta^{-1}(1-\alpha)$. Then

$$F_\zeta^{-1}(\eta) = \operatorname{ess\,inf}(\zeta) + \int_{[1-\eta,1)} \frac{1}{\alpha}\, \mu(d\alpha), \quad \eta \in (0, 1).$$

Recall that each $\zeta \in \mathcal{A}$ is a probability density function, and thus $\operatorname{ess\,inf}(\zeta) \in [0, 1]$. We substitute the last expression into the formula for ϱ and integrate by parts:

$$\begin{aligned}
\varrho(Z) &= \sup_{\zeta\in\mathcal{A}} \int_{(0,1)} F_Z^{-1}(\eta) \left[\operatorname{ess\,inf}(\zeta) + \int_{[1-\eta,1)} \frac{1}{\alpha}\, \mu(\mathrm{d}\alpha) \right] \mathrm{d}\eta \\
&= \sup_{\zeta\in\mathcal{A}} \left\{ \operatorname{ess\,inf}(\zeta) \int_{(0,1)} F_Z^{-1}(\eta)\, \mathrm{d}\eta + \int_{(0,1)} F_Z^{-1}(\eta) \int_{[1-\eta,1)} \frac{1}{\alpha}\, \mu(\mathrm{d}\alpha)\, \mathrm{d}\eta \right\} \\
&= \sup_{\zeta\in\mathcal{A}} \left\{ \operatorname{ess\,inf}(\zeta) \int_{(0,1)} F_Z^{-1}(\eta)\, \mathrm{d}\eta + \int_{(0,1)} \frac{1}{\alpha} \int_{[1-\alpha,1)} F_Z^{-1}(\eta)\, \mathrm{d}\eta\; \mu(\mathrm{d}\alpha) \right\} \\
&= \sup_{\zeta\in\mathcal{A}} \left\{ \operatorname{ess\,inf}(\zeta)\, \mathrm{AVaR}_1^+(Z) + \int_{(0,1)} \mathrm{AVaR}_\alpha^+(Z)\, \mu(\mathrm{d}\alpha) \right\}.
\end{aligned}$$

Defining the measure λ_ζ on $(0, 1]$ by

$$\lambda_\zeta(a, b) = \mu(a, b), \quad \lambda_\zeta(\{1\}) = \operatorname{ess\,inf}(\zeta),$$

we can rewrite the last formula as follows:

$$\varrho(Z) = \max_{\zeta\in\mathcal{A}} \int_{(0,1]} \mathrm{AVaR}_\alpha^+(Z)\, \lambda_\zeta(\mathrm{d}\alpha).$$

It remains to verify that λ_ζ is a probability measure. Since each $\zeta \in \mathcal{A}$ is a probability density function,

$$\int_{(0,1)} F_\zeta^{-1}(\eta)\, \mathrm{d}\eta = \int \zeta(\omega)\, P(\mathrm{d}\omega) = 1.$$

Integrating by parts, we obtain the chain of relations

$$\begin{aligned}
\int_{(0,1)} \mu(\mathrm{d}\alpha) &= -\int_{(0,1)} \alpha\, \mathrm{d}F_\zeta^{-1}(1-\alpha) = -\int_{(0,1)} \int_{(0,\alpha]} \mathrm{d}u\, \mathrm{d}F_\zeta^{-1}(1-\alpha) \\
&= -\int_{(0,1)} \int_{[u,1)} \mathrm{d}F_\zeta^{-1}(1-\alpha)\, \mathrm{d}u = \int_{(0,1)} \left[F_\zeta^{-1}(1-u) - F_\zeta^{-1}(0)\right] \mathrm{d}u \\
&= 1 - F_\zeta^{-1}(0) = 1 - \operatorname{ess\,inf}(\zeta).
\end{aligned}$$

Thus λ_ζ is a probability measure, and (2.40) is true with $\Lambda = \{\lambda_\zeta : \zeta \in \mathcal{A}\}$. The reverse implication, from (2.40) to the coherence of ϱ, follows from the coherence of all $\mathrm{AVaR}^+_\alpha(\cdot)$, $\alpha \in (0, 1]$. □

For distortion risk measures discussed in § 2.2.4, the Kusuoka representation (2.40) is true with the set Λ being a singleton, as proved in Theorem 2.22. Such measures of risk are called *spectral.*

We end this section with a generalization of the classical Jensen inequality to coherent law-invariant measures of risk.

Corollary 2.40. *Let $\mathcal{Z} = \mathcal{L}_p(\Omega, \mathcal{F}, P)$, $p \in [1, \infty)$, where $(\Omega, \mathcal{F}, P)$ is standard and atomless, and let $\varrho : \mathcal{Z} \to \overline{\mathbb{R}}$ be a proper, lower semicontinuous, coherent, and law-invariant risk measure. Then for every $Z \in \mathcal{Z}$ and every σ-subalgebra $\mathcal{G} \subset \mathcal{F}$ we have*

$$\varrho\big(\mathbb{E}[Z|\mathcal{G}]\big) \le \varrho(Z).$$

Proof. The extremal representation of Theorem 2.16 and the classical Jensen inequality yield for every $\alpha \in (0, 1]$ the inequality

$$\begin{aligned}\mathrm{AVaR}^+_\alpha\big(\mathbb{E}[Z|\mathcal{G}]\big) &= \inf_{\eta\in\mathbb{R}} \left\{\eta + \frac{1}{\alpha}\mathbb{E}\big[\big(\mathbb{E}[Z|\mathcal{G}] - \eta\big)_+\big]\right\}\\ &\le \inf_{\eta\in\mathbb{R}} \left\{\eta + \frac{1}{\alpha}\mathbb{E}\big[(Z-\eta)_+\big]\right\} = \mathrm{AVaR}^+_\alpha(Z).\end{aligned}$$

The assertion now follows from Theorem 2.39, because both the integration over α and the maximization with respect to λ preserve the inequality. □

This property is further generalized to consistency with respect to the second order stochastic dominance in Theorem 5.22.

Corollary 2.40 allows us to derive an interesting characterization of the Average Value at Risk. Directly from the definition (2.5), we see that $\mathrm{AVaR}^+_\alpha(Z) \ge \mathrm{VaR}^+_\alpha(Z)$ for every $\alpha \in (0, 1)$ and all $Z \in \mathcal{L}_1(\Omega, \mathcal{F}, P)$. The following lemma shows that it is the smallest convex measure of risk dominating $\mathrm{VaR}^+_\alpha(\cdot)$.

Lemma 2.41. *Suppose the space $(\Omega, \mathcal{F}, P)$ is standard and atomless, and let $\mathcal{Z} = \mathcal{L}_1(\Omega, \mathcal{F}, P)$ and $\alpha \in (0, 1)$. If $\varrho : \mathcal{Z} \to \overline{\mathbb{R}}$ is a convex and law-invariant risk measure such that $\varrho(Z) \ge \mathrm{VaR}^+_\alpha(Z)$ for all $Z \in \mathcal{Z}$, then $\varrho(Z) \ge \mathrm{AVaR}^+_\alpha(Z)$ for all $Z \in \mathcal{Z}$.*

Proof. Suppose $\varrho(\cdot)$ is a convex risk measure that dominates $\mathrm{VaR}^+_\alpha(\cdot)$. As $\mathrm{VaR}^+_\alpha(\cdot)$ is positively homogeneous, with no loss of generality we may assume that $\varrho(\cdot)$ is coherent. Suppose Z has a density. Then $F_Z\big(\mathrm{VaR}^+_\alpha(Z)\big) = 1 - \alpha$. Define the event $A = \big\{Z \ge \mathrm{VaR}^+_\alpha(Z) - \varepsilon\big\}$ and let

$$V = \mathbb{E}[Z \mid A^c]\mathbb{1}_{A^c} + \mathrm{AVaR}_\alpha^+(Z)\mathbb{1}_A,$$
$$W = \mathbb{E}[Z \mid A^c]\mathbb{1}_{A^c} + \mathbb{E}[Z \mid A]\mathbb{1}_A.$$

By construction, $\mathrm{VaR}_\alpha^+(V) = \mathrm{AVaR}_\alpha^+(Z)$, hence $\varrho(V) \geq \mathrm{AVaR}_\alpha^+(Z)$. For the σ-subagebra $\mathcal{G} = \{\emptyset, \Omega, A, A^c\}$, Corollary 2.40 yields

$$\varrho(Z) \geq \varrho\big(\mathbb{E}[Z \mid \mathcal{G}]\big) = \varrho(W).$$

If $\varepsilon \to 0$, then $\|V - W\| \leq \mathrm{AVaR}_\alpha^+(Z) - \mathbb{E}[Z \mid A] \to 0$ as well (see Remark 2.14). This implies that $\varrho(W) \to \varrho(V)$, and thus $\varrho(Z) \geq \mathrm{AVaR}_\alpha^+(Z)$, as claimed.

In an atomless probability space, a general random variable Z can be approximated arbitrarily closely by random variables having densities, and the result follows as well. □

2.5 Risk Measures on the Spaces of Quantile Functions

Since quantile-based risk measures depend only on quantile functions of random variables, we may view them as functionals on the space of the quantile functions. In this way, the original probability space $(\Omega, \mathcal{F}, P)$ does not feature in the analysis at all, and many results can be applied to the situation when the probability measure P is not fixed.

As in subsection 1.3.1, our prospect space is now the set $\mathcal{Q}$ of all nondecreasing and left-continuous functions on the interval $(0, 1)$. We consider functionals $\varrho : \mathcal{Q} \to \overline{\mathbb{R}}$ satisfying the following conditions:

Convexity: $\varrho(\alpha\Phi + (1-\alpha)\Psi) \leq \alpha\varrho(\Phi) + (1-\alpha)\varrho(\Psi),\ \forall \alpha \in [0, 1];$
Monotonicity: $\Phi \leq \Psi \implies \varrho(\Phi) \leq \varrho(\Psi);$
Translation equivariance: $\varrho(\Phi + c\mathbb{1}) = \varrho(\Phi) + c,\ \forall c \in \mathbb{R};$
Positive homogeneity: $\varrho(\gamma\Phi) = \gamma\varrho(\Phi),\ \forall \gamma \geq 0.$

We continue to use the terms *convex* and *coherent* to define risk measures satisfying the first three or all four of these axioms, respectively.

It may be of interest to discuss the meaning of these axioms for preference relations among real random variables. Suppose $(\Omega, \mathcal{F}, P)$ is a standard atomless probability space and $\mathcal{Z}$ is the space of real random variables $Z : \Omega \to \mathbb{R}$.

We can define a risk measure $\varrho_\mathcal{Z} : \mathcal{Z} \to \overline{\mathbb{R}}$ by the formula:

$$\varrho_\mathcal{Z}(Z) = \varrho(F_Z^{-1}),$$

where $F_Z^{-1}(p) = \inf\big\{t : P[Z \leq t] \geq p\big\}$, $p \in (0, 1)$, is the quantile function of Z.

Since the space $(\Omega, \mathcal{F}, P)$ is standard and atomless, a random variable U with the uniform distribution in [0,1] exists. For any such variable, and for every $Z \in \mathcal{Z}$ we have $Z \overset{\mathcal{D}}{\sim} F_Z^{-1}(U)$ and thus $\varrho_{\mathcal{Z}}(Z) = \varrho_{\mathcal{Z}}\big(F_Z^{-1}(U)\big)$. If Z and V are comonotonic then $F^{-1}_{\alpha Z+(1-\alpha)V} = \alpha F_Z^{-1} + (1-\alpha)F_V^{-1}$. Thus, the operation of forming a convex combination of quantile functions in $\mathcal{Q}$ corresponds to forming convex combinations of comonotonic random variables in $\mathcal{Z}$.

The convexity axiom, therefore, can be restated as follows: if Z and V are comonotonic, then $\varrho_{\mathcal{Z}}(\alpha Z + (1-\alpha)V) \leq \alpha\varrho_{\mathcal{Z}}(Z) + (1-\alpha)\varrho_{\mathcal{Z}}(V)$ for all $\alpha \in [0, 1]$. Similarly, the monotonicity axiom can be formulated for pairs of comonotonic random variables.

2.5.1 Integrable Prospects

To expose our ideas, we start from the setting in the space of integrable functions on $(0, 1)$. We restrict the space $\mathcal{Q}$ to these quantile functions which are elements of the Lebesgue space $\mathcal{L}_p(0, 1)$, where $p \in [1, \infty)$:

$$\mathcal{Q}_p = \mathcal{Q} \cap \mathcal{L}_p(0, 1).$$

For any functional $\varrho : \mathcal{Q}_p \to \mathbb{R}$, we define the conjugate functional $\varrho^* : \mathcal{L}_q(0, 1) \to \mathbb{R} \cup \{+\infty\}$, with $\frac{1}{p} + \frac{1}{q} = 1$, as follows:

$$\varrho^*(h) = \sup_{\Phi\in\mathcal{Q}_p} \left\{ \int_0^1 h(p)\Phi(p)\,dp - \varrho(\Phi)\right\}. \tag{2.41}$$

We use here the fact that $\mathcal{L}_q(0, 1)$ is the topological dual to $\mathcal{L}_p(0, 1)$. The following theorem is a direct counterpart of Theorem 2.25.

Theorem 2.42. *Suppose $\varrho : \mathcal{Q}_p \to \mathbb{R}$ is a convex risk measure, which is lower semicontinuous in the topology of $\mathcal{L}_p(0, 1)$. Then*

$$\varrho(\Phi) = \sup_{h\in\mathcal{A}} \left\{ \int_0^1 h(p)\Phi(p)\,dp - \varrho^*(h)\right\}, \tag{2.42}$$

where $\mathcal{A}$ is the convex and closed domain of the conjugate function satisfying the inclusion:

$$\mathcal{A} \subseteq \left\{h \in \mathcal{L}_q(0, 1) : \int_0^a h(p)\,dp \geq 0 \textit{ for all } a \in (0, 1],\ \int_0^1 h(p)\,dp = 1\right\}.$$

If ϱ is coherent, then $\varrho^*(h) = 0$ for all $h \in \mathcal{A}$.

Proof. Observe that the set $\mathcal{Q}_p$ is closed and convex in $\mathcal{L}_p(0,1)$. Therefore, the extension of $\varrho(\cdot)$ constructed by assigning the value $+\infty$ to arguments in $\mathcal{L}(0,1) \setminus \mathcal{Q}_p$ is convex and lower semicontinuous.

In this setting, the formula (2.46) follows from the Fenchel–Moreau theorem of convex analysis. The properties of the set $\mathcal{A}$ are proved in a similar way as in the proof of Theorem 2.25. The only difference is that the monotonicity applies to elements of $\mathcal{Q}_p$ only. Therefore, we can only use the fact that for every $h \in \mathcal{A}$ and every $a \in (0,1]$ we have

$$\varrho(0) \geq \varrho\big(-t\mathbb{1}_{(0,a]}\big) \geq -t\int_0^a h(p)\,\mathrm{d}p - \varrho^*(h), \quad \text{for all} \quad t > 0.$$

Dividing by t and passing to the limit as $t \to \infty$ we conclude that

$$\int_0^a h(p)\,\mathrm{d}p \geq 0, \quad \text{for all} \quad a \in (0,1].$$

The fact that every $h \in \mathcal{A}$ integrates to 1 follows from the translation property, exactly as in the proof of Theorem 2.25. Likewise, the positive homogeneity implies that $\varrho(\cdot)$ is the support function of $\mathcal{A}$. □

Observe that the lower semicontinuity of $\varrho(\cdot)$ must be assumed explicitly, because the domain $\mathcal{Q}_p$ is a strict subset of $\mathcal{L}_p(0,1)$.

The dual representation can be refined under an additional assumption of risk aversion. We discussed this concept in § 1.3.5 in the case of bounded quantile functions. For every quantile function $\Phi \in \mathcal{Q}_p$ and every interval $(a,b] \subseteq (0,1]$, we define, as in (1.16), a new quantile function $\Phi_{(a,b]}$ as follows

$$\Phi_{(a,b)}(p) = \begin{cases} \dfrac{1}{b-a}\displaystyle\int_a^b \Phi(s)\,\mathrm{d}s & \text{if } p \in (a,b), \\ \Phi(p) & \text{otherwise.} \end{cases} \tag{2.43}$$

Formula (2.43) can interpreted as a quantile version of a conditional expectation in the range (a,b). We provide a more detailed discussion at the end of this subsection.

The following definition is the risk-measure counterpart of Definition 1.19.

Definition 2.43. A risk measure ϱ on $\mathcal{Q}_p$ is *risk-averse*, if $\varrho(\Phi_{(a,b)}) \leq \varrho(\Phi)$, for every $\Phi \in \mathcal{Q}_p$ and all $(a,b) \subseteq (0,1)$.

In particular, if $(a,b] = (0,1)$ we infer that the constant function equal to the expected value of the distribution corresponding to Φ is preferred over Φ.

Theorem 2.44. *Suppose $\varrho : \mathcal{Q}_p \to \mathbb{R}$ is a convex risk measure, which is risk-averse and lower semicontinuous in the topology of $\mathcal{L}_p(0,1)$. Then the relation* (2.46) *is true, with*

$$\mathcal{A} \subseteq \left\{ h \in \mathcal{L}_q(0,1) : h(p) \geq 0 \textit{ for all } p \in (0,1), \right.$$

$$\left. h(\cdot) \textit{ is nondecreasing}, \int_0^1 h(p)\,\mathrm{d}p = 1 \right\}. \quad (2.44)$$

If ϱ is coherent, then $\varrho^*(h) = 0$ for all $h \in \mathcal{A}$.

Proof. For any $h \in \mathcal{A}$ we define its distribution function

$$w(p) = \int_0^p h(q)\,\mathrm{d}q, \quad p \in (0,1),$$

and the convex hull (the largest convex minorant)

$$\bar{w} = \mathrm{conv}(w).$$

Define $\bar{h}(p) \in \partial\bar{w}(p)$, $0 < p < 1$. The set of p for which $\bar{w}(p) < w(p)$ can be decomposed into a countable collection of open intervals:

$$\{p \in (0,1) : \bar{w}(p) < w(p)\} = \bigcup_{i \in I} (a_i, b_i).$$

With every function $\Psi \in \mathcal{A}$ we associate the function $\Phi_\Psi \in \mathcal{A}$ as follows:

$$\Phi_\Psi(p) = \begin{cases} \displaystyle\int_{a_i}^{b_i} \Psi(s)\,\mathrm{d}s & \text{if } p \in (a_i, b_i),\ i \in I, \\ \Phi(p) & \text{if } \bar{w}(p) = w(p). \end{cases}$$

Applying Definition 2.43 recursively for $i \in I$ we deduce that $\varrho(\Phi_\Psi) \leq \varrho(\Psi)$. We also observe that $\bar{h}(\cdot)$ is constant on each interval (a_i, b_i) and thus

$$\int_{a_i}^{b_i} \Phi_\Psi(p)\bar{h}(p)\,\mathrm{d}p = \int_{a_i}^{b_i} \Phi_\Psi(p)h(p)\,\mathrm{d}p.$$

These observations allow us to write the following chain of relations:

$$\begin{aligned} \varrho^*(\bar{h}) &= \sup_{\Psi \in \mathcal{Q}_p} \left\{ \int_0^1 \Psi(p)\,\bar{h}(p)\,\mathrm{d}p - \varrho(\bar{h}) \right\} \\ &\leq \sup_{\Psi \in \mathcal{Q}_p} \left\{ \int_0^1 \Phi_\Psi(p)\,\bar{h}(p)\,\mathrm{d}p - \varrho(\Phi_\Psi) \right\} \\ &= \sup_{\Psi \in \mathcal{Q}_p} \left\{ \int_0^1 \Phi_\Psi(p)\,h(p)\,\mathrm{d}p - \varrho(\Phi_\Psi) \right\} \\ &\leq \sup_{\Phi \in \mathcal{Q}_p} \left\{ \int_0^1 \Phi(p)\,h(p)\,\mathrm{d}p - \varrho(\Phi) \right\} = \varrho^*(h), \end{aligned}$$

Now for every $\Phi \in \mathcal{Q}_p$, by the Hardy–Littlewood–Polya inequality we have

$$\int_0^1 h(p)\Phi(p)\, \mathrm{d}p \le \int_0^1 \bar{h}(p)\Phi(p)\, \mathrm{d}p.$$

The last two relations prove that it is sufficient to consider nondecreasing functions in $\mathcal{A}$ in formula (2.42). □

2.5.2 Bounded Prospects

In this section, we focus on bounded quantile functions. Without loss of generality, we may assume that their domain is the interval $(0, 1]$. This corresponds to compactly supported distributions, and the prospect space $\mathcal{Q}_\mathrm{b}$ of all bounded, nondecreasing, and left-continuous functions on $(0, 1]$.[†]

As discussed in § 1.3.3, the set $\mathcal{Q}_\mathrm{b}$ is contained in the Banach space $B\big((0, 1], \Sigma\big)$ of all bounded functions on $(0, 1]$ that can be obtained as uniform limits of sequences of piecewise-constant functions. Here, Σ is the algebra of all sets generated by finite unions and intersections of intervals of the form $(a, b] \subseteq (0, 1]$.

We denote by $\mathcal{M}(\Sigma)$ the space of bounded and finitely additive measures μ on Σ, by $\mathcal{M}$ the space of countably additive finite measures on the Borel σ-algebra on $(0, 1]$, and by $\mathcal{M}_+$ its subset of nonnegative measures. We equip $\mathcal{M}(\Sigma)$ with its weak topology.

For any functional $\varrho : \mathcal{Q}_\mathrm{b} \to \mathbb{R}$, we define the conjugate functional $\varrho^* : \mathcal{M}(\Sigma) \to \mathbb{R} \cup \{+\infty\}$ in the usual way:

$$\varrho^*(\mu) = \sup_{\Phi \in \mathcal{Q}_\mathrm{b}} \left\{ \int_0^1 \Phi(p)\, \mathrm{d}^\Sigma \mu - \varrho(\Phi) \right\}. \tag{2.45}$$

We use here the fact that $\mathcal{M}(\Sigma)$ is the topological dual to $B\big((0, 1], \Sigma\big)$.

In the following duality theorem, we do not need any lower semicontinuity assumption, because it is implied by the monotonicity and convexity of $\varrho(\cdot)$ on the space $\mathcal{Q}$. The cost of it is the esoteric nature of the dual space $\mathcal{M}(\Sigma)$. However, later in this section, under an additional condition of risk aversion, we shall prove that only countably additive measures will matter here.

Theorem 2.45. *Suppose $\varrho : \mathcal{Q}_\mathrm{b} \to \mathbb{R}$ is a convex risk measure. Then*

$$\varrho(\Phi) = \sup_{\mu \in \mathcal{A}} \left\{ \int_0^1 \Phi(p)\, \mathrm{d}^\Sigma \mu - \varrho^*(\mu) \right\}, \tag{2.46}$$

[†] Bounded nondecreasing functions on $(0, 1)$ can be extended to $(0, 1]$ by assigning their left limits as their values at 1.

where

$$\mathcal{A} \subseteq \big\{\mu \in \mathcal{M}(\Sigma) : \mu(0,a] \geq 0 \;\forall\, a \in (0,1],\; \mu(0,1] = 1\big\}$$

is convex and closed. If ϱ is coherent, then $\varrho^*(\mu) = 0$ for all $\mu \in \mathcal{A}$.

Proof. Due to monotonicity and the translation property, for any $\Phi, \Psi \in \mathcal{Q}_{\mathrm{b}}$, such that $\|\Phi - \Psi\| \leq \delta$ we obtain

$$\varrho(\Phi) - \delta = \varrho(\Phi - \delta \mathbb{1}) \leq \varrho(\Psi) \leq \varrho(\Phi + \delta \mathbb{1}) = \varrho(\Phi) + \delta.$$

Therefore, $\varrho(\cdot)$ is continuous on $\mathcal{Q}_{\mathrm{b}}$.

We can formally define $\varrho(F) = +\infty$ for $F \in B\big((0,1], \Sigma\big) \setminus \mathcal{Q}_{\mathrm{b}}$; this will not affect the definition of the conjugate function (2.45). Observe that the set $\mathcal{Q}_{\mathrm{b}}$ is closed and convex in $B\big((0,1], \Sigma\big)$. Therefore, the extension of $\varrho(\cdot)$ is convex and lower semicontinuous.

In this setting the formula (2.46) follows from the Fenchel–Moreau theorem of convex analysis. We need only verify the properties of the set

$$\mathcal{A} = \big\{\mu \in \mathcal{M}(\Sigma) : \varrho^*(\mu) < +\infty\big\}.$$

Suppose the monotonicity axiom is satisfied. Then for every $\mu \in \mathcal{A}$ and every $a \in (0,1]$ we obtain

$$\varrho(0) \geq \varrho\big(-t \mathbb{1}_{(0,a]}\big) \geq -t\, \mu(0,a] - \varrho^*(\mu) \quad \forall\, t > 0.$$

Dividing by t and passing to the limit as $t \to \infty$ we conclude that $\mu(0,a] \geq 0$.

Next, for any $\mu \in \mathcal{A}$ the translation axiom implies that

$$\begin{aligned}
\varrho^*(\mu) &= \sup_{\substack{\varphi \in \mathcal{Q}_{\mathrm{b}} \\ c \in \mathbb{R}}} \left\{ \int_0^1 \big(\Phi(p) + c\big)\, \mathrm{d}^{\Sigma}\mu - \varrho(\Phi + c\mathbb{1}) \right\} \\
&= \sup_{\substack{\varphi \in \mathcal{Q}_{\mathrm{b}} \\ c \in \mathbb{R}}} \left\{ \int_0^1 \Phi(p)\, \mathrm{d}^{\Sigma}\mu - \varrho(\Phi) + c\mu(0,1] - c \right\} \\
&= \varrho^*(\mu) + \sup_{c \in \mathbb{R}} c \big\{\mu(0,1] - 1\big\}.
\end{aligned}$$

It follows that $\mu(0,1] = 1$.

If we assume the positive homogeneity axiom, then $\varrho(\cdot)$ is positively homogeneous and convex. Its conjugate function is then an indicator function of its domain $\mathcal{A}$. □

For every quantile function $\Phi \in \mathcal{Q}_b$ and every interval $(a, b] \in \Sigma$, we can define a new quantile function $\Phi_{(a,b]}$ as in (1.16) in § 1.3.5, or (2.43) in this section. We also adopt Definition 1.19 of risk aversion (it is a minor variation of Definition 2.43 in this section).

The property of risk aversion allows us to substantially strengthen the dual representation (2.46) by including only countably additive measures μ whose distribution functions are convex on (0, 1].

Theorem 2.46. *Suppose a risk measure* $\varrho : \mathcal{Q}_b \to \mathbb{R}$ *is convex and risk-averse. Then*

$$\varrho(\Phi) = \sup_{\mu \in \mathcal{D}} \left\{ \int_0^1 \Phi(p)\, \mathrm{d}\mu - \varrho^*(\mu) \right\}, \tag{2.47}$$

where

$$\mathcal{D} = \{\mu \in \mathcal{A} \cap \mathcal{M}_+ : \mu(0, \cdot] \text{ is convex on } (0, 1]\}.$$

If $\varrho(\cdot)$ *is coherent, then* $\varrho^*(\mu) = 0$ *for all* $\mu \in \mathcal{D}$.

Proof. Let us denote by $\mathcal{S}$ the set of simple functions in $\mathcal{Q}_b$. Due to the continuity of $\varrho(\cdot)$ on $\mathcal{Q}_b$, the conjugate function (2.45) can be equivalently expressed as

$$\varrho^*(\mu) = \sup_{\Phi \in \mathcal{S}} \left\{ \int_0^1 \Phi(p)\, \mathrm{d}^{\Sigma}\mu - \varrho(\Phi) \right\}. \tag{2.48}$$

Together with each measure $\mu \in \mathcal{A}$ we shall consider piecewise unform measures ν with densities with respect to the Lebesgue measure of the following form:

$$h(p) = \sum_{i=1}^{n} \frac{\mu(a_i, a_{i+1}]}{a_{i+1} - a_i} \mathbb{1}_{(a_i, a_{i+1}]}(p), \quad p \in (0, 1], \tag{2.49}$$

where $0 = a_1 < a_2 < \cdots < a_{n+1} = 1$. We denote the set of such measures ν by $\mathcal{U}(\mu)$. Observe that for every $\nu \in \mathcal{U}(\mu)$ we have

$$\begin{aligned} \varrho^*(\nu) &= \sup_{\Phi \in \mathcal{Q}_b} \left\{ \int_0^1 \Phi(p)\, \mathrm{d}^{\Sigma}\nu - \varrho(\Phi) \right\} \\ &= \sup_{\Phi \in \mathcal{Q}_b} \left\{ \sum_{i=1}^{n} \frac{\mu(a_i, a_{i+1}]}{a_{i+1} - a_i} \int_{a_i}^{a_{i+1}} \Phi(p)\, \mathrm{d}p - \varrho(\Phi) \right\}. \end{aligned} \tag{2.50}$$

Define the simple function

$$\tilde{\Phi}(p) = \sum_{i=1}^{n} \frac{\mathbb{1}_{(a_i, a_{i+1}]}(p)}{a_{i+1} - a_i} \int_{a_i}^{a_{i+1}} \Phi(p)\, \mathrm{d}p, \quad p \in (0, 1].$$

Owing to risk aversion, $\varrho(\tilde{\Phi}) \leq \varrho(\Phi)$. This allows us to continue (2.50) as follows:

$$\varrho^*(\nu) \leq \sup_{\Phi \in \mathcal{Q}_b} \left\{ \int_0^1 \tilde{\Phi}(p) \, \mathrm{d}^\Sigma \mu - \varrho(\tilde{\Phi}) \right\} \leq \varrho^*(\mu). \tag{2.51}$$

This proves that for every $\mu \in \mathcal{A} = \mathrm{dom}(\varrho^*)$ the set $\mathcal{U}(\mu) \subseteq \mathcal{A}$.

Now we proceed as in the proof of Theorem 2.44 with minor adjustments. Consider a piecewise uniform measure $\nu \in \mathcal{A}$ and its distribution function $w(\cdot)$ defined as follows:

$$w(p) = \nu(0, p], \quad p \in (0, 1]; \quad w(0) = 0. \tag{2.52}$$

By virtue of Theorem 2.45, $w(p) \geq 0$ for all $p \in [0, 1]$ and $w(1) = 1$. Moreover, it is piecewise linear. We construct its convex envelope:

$$\bar{w} = \mathrm{conv}(w),$$

that is, $\bar{w}(\cdot)$ is the largest convex minorant of $w(\cdot)$ on $[0, 1]$. Observe that $\bar{w}$ is piecewise linear and nondecreasing.

We can now define a new measure $\bar{\nu}$ as follows:

$$\bar{\nu}[0, p] = \bar{w}(p), \quad p \in (0, 1].$$

Clearly, $\bar{\nu}$ is piecewise uniform, and $\bar{\nu} \in \mathcal{M}_+$. We shall show that $\bar{\nu} \in \mathcal{A}$. To this end we shall verify the inequality

$$\varrho^*(\bar{\nu}) \leq \varrho^*(\nu). \tag{2.53}$$

Consider a simple function

$$\Psi(p) = \sum_{i=1}^{n} \psi_i \mathbb{1}_{(a_i, a_{i+1}]}(p), \quad p \in (0, 1],$$

where $a_1 = 0$ and $a_{n+1} = 1$ and $\psi_1 \leq \psi_2 \leq \cdots \leq \psi_n$. With no loss of generality, we may assume that the collection of points a_i, $i = 1, \ldots, n + 1$, contains all break points of $\bar{w}(\cdot)$. Then

$$\int_0^1 \Psi(p) \, \mathrm{d}^\Sigma \bar{\nu} = \sum_{i=1}^{n} \psi_i \bar{\nu}(a_i, a_{i+1}].$$

Let us define an increasing and finite collection of indices in $\{1, 2 \ldots, n + 1\}$:

$$\begin{aligned}
k_1 &= 1,\\
i_1 &= \min\{i > k_1 : \bar{\nu}(0, a_i] < \nu(0, a_i]\},\\
k_2 &= \min\{i > i_1 : \bar{\nu}(0, a_i] = \nu(0, a_i]\},\\
i_2 &= \min\{i > k_2 : \bar{\nu}(0, a_i] < \nu(0, a_i]\},\\
&\vdots\\
i_{L-1} &= \min\{i > k_{L-1} : \bar{\nu}(0, a_i] < \nu(0, a_i]\},\\
k_L &= n + 1.
\end{aligned}$$

Observe that $\bar{w}$ is affine on the intervals $(a_{i_\ell -1}, a_{k_{\ell+1}}]$ for $\ell = 1, \ldots L-1$, and can thus be written as follows:

$$\bar{w}(p) = \bar{w}(a_{i_\ell -1}) + h_\ell \frac{p - a_{i_\ell -1}}{a_{k_{\ell+1}} - a_{i_\ell -1}} \quad \forall p \in (a_{i_\ell -1}, a_{k_{\ell+1}}],$$

where the constants h_ℓ satisfy the inequalities $0 \le h_1 \le h_2 \le \ldots h_L$. Define

$$\Phi_\Psi(p) = \begin{cases} \Psi(p) & \text{if } p \in (a_{k_\ell}, a_{i_\ell -1}], \quad \ell = 1, \ldots, L,\\ C_\ell & \text{if } p \in (a_{i_\ell -1}, a_{k_{\ell+1}}], \quad \ell = 1, \ldots, L, \end{cases}$$

with

$$C_\ell = \begin{cases} \dfrac{1}{\bar{\nu}(a_{i_\ell -1}, a_{k_{\ell+1}}]} \displaystyle\sum_{j=i_\ell -1}^{k_{\ell+1}-1} \psi_j \bar{\nu}(a_j, a_{j+1}] & \text{if } \bar{\nu}(a_{i_\ell -1}, a_{k_{\ell+1}}] > 0,\\ \dfrac{1}{a_{k_{\ell+1}} - a_{i_\ell -1}} \displaystyle\sum_{j=i_\ell -1}^{k_{\ell+1}-1} \psi_j (a_{j+1} - a_j) & \text{otherwise.} \end{cases}$$

From the definition of the indices $k_1, i_1, \ldots k_L$ we deduce the following chain of equations:

$$\begin{aligned}
\int_0^1 \Phi_\Psi(p)\, \mathrm{d}^\Sigma \bar{\nu} &= \sum_{\ell=1}^{L} \left(\sum_{j=k_\ell}^{i_\ell -2} \psi_j \bar{\nu}(a_j, a_{j+1}] + C_\ell \bar{\nu}(a_{i_\ell -1}, a_{k_{\ell+1}}] \right)\\
&= \sum_{\ell=1}^{L} \left(\sum_{j=k_\ell}^{i_\ell -2} \psi_j \nu(a_j, a_{j+1}] + C_\ell \big(\bar{\nu}(0, a_{k_{\ell+1}}] - \bar{\nu}(0, a_{i_\ell -1}]\big) \right)\\
&= \sum_{\ell=1}^{L} \left(\sum_{j=k_\ell}^{i_\ell -2} \psi_j \nu(a_j, a_{j+1}] + C_\ell \big(\nu(0, a_{k_{\ell+1}}] - \nu(0, a_{i_\ell -1}]\big) \right)\\
&= \sum_{\ell=1}^{L} \left(\sum_{j=k_\ell}^{i_\ell -2} \psi_j \nu(a_j, a_{j+1}] + C_\ell \nu(a_{i_\ell -1}, a_{k_{\ell+1}}] \right) = \int_0^1 \Phi_\Psi(p)\, \mathrm{d}^\Sigma \nu.
\end{aligned} \tag{2.54}$$

Using the definition of C_ℓ we can transform the first equation in (2.54) in another way:

$$\int_0^1 \Phi_\Psi(p)\, \mathrm{d}^\Sigma \bar{\nu} = \sum_{\ell=1}^{L} \left(\sum_{j=k_\ell}^{i_\ell - 2} \psi_j \bar{\nu}(a_j, a_{j+1}] + \sum_{j=i_\ell - 1}^{k_{\ell+1}-1} \psi_j \bar{\nu}(a_j, a_{j+1}] \right)$$
$$= \int_0^1 \Psi(p)\, \mathrm{d}^\Sigma \bar{\nu}. \tag{2.55}$$

As $\bar{w}(\cdot)$ is affine on the intervals $(a_{i_\ell - 1}, a_{k_{\ell+1}}]$, we can rewrite the constants C_ℓ as follows:

$$C_\ell = \frac{1}{a_{k_{\ell+1}} - a_{i_\ell - 1}} \sum_{j=i_\ell - 1}^{k_{\ell+1}-1} \psi_j (a_{j+1} - a_j).$$

By virtue of Definition 1.19,

$$\varrho(\Phi_\Psi) \le \varrho(\Psi). \tag{2.56}$$

Combining the relations (2.48), (2.54), (2.55) and (2.56), we prove (2.53), exactly as in the proof of Theorem 2.44 (with $\bar{h}\, \mathrm{d}p$ replaced by $\mathrm{d}^\Sigma \bar{\nu}$).

$$\begin{aligned}
\varrho^*(\bar{\nu}) &= \sup_{\Psi \in \mathcal{S}} \left\{ \int_0^1 \Psi(p)\, \mathrm{d}^\Sigma \bar{\nu} - \varrho(\Psi) \right\} \\
&\le \sup_{\Psi \in \mathcal{S}} \left\{ \int_0^1 \Phi_\Psi(p)\, \mathrm{d}^\Sigma \bar{\nu} - \varrho(\Phi_\Psi) \right\} \\
&= \sup_{\Psi \in \mathcal{S}} \left\{ \int_0^1 \Phi_\Psi(p)\, \mathrm{d}^\Sigma \nu - \varrho(\Phi_\Psi) \right\} \\
&\le \sup_{\Phi \in \mathcal{S}} \left\{ \int_0^1 \Phi(p)\, \mathrm{d}^\Sigma \nu - \varrho(\Phi) \right\} = \varrho^*(\nu),
\end{aligned}$$

Therefore, $\bar{\nu} \in \mathcal{A}$ for every piecewise uniform $\nu \in \mathcal{A}$. Moreover, for any simple function $F \in \mathcal{S}$ with values f_i on $(a_i, a_{i+1}]$ we have

$$\begin{aligned}
\int_0^1 F(p)\,\mathrm{d}^{\Sigma}\bar{\nu} &= \sum_{i=1}^{n} f_i \bar{\nu}(a_i, a_{i+1}] = \sum_{i=1}^{n}\Big(f_1 + \sum_{j=1}^{i-1}(f_{j+1} - f_j)\Big)\bar{\nu}(a_i, a_{i+1}] \\
&= f_1 \sum_{i=1}^{n} \bar{\nu}(a_i, a_{i+1}] + \sum_{j=1}^{n-1}(f_{j+1} - f_j) \sum_{i=j+1}^{n} \bar{\nu}(a_i, a_{i+1}] \\
&= f_1 + \sum_{j=1}^{n-1}(f_{j+1} - f_j)\bar{\nu}(a_{j+1}, 1] \\
&\geq f_1 + \sum_{j=1}^{n-1}(f_{j+1} - f_j)\nu(a_{j+1}, 1] = \int_0^1 F(p)\,\mathrm{d}^{\Sigma}\nu.
\end{aligned}$$

In the last inequality we used the fact that $\bar{\nu}(0, a] \leq \nu(0, a]$ for all $a \in (0, 1]$. From this relation and (2.53) we deduce that for every piecewise uniform $\nu \in \mathcal{A}$ and every simple function F we have

$$\int_0^1 F(p)\,\mathrm{d}^{\Sigma}\nu - \varrho^*(\nu) \leq \int_0^1 F(p)\,\mathrm{d}^{\Sigma}\bar{\nu} - \varrho^*(\bar{\nu}).$$

Using this relation in (2.46), we derive for any simple function F the chain of inequalities:

$$\begin{aligned}
\varrho(F) &= \sup_{\mu \in \mathcal{A}} \left\{ \int_0^1 F(p)\,\mathrm{d}^{\Sigma}\mu - \varrho^*(\mu) \right\} \\
&\leq \sup_{\mu \in \mathcal{A}} \sup_{\nu \in \mathcal{U}(\mu)} \left\{ \int_0^1 F(p)\,\mathrm{d}^{\Sigma}\nu - \varrho^*(\nu) \right\} \\
&\leq \sup_{\mu \in \mathcal{A}} \sup_{\nu \in \mathcal{U}(\mu)} \left\{ \int_0^1 F(p)\,\mathrm{d}^{\Sigma}\bar{\nu} - \varrho^*(\bar{\nu}) \right\} \\
&\leq \sup_{\mu \in \mathcal{D}} \left\{ \int_0^1 F(p)\,\mathrm{d}^{\Sigma}\mu - \varrho^*(\mu) \right\} \leq \varrho(F).
\end{aligned} \tag{2.57}$$

This proves the formula (2.47) for simple functions. The general case follows from the continuity of $\varrho(\cdot)$. Indeed, for every $\Phi \in \mathcal{Q}_{\mathrm{b}}$ and every $\varepsilon > 0$ we can find a simple function F such that $\Phi(p) \leq F(p) \leq \Phi(p) + \varepsilon$. Then, using (2.57) we obtain

$$\begin{aligned}
\varrho(\Phi) \leq \varrho(F) &= \sup_{\mu \in \mathcal{D}} \left\{ \int_0^1 F(p)\,\mathrm{d}^{\Sigma}\mu - \varrho^*(\mu) \right\} \\
&\leq \sup_{\mu \in \mathcal{D}} \left\{ \int_0^1 [\Phi(p) + \varepsilon]\,\mathrm{d}^{\Sigma}\mu - \varrho^*(\mu) \right\} \\
&= \sup_{\mu \in \mathcal{D}} \left\{ \int_0^1 \Phi(p)\,\mathrm{d}^{\Sigma}\mu - \varrho^*(\mu) \right\} + \varepsilon \leq \varrho(\Phi) + \varepsilon,
\end{aligned}$$

because $\mathcal{D} \subseteq \mathcal{A}$ and $\mu(0, 1] = 1$ for $\mu \in \mathcal{D}$. As $\varepsilon > 0$ was arbitrary, formula (2.47) is true with Σ-additive integration.

Finally, for every $\mu \in \mathcal{D}$ its distribution function $w(\cdot) = \mu(0, \cdot]$ is nondecreasing and convex. It is, therefore, continuous on $[0, 1)$. Since every quantile function $\Phi(\cdot)$ is left-continuous at 1, the linear functionals on the right-hand side of the last displayed relation can be written as Lebesgue and Stieltjes integrals:

$$\int_0^1 \Phi(p) \, \mathrm{d}^{\Sigma}\mu = \int_0^1 \Phi(p) \, \mathrm{d}\mu = \int_0^1 \Phi(p) \, dw(p),$$

as postulated. □

As the distribution function $w(\cdot)$ of every measure $\mu \in \mathcal{D}$ is convex on $[0, 1]$, it can be represented as follows:

$$w(p) = \int_0^p h(q) \, \mathrm{d}q, \quad p \in (0, 1),$$

where the density $h(q)$ is the right derivative of $w(\cdot)$ at $q \in [0, 1)$. The function $h(\cdot)$ is nondecreasing and right-continuous. The function $w(\cdot)$ may have a jump of size $\beta \geq 0$ at 1, so that $w(1) = 1$. It follows that

$$w(1) = \int_0^1 h(q) \, \mathrm{d}q + \beta = 1.$$

2.5.3 Kusuoka Representation

Now, we derive the Kusuoka representation of risk measures on the spaces of quantile functions in the two cases: $\mathcal{L}_1(0, 1)$ and $B((0, 1], \Sigma)$. Our derivation focuses on $B((0, 1], \Sigma)$; in the case of the space $\mathcal{L}_1(0, 1)$ we simply need to set $\beta = 0$.

We can rewrite the integral in (2.47) as follows:

$$\int_0^1 \Phi(p) \, \mathrm{d}\mu = \int_0^1 \Phi(p)h(p) \, \mathrm{d}p + \beta\Phi(1). \tag{2.58}$$

As $h(\cdot)$ is right-continuous and nondecreasing, it can be represented as an integral:

$$h(p) = \int_0^p \nu(\mathrm{d}s), \quad p \in (0, 1),$$

where the measure $\nu \in \mathcal{M}_+$. We can continue (2.58) by changing the order of integration and we obtain:

$$\int_0^1 \Phi(p)\,\mathrm{d}\mu = \int_0^1 \Phi(p)\int_0^p \nu(\mathrm{d}s)\,\mathrm{d}p + \beta\Phi(1)$$
$$= \int_0^1\int_s^1 \Phi(p)\,\mathrm{d}p\;\nu(\mathrm{d}s) + \beta\Phi(1).$$

Let us adapt the definition of the Average Value at Risk to the quantile setting:

$$\mathrm{AVaR}^+_\alpha(\Phi) \triangleq \frac{1}{\alpha}\int_{1-\alpha}^1 \Phi(p)\,\mathrm{d}p. \tag{2.59}$$

Then

$$\int_0^1 \Phi(p)\,\mathrm{d}\mu = \int_{(0,1)} \mathrm{AVaR}^+_{1-s}(\Phi)(1-s)\;\nu(\mathrm{d}s) + \beta\Phi(1).$$

We define the measure λ on $(0,1)$ by $\frac{\mathrm{d}\lambda}{\mathrm{d}\nu} = 1-s$. We formally define $\mathrm{AVaR}^+_0(\Phi) = \Phi(1)$ and extend the measure λ to $(0,1]$ by adding to it an atom β at 1. Then for any quantile function Φ we obtain the formula:

$$\int_0^1 \Phi(p)\,\mathrm{d}\mu = \int_0^1 \mathrm{AVaR}^+_{1-s}(\Phi)\;\lambda(\mathrm{d}s). \tag{2.60}$$

This transformation allows us to derive an analog of the Kusuoka representation of law-invariant coherent measures of risk.

Corollary 2.47. *Suppose the risk measure $\varrho : \mathcal{Q}_\mathrm{b} \to \mathbb{R}$ is coherent and risk-averse. Then*

$$\varrho(\Phi) = \sup_{\lambda\in\Lambda}\int_0^1 \mathrm{AVaR}^+_{1-s}(\Phi)\;\lambda(\mathrm{d}s), \tag{2.61}$$

where the set Λ is a convex closed subset of the set of probability measures on $(0,1]$.

Proof. Owing to Theorem 2.46 and the considerations preceding the corollary, we need only prove that every measure λ is a probability measure. This can be verified via integration by parts:

$$\int_0^1 \lambda(\mathrm{d}s) = \int_{(0,1)} (1-s)\,\mathrm{d}h(s) + \beta = \int_0^1 h(s)\,\mathrm{d}s + \beta = \mu(0,1) + \beta = 1.$$

□

The meaning of the concept of risk aversion of Definition 1.19 in the space $\mathcal{Z}$ of bounded real random variables on a standard atomless probability space $(\Omega, \mathcal{F}, P)$ was discussed in section 1.4.2.3. Here we summarize the conclusions in our setting.

If a risk measure $\varrho_{\mathcal{Z}} : \mathcal{Z} \to \mathbb{R}$ is law-invariant, then it defines a risk measure ϱ on the space $\mathcal{Q}_{\mathrm{b}}$ by the formulas:

$$\begin{aligned} \varrho_{\mathcal{Z}}(Z) &= \varrho\big(F_Z^{-1}\big), \quad Z \in \mathcal{Z}, \\ \varrho(\Phi) &= \varrho_{\mathcal{Z}}\big(\Phi(U)\big), \quad \Phi \in \mathcal{Q}_{\mathrm{b}}, \end{aligned}$$

where U is a uniformly distributed element of $\mathcal{Z}$.

According to Theorem 1.33, the convexity of $\varrho_{\mathcal{Z}}$ on the entire space $\mathcal{Z}$ implies the property of risk aversion.

Corollary 2.48. *If a risk measure $\varrho_{\mathcal{Z}} : \mathcal{Z} \to \mathbb{R}$ is law-invariant and convex, then $\varrho : \mathcal{Q}_{\mathrm{b}} \to \mathbb{R}$ is convex and risk-averse in the sense of Definition* 1.19.

A similar observation is true for $\mathcal{Z} = \mathcal{L}_p(\Omega, \mathcal{F}, P)$, with $p \in [1, \infty)$, and the corresponding space of the quantile functions $\mathcal{Q}_p$.

Finally, we can obtain the dual representation of law-invariant coherent measures of risk on $\mathcal{L}_\infty(\Omega, \mathcal{F}.P)$. Observe that only elements of $\mathcal{L}_1(\Omega, \mathcal{F}, P)$ feature in this representation.

Theorem 2.49. *Suppose the probability space $(\Omega, \mathcal{F}, P)$ is standard and atomless and $\varrho_{\mathcal{Z}} : \mathcal{L}_\infty(\Omega, \mathcal{F}, P) \to \mathbb{R}$ is coherent and law-invariant. Then a convex and closed set*

$$\mathcal{A} \subset \Big\{(\zeta, \beta) \in \mathcal{L}_1(\Omega, \mathcal{F}, P) \times \mathbb{R} : \int_\Omega \zeta(\omega)\, P(d\omega) + \beta = 1,\ \zeta \ge 0,\ \beta \ge 0\Big\}$$

exists, such that for all $Z \in \mathcal{L}_\infty(\Omega, \mathcal{F}, P)$ we have

$$\varrho_{\mathcal{Z}}(Z) = \max_{(\zeta,\beta)\in\mathcal{A}} \left\{ \int_\Omega \zeta(\omega) Z(\omega)\, P(d\omega) + \beta \operatorname{ess\,sup}(Z) \right\}. \tag{2.62}$$

Proof. Define the risk measure $\varrho : \mathcal{Q}_{\mathrm{b}} \to \mathbb{R}$ as

$$\varrho(\Phi) = \varrho_{\mathcal{Z}}\big(\Phi(U)\big), \quad \Phi \in \mathcal{Q}_{\mathrm{b}}.$$

As discussed before the theorem, it is coherent and risk averse. Therefore, it has the unique dual representation (2.47) with $\varrho^*(\mu) = 0$ on $\mathcal{D}$. This, in turn, yields the Kusuoka representation (2.61). Since $\mathrm{AVaR}_\alpha^+(\Phi)$ is defined in the same way as $\mathrm{AVaR}_\alpha^+(Z)$, we get

$$\varrho_{\mathcal{Z}}(Z) = \max_{\lambda\in\Lambda} \left\{ \int_{(0,1)} \mathrm{AVaR}_{1-s}^+(Z)\, \lambda(ds) + \lambda(\{1\}) \operatorname{ess\,sup}(Z) \right\}.$$

From Example 2.29 with $p = 1$, we know the dual representation of each $\mathrm{AVaR}_{1-s}^+(Z)$ under the integral:

$$\partial\,\mathrm{AVaR}^{+}_{1-s}(0) = \left\{ \xi_s \in \mathcal{L}_\infty(\Omega, \mathcal{F}, P) : \ \mathbb{E}[\xi_s] = 1,\ 0 \le \xi_s \le \frac{1}{1-s} \right\}.$$

Using Strassen's theorem 3.13, we conclude that the dual representation (2.62) is true with the set

$$\mathcal{A} = \mathrm{cl}\Big\{ (\zeta, \beta) \in \mathcal{L}_1(\Omega, \mathcal{F}, P) \times \mathbb{R} : \zeta = \int_{(0,1)} \xi_s \, \lambda(\mathrm{d}s),$$
$$\xi_s \in \partial\,\mathrm{AVaR}^{+}_{1-s}(0),\ \forall s \in (0,1),\ \beta = \lambda(\{1\}),\ \lambda \in \Lambda \Big\}.$$

As before, the integral above is understood in the weak* sense. □

The only singular element in the dual representation is the finitely additive measure associated with $\operatorname{ess\,sup}(Z)$.

2.6 Systemic Risk Measures

In this section, we propose a set of axioms of measures of risk for random vectors with realizations in $\mathbb{R}^m$. This framework is analogous to the coherent risk measures properties for scalar-valued random variables. The term "systemic" refers to the interpretation that the components of the random vector represent "losses" in parts of a system.

If $m = 1$, the proposed set of axioms exactly coincides with those in § 2.1.1. Throughout the section, we consider a risk measure for random vectors in the space $\mathcal{Z}_m = \mathcal{L}_p(\Omega, \mathcal{F}, P; \mathbb{R}^m)$ to be a lower semicontinuous functional $\varrho : \mathcal{Z}_m \to \overline{\mathbb{R}}$ with a non-empty domain. We denote by I_m the m-dimensional vector, whose components are all equal to one, and by $\mathbb{1}_m$ the random vector with realizations equal to I_m.

Definition 2.50. A functional $\varrho : \mathcal{Z}_m \to \overline{\mathbb{R}}$ is a coherent risk measure if it satisfies for all $Z, V \in \mathcal{Z}_m$ the following axioms:

Convexity: $\varrho(\alpha Z + (1-\alpha)V) \le \alpha\varrho(Z) + (1-\alpha)\varrho(V),\ \forall \alpha \in [0,1]$;
Monotonicity: $Z_i \le V_i,\ i = 1, \dots, m \implies \varrho(Z) \le \varrho(V)$;
Translation equivariance: $\varrho(Z + c\mathbb{1}_m) = \varrho(Z) + c\varrho(\mathbb{1}_m),\ \forall c \in \mathbb{R}$;
Positive homogeneity: $\varrho(\gamma Z) = \gamma\varrho(Z),\ \forall \gamma \ge 0$.

A functional $\varrho : \mathcal{Z}_m \to \overline{\mathbb{R}}$ is a convex risk measure if it satisfies the axioms of convexity, monotonicity, and translation equivariance.

The main difference from the one-dimensional case is the translation equivariance axiom.

2.6.1 Dual Representation

In order to derive a dual representation of a systemic risk measure, we pair the space of random vectors $\mathcal{Z}_m = \mathcal{L}_p(\Omega, \mathcal{F}, P; \mathbb{R}^m)$, $p \in [1, \infty)$, with the space $\mathcal{Z}_m^* = \mathcal{L}_q(\Omega, \mathcal{F}, P; \mathbb{R}^m)$, where $q \in (1, \infty]$ is such that $\frac{1}{p} + \frac{1}{q} = 1$. For $Z \in \mathcal{Z}_m$ and $\zeta \in \mathcal{Z}_m^*$ the bilinear form $\langle \cdot, \cdot \rangle$ on the product space $\mathcal{Z}_m^* \times \mathcal{Z}_m$ is defined as follows:

$$\langle \zeta, Z \rangle = \int_\Omega \big\langle \zeta(\omega), Z(\omega) \big\rangle \, P(d\omega).$$

The Fenchel conjugate function $\varrho^* : \mathcal{Z}_m^* \to \overline{\mathbb{R}}$ of a risk measure ϱ is given by

$$\varrho^*(\zeta) = \sup_{Z \in \mathcal{Z}_m} \big\{ \langle \zeta, Z \rangle - \varrho(Z) \big\},$$

and the conjugate of ϱ^* (the biconjugate function) is

$$\varrho^{**}(Z) = \sup_{\zeta \in \mathcal{Z}_m^*} \big\{ \langle \zeta, Z \rangle - \varrho^*(\zeta) \big\}.$$

The Fenchel–Moreau theorem implies that if $\varrho(\cdot)$ is convex and lower semicontinuous, then $\varrho^{**} = \varrho$, that is,

$$\varrho(Z) = \sup_{\zeta \in \mathcal{A}} \big\{ \langle \zeta, Z \rangle - \varrho^*(\zeta) \big\}, \tag{2.63}$$

where $\mathcal{A} = \mathrm{dom}(\varrho^*)$ is the domain of the conjugate function ϱ^*. Then, based on the Fenchel–Moreau theorem and the axioms formulated above, we can show the following theorem, which parallels Theorem 2.25.

Theorem 2.51. *Suppose $\varrho : \mathcal{Z}_m \to \overline{\mathbb{R}}$ is a convex and lower semicontinuous risk functional. Then the following holds:*

(i) *The monotonicity axiom is satisfied if and only if all $\zeta \in \mathcal{A}$ are nonnegative;*

(ii) *The translation equivariance axiom is satisfied if and only if $\varrho(\mathbb{1}_m) = \langle 1_m, \mu_\zeta \rangle$ for all $\zeta \in \mathcal{A}$, where $\mu_\zeta = \int_\Omega \zeta(\omega) \, P(d\omega)$;*

(iii) *The positive homogeneity axiom is satisfied if and only if ϱ^* is the indicator function of $\partial\varrho(0)$, i.e.*

$$\varrho(Z) = \sup_{\zeta \in \mathcal{A}} \langle \zeta, Z \rangle, \quad \mathcal{A} = \partial\varrho(0). \tag{2.64}$$

Proof. Since $\varrho(\cdot)$ is convex, lower semicontinuous, and has a non-empty domain, the representation (2.63) holds by virtue of the Fenchel–Moreau theorem.

(i) Suppose ϱ satisfies the monotonicity condition. Assume that $\zeta_i(\omega) < 0$ for $\omega \in \Delta \in \mathcal{F}$ with $P(\Delta) > 0$ for some component $i = 1, \ldots, m$. Define $X_i = \mathbb{1}_\Delta$, and $X_j = 0$ for $j \neq i$, $j = 1, \ldots, m$. Take any Z with support in Δ such that $\varrho(Z)$ is finite and define $Z_t := Z - tX$. Then for $t \geq 0$, we have that $Z \geq Z_t$, and $\varrho(Z) \geq \varrho(Z_t)$ by monotonicity. Consequently,

$$\begin{aligned}\varrho^*(\zeta) &\geq \sup_{t \in \mathbb{R}_+} \Big\{\langle \zeta, Z_t\rangle - \varrho(Z_t)\Big\} \geq \sup_{t \in \mathbb{R}_+} \Big\{\langle \zeta, Z\rangle - t\langle \zeta, X\rangle - \varrho(Z)\Big\} \\ &= \sup_{t \in \mathbb{R}_+} \Big\{\langle \zeta, Z\rangle - t\int_\Delta \zeta_i(\omega)\, P(\mathrm{d}\omega) - \varrho(Z)\Big\} = +\infty.\end{aligned}$$

It follows that $\varrho^*(\zeta) = +\infty$ for every $\zeta \in \mathcal{Z}_m^*$ with at least one negative component and, thus $\zeta \notin \operatorname{dom} \varrho^*$. Conversely, suppose $\zeta \in \mathcal{Z}_m^*$ has realizations in $\mathbb{R}^n$ with nonnegative components P-a.s. Then whenever $Z \geq Z'$, we have:

$$\langle \zeta, Z\rangle = \int_\Omega \langle \zeta(\omega), Z(\omega)\rangle\, P(\mathrm{d}\omega) \geq \int_\Omega \langle \zeta(\omega), Z'(\omega)\rangle\, P(\mathrm{d}\omega) = \langle \zeta, Z'\rangle.$$

Consequently,

$$\varrho(Z) = \sup_{\zeta \in \mathcal{Z}_m^*} \Big\{\langle \zeta, Z\rangle - \varrho^*(\zeta)\Big\} \geq \sup_{\zeta \in \mathcal{Z}_m^*} \Big\{\langle \zeta, Z'\rangle - \varrho^*(\zeta)\Big\} = \varrho(Z').$$

Hence, the monotonicity condition holds.

(ii) Suppose the translation property is satisfied. Then for any $c \in \mathbb{R}$ and $\zeta \in \mathcal{Z}_m^*$, we get:

$$\begin{aligned}\varrho^*(\zeta) &= \sup_{Z \in \mathcal{Z}_m} \Big\{\langle \zeta, Z + c\mathbb{1}_m\rangle - \varrho(Z + c\mathbb{1}_m)\Big\} \\ &= \sup_{Z \in \mathcal{Z}_m} \Big\{\langle \zeta, Z\rangle + c\int_\Omega \langle I_m, \zeta(\omega)\rangle\, P(\mathrm{d}\omega) - \varrho(Z) - c\varrho(\mathbb{1}_m)\Big\} \\ &= \varrho^*(\zeta) + c\Big(\int_\Omega \langle I_m, \zeta(\omega)\rangle\, P(\mathrm{d}\omega) - \varrho(\mathbb{1}_m)\Big).\end{aligned}$$

If $\varrho^*[\zeta]$ is finite, and since c may be arbitrary, then

$$\varrho(\mathbb{1}_m) = \Big\langle I_m, \int_\Omega \zeta(\omega) P(d\omega)\Big\rangle = \langle I_m, \mu_\zeta\rangle \quad \text{for all } \zeta \in \mathcal{A}. \tag{2.65}$$

Conversely, suppose $\varrho(\mathbb{1}_m) = \langle I_m, \mu_\zeta\rangle$ for all $\zeta \in \mathcal{A}$. Then for any $Z \in \mathcal{Z}_m$ and $c \in \mathbb{R}$:

$$\begin{aligned}\varrho(Z + c\mathbb{1}_m) &= \sup_{\zeta \in \mathcal{Z}_m^*} \Big\{ \langle \zeta, Z + c\mathbb{1}_m \rangle - \varrho^*(\zeta) \Big\} \\ &= \sup_{\zeta \in \mathcal{Z}_m^*} \Big\{ \langle \zeta, Z \rangle + c \int_\Omega \langle \mathbb{1}_m, \zeta(\omega) \rangle \, P(\mathrm{d}\omega) - \varrho^*(\zeta) \Big\} \\ &= \sup_{\zeta \in \mathcal{Z}_m^*} \Big\{ \langle \zeta, Z \rangle - \varrho^*(\zeta) + c\varrho(\mathbb{1}_m) \Big\} = \varrho(Z) + c\varrho(\mathbb{1}_m).\end{aligned}$$

Hence, the translation property is satisfied.

(iii) The proof is identical to the proof of Theorem 2.25(iii). □

It follows from Theorem 2.51 that if a systemic risk measure ϱ is lower semicontinuous and satisfies the axioms of monotonicity, convexity and translation equivariance, then the representation (2.64) holds with the set $\mathcal{A}$ defined as:

$$\mathcal{A} = \Big\{ \zeta \in \mathcal{Z}_m^* \ : \ \Big\langle \mathbb{1}_m, \int_\Omega \zeta(\omega) \, P(\mathrm{d}\omega) \Big\rangle = \varrho(\mathbb{1}_m), \ \zeta \succeq 0 \Big\}.$$

Definition 2.52. A coherent systemic measure of risk $\varrho : \mathcal{Z}_m \to \overline{\mathbb{R}}$ is *normalized* if $\varrho(\mathbb{1}_m) = 1$.

Corollary 2.53. *For a normalized systemic coherent measure of risk* $\varrho(\cdot)$*, we have* $\int_\Omega \langle \mathbb{1}_m, \zeta \rangle \, P(\mathrm{d}\omega) = 1$*, for all* $\zeta \in \mathrm{dom}(\varrho^*)$.

Proof. It follows from equation (2.65) that $\langle \mathbb{1}_m, \mu_\zeta \rangle = 1$, as stated. This entails that for all $\zeta \in \mathcal{A}$ the measure $\zeta \cdot \mathrm{d}P$ is a probability measure on the space $\Omega \times \{1, 2, \ldots, m\}$. □

2.6.2 Linear Scalarizations

In this section, we propose a particular construction of a systemic risk measure using a scalarization of the components of the systemic losses.

Suppose we have a random vector $Z \in \mathcal{Z}_m$ with a right-continuous distribution function $F(Z; \cdot)$ and marginal distribution functions $F_i(Z_i; \cdot)$ of each component $i = 1, \ldots, m$. We consider linear scalarization using vectors taken from the simplex $\mathbb{S}^m = \{c \in \mathbb{R}^m : \sum_{i=1}^m c_i = 1, \ c_i \geq 0, \ i = 1, \ldots, m\}$. Let $\varrho : \mathcal{L}_p(\Omega, \mathcal{F}, P) \to \overline{\mathbb{R}}$ be a lower semicontinuous risk measure. For a fixed closed convex set $S \subseteq \mathbb{S}^m$, we define the risk measure

$$\varrho_S(Z) = \varrho(Z_S), \quad \text{where } Z_S(\omega) = \max_{c \in S} c^\top Z(\omega), \quad \omega \in \Omega. \tag{2.66}$$

It is straightforward to see that $Z_S \in \mathcal{L}_p(\Omega, \mathcal{F}, P)$ and hence, the risk measure $\varrho_S(\cdot)$ is well-defined on $\mathcal{Z}_m$.

Theorem 2.54. *If $\varrho : \mathcal{L}_p(\Omega, \mathcal{F}, P) \to \overline{\mathbb{R}}$ is a coherent (convex) risk measure, then for any closed convex set $S \subseteq \mathbb{S}^m$, the systemic risk measure $\varrho_S(Z)$ is coherent (convex) and normalized.*

Proof. For two random vectors $Z, V \in \mathcal{Z}_m$ with $Z \leq V$ component-wise a.s., we have $c^\top Z \leq c^\top V$ a.s. for all $c \in \mathbb{S}^m$. This implies that $\max_{c\in S} c^\top Z \leq \max_{c\in S} c^\top V$ a.s. and, hence, $\varrho(Z_S) \leq \varrho(V_S)$. Thus, the monotonicity axiom is satisfied. Given two random vectors $Z, V \in \mathcal{Z}_m$ and $\alpha \in [0, 1]$, consider their convex combination $\alpha Z + (1-\alpha)V$. Due to the convexity and monotonicity of $\varrho(\cdot)$, we have

$$\begin{aligned}\varrho_S(\alpha Z + (1-\alpha)V) &= \varrho[\max_{c\in S} c^\top(\alpha Z + (1-\alpha)V)]\\ &\leq \varrho\big(\alpha \max_{c\in S} c^\top Z + (1-\alpha)\max_{c\in S} c^\top V\big)\\ &\leq \alpha\varrho\big(\max_{c\in S} c^\top Z\big) + (1-\alpha)\varrho\big(\max_{c\in S} c^\top V\big)\\ &= \alpha\varrho_S(Z) + (1-\alpha)\varrho_S(V).\end{aligned}$$

Thus, the convexity axiom is satisfied. Given a random vector $Z \in \mathcal{Z}_m$ and a constant $t \in \mathbb{R}$, it follows:

$$\begin{aligned}\varrho_S(Z + t\mathbb{1}_m) &= \varrho\big(\max_{c\in S} c^\top(Z + t\mathbb{1}_m)\big)\\ &= \varrho\big(\max_{c\in S} c^\top Z + t\big) = \varrho(Z_S) + t.\end{aligned}$$

Furthermore

$$\varrho_S(\mathbb{1}_m) = \varrho\big(\max_{c\in S} c^\top \mathbb{1}_m\big) = \varrho(1) = 1.$$

This verifies the translation axiom and the normalization. The positive homogeneity follows straightforwardly. □

If $\varrho(\cdot)$ is a coherent measure of risk, we can calculate the dual representation of the systemic risk measure $\varrho_S(\cdot)$ explicitly. Every $\xi \in \operatorname{dom}(\varrho^*)$ is nonnegative, and thus

$$\begin{aligned}\varrho_S(Z) = \varrho\big(\max_{c\in S} c^\top Z\big) &= \sup_{\xi\in\partial\varrho(0)} \int_\Omega \xi(\omega) \max_{c\in S} c^\top Z(\omega)\, P(d\omega)\\ &= \sup_{\xi\in\partial\varrho(0)} \int_\Omega \max_{c\in S} c^\top \xi(\omega) Z(\omega)\, P(d\omega).\end{aligned}$$

Define the set $\mathbb{C}_S$ of measurable functions $C : \Omega \to S$. Then, due to the measurable selection theorem, the last chain of relations may be continued as follows:

$$\varrho\big(\max_{c\in S} c^\top Z\big) = \sup_{\xi\in\partial\varrho(0)} \sup_{C\in\mathbb{C}_S} \int_\Omega C(\omega)^\top \xi(\omega) Z(\omega)\ P(\mathrm{d}\omega)$$
$$= \sup_{\zeta\in\mathcal{A}} \int_\Omega \zeta(\omega) Z(\omega)\ P(\mathrm{d}\omega),$$

with

$$\mathcal{A} = \{\xi C : \xi \in \partial\varrho(0),\ C \in \mathbb{C}_S\}.$$

Notice that the representations just derived have the form of the dual representation (2.64).

The convexity of $\varrho(\cdot)$ yields the following property of the aggregation by a single linear scalarization.

Remark 2.55. Given a coherent risk measure $\varrho : \mathcal{Z}_m \to \mathbb{R}$, and a scalarization vector $c \in \mathbb{S}^m$, the following holds for any random vector $Z \in \mathcal{Z}_m$:

$$\varrho\big(c^\top Z\big) \le \sum_{i=1}^m c_i \varrho(Z_i).$$

□

2.6.3 Nonlinear Scalarization

The second aggregation method that falls within the scope of our axiomatic framework is the approach that evaluates the risk of each component first, and then scalarizes the vector of risks.

We define $\Omega_m = \{1, \ldots, m\}$ and consider a probability space $(\Omega_m, \mathcal{F}_m, c)$, where $c \in \mathbb{S}^m$ and $\mathcal{F}_m$ contains all subsets of Ω_m. Given an m-dimensional random vector $Z \in \mathcal{L}_p(\Omega, \mathcal{F}, P; \mathbb{R}^m)$ and a collection of m univariate measures of risk $\varrho_i : \mathcal{L}_p(\Omega, \mathcal{F}, P) \to \mathbb{R}$, $i = 1, \ldots, m$, we define the random variable R_Z on the space $(\Omega_m, \mathcal{F}_m, c)$ as follows:

$$R_Z(i) = \varrho_i(Z_i), \quad i = 1, \ldots m. \tag{2.67}$$

Then, with a convex or coherent measure of risk $\varrho_0 : \mathcal{L}_0(\Omega_m, \mathcal{F}_m, c) \to \mathbb{R}$ we define the systemic measure of risk as follows:

$$\hat{\varrho}(Z) = \varrho_0(R_Z). \tag{2.68}$$

This amounts to a nonlinear aggregation of the individual risks $\varrho_i(Z_i)$. The measure $\hat{\varrho}(\cdot)$ satisfies the axioms postulated for systemic risk measures.

Theorem 2.56.

(i) *If the univariate measures of risk* $\varrho_i(\cdot)$, $i = 1, \ldots, m$, *are convex and* $\varrho_0(\cdot)$ *is convex and monotonic, then* $\hat{\varrho}(\cdot)$ *is convex;*

(ii) *If* $\varrho_i(\cdot)$, $i = 0, 1, \ldots, m$, *are nondecreasing, then* $\hat{\varrho}(\cdot)$ *is nondecreasing as well;*

(iii) *If* $\varrho_i(\cdot)$, $i = 0, 1, \ldots, m$, *satisfy the translation property, then* $\hat{\varrho}(\cdot)$ *satisfies this property as well;*

(iv) *If* $\varrho_i(\cdot)$, $i = 0, 1, \ldots, m$, *are positively homogeneous, so is* $\hat{\varrho}(\cdot)$.

Proof. (i) Given any $Z, V \in \mathcal{L}_p(\Omega, \mathcal{F}, P; \mathbb{R}^m)$ and $\alpha \in (0, 1)$, we consider the random vector $W = \alpha Z + (1-\alpha)V$. We have $\varrho_i(W_i) \leq \alpha\varrho_i(Z_i) + (1-\alpha)\varrho_i(V_i)$, $i = 1, \ldots, m$. Defining a random variable W' on Ω_m by setting $W'(i) = \alpha\varrho_i(Z_i) + (1-\alpha)\varrho_i(V_i)$, we obtain that $W \leq W'$. Using the monotonicity and convexity of ϱ_0, we obtain

$$\varrho_0(W) \leq \varrho_0(W') \leq \alpha\varrho_0(R_Z) + (1-\alpha)\varrho_0(R_V).$$

Hence, $\hat{\varrho}\big(\alpha Z + (1-\alpha)V\big) \leq \alpha\hat{\varrho}(Z) + (1-\alpha)\hat{\varrho}(V)$.

(ii) Suppose the vectors $Z, V \in \mathcal{L}_p(\Omega, \mathcal{F}, P; \mathbb{R}^m)$ satisfy $Z_i \leq V_i$, $i = 1, \ldots, m$, a.s.. Hence, $\varrho_i(Z_i) \leq \varrho_i(V_i)$ for all $i = 1, \ldots, m$, by the monotonicity property of ϱ_i. This further implies that $R_Z \leq R_V$, entailing that $\varrho_0(R_Z) \leq \varrho_0(R_V)$, in view of the monotonicity of $\varrho_0(\cdot)$.

(iii) Given a random vector $Z \in \mathcal{L}_p(\Omega, \mathcal{F}, P; \mathbb{R}^m)$ and a constant a, we have $R_{Z+a\mathbb{1}_m}(i) = \varrho_i(Z_i + a) = \varrho_i(Z_i) + a$. Hence $\varrho_0\big(R_{Z+a\mathbb{1}_m}\big) = \varrho_0(R_Z) + a$.

(iv) It is sufficient to verify the positive homogeneity. Given a random vector $Z \in \mathcal{L}_p(\Omega, \mathcal{F}, P; \mathbb{R}^m)$ and $t > 0$, we have

$$\hat{\varrho}(tZ) = \varrho_0(R_{tZ}) = \varrho_0\big(tR_Z\big) = t\varrho_0(R_Z) = t\hat{\varrho}(Z),$$

where we have used the positive homogeneity property of $\varrho_i(\cdot)$ for all $i = 0, 1, \ldots, m$. □

We may remark here that the construction of a systemic risk measure in this subsection is similar to a time-consistent dynamic risk measure in a two-stage setting: stage 2 evaluates the conditional risk of the individual components of Z, and stage 1 evaluates the risk of these conditional risks. There is, however, a significant difference between the systemic setting and the two-stage dynamic setting. In the dynamic setting, for every ω, only one of the possible "futures" is observed, while in the systemic setting, we need to deal with all components of the system.

2.7 Risk Forms

In this section, instead of considering an abstract probability space $(\Omega, \mathcal{F}, P)$, with the probability measure P fixed, we focus on a Polish space $\mathcal{X}$ with its Borel σ-algebra $\mathcal{B}(\mathcal{X})$ as the base measure space, but we allow arbitrary probability measures on this space.

2.7.1 Risk Models with Variable Probability Measures

Let $\mathcal{P}(\mathcal{X})$ be the set of probability measures on $\big(\mathcal{X}, \mathcal{B}(\mathcal{X})\big)$. The space of all real-valued bounded measurable functions on $\mathcal{X}$ is denoted by $\mathbb{B}(\mathcal{X})$. We use x to denote an element of $\mathcal{X}$ and δ_x to denote the Dirac measure concentrated at x. As before, the symbol $\mathbb{1}$ stands for the constant function on $\mathcal{X}$ equal to 1.

A *probabilistic model* is a pair $[Z, P] \in \mathbb{B}(\mathcal{X}) \times \mathcal{P}(\mathcal{X})$. For two probabilistic models $[V, P]$ and $[W, Q]$ the notation $[V, P] \overset{\mathcal{D}}{\sim} [W, Q]$ means that $P\{V \le \eta\} = Q\{W \le \eta\}$ for all $\eta \in \mathbb{R}$ (both models have the same distribution). The inequality $Z \le V$ between elements of $\mathbb{B}(\mathcal{X})$ is always understood pointwise.

Our goal is to propose an approach to risk evaluation of a family of probabilistic models. In the definition below, the first four properties are the same as for a usual measure of risk, with the second argument of the risk form fixed. The last two properties are specific to our model with two arguments.

Definition 2.57. A measurable functional $\varrho : \mathbb{B}(\mathcal{X}) \times \mathcal{P}(\mathcal{X}) \to \mathbb{R}$ is called a *risk form*.

(i) It is *monotonic*, if $V \le W \implies \varrho[V, P] \le \varrho[W, P]$ for all $P \in \mathcal{P}(\mathcal{X})$;

(ii) It is *normalized* if $\varrho[0, P] = 0$ for all $P \in \mathcal{P}(\mathcal{X})$;

(iii) It is *translation equivariant* if for all $V \in \mathbb{B}(\mathcal{X})$, all $a \in \mathbb{R}$, and all $P \in \mathcal{P}(\mathcal{X})$, $\varrho[a\mathbb{1} + V, P] = a + \varrho[V, P]$;

(iv) It is *positively homogeneous*, if for all $V \in \mathbb{B}(\mathcal{X})$, all $\beta \in \mathbb{R}_+$, and all $P \in \mathcal{P}(\mathcal{X})$, $\varrho[\beta V, P] = \beta\varrho[V, P]$;

(v) It is *law-invariant* if $[V, P] \overset{\mathcal{D}}{\sim} [W, Q] \implies \varrho[V, P] = \varrho[W, Q]$;

(vi) It has the *support property*, if $\varrho\big[\mathbb{1}_{\text{supp}(P)} V, P\big] = \varrho[V, P]$ for all $(V, P) \in \mathbb{B}(\mathcal{X}) \times \mathcal{P}(\mathcal{X})$.

An example of a risk form is the expected value, which is a well-understood bilinear form:

$$\mathbb{E}[Z, P] = \int_{\mathcal{X}} Z(x)\; P(\mathrm{d}x).$$

In our analysis, we are interested mainly in forms depending on one or both arguments in a nonlinear way. All examples discussed in § 2.1.1 can be viewed as risk forms, if they allow the probability measure P to vary. One of them is the *mean–semideviation model* (see Example 2.4):

$$\varrho[Z, P] = \int_{\mathcal{X}} Z(u)\ P(\mathrm{d}u) + \varkappa \int_{\mathcal{X}} \Big[Z(u) - \int_{\mathcal{X}} Z(v)\ P(\mathrm{d}v) \Big]_+\ P(\mathrm{d}u), \quad (2.69)$$

with $\varkappa \in [0, 1]$. Yet another example, rarely used in the risk measure theory, due to its conservative nature, but very relevant for us, is the *worst-case risk form*:

$$\varrho[Z, P] = \operatorname{ess\,sup}(Z; P) = \inf \big\{ b \in \mathbb{R} : P[v : Z(v) \le b] = 1 \big\}. \quad (2.70)$$

We mark the dependence on the measure P by the second argument of the "ess sup" operator.

All three examples above are coherent and law-invariant risk forms having the support property.

Our concept of law invariance is broader than that used in the previous section because it allows for the probability measure to vary. If the risk form is law-invariant, then it has the support property, because $[V, P] \overset{\mathcal{D}}{\sim} [\mathbb{1}_{\operatorname{supp}(P)} V, P]$.

Lemma 2.58. *If a risk form $\varrho : \mathcal{B}(\mathcal{X}) \times \mathcal{P}(\mathcal{X}) \to \mathbb{R}$ has the normalization, translation equivariance, and support properties then for every $Z \in \mathcal{B}(\mathcal{X})$ and every $x \in \mathcal{X}$*

$$\varrho\big[Z, \delta_x\big] = Z(x). \quad (2.71)$$

Proof. Using the support property twice, the translation property, and the normalization property, we obtain the chain of equations:

$$\varrho\big[Z, \delta_x\big] = \varrho\big[\mathbb{1}_x Z, \delta_x\big] = \varrho\big[Z(x)\mathbb{1}, \delta_x\big] = Z(x) + \varrho[0, \delta_x] = Z(x).$$

□

2.7.2 Dual and Kusuoka Representations

In this section, we generalize the dual representation of Theorem 2.25 and the Kusuoka representation of law-invariant risk measures (Theorem 2.39) to risk forms. In sections 2.3 and 2.4.2, these representations were derived under the assumption that the probability measure was fixed. We show that a more general result using variable probability measures is true.

With every probabilistic model $[Z, P]$, we associate its distribution function,

$$F[Z, P](z) = P[Z \le z], \quad z \in \mathbb{R},$$

and its quantile function,

$$\Phi[Z,P](p) = \inf\{\eta : P[Z \le \eta] \ge p\}, \quad p \in (0,1].$$

The quantile functions are elements of the space $\mathcal{Q}_b$ of bounded, nondecreasing, and left-continuous functions on $(0, 1]$.

We recall that a risk form $\varrho : \mathbb{B}(\mathcal{X}) \times \mathcal{P}(\mathcal{X}) \to \mathbb{R}$ is *comonotonically convex*, if for any $P \in \mathcal{P}(\mathcal{X})$ and any two comonotonic functions $Z, V \in \mathbb{B}(\mathcal{X})$ we have

$$\varrho[\alpha Z + (1-\alpha)V, P] \le \alpha\varrho[Z, P] + (1-\alpha)\varrho[V, P], \quad \forall\, \alpha \in [0.1].$$

Definition 2.59. A probabilistic model $[Z, P]$ is smaller than a probabilistic model $[Z', P']$ in the *increasing convex order*, written $[Z, P] \preceq_{(\mathrm{ic})} [Z', P']$, if for all $\eta \in \mathbb{R}$

$$\int_{\mathcal{X}} \big[Z(x) - \eta\big]_+ \, P(dx) \le \int_{\mathcal{X}} \big[Z'(x) - \eta\big]_+ \, P'(dx).$$

Here, $[a]_+ = \max(0, a)$.

This concept allows us to consider risk forms consistent with the increasing order.

Definition 2.60. A risk form $\varrho : \mathbb{B}(\mathcal{X}) \times \mathcal{P}(\mathcal{X}) \to \mathbb{R}$ is *consistent with the increasing convex order*, if

$$[Z, P] \preceq_{(\mathrm{ic})} [Z', P'] \implies \varrho[Z, P] \le \varrho[Z', P'].$$

Evidently, the consistency with the increasing convex order implies the monotonicity and the law invariance.

Now we are ready to develop the dual representation of risk forms. We denote by $\mathcal{M}$ the set of countably additive finite measures on $(0, 1]$. For every risk form $\varrho : \mathbb{B}(\mathcal{X}) \times \mathcal{P}(\mathcal{X}) \to \mathbb{R}$, we define the conjugate functional $\varrho^* : \mathcal{M} \to \mathbb{R} \cup \{+\infty\}$ as follows:

$$\varrho^*(\mu) = \sup_{[Z,P] \in \mathbb{B}(\mathcal{X}) \times \mathcal{P}(\mathcal{X})} \left\{ \int_0^1 \Phi[Z, P](p)\, \mu(\mathrm{d}p) - \varrho[Z, P] \right\}. \tag{2.72}$$

We obtain the following dual representation.

Theorem 2.61. *If the space $\mathcal{X}$ is uncountable and a risk form $\varrho : \mathbb{B}(\mathcal{X}) \times \mathcal{P}(\mathcal{X}) \to \mathbb{R}$ is normalized, translation equivariant, comonotonically convex, and consistent with the increasing convex order, then a uniquely defined closed convex set*

$$\mathcal{D}_\varrho \subseteq \big\{\mu \in \mathcal{M} : \mu(0, \cdot] \text{ is nondecreasing and convex on } (0, 1],\ \mu(0, 1] = 1\big\} \tag{2.73}$$

exists, such that for all $[Z, P] \in \mathbb{B}(\mathcal{X}) \times \mathcal{P}(\mathcal{X})$

$$\varrho[Z, P] = \sup_{\mu \in \mathcal{D}_\varrho} \left\{ \int_0^1 \Phi[Z, P](p)\, \mu(\mathrm{d}p) - \varrho^*(\mu) \right\}. \tag{2.74}$$

If, additionally, ϱ *is positively homogeneous, then* $\varrho^*(\mu) = 0$ *for all* $\mu \in \mathcal{D}_\varrho$.

Proof. First, we show that the risk form $\varrho[\cdot, \cdot]$ defines a functional $R(\cdot)$ on the space of the quantile functions $\mathcal{Q}_{\mathrm{b}}$ on $(0, 1]$ by the identity:

$$R\big(\Phi[Z, P]\big) = \varrho[Z, P]. \tag{2.75}$$

Indeed, $\varrho[\cdot, \cdot]$ is law-invariant, due to its consistency with the increasing convex order. Therefore, if $\Phi[Z, P] = \Phi[Z', P']$ then $\varrho[Z, P] = \varrho[Z', P']$. Thus the functional R is well-defined on the set of quantile functions $\big\{\Phi[Z, P] : Z \in \mathbb{B}(\mathcal{X}),\ P \in \mathcal{P}(\mathcal{X})\big\}$.

Since $\mathcal{X}$ is an uncountable Polish space, it is isomorphic to [0,1] equipped with the Borel σ-algebra. Denote by $Z_0 : \mathcal{X} \to [0, 1]$ the said isomorphism. Since both Z_0 and Z_0^{-1} are measurable, we can define a probability measure P_0 on $\mathcal{B}(\mathcal{X})$ by $\lambda \circ Z_0$, where λ is the Lebesgue measure on [0,1]. By construction, $P_0\big[Z_0 \le p\big] = p$ and thus $\Phi[Z_0, P_0](p) = p$, for all $p \in [0, 1]$. Moreover, for every function $\Psi \in \mathcal{Q}_{\mathrm{b}}$, we can define $Z(x) = \Psi(Z_0(x))$, $x \in \mathcal{X}$, and $\Psi^{-1}(z) = \sup\{p \in [0, 1] : \Psi(p) \le z\}$. Then, for every $z \in \mathbb{R}$,

$$\begin{aligned} P_0\big\{x : Z(x) \le z\big\} &= P_0\big\{x : \Psi(Z_0(x)) \le z\big\} \\ &= P_0\big\{x : Z_0(x) \le \Psi^{-1}(z)\big\} = \Psi^{-1}(z). \end{aligned} \tag{2.76}$$

Consequently, the distribution function of Z under P_0 is the inverse of Ψ, and thus $\Phi[Z, P_0] = \Psi$. This means that the domain of R is the entire space $\mathcal{Q}_{\mathrm{b}}$.

We verify the assumptions of Theorem 2.46. We define a preference relation $\trianglelefteq$ on $\mathcal{Q}_{\mathrm{b}}$ by setting

$$\Psi_1 \trianglelefteq \Psi_2 \ \text{ if and only if } \ R(\Psi_1) \le R(\Psi_2).$$

Clearly, the relation $\trianglelefteq$ is a total preorder with R being its numerical representation. Since $\varrho[\cdot, \cdot]$ is normalized and translation equivariant, the identity (2.75) implies that R is normalized and translation equivariant as well. We observe that R is monotonic, i.e., if $\Psi_1 \le \Psi_2$ (pointwise), then $R(\Psi_1) \le R(\Psi_2)$. Indeed, let $\Psi_1 \le \Psi_2$ and set $Z_1(x) = \Psi_1(Z_0(x))$, $Z_2(x) = \Psi_2(Z_0(x))$ for all $x \in \mathcal{X}$. Similar to (2.76),

$$P_0(Z_1 \le z) = \Psi_1^{-1}(z) \ge \Psi_2^{-1}(z) = P_0(Z_2 \le z)$$

for all $z \in \mathbb{R}$. The last relation implies that $[Z_1, P_0] \preceq_{(\text{ic})} [Z_2, P_0]$. The consistency of ϱ with the increasing convex order entails

$$R(\Psi_1) = \varrho[Z_1, P_0] \leq \varrho[Z_2, P_0] = R(\Psi_2),$$

which is the desired monotonicity. The properties of R further imply that the order $\trianglelefteq$ is directed, monotonic, and satisfies the dual translation property.

For any two comonotonic functions Z_1 and Z_2 in $\mathcal{B}(\mathcal{X})$, any $\lambda \in [0, 1]$, and any $P \in \mathcal{P}(\mathcal{X})$,

$$\Phi\big[\lambda Z_1 + (1-\lambda)Z_2, P\big] = \lambda\Phi[Z_1, P] + (1-\lambda)\Phi[Z_2, P].$$

The comonotonic convexity assumption implies that

$$\begin{aligned} &R\big(\lambda\Phi[Z_1, P] + (1-\lambda)\Phi[Z_2, P]\big) = \varrho\big[\lambda Z_1 + (1-\lambda)Z_2, P\big] \\ &\leq \lambda\varrho[Z_1, P] + (1-\lambda)\varrho[Z_2, P] = \lambda R\big(\Phi[Z_1, P]\big) + (1-\lambda)R\big(\Phi[Z_2, P]\big). \end{aligned}$$

Since any two functions $\Psi_1, \Psi_2 \in \mathcal{Q}_{\text{b}}$ can be represented as

$$\begin{aligned} \Psi_1 &= \Phi\big[Z_1, P_0\big], \quad \text{with} \quad Z_1(x) = \Psi_1(Z_0(x)),\ x \in \mathcal{X}, \\ \Psi_2 &= \Phi\big[Z_2, P_0\big], \quad \text{with} \quad Z_2(x) = \Psi_2(Z_0(x)),\ x \in \mathcal{X}, \end{aligned}$$

and the functions Z_1 and Z_2 are comonotonic by construction, the functional R is convex.

Consider a function $\Psi \in \mathcal{Q}_{\text{b}}$. For $(a, b] \subseteq (0, 1]$ we define $\Psi_{(a,b]}(\cdot)$ by formula (1.16) in § 1.3.5. Directly from (1.16) we observe that for every $\alpha \in (0, 1]$

$$\int_{1-\alpha}^{1} \Psi_{(a,b]}(p)\,\mathrm{d}p \leq \int_{1-\alpha}^{1} \Psi(p)\,\mathrm{d}p.$$

Therefore, for any $[Z, P]$ and $[V, Q]$ such that $\Psi_{(a,b]} = \Phi[Z, P]$ and $\Psi = \Phi[V, Q]$, we have $[Z, P] \preceq [V, Q]$. Due to the consistency of $\varrho[\cdot, \cdot]$ with the increasing convex order,

$$R\big(\Psi_{(a,b]}\big) = \varrho[Z, P] \leq \varrho[V, Q] = R\big(\Psi\big).$$

Therefore, the preorder $\trianglelefteq$ is risk averse. It follows from Theorem 2.46 that a set $\mathcal{D}_\varrho$ satisfying (2.73) exists, such that

$$R(\Psi) = \sup_{\mu \in \mathcal{D}_\varrho} \left\{ \int_0^1 \Psi(p)\,\mu(\mathrm{d}p) - R^*(\mu) \right\},$$

with

$$R^*(\mu) = \sup_{\Psi \in \mathcal{Q}_b} \left\{ \int_0^1 \Psi(p)\, \mu(\mathrm{d}p) - R(\Psi) \right\}.$$

Moreover, $R^*(\mu) = 0$ for $\mu \in \mathcal{D}_\varrho$, if R is positively homogeneous. The assertion of the theorem follows now from the substitution (2.75). □

The dual representation (2.74) can be written with the use of the Stjelties integral with respect to the distribution function $w(\cdot) = \mu(0, \cdot]$. It is particularly revealing in the homogeneous case:

$$\varrho[Z, P] = \sup_{w \in \mathcal{W}_\varrho} \int_0^1 \Phi[Z, P](p)\, \mathrm{d}w(p), \tag{2.77}$$

where each $w(\cdot) \in \mathcal{W}_\varrho$ can be interpreted as a dual (rank dependent) utility function. The set $\mathcal{W}_\varrho$ is the set of distribution functions of measures $\mu \in \mathcal{D}_\varrho$: a convex subset of convex and nondecreasing functions from $[0, 1]$ to $[0, 1]$. The relation (2.77) suggests an intriguing relation of law-invariant risk forms and the dual utility theory discussed in § 1.3.

Theorem 2.61 allows us to derive a generalization of the Kusuoka representation of law-invariant coherent measures of risk.

Definition 2.62. The Average Value at Risk at level $\alpha \in [0, 1]$ of a probabilistic model $[Z, P]$ is defined as follows:

$$\mathrm{AVaR}_\alpha^+[Z, P] = \begin{cases} \frac{1}{\alpha}\int_{1-\alpha}^1 \Phi[Z, P](p)\, \mathrm{d}p & \text{if } \alpha \in (0, 1), \\ \Phi[Z, P](1) & \text{if } \alpha = 0, \\ \mathbb{E}[Z, P] & \text{if } \alpha = 1. \end{cases}$$

Then, repeating the considerations leading to Corollary 2.47 *verbatim*, we obtain the following result.

Corollary 2.63. *Suppose the conditions of Theorem* 2.61 *are satisfied and the risk form* $\varrho[\cdot, \cdot]$ *is positively homogeneous. Then a convex subset* Λ_ϱ *of the set of probability measures on* $[0, 1]$ *exists, such that for all* $[Z, P] \in \mathbb{B}(\mathcal{X}) \times \mathcal{P}(\mathcal{X})$

$$\varrho[Z, P] = \sup_{\lambda \in \Lambda_\varrho} \int_0^1 \mathrm{AVaR}_s^+[Z, P]\, \lambda(\mathrm{d}s).$$

It it worth stressing that in the extant literature, the Kusuoka representations were derived for probabilistic models with a fixed atomless probability measure P. If P has atoms, additional conditions are needed. Our approach proves the validity of the Kusuoka representations for probabilistic models with an arbitrary probability measure P. The set Λ_ϱ is determined by the risk form alone; it is the same for all probabilistic models $[Z, P]$. This universal property is due to two key conditions: the consistency with the increasing convex order with *both* Z and P varying, and the requirement that $\mathcal{X}$ be Polish and uncountable.

2.7.3 Risk Disintegration

Our interest in this section is in measuring risk in product spaces. Consider two Polish spaces $\mathcal{X}$ and $\mathcal{Y}$ and their corresponding Borel σ-algebras $\mathcal{B}(\mathcal{X})$ and $\mathcal{B}(\mathcal{Y})$. We can disintegrate any $P \in \mathcal{P}(\mathcal{X} \times \mathcal{Y})$ into its marginal $P_{\mathcal{X}} \in \mathcal{P}(\mathcal{X})$ and a transition kernel $P_{\mathcal{Y}|\mathcal{X}} : \mathcal{X} \to \mathcal{P}(\mathcal{Y})$ as follows: $P(\mathrm{d}x, \mathrm{d}y) = P_{\mathcal{X}}(\mathrm{d}x)\, P_{\mathcal{Y}|\mathcal{X}}(\mathrm{d}y|x)$, which we symbolically write as $P = P_{\mathcal{X}} \otimes P_{\mathcal{Y}|\mathcal{X}}$.

Let $\mathcal{Q}(\mathcal{Y}|\mathcal{X})$ be the space of all stochastic kernels $Q : \mathcal{X} \to \mathcal{P}(\mathcal{Y})$. For any measure $\lambda \in \mathcal{P}(\mathcal{X})$ and any kernel $Q \in \mathcal{Q}(\mathcal{Y}|\mathcal{X})$, the measure $\lambda \otimes Q$, defined as

$$\big[\lambda \otimes Q\big](\mathrm{d}x, \mathrm{d}y) = \lambda(\mathrm{d}x)\, Q(\mathrm{d}y|x),$$

is an element of $\mathcal{P}(\mathcal{X} \times \mathcal{Y})$.

Suppose the risk form $\varrho : \mathbb{B}(\mathcal{X} \times \mathcal{Y}) \times \mathcal{P}(\mathcal{X} \times \mathcal{Y}) \to \mathbb{R}$ is monotonic, translation equivariant, and normalized. Then it induces a mapping $\varrho_{\mathcal{Y}|\mathcal{X}} : \mathbb{B}(\mathcal{X} \times \mathcal{Y}) \times \mathcal{Q}(\mathcal{Y}|\mathcal{X}) \to \mathbb{B}(\mathcal{X})$ defined as follows:

$$\varrho_{\mathcal{Y}|\mathcal{X}}[Z, Q](x) = \varrho[Z, \delta_x \otimes Q], \quad x \in \mathcal{X}. \tag{2.78}$$

We call the mapping $\varrho_{\mathcal{Y}|\mathcal{X}}[\cdot, \cdot]$ the *conditional risk operator* associated with $\varrho[\cdot, \cdot]$.

To verify that the values are indeed elements of $\mathbb{B}(\mathcal{X})$, let $c \in \mathbb{R}$ be such that $Z \le c\mathbb{1}$. Then, by monotonicity, translation equivariance, and normalization,

$$\varrho[Z, \delta_x \otimes Q] \le \varrho[c\mathbb{1}, \delta_x \otimes Q] = c.$$

The lower bound is similar, and thus the function $\varrho_{\mathcal{Y}|\mathcal{X}}[Z, Q](\cdot)$ is bounded. It is measurable, as a composition of measurable mappings.

If the risk form ϱ has the support property, then for each $x \in \mathcal{X}$ the value of (2.78) depends only on the function $Z(x, \cdot) \in \mathbb{B}(\mathcal{Y})$ and the measure $Q(x) \in \mathcal{P}(\mathcal{Y})$. We can, therefore, define the functionals $\varrho_{\mathcal{Y}|x} : \mathbb{B}(\mathcal{Y}) \times \mathcal{P}(\mathcal{Y}) \to \mathbb{R}$, $x \in \mathcal{X}$, as follows:

$$\varrho_{\mathcal{Y}|x}[Z(x, \cdot), Q(x)] = \varrho_{\mathcal{Y}|\mathcal{X}}[Z, Q](x), \quad x \in \mathcal{X}. \tag{2.79}$$

We call them *conditional risk forms* associated with $\varrho[\cdot, \cdot]$. Observe that any function from $\mathbb{B}(\mathcal{Y})$ and any measure from $\mathcal{P}(\mathcal{Y})$ may feature as arguments of $\varrho_{\mathcal{Y}|x}[\cdot, \cdot]$.

From now on, we always assume that the risk forms in question have the support property. The inequalities "$\le$" between functions are always understood pointwise.

Lemma 2.64. *If the risk form $\varrho[\cdot,\cdot]$ is monotonic (normalized, translation equivariant), then, for every $x \in \mathcal{X}$, the conditional risk form $\varrho_{\mathcal{Y}|x}$ is monotonic (normalized, translation equivariant).*

Proof. All the properties follow directly from the equation

$$\varrho_{\mathcal{Y}|x}[Z(x,\cdot), Q(x)] = \varrho[Z, \delta_x \otimes Q],$$

which defines the conditional risk form. □

Definition 2.65. A risk form $\varrho : \mathcal{B}(\mathcal{X}\times\mathcal{Y}) \times \mathcal{P}(\mathcal{X}\times\mathcal{Y}) \to \mathbb{R}$ is *conditionally consistent* if for all $Z, Z' \in \mathcal{B}(\mathcal{X}\times\mathcal{Y})$ and all $Q, Q' \in \mathcal{Q}(\mathcal{Y}|\mathcal{X})$ such that

$$\varrho_{\mathcal{Y}|\mathcal{X}}[Z, Q] \le \varrho_{\mathcal{Y}|\mathcal{X}}[Z', Q'],$$

we also have

$$\varrho[Z, \lambda\otimes Q] \le \varrho[Z', \lambda\otimes Q'], \quad \forall\, \lambda \in \mathcal{P}(\mathcal{X}).$$

The following result is the foundation of our further considerations.

Theorem 2.66. *Suppose a risk form $\varrho : \mathcal{B}(\mathcal{X}\times\mathcal{Y}) \times \mathcal{P}(\mathcal{X}\times\mathcal{Y}) \to \mathbb{R}$ is monotonic, normalized, translation equivariant, has the support property, and is conditionally consistent. Then a risk form $\varrho_{\mathcal{X}} : \mathcal{B}(\mathcal{X}) \times \mathcal{P}(\mathcal{X}) \to \mathbb{R}$ exists, such that for all $[Z, P] \in \mathcal{B}(\mathcal{X}\times\mathcal{Y}) \times \mathcal{P}(\mathcal{X}\times\mathcal{Y})$ the following formula is true:*

$$\varrho[Z, P] = \varrho_{\mathcal{X}}\big[\varrho_{\mathcal{Y}|\mathcal{X}}[Z, P_{\mathcal{Y}|\mathcal{X}}], P_{\mathcal{X}}\big]. \tag{2.80}$$

The risk form $\varrho_{\mathcal{X}}$ is uniquely defined by the equation

$$\varrho_{\mathcal{X}}[f, P_{\mathcal{X}}] = \varrho[\underline{f}, P], \quad \text{with} \quad \underline{f}(x,y) \equiv f(x). \tag{2.81}$$

It is monotonic, normalized, translation equivariant, and has the support property.

Proof. Let us verify (2.80). Suppose $[Z, P]$ and $[Z', P']$ are such that

$$\varrho_{\mathcal{Y}|\mathcal{X}}[Z, P_{\mathcal{Y}|\mathcal{X}}] = \varrho_{\mathcal{Y}|\mathcal{X}}[Z', P'_{\mathcal{Y}|\mathcal{X}}].$$

Then it follows from Definition 2.65 that $\varrho[Z, \lambda\otimes P_{\mathcal{Y}|\mathcal{X}}] = \varrho[Z', \lambda\otimes P'_{\mathcal{Y}|\mathcal{X}}]$ for all $\lambda \in \mathcal{P}(\mathcal{X})$. If, additionally, the marginal measures $P_{\mathcal{X}}$ and $P'_{\mathcal{X}}$ are identical, by setting $\lambda = P_{\mathcal{X}} = P'_{\mathcal{X}}$ we conclude that

$$\varrho[Z, P] = \varrho[Z, P_{\mathcal{X}}\otimes P_{\mathcal{Y}|\mathcal{X}}] = \varrho[Z', P'_{\mathcal{X}}\otimes P'_{\mathcal{Y}|\mathcal{X}}] = \varrho[Z', P'].$$

It follows that the value of $\varrho[Z, P]$ is fully determined by the value of the conditional risk operator $\varrho_{\mathcal{Y}|\mathcal{X}}[Z, P_{\mathcal{Y}|\mathcal{X}}]$ and the marginal measure $P_{\mathcal{X}}$. Therefore, the disintegration formula (2.80) is true.

It remains to verify the properties of $\varrho_{\mathcal{X}}$. Set $Z(x, y) = \underline{f}(x, y) \equiv f(x)$ in (2.80). Then, by the support, translation equivariance, and normalization properties of $\varrho[\cdot, \cdot]$, for every $x \in \mathcal{X}$ we obtain

$$\varrho_{\mathcal{Y}|\mathcal{X}}[\underline{f}, P_{\mathcal{Y}|\mathcal{X}}](x) = \varrho[f(x)\mathbb{1}, \delta_x \otimes P_{\mathcal{Y}|\mathcal{X}}] = f(x).$$

Combining with (2.80), we observe that the identity (2.81) is true. All the postulated properties of $\varrho_{\mathcal{X}}[\cdot, \cdot]$ follow from the corresponding properties of $\varrho[\cdot, \cdot]$. □

We call the identity (2.80) the *risk disintegration formula*. It represents the risk form $\varrho[\cdot, \cdot]$ by its marginal risk form $\varrho_{\mathcal{X}}[\cdot, \cdot]$ and its conditional risk operator $\varrho_{\mathcal{Y}|\mathcal{X}}[\cdot, \cdot]$.

Example 2.67. Consider the risk form $\varrho : \mathcal{B}(\mathcal{X} \times \mathcal{Y}) \times \mathcal{P}(\mathcal{X} \times \mathcal{Y}) \to \mathbb{R}$ defined as follows:

$$\varrho[Z, P] = \min_{\eta(\cdot)} \mathbb{E}_P\Big[\eta(x) + \frac{1}{\alpha}\big(Z(x, y) - \eta(x)\big)_+\Big],$$

where the minimization is over all measurable mappings $\eta : \mathcal{X} \to \mathbb{R}$. Directly from Definition 2.57 we verify that it is normalized, monotonic, translation invariant, and has the support property. To verify the conditional consistency, observe that

$$\varrho_{\mathcal{Y}|x}[Z, Q] = \min_{\eta} \Big\{\eta + \frac{1}{\alpha}\mathbb{E}_{Q(x)}\big[\big(Z(x, y) - \eta\big)_+\big]\Big\}. \tag{2.82}$$

Therefore, the relation

$$\varrho[Z, \delta_x \otimes Q] \le \varrho[Z', \delta_x \otimes Q'], \quad \forall\, x \in \mathcal{X},$$

means that

$$\min_{\eta} \Big\{\eta + \frac{1}{\alpha}\mathbb{E}_{Q(x)}\big[\big(Z(x, y) - \eta\big)_+\big]\Big\} \le \min_{\eta} \Big\{\eta + \frac{1}{\alpha}\mathbb{E}_{Q'(x)}\big[\big(Z'(x, y) - \eta\big)_+\big]\Big\},$$

for all $x \in \mathcal{X}$. Using the interchangeability of $\min_{\eta(\cdot)}$ and $\mathbb{E}_\lambda$, for every $\lambda \in \mathcal{P}(\mathcal{X})$ we obtain:

$$
\begin{aligned}
\varrho[Z, \lambda \otimes Q] &= \min_{\eta(\cdot)} \mathbb{E}_\lambda\Big\{\eta(x) + \frac{1}{\alpha}\mathbb{E}_{Q(x)}\big[\big(Z(x,y) - \eta(x)\big)_+\big]\Big\} \\
&= \mathbb{E}_\lambda \min_\eta \Big\{\eta + \frac{1}{\alpha}\mathbb{E}_{Q(x)}\big[\big(Z(x,y) - \eta\big)_+\big]\Big\} \\
&\le \mathbb{E}_\lambda \min_\eta \Big\{\eta + \frac{1}{\alpha}\mathbb{E}_{Q'(x)}\big[\big(Z'(x,y) - \eta\big)_+\big]\Big\} = \varrho[Z', \lambda \otimes Q'].
\end{aligned}
$$

Therefore, the assumptions of Theorem 2.66 are satisfied, and the risk form $\varrho[\cdot,\cdot]$ can be disintegrated into a marginal and conditional form. Using the dual representation of the Average Value at Risk in (2.82), we obtain the explicit formula:

$$
\varrho[Z, P] = \mathbb{E}_{P_{\mathfrak{X}}}\Big[\operatorname{AVaR}_\alpha^+\big[Z(x,y), P_{\mathfrak{Y}|\mathfrak{X}}\big]\Big].
$$

This means that $\varrho_{\mathfrak{X}}[\cdot, P_{\mathfrak{X}}] = \mathbb{E}_{P_{\mathfrak{X}}}[\cdot]$ and $\varrho_{\mathfrak{Y}|\mathfrak{X}}[\cdot, P_{\mathfrak{Y}|\mathfrak{X}}] = \operatorname{AVaR}_\alpha^+[\cdot, P_{\mathfrak{Y}|\mathfrak{X}}]$.

The risk form $\varrho[\cdot,\cdot]$ is not law-invariant on the product space, because pairs $[Z, P]$ having identical distribution may have different conditional and marginal distributions. This can be seen on the case of $\mathfrak{X} = \mathfrak{Y} = [0,1]$ with the Lebesgue measure P on $\mathfrak{X} \times \mathfrak{Y}$ and two functions: $Z(x,y) = x$ and $Z'(x,y) = y$. Then $[Z, P] \overset{\mathfrak{D}}{\sim} [Z', P]$. However,

$$
\begin{aligned}
\varrho[Z, P] &= \min_{\eta(\cdot)} \int_0^1 \int_0^1 \Big[\eta(x) + \frac{1}{\alpha}\big(x - \eta(x)\big)_+\Big]\, dy\, dx \\
&= \int_0^1 \min_\eta \Big[\eta + \frac{1}{\alpha}(x - \eta)_+\Big]\, dx = \int_0^1 x\, dx = \frac{1}{2},
\end{aligned}
$$

and

$$
\begin{aligned}
\varrho[Z', P] &= \min_{\eta(\cdot)} \int_0^1 \int_0^1 \Big[\eta(x) + \frac{1}{\alpha}\big(y - \eta(x)\big)_+\Big]\, dy\, dx \\
&\ge \int_0^1 \min_\eta \int_0^1 \Big[\eta + \frac{1}{\alpha}(y - \eta)_+\Big]\, dy\, dx \\
&= \int_0^1 \operatorname{AVaR}_\alpha^+(Y)\, dx = \operatorname{AVaR}_\alpha^+(Y).
\end{aligned}
$$

We observe that the minimizing η in the second line of the formula above is in fact constant and equal to $1 - \alpha$. Therefore, it can be used in the first line, the inequality becomes an equation, and $\varrho[Z', P] = 1 - \alpha/2$. We conclude that $\varrho[Z, P] \neq \varrho[Z', P]$, unless $\alpha = 1$. This example illustrates the fact that the concept of law invariance on product spaces is excessively demanding. Therefore, we do not assume law invariance of risk forms on product spaces. □

2.7.4 Risk in Two-Stage Partially Observable Systems

2.7.4.1 Fixed Observation Distribution

Let us start with the following simple setting. A random vector (X, Y) is distributed in the product of Polish spaces $\mathcal{X} \times \mathcal{Y}$ according to a measure P. For a bounded measurable function $c : \mathcal{X} \times \mathcal{Y} \to \mathbb{R}$, we can evaluate the risk of $c(X, Y)$ by a risk form $\varrho[c, P]$.

However, we know that we shall be able to observe the value of X. After X is observed, we might refine our risk evaluation of $c(X, Y)$. Thus, a question arises: if a future possibility to observe X exists, what should be our present evaluation of the risk of $c(X, Y)$, *before* X is observed. Note that Y is never observed.

Let us start with the problem of risk evaluation after X is observed. We can disintegrate P into its marginal $P_{\mathcal{X}}$ on $\mathcal{X}$ and a transition kernel $P_{\mathcal{Y}|\mathcal{X}}$ from $\mathcal{X}$ to $\mathcal{P}(\mathcal{Y})$:

$$P(\mathrm{d}x, \mathrm{d}y) = P_{\mathcal{X}}(\mathrm{d}x) P_{\mathcal{Y}|\mathcal{X}}(\mathrm{d}y|x).$$

Suppose the risk form ϱ is monotonic, normalized, and has the translation property and the support property. Then the correct evaluation of the risk after $X = x$ is observed is

$$\varrho_{\mathcal{Y}|x}\big[c(x, \cdot), P_{\mathcal{Y}|\mathcal{X}}(x)\big] = \varrho\big[c, \delta_x \otimes P_{\mathcal{Y}|\mathcal{X}}\big].$$

This is nothing else, but the conditional risk form defined in (2.79). As a function of x, we obtain the conditional risk operator $\varrho_{\mathcal{Y}|\mathcal{X}}\big[c, P_{\mathcal{Y}|\mathcal{X}}\big]$. Now, to evaluate the overall risk, we calculate $\varrho_{\mathcal{X}}\Big[\varrho_{\mathcal{Y}|\mathcal{X}}\big[c, P_{\mathcal{Y}|\mathcal{X}}\big], P_{\mathcal{X}}\Big]$. We thus arrive to the following conclusion from Theorem 2.66. If the risk form $\varrho : \mathcal{B}(\mathcal{X} \times \mathcal{Y}) \times \mathcal{P}(\mathcal{X} \times \mathcal{Y}) \to \mathbb{R}$ is normalized, translation equivariant, conditionally consistent, and has the support property, then

$$\varrho[c, P] = \varrho_{\mathcal{X}}\Big[\varrho_{\mathcal{Y}|\mathcal{X}}\big[c, P_{\mathcal{Y}|\mathcal{X}}\big], P_{\mathcal{X}}\Big].$$

It follows that the two risk evaluations: without and with the perspective of inspection, are identical. The mere existence of inspection does not affect risk.

When a possibility of control exists, the situation is different. Suppose there are two Polish spaces $\mathcal{U}_1$ and $\mathcal{U}_2$, which we call *control spaces*. At stage one, a control $u \in U_1 \subset \mathcal{U}_1$ is chosen, where U_1 is a subset of $\mathcal{U}_1$. Then an observation of X is made. After observing the value of X, we choose control $u_2 \in U_2(X, u_1) \subset \mathcal{U}_2$ to minimize the risk of $c(X, Y, u_1, u_2)$. The risk is measured by the form $\varrho[\cdot, \cdot]$. Here $U_2 : \mathcal{X} \times \mathcal{U}_1 \rightrightarrows \mathcal{U}_2$ is a measurable multifunction representing the feasible set at the second stage. We shall use the symbol

$\pi(\cdot) \lessdot U_2(\cdot, u_1)$ to indicate that the function π is a measurable selection of $U_2(\cdot, u_1)$.

We may look at this problem from two perspectives. Let us start from the functional perspective. Since u_2 can be chosen after X is observed, we may represent it as a measurable function: $u_2 = \pi(x)$, $x \in \mathfrak{X}$. Therefore, the overall cost has the form:

$$Z^{u_1,\pi}(x, y) = c(x, y, u_1, \pi(x)), \quad (x, y) \in \mathfrak{X} \times \mathfrak{Y}. \tag{2.83}$$

The distribution of (X, Y) is P. The problem takes on the form

$$\begin{aligned} \min_{u_1,\pi} \; & \varrho\big[Z^{u_1,\pi}, P\big] \\ \text{s.t. } & u_1 \in U_1, \\ & \pi(\cdot) \lessdot U_2(\cdot, u_1). \end{aligned} \tag{2.84}$$

We now derive a two-stage representation of this problem.

Theorem 2.68. *We assume the following:*

(i) *The risk form ϱ is monotonic, normalized, translation equivariant, has the support property, and is conditionally consistent;*

(ii) *The multifunction U_2 is upper-semicontinuous and has nonempty and compact values;*

(iii) *The function c is uniformly bounded, measurable, and lower-semicontinuous with respect to its second argument.*

Then, problem (2.84) *is equivalent to the two-stage problem:*

$$\min_{u_1 \in U_1} \varrho_{\mathfrak{X}}\big[V(\cdot, u_1), P_{\mathfrak{X}}\big], \tag{2.85}$$

where $V(\cdot, \cdot)$ is the optimal value of the following second-stage problem:

$$V(x, u_1) = \min_{u_2 \in U_2(x,u_1)} \varrho_{\mathfrak{Y}|x}\big[c(x, \cdot, u_1, u_2), P_{\mathfrak{Y}|\mathfrak{X}}(x)\big], \quad x \in \mathfrak{X}, \quad u_1 \in U_1. \tag{2.86}$$

Proof. Since $\varrho[\cdot, \cdot]$ satisfies the assumptions of Theorem 2.66, the risk of the function $Z^{u_1,\pi}$ can be calculated by the risk disintegration formula:

$$\begin{aligned} \varrho\big[Z^{u_1,\pi}, P\big] &= \varrho_{\mathfrak{X}}\Big[\varrho_{\mathfrak{Y}|\mathfrak{X}}\big[Z^{u_1,\pi}, P_{\mathfrak{Y}|\mathfrak{X}}\big], P_{\mathfrak{X}}\Big] \\ &= \varrho_{\mathfrak{X}}\Big[x \mapsto \varrho_{\mathfrak{Y}|x}\big[c(x, \cdot, u_1, \pi(x)), P_{\mathfrak{Y}|\mathfrak{X}}(x)\big], P_{\mathfrak{X}}\Big]. \end{aligned}$$

Then problem (2.84) takes on the form:

$$\min_{u_1,\pi} \varrho_{\mathfrak{X}}\Big[x \mapsto \varrho_{\mathfrak{Y}|x}\big[c(x,\cdot,u_1,\pi(x)), P_{\mathfrak{Y}|\mathfrak{X}}(x)\big], P_{\mathfrak{X}}\Big],$$

subject to the same constraints. Due to the monotonicity of the marginal risk form $\varrho_{\mathfrak{X}}$, the smaller the values of the function $x \mapsto \varrho_{\mathfrak{Y}|x}\big[c(x,\cdot,u_1,\pi(x))$, the smaller the value of $\varrho_{\mathfrak{X}}$. Since $u_2 = \pi(x)$ may depend on x, we may carry out the minimization with respect to u_2 inside the argument of $\varrho_{\mathfrak{X}}$:

$$\begin{aligned}\inf_{u_1\in U_1}\ \inf_{\pi(\cdot)\lessdot U_2(\cdot,u_1)} &\varrho_{\mathfrak{X}}\Big[x \mapsto \varrho_{\mathfrak{Y}|x}\big[c(x,\cdot,u_1,\pi(x)), P_{\mathfrak{Y}|\mathfrak{X}}(x)\big], P_{\mathfrak{X}}\Big]\\ &= \inf_{u_1\in U_1} \varrho_{\mathfrak{X}}\Big[x \mapsto \inf_{u_2\in U_2(x,u_1)} \varrho_{\mathfrak{Y}|x}\big[c(x,\cdot,u_1,u_2), P_{\mathfrak{Y}|\mathfrak{X}}(x)\big], P_{\mathfrak{X}}\Big].\end{aligned}$$

The only condition for the validity of this transformation is the measurability and boundedness of the optimal value function (2.86). This follows from Berge's theorem, which can be applied due to the assumptions (ii) and (iii). In fact, they also guarantee that the optimal value function is lower semicontinuous with respect to u_1.

We conclude that problem (2.84) reduces to the marginal risk optimization (2.85). □

Theorem 2.68 provides us with the second perspective on the problem. It has a hierarchical structure, similar to its expected-value full information counterpart: after $X = x$ is observed, the second stage problem (2.86) is to minimize the conditional risk. Then, the first-stage problem takes on the form of minimizing the marginal risk (2.85) of the second-stage optimal value. The most important conclusion is that the risk disintegration formula allows us to write the overall problem in a hierarchical structure.

2.7.4.2 Controlled Observation Distribution

Now we consider a more complex situation. There are still two Polish control spaces $\mathcal{U}_1$ and $\mathcal{U}_2$. However, after a control $u_1 \in U_1 \subset \mathcal{U}_1$ is chosen, the distribution of the observation X depends on u_1. The dependence is described by a controlled kernel $K : \mathfrak{Y} \times \mathcal{U}_1 \to \mathcal{P}(\mathfrak{X})$. After observing X, we choose control $u_2 \in U_2(X,u_1) \subset \mathcal{U}_2$ to minimize the risk of $c(X,Y,u_1,u_2)$. The risk is measured by the form $\varrho[\cdot,\cdot]$.

Assume the same conditions on U_1, U_2, and ϱ as in the previous subsection. Let P_Y be the marginal distribution of Y. After the first decision u_1 will be chosen, the joint distribution of (Y,X) will become

$$M(u_1) = P_{\mathfrak{Y}} \otimes K(\cdot,u_1),$$

that is, $M(\mathrm{d}y, \mathrm{d}x|u_1) = P_{\mathfrak{Y}}(\mathrm{d}y)K(\mathrm{d}x|y, u_1)$. Therefore, denoting the second stage decision by $u_2 = \pi(x)$ (it may depend on x), our problem is to find

$$\begin{aligned} \min_{u_1,\pi}\ & \varrho\big[Z^{u_1,\pi}, M(u_1)\big],\\ \text{s.t. } & u_1 \in U_1, \\ & \pi(\cdot) \lessdot U_2(\cdot, u_1). \end{aligned} \tag{2.87}$$

where the function $Z^{u_1,\pi}(\cdot,\cdot)$ is given by (2.83).

Let us develop a two-stage version of the functional problem (2.87). The marginal distribution of the observation result is

$$M_{\mathfrak{X}}(u_1) = \int_{\mathfrak{Y}} K(y, u_1)\ P_{\mathfrak{Y}}(\mathrm{d}y)$$

where the integral is understood in the weak sense. Since the space $\mathfrak{Y}$ is standard, the measure $M(u_1)$ can be disintegrated into the marginal $M_{\mathfrak{X}}(u_1)$ and a transition kernel $\Gamma : \mathfrak{X} \times \mathcal{U}_1 \to \mathcal{P}(\mathfrak{Y})$ as follows

$$M(u_1) = M_{\mathfrak{X}}(u_1) \otimes \Gamma(u_1),$$

which reads $M(\mathrm{d}x, \mathrm{d}y|u_1) = M_{\mathfrak{X}}(\mathrm{d}x|u_1)\Gamma(\mathrm{d}y|x, u_1)$. The transition kernel Γ is called the *Bayes operator*.

Example 2.69. Assume that the joint distribution $M(u_1)$ of (X, Y) has a density $q(\cdot,\cdot \mid u_1)$ with respect to a finite product measure $\mu_{\mathfrak{X}} \otimes \mu_{\mathfrak{Y}}$ on $\mathfrak{X} \times \mathfrak{Y}$. Then the Bayes operator has the form

$$\Gamma(A|x, u_1) = \frac{\int_A \int_{\mathfrak{Y}} q(x', y' \mid u_1)\ M_{\mathfrak{Y}}(\mathrm{d}y)\ \mu_{\mathfrak{Y}}(\mathrm{d}y')}{\int_{\mathfrak{Y}} \int_{\mathfrak{Y}} q(x', y' \mid u_1)\ M_{\mathfrak{Y}}(\mathrm{d}y)\ \mu_{\mathfrak{Y}}(\mathrm{d}y')}, \quad \forall\, A \in \mathcal{B}(\mathfrak{Y}).$$

If the formula above has a zero denominator, we can formally define $\Gamma(x, u_1)$ to be an arbitrarily selected distribution on $\mathfrak{Y}$. □

With the use of the Bayes operator, we can equivalently write problem (2.87) as a two-stage problem.

2.7.5 Mini-Batch Risk Measures

Risk forms are convenient tools for defining measures of risk which are easier to deal with in numerical optimization methods.

Suppose $\varrho : \mathcal{B}(\mathfrak{X}) \times \mathcal{P}(\mathfrak{X}) \to \mathbb{R}$ is a risk form and $P \in \mathcal{P}(\mathfrak{X})$. If we draw a sample $X^{1:N} \triangleq (X^1, \ldots, X^N)$ with N independent components distributed according to P in $\mathfrak{X}$, we obtain a random empirical measure

$$P^{(N)} = \sum_{i=1}^{N} \delta_{X^i}.$$

Using it as the second argument of the risk form $\varrho[\cdot,\cdot]$, we obtain a random risk form $(Z,P) \mapsto \varrho[Z,P^{(N)}]$. Finally, we define a new risk form

$$\varrho^{(N)}[Z,P] \triangleq \mathbb{E}\big\{\varrho[Z,P^{(N)}]\big\}. \tag{2.88}$$

We call it the *mini-batch risk form.*

The properties of the original risk form are inherited by the mini-batch version.

Lemma 2.70. *If the risk form $\varrho[\cdot,\cdot]$ is convex (monotonic, translation equivariant, positively homogeneous, law-invariant, or has the support property) then the mini-batch risk form $\varrho^{(N)}[\cdot,\cdot]$ has the corresponding property as well.*

Proof. Only the inheritance of law invariance requires proof; other properties are evident. Consider two pairs:

$$\Big[Z, \frac{1}{N}\sum_{i=1}^{N} \delta_{v^i}\Big] \quad \text{and} \quad \Big[U, \frac{1}{N}\sum_{i=1}^{N} \delta_{w^i}\Big].$$

If, for some permutation π of $\{1,\dots,N\}$,

$$\big(Z(v^1),\dots,Z(v^N)\big) = \big(U(w^{\pi(1)}),\dots,U(w^{\pi(N)})\big),$$

then the two pairs have identical distribution functions. By the law invariance of $\varrho[\cdot,\cdot]$,

$$\varrho\Big[Z, \frac{1}{N}\sum_{i=1}^{N} \delta_{v^i}\Big] = \varrho\Big[U, \frac{1}{N}\sum_{i=1}^{N} \delta_{w^i}\Big].$$

This implies that a measurable function $\Psi : \mathbb{R}^n \to \mathbb{R}$ exists, such that

$$\varrho[Z,P^{(N)}] = \Psi\big(Z(X^1),\dots,Z(X^N)\big), \tag{2.89}$$

and for every permutation π of $\{1,\dots,N\}$,

$$\Psi\big(Z(X^1),\dots,Z(X^N)\big) = \Psi\big(Z(X^{\pi(1)}),\dots,Z(X^{\pi(N)})\big). \tag{2.90}$$

Now, if $[Z,P] \overset{\mathcal{D}}{\sim} [U,Q]$, then for $X^{1:N} \sim P^N$ and $Y^{1:N} \sim Q^N$, the vectors $\big(Z(X^1),\dots,Z(X^N)\big)$ and $\big(U(Y^1),\dots,U(Y^N)\big)$ have the same distribution. Therefore,

$$\begin{aligned}\varrho^{(N)}[Z,P] &= \mathbb{E}_{X^{1:N}\sim P^N}\{\Psi(Z(X^1),\dots,Z(X^N))\}\\ &= \mathbb{E}_{Y^{1:N}\sim Q^N}\{\Psi(U(Y^1),\dots,U(Y^N))\} = \varrho^{(N)}[U,Q].\end{aligned}$$

□

Example 2.71. Consider the mini-batch risk form derived from (2.70):

$$\varrho^{(N)}[Z,P] = \mathbb{E}_{X^{1:N}\sim P^N}\Big[\max_{1\le i\le N} Z(X^i)\Big]. \tag{2.91}$$

In the special case of $N = 2$ it has the mean–risk form

$$\begin{aligned}\varrho^{(2)}[Z,P] &= \mathbb{E}_{(X^1,X^2)\sim P^2}\Big[\max\big(Z(X^1), Z(X^2)\big)\Big]\\ &= \mathbb{E}_{(X^1,X^2)\sim P^2}\Big[Z(X^1) + \max\big(0, Z(X^2) - Z(X^1)\big)\Big]\\ &= \mathbb{E}\big[Z\big] + \Gamma[Z],\end{aligned}$$

with the Gini index (2.13):

$$\Gamma[Z] = \frac{1}{2}\mathbb{E}_{(X^1,X^2)\sim P^2}\Big[\big|Z(X^1) - Z(X^2)\big|\Big].$$

For $N \ge 2$, (2.91) is a generalized mean–Gini model, with an increasing degree of risk aversion. Its explicit form allows for more detailed analysis. The distribution function of the sample maximum is

$$P^N\{\varrho^{(N)}[Z,P] \le v\} = \big(P[Z\le v]\big)^N, \quad v\in\mathbb{R}.$$

Therefore, the mini-batch risk form (2.91) can be re-written by changing the variables:

$$\varrho^{(N)}[Z,P] = \int_{-\infty}^{\infty} v\ \mathrm{d}\Big(\big(P[Z\le v]\big)^N\Big) = \int_0^1 F_Z^{-1}(\alpha)\ \mathrm{d}(\alpha^N),$$

where $F_Z^{-1}(\cdot)$ is the quantile function of Z. The last expression is a dual (rank-dependent) utility functional (see § 1.3.3)

$$\varrho^{(N)}[Z,P] = \int_0^1 F_Z^{-1}(\alpha)\ \mathrm{d}w(\alpha), \tag{2.92}$$

with the rank-dependent utility function

$$w(\alpha) = \alpha^N.$$

For a fixed P, the functional (2.91) as a function of Z is a coherent measure of risk. We observe that it is well-defined and finite on $\mathcal{L}_1(\mathcal{X},\mathcal{B}(\mathcal{X}),P)$, while

(2.70) was finite on bounded functions only. We return to this issue in the next section in a general setting.

The expression (2.92) allows for the derivation of the Kusuoka representation of the mini-batch risk form (2.91). Changing the order of integration, we obtain

$$\begin{aligned}\varrho^{(N)}[Z,P] &= N\int_0^1 F_Z^{-1}(\alpha)\alpha^{N-1}\,\mathrm{d}\alpha\\ &= N(N-1)\int_0^1 F_Z^{-1}(\alpha)\int_0^\alpha \beta^{N-2}\,\mathrm{d}\beta\,\mathrm{d}\alpha\\ &= N(N-1)\int_0^1 \beta^{N-2}\int_\beta^1 F_Z^{-1}(\alpha)\,\mathrm{d}\alpha\,\mathrm{d}\beta\\ &= N(N-1)\int_0^1 \alpha(1-\alpha)^{N-2}\,\mathrm{AVaR}_\alpha^+[Z]\,\mathrm{d}\alpha,\end{aligned} \tag{2.93}$$

where

$$\mathrm{AVaR}_\alpha^+[Z] = \frac{1}{\alpha}\int_{1-\alpha}^1 F_Z^{-1}(1-\alpha)\,\mathrm{d}\alpha$$

is the Average Value at Risk at level $\alpha \in (0,1]$ (recall that Z represents losses). It turns out that (2.91) is a spectral risk measure. Evidently,

$$N(N-1)\int_0^1 \alpha(1-\alpha)^{N-2}\,\mathrm{d}\alpha = 1,$$

as required from the spectrum. □

Example 2.72. Consider the mini-batch mean–semideviation risk form derived from (2.69):

$$\varrho^{(N)}[Z,P] = \\ \mathbb{E}_{X^{1:N}\sim P^N}\left\{\frac{1}{N}\sum_{j=1}^N Z(X^j) + \frac{\varkappa}{N}\sum_{j=1}^N\Big(Z(X^j) - \frac{1}{N}\sum_{k=1}^N Z(X^k)\Big)_+\right\}. \tag{2.94}$$

In the special case of $N = 2$ it has a mean–risk form similar to the previous example:

$$\begin{aligned}\varrho^{(2)}[Z,P] &= \mathbb{E}\big[Z\big] + \frac{\varkappa}{4}\mathbb{E}_{(X^1,X^2)\sim P^2}\Big[\big|Z(X^1) - Z(X^2)\big|\Big]\\ &= \mathbb{E}\big[Z\big] + \frac{\varkappa}{2}\Gamma[Z].\end{aligned}$$

Again, for a fixed P, the functional (2.94) as a function of Z is a coherent measure of risk which is well-defined and finite on $\mathcal{L}_1(\mathcal{X},\mathcal{B}(\mathcal{X}),P)$. □

We now develop the dual representation of mini-batch risk forms. Suppose the risk form $\varrho : \mathbb{B}(\mathcal{X}) \times \mathcal{P}(\mathcal{X}) \to \mathbb{R}$ is coherent and law-invariant. Then, for a fixed probability measure P, the function $\varrho^{(N)}[\cdot, P]$ is a coherent measure of risk on the space of bounded functions $\mathbb{B}(\mathcal{X})$. We plan to derive its dual representation. In general, such a representation would have to involve elements from the space of finitely additive measures $\text{ba}(\mathcal{B}(\mathcal{X}))$, which is the topological dual of $\mathbb{B}(\mathcal{X})$. In Theorem 2.61 we have shown, however, that the dual representation of coherent and law-invariant risk forms involves only countably additive measures. Now, we advance the analysis for the mini-batch risk forms and we derive a more explicit representation in terms of measures that are absolutely continuous with respect to P.

Owing to (2.89)–(2.90), the random risk form $\varrho[\cdot, P^{(N)}]$ is a function $\Psi(\cdot)$ of the vector $\big(Z(X^1), \dots, Z(X^N)\big)$. The function $\Psi : \mathbb{R}^N \to \mathbb{R}$ satisfies the axioms of a coherent measure of risk. By the dual representation of risk measures on a finite probability space, a closed convex set

$$\mathcal{A}^{(N)} \subseteq S^{(N)} = \left\{ \xi \in \mathbb{R}^N_+ : \sum_{j=1}^{N} \xi_j = 1 \right\} \tag{2.95}$$

exist, such that for all $Z \in \mathbb{B}(\mathcal{X})$

$$\varrho[Z, P^{(N)}] = \max_{\xi \in \mathcal{A}^{(N)}} \sum_{j=1}^{N} \xi_j Z(X^j). \tag{2.96}$$

The set $\mathcal{A}^{(N)}$ depends on N, but not on the specific sample $X^{1:N}$, and is law-invariant as well, in the sense that for every $\xi \in \mathcal{A}^{(N)}$ and for every permutation π of $\{1, \dots, N\}$, the vector $\big(\xi_{\pi(1)}, \dots, \xi_{\pi(N)}\big)$ is an element of $\mathcal{A}^{(N)}$ as well.†

Therefore,

$$\varrho^{(N)}[Z, P] = \mathbb{E}_{X^{1:N} \sim P^N} \left\{ \max_{\xi \in \mathcal{A}^{(N)}} \sum_{j=1}^{N} \xi_j Z(X^j) \right\}. \tag{2.97}$$

We now transform formula (2.97) into the standard dual representation (2.28) of a coherent measure of risk. Consider the cylindrical multifunction $H : \mathcal{X}^N \rightrightarrows S^{(N)}$ defined as

$$H(X^{1:N}) = \mathcal{A}^{(N)}, \quad \text{for all} \quad X^{1:N} \in \mathcal{X}^N.$$

† In § 2.3.1 we derived the dual representation in terms of the densities with respect to the reference measure. In the case considered here, the reference measure is the empirical measure with all probabilities equal to $\frac{1}{N}$. We skip the normalization by N here, and thus $\mathcal{A}^{(N)} = \frac{1}{N}\mathcal{A}$ with the set $\mathcal{A}$ defined in (2.28) for the risk measure $\varrho[\cdot, P^{(N)}]$. The elements of $\mathcal{A}^{(N)}$ are probability measures.

By interchangeability of the maximization and expectation operators, formula (2.97) can be rewritten as follows:

$$\varrho^{(N)}[Z,P] = \max_{\xi(\cdot)\lessdot H} \mathbb{E}_{X^{1:N}\sim P^N}\left\{\sum_{j=1}^{N} \xi_j(X^{1:N})\, Z(X^j)\right\}. \tag{2.98}$$

Here, the notation $\xi(\cdot) \lessdot H$ means that $\xi(\cdot)$ is a measurable selector of $H(\cdot)$. For each $X^{1:N}$ the maximizer in the braces in (2.97) exists; with no loss of generality we may assume that it is a measurable function of the sample and thus creates a maximizer in (2.98). This follows from Berge's theorem about the measurability of the optimal value, and from the existence of measurable selectors.

Define linear continuous operators

$$\Pi_j^{(N)} : \mathcal{L}_\infty(\mathcal{X}^N, \mathcal{B}(\mathcal{X}^N), P^N) \to \mathcal{L}_\infty(\mathcal{X}, \mathcal{B}(\mathcal{X}), P), \quad j = 1,\dots,N,$$

as follows:

$$\big[\Pi_j^{(N)}(\xi)\big](v) = \mathbb{E}_{X^{1:N}\sim P^N}\big[\xi_j(X^{1:N})\,\big|\, X^j = v\big], \quad v \in \mathcal{X}. \tag{2.99}$$

By using the fact that all X^j are distributed according to P, we obtain

$$\begin{aligned}\varrho^{(N)}[Z,P] &= \max_{\xi(\cdot)\lessdot H} \mathbb{E}_{X^{1:N}\sim P^N}\left\{\sum_{j=1}^{N} \big[\Pi_j^{(N)}(\xi)\big](X^j)\, Z(X^j)\right\} \\ &= \max_{\xi(\cdot)\lessdot H} \int_{\mathcal{X}} \sum_{j=1}^{N} \big[\Pi_j^{(N)}(\xi)\big](v)\, Z(v)\, P(dv). \end{aligned} \tag{2.100}$$

In this way, we established the following result.

Theorem 2.73. *If the risk form $\varrho : \mathbb{B}(\mathcal{X}) \times \mathcal{P}(\mathcal{X}) \to \mathbb{R}$ is coherent and law-invariant, then for every $N \geq 1$ the corresponding mini-batch risk form has the following dual representation*

$$\varrho^{(N)}[Z,P] = \sup_{Q\in\partial\varrho^{(N)}[0,P]} \int_{\mathcal{X}} Z(v)\, Q(dv) \tag{2.101}$$

where

$$\partial\varrho^{(N)}[0,P] = \left\{Q \in \mathcal{P}(\mathcal{X}) : \exists\,(\xi(\cdot) \lessdot H)\ \ \frac{dQ}{dP} = \sum_{j=1}^{N} \big[\Pi_j^{(N)}(\xi)\big]\right\}, \tag{2.102}$$

with the operators $\Pi_j^{(N)}$ defined in (2.99).

We may double-check the essential properties of $\partial\varrho^{(N)}[0, P]$. Since $\mathcal{A}^{(N)}$ is convex, then the set of measurable selectors $\xi(\cdot) \lessdot H$ is convex as well. As the mapping $\xi \mapsto \sum_{j=1}^{N} \Pi_j^{(N)}(\xi)$ is linear, then the resulting set of densities is indeed convex. Furthermore, for every measure Q in the set (2.102), using (2.99) and (2.95), we obtain

$$Q(\mathcal{X}) = \sum_{j=1}^{N} \mathbb{E}_{X^{1:N}\sim P^N}[\Pi_j^{(N)}(\xi)] \\ = \sum_{j=1}^{N} \mathbb{E}_{X^{1:N}\sim P^N}[\xi_j] = \mathbb{E}_{X^{1:N}\sim P^N}\Big[\sum_{j=1}^{N} \xi_j\Big] = 1.$$

Nonnegativity is evident by construction and thus indeed $Q \in \mathcal{P}(\mathcal{X})$, $Q \ll P$.

We conclude that for a fixed probability measure P, our mini-batch risk form $\varrho^{(N)}[\cdot, P]$, although defined on the set of bounded functions, has a dual representation in terms of countably additive measures, without any singular components. Furthermore, all probability measures in the dual representation are absolutely continuous with respect to the base measure P. Their densities are bounded by N, because all $\big[\Pi_j^{(N)}(\xi)\big](\cdot) \in [0, 1]$ due to (2.95). Therefore, for a fixed probability measure P, we can consider the following mini-batch risk measure $r^{(N)} : \mathcal{L}_1(\mathcal{X}, \mathcal{B}(\mathcal{X}), P) \to \mathbb{R}$:

$$r^{(N)}[Z] = \varrho^{(N)}[Z, P], \quad Z \in \mathcal{L}_1(\mathcal{X}, \mathcal{B}(\mathcal{X}), P). \tag{2.103}$$

Corollary 2.74. *If the risk form $\varrho : \mathbb{B}(\mathcal{X}) \times \mathcal{P}(\mathcal{X}) \to \mathbb{R}$ is coherent and law-invariant, then for every $N \geq 1$ the functional* (2.103) *is a coherent measure of risk on the space $\mathcal{L}_1(\mathcal{X}, \mathcal{B}(\mathcal{X}), P)$ and its subdifferential is given by the equation:*

$$\partial r^{(N)}[Z] = \Big\{\zeta \in \mathcal{L}_\infty(\mathcal{X}, \mathcal{B}(\mathcal{X}), P) : \exists\, (\hat{\xi}(\cdot) \lessdot \hat{H})\ \zeta = \sum_{j=1}^{N} \Pi_j^{(N)}(\hat{\xi})\Big\}, \tag{2.104}$$

where the multifunction $\hat{H} : \mathcal{X}^N \rightrightarrows S^{(N)}$ is defined as

$$\hat{H}(X^{1:N}) = \operatorname*{Arg\,max}_{\xi\in\mathcal{A}^{(N)}} \sum_{j=1}^{N} \xi_j Z(X^j), \quad X^{1:N} \in \mathcal{X}^N.$$

Example 2.75. Consider the mini-batch max risk form (2.91) of Example 2.71:

$$r^{(N)}[Z] = \mathbb{E}_{X^{1:N}\sim P^N}\Big[\max_{1\leq i\leq N} Z(X^i)\Big]; \tag{2.105}$$

the measure P is fixed here. We have $\mathcal{A}^{(N)} = \mathcal{S}^{(N)}$ in this case, and the formula (2.104) applies directly.

An alternative way to derive the dual representation in this case follows from the Kusuoka representation. Formula (2.93) and Strassen's theorem 3.13 imply that every subgradient of $r^{(N)}[0]$ has the density of the form

$$\zeta(v) = N(N-1) \int_0^1 \alpha(1-\alpha)^{N-2} \eta(\alpha, v)\, \mathrm{d}\alpha, \quad v \in \mathcal{X}, \tag{2.106}$$

where the function $\eta : (0,1] \times \mathcal{X} \to \mathbb{R}_+$ satisfies the conditions

$$0 \le \eta(\alpha, \cdot) \le \frac{1}{\alpha}, \quad \alpha \in (0,1],$$
$$\int_{\mathcal{X}} \eta(\alpha, v)\, P(\mathrm{d}v) = 1, \quad \alpha \in (0,1],$$

which express the requirement that $\eta(\alpha, \cdot) \in \partial\, \mathrm{AVaR}_\alpha^+[0]$. The collection of functions (2.106) is identical to (2.104) at $Z = 0$ with $\mathcal{A}^{(N)} = \mathcal{S}^{(N)}$. □

Chapter 3
Optimization of Measures of Risk

In this chapter, we are concerned with the optimization problem

$$\min_{x \in X} \varrho\big(F(x)\big), \tag{3.1}$$

where X is a subset of a decision space $\mathcal{X}$, $F : \mathcal{X} \to \mathcal{Z}$ is an operator assigning to a decision $x \in \mathcal{X}$ a random variable $Z = F(x)$, and $\varrho : \mathcal{Z} \to \mathbb{R}$ is a measure of risk. In successive sections, we address the questions associated with the subdifferentiability of the composition $\varrho \circ F$, the optimality conditions and duality theory, and we present specialized stochastic algorithms for solving problem (3.1).

3.1 Differentiation of Composite Risk Functions

Suppose a mapping $F : \mathcal{X} \to \mathcal{L}_p(\Omega, \mathcal{F}, P)$, where $p \geq 1$, describes the dependence of a random "cost" in a stochastic system on decision variables $x \in \mathcal{X}$, that is, $\big[F(x)\big](\omega)$ is the cost associated with decision x and elementary event $\omega \in \Omega$. We assume that the decision space $\mathcal{X}$ is a Banach space. Frequently, $\mathcal{X} = \mathbb{R}^n$, but it is convenient to consider more general cases, in view of applications to dynamic models. Usually, the mapping $F(\cdot)$ results from a composition of a cost function $c : \mathcal{X} \times \mathfrak{D} \to \mathbb{R}$ and a random data vector $D : \Omega \to \mathfrak{D}$, where $\mathfrak{D}$ is a space of the data vector realizations. Then $\big[F(x)\big](\omega) \triangleq c(x, D(\omega))$, $\omega \in \Omega$. In another setting, we may allow the decision x to be random itself. In this case, we let $\mathcal{X} = \mathcal{L}_p(\Omega, \mathcal{F}, P; U)$, where U is a separable Banach space, and $\big[F(x)\big](\omega) \triangleq c(x(\omega), D(\omega))$, $\omega \in \Omega$. The second case, although it appears more advanced, is a little easier to analyze, and we will start our considerations from it.

D. Dentcheva, A. Ruszczyński, *Risk-Averse Optimization and Control*, Springer Series in Operations Research and Financial Engineering,
https://doi.org/10.1007/978-3-031-57988-2_3

The main topic of this section is the calculation of the subdifferential of the composite risk function $\varphi(\cdot) = \varrho(F(\cdot))$, where $\varrho : \mathcal{L}_p(\Omega, \mathcal{F}, P) \to \mathbb{R}$ is a measure of risk.

3.1.1 Composition with a Convex Operator

In what follows, all equations and inequalities between elements of the space $\mathcal{L}_p(\Omega, \mathcal{F}, P)$, with $p \geq 1$, are understood in the almost sure sense.

Definition 3.1. A mapping $F : \mathcal{X} \to \mathcal{L}_p(\Omega, \mathcal{F}, P)$ is *convex* if for all $x, y \in \mathcal{X}$ and all $\alpha \in [0, 1]$ we have

$$F(\alpha x + (1-\alpha)y) \leq \alpha F(x) + (1-\alpha)F(y). \tag{3.2}$$

For a convex mapping $F : \mathcal{X} \to \mathcal{L}_p(\Omega, \mathcal{F}, P)$, at each $x \in \mathcal{X}$ and for any $h \in \mathcal{X}$, we can construct the differential quotients:

$$Q_t(x; h) \triangleq \frac{1}{t}\big[F(x+th) - F(x)\big], \quad t > 0. \tag{3.3}$$

They are nondecreasing (in the lattice sense) functions of t, and satisfy for $t \in (0, 1]$ the inequalities

$$F(x) - F(x-h) \leq Q_t(x; h) \leq F(x+h) - F(x).$$

Thus the directional derivative

$$F'(x; h) = \lim_{t \downarrow 0} Q_t(x; h), \quad h \in \mathcal{X}, \tag{3.4}$$

is well-defined and belongs to $\mathcal{L}_p(\Omega, \mathcal{F}, P)$, as the limit in the strong and the order sense. By construction, $F'(x; \cdot)$ is convex and positively homogeneous. If $F(\cdot)$ is continuous at x, so is $F'(x; \cdot)$ at 0.

Analogously to the scalar case, we can define subgradients of vector-valued convex mappings.

Definition 3.2. Suppose $F : \mathcal{X} \to \mathcal{L}_p(\Omega, \mathcal{F}, P)$ is a convex mapping. A continuous linear operator $S : \mathcal{X} \to \mathcal{L}_p(\Omega, \mathcal{F}, P)$ such that for all $h \in \mathcal{X}$

$$Sh \leq F(x+h) - F(x) \tag{3.5}$$

is called the *subgradient* of $F(\cdot)$ at x. The set of all such operators is called the *subdifferential* of $F(\cdot)$ at x and is denoted by $\partial F(x)$.

Directly from the construction, we observe that the set $\partial F(x)$ is convex and compact in the weak operator topology.

It is evident from the definition, that $S \in \partial F(x)$ if and only if for all $h \in \mathcal{X}$

$$Sh \le F'(x; h). \tag{3.6}$$

Immediately from (3.6) we obtain the following observation.

Lemma 3.3. *For all $x \in \mathcal{X}$, the set $\partial F(x)$ is convex in the following generalized sense: if $S_1, S_2 \in \partial F(x)$ and $\alpha : \Omega \to [0, 1]$ is measurable, then $\alpha S_1 + (1 - \alpha)S_2 \in \partial F(x)$ as well.*

The following fundamental fact is essential for our further considerations.

Theorem 3.4. *Suppose $F : \mathcal{X} \to \mathcal{L}_p(\Omega, \mathcal{F}, P)$ is convex and continuous at x. Then $\partial F(x)$ is nonempty and for every $h \in \mathcal{X}$ we can find $S \in \partial F(x)$ such that $F'(x; h) = Sh$.*

Proof. Fix $h \in \mathcal{X}$ and consider the subspace $\mathcal{N} \triangleq \{th : t \in \mathbb{R}\}$ in $\mathcal{X}$. The linear operator $S : \mathcal{N} \to \mathcal{L}_p(\Omega, \mathcal{F}, P)$ defined as $S(th) \triangleq tF'(x; h)$ satisfies the inequality:

$$S(d) \le F'(x; d), \quad d \in \mathcal{N}.$$

By the generalization of the Hahn-Banach theorem to lattice-valued operators, the operator S can be extended to the whole space $\mathcal{X}$ with the above inequality satisfied for all $d \in \mathcal{X}$. Due to (3.6), $S \in \partial F(x)$. Since $F(\cdot)$ is continuous at x, S is continuous as well. By construction, $Sh = F'(x; h)$. □

In view of the above theorem, not only we have

$$F'(x; h) = \operatorname*{ess\,sup}_{S \in \partial F(x)} Sh,$$

but the essential supremum is attained at one of the subgradients.

We pass now to the analysis of the composition

$$\varphi(x) \triangleq \varrho\big(F(x)\big), \quad x \in \mathcal{X}, \tag{3.7}$$

where $F : \mathcal{X} \to \mathcal{L}_p(\Omega, \mathcal{F}, P)$ is a convex operator, and $\varrho : \mathcal{L}_p(\Omega, \mathcal{F}, P) \to \mathbb{R}$ is a measure of risk.

Lemma 3.5. *If the mapping $F : \mathcal{X} \to \mathcal{L}_p(\Omega, \mathcal{F}, P)$ is convex and $\varrho : \mathcal{L}_p(\Omega, \mathcal{F}, P) \to \mathbb{R}$ satisfies the conditions of convexity and monotonicity, then the composite function $\varrho \circ F$ is convex.*

Proof. For any $x, y \in \mathcal{X}$ and $\alpha \in [0, 1]$, the convexity of $F(\cdot)$, the monotonicity of $\varrho(\cdot)$, and the convexity of $\varrho(\cdot)$ imply that

$$\varrho\big(F(\alpha x+(1-\alpha)y)\big) \le \varrho\big(\alpha F(x)+(1-\alpha)F(y)\big) \le \alpha\varrho\big(F(x)\big)+(1-\alpha)\varrho\big(F(y)\big).$$

This proves the convexity of $\varrho \circ F$. □

To calculate the subdifferential of the composite function (3.7), we first consider its directional derivatives.

Lemma 3.6. *Suppose the operator* $F : \mathcal{X} \to \mathcal{L}_p(\Omega, \mathcal{F}, P)$ *is convex and continuous at* x, *and the measure of risk* $\varrho : \mathcal{L}_p(\Omega, \mathcal{F}, P) \to \mathbb{R}$ *satisfies the conditions of convexity and monotonicity. Then for all* $x, h \in \mathcal{X}$

$$\varphi'(x;h) = \varrho'\big(F(x); F'(x;h)\big). \tag{3.8}$$

Proof. Recalling the definition (3.4), for all $\tau \geq 0$ we have

$$F(x + \tau h) = F(x) + \tau F'(x;h) + \tau r(\tau, h),$$

with $r(\tau, h) \downarrow 0$ when $\tau \downarrow 0$. Furthermore, by virtue of Theorem 2.32, $\varrho(\cdot)$ is continuous and subdifferentiable everywhere, and for any $d \in \mathcal{L}_p(\Omega, \mathcal{F}, P)$ we have

$$\varrho\big(F(x) + \tau d\big) = \varrho\big(F(x)\big) + \tau\varrho'\big(F(x); d\big) + \tau R(\tau, d),$$

where $R(\tau, d) \downarrow 0$ when $\tau \downarrow 0$. Suppose $0 < \tau \leq \bar{\tau}$. Since $r(\cdot, h)$ is nondecreasing, the two expansions yield

$$\begin{aligned}
\varphi(x + \tau h) - \varphi(x) &= \varrho\big(F(x + \tau h)\big) - \varrho\big(F(x)\big)\\
&= \varrho\big(F(x) + \tau F'(x;h) + \tau r(\tau, h)\big) - \varrho\big(F(x)\big)\\
&\leq \varrho\big(F(x) + \tau F'(x;h) + \tau r(\bar{\tau}, h)\big) - \varrho\big(F(x)\big)\\
&= \tau\varrho'\big(F(x); F'(x;h) + r(\bar{\tau}, h)\big) + \tau R\big(\tau, F'(x;h) + r(\bar{\tau}, h)\big).
\end{aligned}$$

The directional derivative $\varrho'\big(F(x); \cdot\big)$ is convex and positively homogeneous, hence subadditive, and thus

$$\begin{aligned}
&\varphi(x + \tau h) - \varphi(x) \leq\\
&\quad \tau\varrho'\big(F(x); F'(x;h)\big) + \tau\Big[\varrho'\big(F(x); r(\bar{\tau}, h)\big) + R\big(\tau, F'(x;h) + r(\bar{\tau}, h)\big)\Big].
\end{aligned}$$

On the other hand, the convexity of both functions and the monotonicity of $\varrho(\cdot)$ imply that

$$\varphi(x + \tau h) - \varphi(x) \geq \varrho\big(F(x) + \tau F'(x;h)\big) - \varrho\big(F(x)\big) \geq \tau\varrho'\big(F(x); F'(x;h)\big).$$

Combining the last two inequalities and dividing by τ we obtain

$$\begin{aligned}
0 \leq \frac{1}{\tau}\big[\varphi(x + \tau h) - \varphi(x)\big] - \varrho'\big(F(x); F'(x;h)\big)&\\
\leq \varrho'\big(F(x); r(\bar{\tau}, h)\big) + R\big(\tau, F'(x;h) + r(\bar{\tau}, h)\big)&.
\end{aligned}$$

After passing to the limit with $\tau \downarrow 0$, we get

$$0 \leq \varphi'(x; h) - \varrho'\big(F(x); F'(x; h)\big) \leq \varrho'\big(F(x); r(\bar{\tau}, h)\big).$$

As $\varrho'\big(F(x); r(\bar{\tau}, h)\big) \downarrow 0$, when $\bar{\tau} \downarrow 0$, equation (3.8) is true. □

This allows us to calculate the subdifferential of the composition.

Theorem 3.7. *Suppose the operator $F : \mathcal{X} \to \mathcal{L}_p(\Omega, \mathcal{F}, P)$, where $\mathcal{X}$ is a Banach space and $p \in [1, \infty)$, is convex and continuous at x, and the measure of risk $\varrho : \mathcal{L}_p(\Omega, \mathcal{F}, P) \to \mathbb{R}$ satisfies the conditions of convexity and monotonicity. Then the composite function $\varphi = \varrho \circ F$ is subdifferentiable at x and*

$$\partial\varphi(x) = \bigcup_{\substack{\zeta \in \partial\varrho(F(x)) \\ S \in \partial F(x)}} S^*\zeta. \tag{3.9}$$

Proof. Using Lemma 3.6 and Theorem 3.4, we obtain the equation

$$\varphi'(x; h) = \max_{\zeta \in \partial\varrho(F(x))} \mathbb{E}\big[\zeta\, F'(x; h)\big] = \max_{\zeta \in \partial\varrho(F(x))} \mathbb{E}\big[\zeta \operatorname*{ess\,sup}_{S \in \partial F(x)} Sh\big].$$

Since every $\zeta \in \partial\varrho(F(x))$ is nonnegative, and the "ess sup" above is attained, we can continue this chain of equations as follows:

$$\begin{aligned} \varphi'(x; h) &= \max_{\zeta \in \partial\varrho(F(x))} \max_{S \in \partial F(x)} \mathbb{E}\big[\zeta\, Sh\big] \\ &= \max_{\zeta \in \partial\varrho(F(x))} \max_{S \in \partial F(x)} \langle S^*\zeta, h\rangle = \max_{g \in \mathcal{D}} \langle g, h\rangle, \end{aligned} \tag{3.10}$$

where $\mathcal{D}$ is the set on the right hand side of (3.9).

We can verify that $\mathcal{D}$ is convex. Suppose $g_1 = S_1^*\zeta_1$, $g_2 = S_2^*\zeta_2$, where $\zeta_1, \zeta_2 \in \partial\varrho(F(x))$, and $S_1, S_2 \in \partial F(x)$. For $0 < \alpha < 1$ we verify whether $g = \alpha g_1 + (1 - \alpha)g_2 \in \mathcal{D}$. By the convexity od $\partial\varrho(\cdot)$,

$$\zeta = \alpha\zeta_1 + (1 - \alpha)\zeta_2 \in \partial\varrho(F(x)).$$

Furthermore,

$$g = \big(\theta S_1 + (1 - \theta)S_2\big)^*\zeta,$$

with (for $\zeta(\omega) > 0$)

$$\theta(\omega) = \frac{\alpha\zeta_1(\omega)}{\zeta(\omega)}, \quad \omega \in \Omega.$$

If $\zeta(\omega) = 0$, then both $\zeta_1(\omega) = 0$ and $\zeta_2(\omega) = 0$, and we can set $\theta(\omega) = 0$. In any case, since $\zeta_1, \zeta_2 \geq 0$, we have $\theta \in [0, 1]$ a.s.. By the generalized convexity of $\partial F(x)$ (see Lemma 3.3),

$$S = \theta S_1 + (1 - \theta)S_2 \in \partial F(x),$$

and thus $g \in \mathcal{D}$. Therefore, the set $\mathcal{D}$ is convex.

The closedness of $\mathcal{D}$ follows from the fact that for every $h \in \mathcal{X}$ the maximum of $\langle g, h\rangle$ over $g \in \mathcal{D}$ is attained in (3.10). Denote a maximizer by $g(h)$. If a limit $\bar{g}$ of a sequence $\{g^k\}_{k\in N} \subset \mathcal{D}$ was not an element of $\mathcal{D}$, then it could be separated from $\mathcal{D}$ by a linear functional $\bar{h}$:

$$\langle \bar{g}, \bar{h}\rangle > \langle g, \bar{h}\rangle, \quad \forall\, g \in \mathcal{D}.$$

This would lead to a contradiction:

$$\langle \bar{g}, \bar{h}\rangle > \langle g(\bar{h}), \bar{h}\rangle = \max_{g\in\mathcal{D}} \langle g, \bar{h}\rangle \geq \lim_{k\to\infty} \langle g^k, \bar{h}\rangle = \langle \bar{g}, \bar{h}\rangle.$$

Summing up, it follows from (3.10) that $\varphi'(x;\cdot)$ is the support function of a convex closed set $\mathcal{D}$, and thus $\mathcal{D} = \partial\varphi'(x;0) = \partial\varphi(x)$. □

Further analysis of the composition $\rho \circ F$ is dependent on the structure of the mapping $F(\cdot)$. We address this issue in the next two subsections.

3.1.2 Composition with a Convex Integrand

We assume that $\mathcal{X} = \mathcal{L}_{p'}(\Omega, \mathcal{F}, P; U)$, where U is a separable Banach space, and $p \in [1,\infty)$, $p' \in [p,\infty]$. We include the case of $p' = \infty$ in view of the application to the analysis in the next section. In some intermediate results, we also allow $p = \infty$.

The mapping $F : \mathcal{L}_{p'}(\Omega, \mathcal{F}, P; U) \to \mathcal{L}_p(\Omega, \mathcal{F}, P)$ is defined as follows:

$$\big[F(x)\big](\omega) = f(x(\omega), \omega), \quad \omega \in \Omega, \tag{3.11}$$

where $f : U \times \Omega \to \mathbb{R}$ is a convex and continuous function of the first argument, for almost all values of the second argument, and measurable with respect to the second argument, for all values of the first argument. We call such functions *convex integrands*.

Evidently, such a mapping $F(\cdot)$ is convex in the sense of Definition 3.1, but our setting is more specific because it excludes a null set *before* all $x, y \in \mathcal{X}$ are considered. This allows for an explicit description of the subgradients of $F(\cdot)$.

The differential quotients (3.3) of the mapping (3.11) have a specific form: for almost all $\omega \in \Omega$

$$\big[Q_t(x;h)\big](\omega) = \frac{1}{t}\big[f\big(x(\omega) + th(\omega), \omega\big) - f\big(x(\omega), \omega\big)\big], \quad t > 0.$$

By the convexity of $f(\cdot, \omega)$ we conclude that for all $h \in \mathcal{X}$ and for almost all $\omega \in \Omega$

$$\big[F'(x;h)\big](\omega) = \lim_{t\downarrow 0}\big[Q_t(x;h)\big](\omega) = f'\big(x(\omega),\omega;h(\omega)\big).$$

We can now establish the decomposable structure of the subdifferential.

Lemma 3.8. *If $L \in \partial F(x)$ then L is* local *in the following sense: for all $B \in \mathcal{F}$ and all $h \in \mathcal{X}$ we have*

$$L(\mathbb{1}_B h) = \mathbb{1}_B Lh.$$

Proof. For all events $B \in \mathcal{F}$ and all $h \in \mathcal{X}$ we have $F'(x;\mathbb{1}_B h) = \mathbb{1}_B F'(x;h)$, and thus every $L \in \partial F(x)$ satisfies the inequalities:

$$-\mathbb{1}_B F'(x;-h) \le L(\mathbb{1}_B h) \le \mathbb{1}_B F'(x;h). \tag{3.12}$$

Consequently, $L(\mathbb{1}_B h)$ vanishes outside of B. As L is linear,

$$\mathbb{1}_B L(\mathbb{1}_B h) = \mathbb{1}_B Lh - \mathbb{1}_B L(\mathbb{1}_{B^c} h) = \mathbb{1}_B Lh,$$

because the second term disappears due to (3.12) for B^c. Therefore, $L(\mathbb{1}_B h)$ coincides with Lh on B. □

The local property is essential in the following theorem.

Theorem 3.9. *Suppose $F : \mathcal{L}_{p'}(\Omega,\mathcal{F},P;U) \to \mathcal{L}_p(\Omega,\mathcal{F},P)$ is given by (3.11), $1 \le p \le p' \le \infty$, U is a separable Banach space, $f(\cdot,\omega)$ is convex and continuous at $x(\omega)$ for almost all $\omega \in \Omega$, and $\mathbb{E}[F(\cdot)]$ is continuous at x as well. Then $S \in \partial F(x)$ if and only if a function $s \in \mathcal{L}_r(\Omega,\mathcal{F},P;U^*)$ exists, with $1/p' + 1/r = 1/p$, such that for almost all $\omega \in \Omega$ we have $s(\omega) \in \partial f(x(\omega),\omega)$, and for all $h \in \mathcal{L}_p(\Omega,\mathcal{F},P;U)$*

$$\big[Sh\big](\cdot) = \langle s(\cdot),h(\cdot)\rangle, \quad \textit{a.s.}. \tag{3.13}$$

Proof. Every operator of the form (3.13), with a measurable selector $s(\cdot) \in \partial f(x(\cdot),\cdot)$, is a subgradient of F at x, because for all $h \in \mathcal{X}$

$$\langle s(\omega),h(\omega)\rangle \le f'\big(x(\omega),\omega;h(\omega)\big) = \big[F'(x;h)\big](\omega),$$

for almost all $\omega \in \Omega$, and thus Definition 3.2 is satisfied.

Inequality (3.6) implies that for all $h \in \mathcal{X}$

$$-F'(x;-h) \le \langle s,h\rangle \le F'(x;h) \quad \text{a.s.},$$

and thus $\langle s(\cdot),h(\cdot)\rangle$ is an element of $\mathcal{L}_p(\Omega,\mathcal{F},P)$. By the Hölder inequality, $s \in \mathcal{L}_r(\Omega,\mathcal{F},P;U^*)$, with $1/p' + 1/r = 1/p$.

We need to prove that no other subgradients exist. Suppose $L \in \partial F(x)$. Consider the functional $\ell : \mathcal{X} \to \mathbb{R}$ defined as

$$\ell(h) = \mathbb{E}[Lh], \quad h \in \mathcal{X}.$$

As L is linear and continuous, and so is $\ell(\cdot)$. By the local property of L established in Lemma 3.8,

$$\ell(\mathbb{1}_B h) = \mathbb{E}\big[L(\mathbb{1}_B h)\big] = \mathbb{E}\big[\mathbb{1}_B Lh\big], \quad \forall B \in \mathcal{F}. \tag{3.14}$$

If $p' < \infty$, by the Riesz representation theorem, we can identify $\ell(\cdot)$ with an element $v \in \mathcal{L}_q(\Omega, \mathcal{F}, P; U^*)$, $1/p' + 1/q = 1$, such that

$$\ell(h) = \mathbb{E}\big[\langle v(\cdot), h(\cdot)\rangle\big], \quad \forall\, h \in \mathcal{X}.$$

This entails the equation

$$\mathbb{E}\big[\mathbb{1}_B Lh\big] = \mathbb{E}\big[\mathbb{1}_B \langle v(\cdot), h(\cdot)\rangle\big], \quad \forall B \in \mathcal{F}.$$

Thus, for all $h \in \mathcal{X}$,

$$[Lh](\cdot) = \langle v(\cdot), h(\cdot)\rangle, \quad \text{a.s..} \tag{3.15}$$

If $p' = \infty$, then linear continuous functionals on $\mathcal{L}_\infty(\Omega, \mathcal{F}, P; U)$ may contain singular parts. However, in our case, (3.14) implies that the set function $B \mapsto \ell(\mathbb{1}_B h)$ is countably additive on $\mathcal{F}$. Consequently, no singular parts occur, $\ell(\cdot)$ can be identified with an element $v \in \mathcal{L}_1(\Omega, \mathcal{F}, P; U^*)$, and the representation (3.15) remains valid.

The representation (3.15) implies that the function $\langle v(\cdot), h(\cdot)\rangle$ is an element of $\mathcal{L}_p(\Omega, \mathcal{F}, P)$, for all $h \in \mathcal{X}$. Therefore, $v \in \mathcal{L}_r(\Omega, \mathcal{F}, P; U^*)$.

Let

$$\mathcal{S} = \big\{s \in \mathcal{L}_r(\Omega, \mathcal{F}, P; U^*) : s(\cdot) \in \partial f(x(\cdot), \cdot)\big\}.$$

Suppose $v \notin \mathcal{S}$. Because $\mathcal{S}$ is convex and closed, by virtue of the separation theorem, a function $\bar{h} \in \mathcal{X}$ and a real number $\varepsilon > 0$ exist, such that for all $s \in \mathcal{S}$

$$\mathbb{E}\big[\langle v, \bar{h}\rangle\big] \geq \mathbb{E}\big[\langle s, \bar{h}\rangle\big] + \varepsilon. \tag{3.16}$$

Let $\bar{s}$ be an element of $\mathcal{S}$ such that

$$\langle \bar{s}(\cdot), \bar{h}(\cdot)\rangle = f'\big(x(\cdot), \cdot; \bar{h}(\cdot)\big); \tag{3.17}$$

such a function can be found, owing to the measurable selection theorem. Indeed, for almost all $\omega \in \Omega$ we have

$$f'(x(\omega), \omega; \bar{h}(\omega)) = \max_{s(\omega) \in \partial f(\bar{x}(\omega), \omega)} \langle s(\omega), \bar{h}(\omega)\rangle;$$

the maximum is attained at some $\bar{s}(\omega)$, due to the weak* compactness of $\partial f(\bar{x}(\omega), \omega)$. The relations (3.16)–(3.17) imply that

$$\mathbb{E}\big[L\bar{h}\big] = \mathbb{E}\big[\langle v, h\rangle\big] \geq \mathbb{E}\big[F'(x; \bar{h})\big] + \varepsilon.$$

which contradicts (3.6). Consequently, $v \in \mathcal{S}$. □

If $U = \mathbb{R}^n$, we do not need the condition about the continuity of the functions $f(\cdot, \omega)$, because a convex function in finite dimensions is continuous on the interior of its domain.

We can now calculate the subdifferential of the composition $\varphi(\cdot) = \varrho(F(\cdot))$, where $\varrho(\cdot)$ is a measure of risk.

Theorem 3.10. *Suppose $F : \mathcal{L}_{p'}(\Omega, \mathcal{F}, P; U) \to \mathcal{L}_p(\Omega, \mathcal{F}, P)$ is given by (3.11), with $p \in [1, \infty)$ and $p' \in [p, \infty]$, $f(\cdot, \omega)$ is convex and continuous at $x(\omega)$ for almost all $\omega \in \Omega$, U is a separable Banach space, and $\mathbb{E}[F(\cdot)]$ is continuous at x as well. Further, suppose $\varrho : \mathcal{L}_p(\Omega, \mathcal{F}, P) \to \mathbb{R}$ satisfies the conditions of convexity and monotonicity. Then the composite function $\varphi = \varrho \circ F$ is subdifferentiable at x and*

$$\partial\varphi(x) = \bigcup_{\zeta \in \partial\varrho(F(x))} \big\{ g \in \mathcal{L}_{q'}(\Omega, \mathcal{F}, P; U^*) : g(\omega) \in \zeta(\omega)\, \partial f(x(\omega), \omega),\ a.s. \big\}, \tag{3.18}$$

where $1/p' + 1/q' = 1$.

Proof. We use the representation of the subdifferential established in Theorem 3.7 and the characterizations of the subgradients of $F(\cdot)$ established in Theorem 3.9. For any $h \in \mathcal{X}$, any $\zeta \in \partial\varrho(F(x))$ and any $S \in \partial F(x)$ we have

$$\langle S^*\zeta, h\rangle = \langle \zeta, Sh\rangle = \int_\Omega \big\langle \zeta(\omega)s(\omega), h(\omega)\big\rangle\, P(\mathrm{d}\omega),$$

where $s(\cdot) \in \partial f(x(\cdot), \cdot)$. This verifies (3.18). □

3.1.3 Composition with a Convex Function

We now consider the case when $\mathcal{X}$ is a separable Banach space, and the mapping $F : \mathcal{X} \to \mathcal{L}_p(\Omega, \mathcal{F}, P)$ is such that for almost all $\omega \in \Omega$

$$\big[F(x)\big](\omega) = f(x, \omega), \quad x \in \mathcal{X}, \tag{3.19}$$

where $f(\cdot, \omega)$ is convex on $\mathcal{X}$.

Again, the mapping $F(\cdot)$ is convex and we plan to analyze the structure of its subdifferential.

The differential quotients (3.3) of the mapping (3.19) again have a specific form: for almost all $\omega \in \Omega$

$$\big[Q_t(x;h)\big](\omega) = \frac{1}{t}\big[f\big(x + th, \omega\big) - f\big(x, \omega\big)\big], \quad t > 0.$$

By the convexity of $f(\cdot, \omega)$ we conclude that for all $h \in \mathcal{X}$ and for almost all $\omega \in \Omega$

$$\big[F'(x;h)\big](\omega) = f'\big(x, \omega; h\big).$$

We will use this form to disintegrate the subdifferential of $F(\cdot)$.

Theorem 3.11. *Suppose $F : \mathcal{X} \to \mathcal{L}_p(\Omega, \mathcal{F}, P)$, with $p \in [1, \infty]$, is given by (3.19), $\mathcal{X}$ is a separable Banach space, $f(\cdot, \omega)$ is convex and continuous at x for almost all $\omega \in \Omega$, and $\mathbb{E}[F(\cdot)]$ is continuous at x as well. Then $S \in \partial F(x)$ if and only if a function $s \in \mathcal{L}_p(\Omega, \mathcal{F}, P; \mathcal{X}^*)$ exists, such that $s(\omega) \in \partial f(x, \omega)$ for almost all $\omega \in \Omega$, and for all $h \in \mathcal{X}$*

$$\big[Sh\big](\cdot) = \langle s(\cdot), h\rangle \quad a.s. \tag{3.20}$$

Proof. It is evident that every operator S of the form (3.20) is a subgradient of $F(\cdot)$. We need to prove that no other subgradients exist.

Our idea is to allow x and h to be random and to leverage Theorem 3.9. Define the space $\mathbb{X} = \mathcal{L}_\infty(\Omega, \mathcal{F}, P; \mathcal{X})$ and denote by $\mathbb{x}$ and $\mathbb{h}$ its elements. Let the mapping $\mathbb{F} : \mathbb{X} \to \mathcal{L}_p(\Omega, \mathcal{F}, P)$ be defined as follows:

$$\big[\mathbb{F}(\mathbb{x})\big](\omega) = f(\mathbb{x}(\omega), \omega), \quad \omega \in \Omega.$$

Suppose $\mathbb{x} \in \mathbb{X}$ is equal to x a.s.. We have, for almost all $\omega \in \Omega$,

$$\big[\mathbb{F}'(\mathbb{x};\mathbb{h})\big](\omega) = f'\big(x, \omega; \mathbb{h}(\omega)\big).$$

The function $\mathbb{F}'(\mathbb{x};\cdot)$ is convex, positively homogeneous, and continuous on $\mathbb{X}$. Suppose $S \in \partial F(x)$. Then $S(\cdot) \le F'(x;\cdot)$ on $\mathcal{X}$. This means that $S(\cdot)$ is a minorant of $\mathbb{F}'(\mathbb{x};\cdot)$ on the subspace of constant functions $\mathcal{N} = \big\{\mathbb{h} \in \mathbb{X} : \mathbb{E}[\mathbb{h}] = h\big\}$. By the generalization of the Hahn–Banach theorem to lattice-valued operators, S can be extended to a linear functional $\mathbb{S}$ on $\mathbb{X}$ which is a minorant of $\mathbb{F}'(\mathbb{x};\cdot)$ and which coincides with S on $\mathcal{N}$. As $\mathbb{F}'(\mathbb{x};\cdot)$ is continuous, $\mathbb{S} \in \partial\mathbb{F}(\mathbb{x})$. By Theorem 3.11 at the point $\mathbb{x}$, a function $s \in \mathcal{L}_p(\Omega, \mathcal{F}, P; \mathcal{X}^*)$ exists, such that for almost all $\omega \in \Omega$ we have $s(\omega) \in \partial f(x, \omega)$, and for all $\mathbb{h} \in \mathbb{X}$

$$\big[\mathbb{S}\mathbb{h}\big](\cdot) = \langle s(\cdot), \mathbb{h}(\cdot)\rangle, \quad \text{a.s..}$$

On the subspace $\mathcal{N}$, when $\mathbb{h}(\cdot) = h$ a.s., we get $[Sh](\cdot) = [\mathbb{S}\mathbb{h}](\cdot) = \langle s(\cdot), h\rangle$ a.s.. □

The subdifferential of $\varphi(x) = \varrho(F(x))$ can be calculated in a similar way to Theorem 3.10.

Theorem 3.12. *Suppose $\mathcal{X}$ is a separable Banach space, $F : \mathcal{X} \to \mathcal{L}_p(\Omega, \mathcal{F}, P)$ is given by (3.19), $p \in [1, \infty)$, $f(\cdot, \omega)$ is convex and continuous at x for almost all $\omega \in \Omega$, $\mathbb{E}[F(\cdot)]$ is continuous at x as well, and $\varrho : \mathcal{L}_p(\Omega, \mathcal{F}, P) \to \mathbb{R}$ satisfies the conditions of convexity and monotonicity. Then the composite function*

$\varphi = \varrho \circ F$ is subdifferentiable at x and[†]

$$\partial\varphi(x) = \bigcup_{\zeta \in \partial\varrho(F(x))} \int_{\Omega} \partial f(x,\omega)\, \zeta(\omega)\; P(\mathrm{d}\omega). \tag{3.21}$$

Proof. We use the representation of the subdifferential established in Theorem 3.7 and the characterizations of the subgradients of $F(\cdot)$ established in Theorem 3.11. For any $h \in \mathcal{X}$, any $\zeta \in \partial\varrho(F(x))$ and any $S \in \partial F(x)$ we have

$$\langle S^*\zeta, h\rangle = \langle \zeta, Sh\rangle = \int_{\Omega} \zeta(\omega)\big\langle s(\omega), h\big\rangle\; P(\mathrm{d}\omega) = \Big\langle \int_{\Omega} s(\omega)\zeta(\omega)\; P(\mathrm{d}\omega), h\Big\rangle,$$

where $s(\cdot) \in \partial f(x,\cdot)$. This verifies (3.21). □

The following conclusion is known as the *Strassen Theorem.*

Theorem 3.13. *Suppose $\mathcal{X}$ is a separable Banach space, $F : \mathcal{X} \to \mathcal{L}_1(\Omega, \mathcal{F}, P)$ is given by* (3.19)*, $f(\cdot,\omega)$ is convex and continuous at x for almost all $\omega \in \Omega$, and $\mathbb{E}[F(\cdot)]$ is continuous at x as well. Then the function $\mathbb{E}[F(\cdot)]$ is subdifferentiable at x and*

$$\partial\mathbb{E}[F(x)] = \mathbb{E}[\partial F(x)] = \int_{\Omega} \partial f(x,\omega)\; P(\mathrm{d}\omega).$$

Our method can be easily adapted when the decision x is allowed to depend on additional information, modeled by a σ-subalgebra $\mathcal{G} \subset \mathcal{F}$.

Remark 3.14. Suppose $\mathcal{X} = \mathcal{L}_{p'}(\Omega, \mathcal{G}, P; U)$ when $\mathcal{G} \subset \mathcal{F}$, U is a separable Banach space, and $p' \in [1,\infty)$. In this case, under other assumptions as before

$$\partial\varphi(x) = \bigcup_{\substack{\zeta \in \partial\varrho(F(x)) \\ s(\cdot) \lessdot \partial f(x(\cdot),\cdot)}} \mathbb{E}[s\zeta|\mathcal{G}].$$

The derivation is almost the same as Theorem 3.12, we only need to make sure that the subgradients live in $\mathcal{X}^* = \mathcal{L}_{q'}(\Omega, \mathcal{G}, P; U^*)$. □

3.1.4 Composition with a Regular Lipschitz Continuous Function

Another case of interest in applications is when the mapping $F : \mathcal{X} \to \mathcal{L}_p(\Omega, \mathcal{F}, P)$, $p \in [1,\infty)$, is such that for almost all $\omega \in \Omega$

$$\big[F(x)\big](\omega) = f(x,\omega), \quad x \in \mathcal{X}, \tag{3.22}$$

[†] The integral in (3.21) is understood as the set of all weak* integrals of measurable selectors of the integrated multifunction.

where $f : \mathcal{X} \times \Omega \to \mathbb{R}$ is locally Lipschitz continuous with respect to the first argument, and measurable with respect to the second argument. The decision space $\mathcal{X}$ is a separable Banach space.

In this case, the composition $\varphi = \varrho \circ F$ is neither convex nor differentiable. Fortunately, its special structure allows for direct analysis, under additional assumptions about the functions involved.

Before we present the main findings, we recall some basic definitions.

Definition 3.15. For a locally Lipschitz continuous function $f : \mathcal{X} \to \mathbb{R}$, its *generalized directional derivative* at a point $x \in \mathcal{X}$ in the direction $h \in \mathcal{X}$ is defined as

$$f^0(x;h) \triangleq \limsup_{\substack{y \to x \\ \tau \downarrow 0}} \frac{f(y + \tau h) - f(y)}{\tau}.$$

It can be easily verified that $f^0(x, \cdot)$ is convex, positively homogeneous, and Lipschitz continuous.

Definition 3.16. For a locally Lipschitz continuous function $f : \mathcal{X} \to \mathbb{R}$, its *Clarke subdifferential* at a point $x \in \mathcal{X}$ is the closed convex set

$$\partial f(x) \triangleq \{g \in \mathcal{X}^* : \langle g, \cdot \rangle \le f^0(x; \cdot)\}.$$

The Clarke subdifferential coincides with the usual subdifferential when $f(\cdot)$ is convex and with the gradient $\{\nabla f(x)\}$ when $f(\cdot)$ is Fréchet differentiable.

Definition 3.17. A locally Lipschitz continuous function $f : \mathcal{X} \to \mathbb{R}$ is *subdifferentially regular* at a point $x \in \mathcal{X}$, if it has the usual directional derivative $f'(x; h)$ in every direction $h \in \mathcal{X}$, and

$$f^0(x; \cdot) = f'(x; \cdot).$$

Now, we pass to *random* functions $f : \mathcal{X} \times \Omega \to \mathbb{R}$, that is, such functions that for each $x \in \mathcal{X}$ the mapping $f(x, \cdot)$ is measurable. We make the following general assumptions.

Assumption 3.18. *For every bounded set $U \subset \mathcal{X}$, a function $C \in \mathcal{L}_p(\Omega, \mathcal{F}, P)$, $p \in [1, \infty)$, exists, such that for all $x, y \in U$*

$$|f(x, \omega) - f(y, \omega)| \le C(\omega)\|x - y\|, \quad \omega \in \Omega. \tag{3.23}$$

Under this condition, the operator (3.22) is Lipschitz continuous as well, because

$$\|F(x) - F(y)\| = \left(\int |f(x, \omega) - f(y, \omega)|^p \, P(\mathrm{d}\omega)\right)^{1/p} \le \|C\| \|x - y\|.$$

We also assume the regularity.

Assumption 3.19. *The functions $f(\cdot,\omega)$, $\omega \in \Omega$, are subdifferentially regular.*

These properties allow us to calculate the directional derivative of the composite risk function.

Lemma 3.20. *Suppose $F : \mathcal{X} \to \mathcal{L}_p(\Omega, \mathcal{F}, P)$, with $p \in [1,\infty)$, is defined by* (3.22) *and satisfies Assumptions* 3.18 *and* 3.19, *and $\mathcal{X}$ is a Banach space. Furthermore, suppose $\varrho : \mathcal{L}_p(\Omega, \mathcal{F}, P) \to \mathbb{R}$ satisfies the conditions of convexity and monotonicity. Then for all $x, h \in \mathcal{X}$*

$$\varphi'(x;h) = \max_{\zeta\in\partial\varrho(F(x))} \big\langle \zeta, F'(x;h)\big\rangle, \tag{3.24}$$

where

$$\big[F'(x;h)\big](\omega) = f'(x,\omega;h), \quad \omega \in \Omega. \tag{3.25}$$

Proof. Define $D(x;h)(\omega)$ as the right-hand side of (3.25). Due to Assumption 3.19, it is well defined and measurable, as a function of $\omega \in \Omega$. By Assumption 3.18, $\|D(x,h)(\omega)\| \le C(\omega)\|h\|$ a.s., and thus $D(x,h) \in \mathcal{L}_p(\Omega, \mathcal{F}, P)$. Consider a sequence $\{\tau_k\} \downarrow 0$. By (3.25), we have

$$\frac{1}{\tau_k}\big[F(x+\tau_k h) - F(x)\big] - D(x,h) \to 0, \quad \text{a.s.}, \tag{3.26}$$

and

$$\left\| \frac{1}{\tau_k}\big[F(x+\tau_k h) - F(x)\big] - D(x,h)\right\| \le 2C\,\|h\|.$$

By the Lebesgue dominated convergence theorem, the limit in (3.26) is zero the space $\mathcal{L}_p(\Omega, \mathcal{F}, P)$ and thus the formula (3.25) is true. Consequently, we can write the expansion

$$F(x+\tau h) = F(x) + \tau F'(x;h) + r(\tau,h), \quad \lim_{\tau\downarrow 0}\frac{r(\tau,h)}{\tau} = 0. \tag{3.27}$$

Furthermore, by virtue of Theorem 2.32, $\varrho(\cdot)$ is continuous and subdifferentiable everywhere. Thus, it is Hadamard directionally differentiable:

$$\lim_{\substack{\tau\downarrow 0\\ d'\to d}} \frac{1}{\tau}\big[\varrho\big(F(x)+\tau d'\big) - \varrho\big(F(x)\big)\big] = \varrho'\big(F(x);d\big).$$

The two expansions yield

$$\begin{aligned}\varphi'(x;h) &= \lim_{\tau\downarrow 0}\frac{1}{\tau}\big[\varphi(x+\tau h)-\varphi(x)\big]\\ &= \lim_{\tau\downarrow 0}\frac{1}{\tau}\big[\varrho\big(F(x)+\tau F'(x;h)+r(\tau,h)\big)-\varrho\big(F(x)\big)\big]\\ &= \lim_{\tau\downarrow 0}\frac{1}{\tau}\Big[\varrho\big(F(x)+\tau\big(F'(x;h)+\frac{r(\tau,h)}{\tau}\big)\big)-\varrho\big(F(x)\big)\Big]\\ &= \varrho'\big(F(x);F'(x;h)\big).\end{aligned}$$

Consequently, formula (3.24) is true. □

We can now derive the formula for the Clarke subdifferential of the composite function $\varphi(\cdot)$.

Theorem 3.21. *Suppose $F : \mathcal{X} \to \mathcal{L}_p(\Omega, \mathcal{F}, P)$, with $p \in [1,\infty)$, is defined by* (3.22) *and satisfies Assumptions* 3.18 *and* 3.19, *and $\mathcal{X}$ is a separable Banach space. Furthermore, suppose $\varrho : \mathcal{L}_p(\Omega, \mathcal{F}, P) \to \mathbb{R}$ is nondecreasing and convex. Then the composite function $\varphi = \varrho \circ F$ is Lipschitz continuous and subdifferentially regular at every point $x \in \mathcal{X}$, and its Clarke subdifferential is given by the formula*

$$\partial\varphi(x) = \bigcup_{\zeta\in\partial\varrho(F(x))}\int_\Omega \zeta(\omega)\,\partial f(x,\omega)\;P(\mathrm{d}\omega). \tag{3.28}$$

Proof. Due to the monotonicity of $\varrho(\cdot)$, every $\zeta \in \partial\varrho(F(x))$ is nonnegative. Therefore, by virtue of Lemma 3.20,

$$\begin{aligned}\varphi'(x;h) &= \max_{\zeta\in\partial\varrho(F(x))}\int \zeta(\omega)\max_{s\in\partial f(x,\omega)}\langle s,h\rangle\;P(\mathrm{d}\omega)\\ &= \max_{\zeta\in\partial\varrho(F(x))}\int \max_{s\in\partial f(x,\omega)}\zeta(\omega)\,\langle s,h\rangle\;P(\mathrm{d}\omega).\end{aligned} \tag{3.29}$$

The multifunction $\partial f(x,\cdot)$ has closed, convex, and bounded values in $\mathcal{X}^*$. Therefore, they are weakly* compact, and the maximizer

$$s^*(\omega) \in \operatorname*{Arg\,max}_{s\in\partial f(x,\omega)}\langle s,h\rangle,\qquad \omega\in\Omega,$$

exists. Since $\partial f(x,\cdot)$ is measurable and closed and convex valued, the maximizer $s^*(\cdot)$ may be chosen to be measurable, by virtue of the measurable selection theorem. Consequently, we may exchange the integral and the "max" operator in (3.29), obtaining[†]

[†] The symbol "$\lessdot$" means "is a measurable selector of."

$$\varphi'(x;h) = \max_{\zeta \in \partial\varrho(F(x))} \max_{s(\cdot) \lessdot \partial f(x,\cdot)} \int \zeta(\omega)\langle s(\omega), h\rangle \; P(\mathrm{d}\omega). \tag{3.30}$$

Thus, the directional derivative $\varphi'(x;\cdot)$ is the support function of the set $\mathcal{D}$ on the right-hand side of (3.28). Furthermore,

$$\varphi'(x;h) = \max_{\zeta \in \partial\varrho(F(x))} \int \zeta(\omega) f'(x,\omega;h) \; P(\mathrm{d}\omega).$$

Again, the set $\partial\varrho(F(x))$ is weakly* compact and a maximizer exists, for every $h \in \mathcal{X}$.

Using the nonnegativity of each $\zeta \in \partial\varrho(F(x))$ and the convexity of the sets $\partial f(x,\omega)$, $\omega \in \Omega$, in exactly the same way as in the proof of Theorem 3.7, we can verify that the set $\mathcal{D}$ is convex and closed. Therefore, $\mathcal{D} = \partial\varphi'(x;0)$.

We shall estimate the generalized directional derivative of $\varphi(\cdot)$ at a point x in a direction h. Due to the subdifferentiability of $\varrho(\cdot)$, we have the inequality

$$\varrho\big(F(y+\tau h)\big) \le \varrho\big(F(y)\big) + \max_{\zeta \in \partial\varrho(F(y+\tau h))} \big\langle \zeta, F(y+\tau h) - F(y)\big\rangle.$$

Hence,

$$\frac{\varphi(y+\tau h) - \varphi(y)}{\tau} \le \frac{1}{\tau} \max_{\zeta \in \partial\varrho(F(y+\tau h))} \big\langle \zeta, F(y+\tau h) - F(y)\big\rangle.$$

Therefore,

$$\varphi^0(x;h) \le \limsup_{\substack{y\to x \\ \tau\downarrow 0}} \max_{\zeta \in \partial\varrho(F(y+\tau h))} \frac{1}{\tau}\big\langle \zeta, F(y+\tau h) - F(y)\big\rangle.$$

Consider sequences $\{y_k\}$ converging to x and $\{\tau_k\}$ converging to 0 from the right, and $\zeta_k \in \partial\varrho(F(y_k + \tau_k h))$, at which the "lim sup" above is realized. We have

$$\begin{aligned}
\varphi^0(x;h) &\le \lim_{k\to\infty} \int \zeta_k(\omega) \frac{f(y_k + \tau_k h, \omega) - f(y_k,\omega)}{\tau_k} \; P(\mathrm{d}\omega) \\
&\le \lim_{k\to\infty} \int \zeta_k(\omega) \left[\frac{f(y_k + \tau_k h, \omega) - f(y_k,\omega)}{\tau_k} - f'(x;h)\right] P(\mathrm{d}\omega) \\
&\quad + \lim_{k\to\infty} \int \zeta_k(\omega) f'(x;h) \; P(\mathrm{d}\omega)
\end{aligned} \tag{3.31}$$

Due to the subdifferential regularity of $f(\cdot,\omega)$,

$$\lim_{k\to\infty} \frac{f(y_k + \tau_k h, \omega) - f(y_k,\omega)}{\tau_k} = f'(x,\omega;h), \quad \text{a.s.}$$

As the sequence $\{\zeta_k\}$ is bounded, by the Lebesgue dominated convergence theorem, the first term on the right-hand side of (3.31) converges to 0. Next, due to the weak* upper semicontinuity of $\partial\varrho(\cdot)$, we can continue (3.31) as follows

$$\varphi^0(x;h) \leq \lim_{k\to\infty} \int \zeta_k(\omega) f'(x;h)\ P(\mathrm{d}\omega)$$
$$\leq \max_{\zeta\in\partial\varrho(F(x))} \int \zeta(\omega) f'(x;h)\ P(\mathrm{d}\omega) = \varphi'(x;h).$$

As $\varphi^0(x;h) \geq \varphi'(x;h)$ by construction, we conclude that $\varphi^0(x;h) = \varphi'(x;h)$ for all $h \in \mathcal{X}$. This proves the subdifferential regularity of $\varphi(\cdot)$ and justifies the representation (3.28). □

We may remark that the subdifferential regularity of $\varphi(\cdot)$ implies its Hadamard directional differentiability.

3.1.5 Composition with a Regular Lipschitz Continuous Integrand

Now, we consider the case when $\mathcal{X} = \mathcal{L}_{p'}(\Omega, \mathcal{F}, P; U)$, where U is a separable Banach space, and $\mathcal{Z} = \mathcal{L}_p(\Omega, \mathcal{F}, P)$, with $1 \leq p \leq p' < \infty$. The operator $F : \mathcal{X} \to \mathcal{L}_p(\Omega, \mathcal{F}, P)$ is given by

$$\big[F(x)\big](\omega) = f(x(\omega), \omega), \quad \omega \in \Omega, \tag{3.32}$$

where $f : U \times \Omega \to \mathbb{R}$ is a locally Lipschitz continuous function of the first argument, for almost all values of the second argument.

We make the following assumption.

Assumption 3.22. *A function $C \in \mathcal{L}_r(\Omega, \mathcal{F}, P)$, with $1/r + 1/p' = 1/p$ exists, such that for all $x, y \in \mathcal{X}$ and almost all $\omega \in \Omega$,*

$$|f(x,\omega) - f(y,\omega)| \leq C(\omega)\|x - y\|. \tag{3.33}$$

Furthermore, for almost all $\omega \in \Omega$, the function $f(\cdot,\omega)$ is subdifferentially regular.

Under this condition, we can still prove the existence and calculate the Clarke subdifferential of the composition

$$\varphi(\cdot) = \varrho\big(F(\cdot)\big), \tag{3.34}$$

with a risk measure $\varrho(\cdot)$.

We start from the directional derivative of the composite function.

Lemma 3.23. *Suppose $F : \mathcal{X} \to \mathcal{L}_p(\Omega, \mathcal{F}, P)$, with $p \in [1, \infty)$, is defined by* (3.32) *and satisfies Assumption* 3.22. *Furthermore, suppose $\varrho : \mathcal{L}_p(\Omega, \mathcal{F}, P) \to \mathbb{R}$ satisfies the conditions of convexity and monotonicity. Then for all $x, h \in \mathcal{X}$*

$$\varphi'(x; h) = \max_{\zeta \in \partial\varrho(F(x))} \langle \zeta, F'(x; h) \rangle, \tag{3.35}$$

where

$$\big[F'(x; h)\big](\omega) = f'(x(\omega), \omega; h(\omega)), \quad \omega \in \Omega. \tag{3.36}$$

Proof. Define $D(x; h)(\omega)$ as the right-hand side of (3.36). Due to Assumption 3.22, it is well defined and measurable, as a function of $\omega \in \Omega$, and $\|D(x, h)(\omega)\| \le C(\omega)\|h(\omega)\|$ a.s., and thus $D(x, h) \in \mathcal{L}_p(\Omega, \mathcal{F}, P)$. The remaining proof is almost identical to the proof of Lemma 3.20. □

We can now formulate the main result of this section.

Theorem 3.24. *Suppose $F : \mathcal{L}_{p'}(\Omega, \mathcal{F}, P; U) \to \mathcal{L}_p(\Omega, \mathcal{F}, P)$ with $1 \le p \le p' < \infty$, is an operator given by* (3.32) *satisfying Assumption* 3.22, *U is a separable Banach space, and $\varrho : \mathcal{L}_p(\Omega, \mathcal{F}, P) \to \mathbb{R}$ is a convex nondecreasing functional. Then the composite functional $\varphi = \varrho \circ F$ is subdifferentially regular, and its Clarke subdifferential has the following form:*

$$\partial\varphi(x) = \bigcup_{\zeta \in \partial\varrho(F(x))} \big\{g \in \mathcal{L}_q(\Omega, \mathcal{F}, P; U^*) : g(\omega) \in \zeta(\omega)\, \partial f(x(\omega), \omega),\ a.s.\big\}, \tag{3.37}$$

where $1/q + 1/p = 1$.

Proof. Since $\varrho(\cdot)$ is convex and continuous, it is subdifferentiable, and we can use the subgradient inequality:

$$\varrho\big(F(y + \tau h)\big) - \rho\big(F(y)\big) \le \big\langle \zeta, F(y + \tau h) - F(y)\big\rangle,$$

for any $\zeta \in \partial\varrho\big(F(y + \tau h)\big)$. The generalized directional derivative of $\varphi(\cdot)$ at x in the direction h can thus be bounded from above as follows:

$$\begin{aligned} \varphi^0(x; h) &\triangleq \limsup_{\substack{y \to x \\ \tau \downarrow 0}} \frac{\varrho\big(F(y + \tau h)\big) - \varrho\big(F(y)\big)}{\tau} \\ &\le \limsup_{\substack{y \to x \\ \tau \downarrow 0}} \Big\langle \zeta(y, \tau), \frac{F(y + \tau h) - F(y)}{\tau} \Big\rangle, \end{aligned} \tag{3.38}$$

for any selection $\zeta(y, \tau) \in \partial\varrho(F(y + \tau h))$.

Consider the sequences $\{y_k\} \to x$, $\{\tau_k\} \downarrow 0$, as $k \to \infty$, for which the "lim sup" in the last displayed relation is realized, with some $\zeta_k = \zeta(y_k, \tau_k)$, $k = 1, 2, \ldots$:

$$\begin{aligned}\varphi^0(x;h) &\le \lim_{k\to\infty} \int \zeta_k(\omega) \frac{f(y_k(\omega)+\tau_k h(\omega),\omega) - f(y_k(\omega),\omega)}{\tau_k}\, P(\mathrm{d}\omega)\\ &= \lim_{k\to\infty} \int \zeta_k(\omega) f'\big(x(\omega),\omega;h(\omega)\big)\, P(\mathrm{d}\omega) + \Delta_k,\end{aligned}$$

where

$$\begin{aligned}\Delta_k = \int \zeta_k(\omega) \Big[&\frac{f(y_k(\omega)+\tau_k h(\omega),\omega) - f(y_k(\omega),\omega)}{\tau_k} \\ &\qquad - f'\big(x(\omega),\omega;h(\omega)\big)\Big]\, P(\mathrm{d}\omega).\end{aligned}$$

Since $\{y_k\} \to x$ in $\mathcal{L}_{p'}(\Omega, \mathcal{F}, P)$, then an infinite subsequence $\{y_k\}_{k\in\mathcal{K}}$ exists, which is convergent to x almost surely. Hence, for almost all $\omega \in \Omega$, due to Assumption 3.22,

$$\lim_{\substack{k\to\infty\\ k\in\mathcal{K}}} \frac{f(y_k(\omega)+\tau_k h(\omega),\omega) - f(y_k(\omega),\omega)}{\tau_k} = f'\big(x(\omega),\omega;h(\omega)\big).$$

Since ζ_k are uniformly bounded in $\mathcal{L}_q(\Omega, \mathcal{F}, P)$, and the differential quotients in the definition of Δ_k are bounded by $C(\omega)\|h(\omega)\|$, which is an element of $\mathcal{L}_p(\Omega, \mathcal{F}, P)$, we can use the Lebesgue dominated convergence theorem to conclude that $\Delta_k \to 0$ when $k \in \mathcal{K}$.

By the Banach–Alaoglu theorem, we can choose a sub-subsequence $\{\zeta_k\}_{k\in\mathcal{K}_1}$ which is weakly* convergent to some ζ in $\mathcal{Z}^*$. By the norm-to-weak* upper semicontinuity of the subdifferential, $\zeta \in \partial\varrho(F(x))$. Therefore,

$$\varphi^0(x;h) \le \max_{\zeta\in\partial\varrho(F(x))} \int \zeta(\omega) f'\big(x(\omega),\omega;h(\omega)\big)\, P(\mathrm{d}\omega).$$

The converse inequality follows in the same way as Lemma 3.20, and thus the equality is true in the last displayed relation. This proves that $\varphi(\cdot)$ is subdifferentially regular and establishes (3.37). The convexity of the set (3.37) follows from the nonnegativity of each $\zeta \in \partial\varrho(F(x))$ and the convexity of the sets $\partial f(x,\omega)$, $\omega \in \Omega$, in the same way as in the proof of Theorem 3.7. □

3.1.6 Composition with a Vector-Valued Integrand

Finally, in view of later application to causal operators in dynamical systems, we consider the case when $F : \mathcal{X} \to \mathcal{L}_p(\Omega, \mathcal{F}, P; \mathcal{Y})$, where $\mathcal{Y}$ is a separable Banach space, and when the functions $f(\cdot,\omega)$ in (3.32) are continuously

Fréchet differentiable. Even under these rather strong conditions, the analysis of the composite functional $\varphi(\cdot) = \varrho(F(\cdot))$ is difficult.

We first establish the Gâteaux differentiability of $F(\cdot)$.

Lemma 3.25. *An operator $F(\cdot)$ satisfying Assumption 3.22 with continuously Fréchet differentiable functions $f(\cdot, \omega)$, $\omega \in \Omega$, is Gâteaux differentiable with the derivative $F'(x)$ defined by*

$$[F'(x)\,h](\omega) = f'(x(\omega), \omega)\,h(\omega), \quad \omega \in \Omega. \tag{3.39}$$

Proof. We define $J(x) : \mathcal{L}_{p'}(\Omega, \mathcal{F}, P; U) \to \mathcal{L}_p(\Omega, \mathcal{F}, P)$ by using the right hand side of formula (3.39):

$$[J(x)\,h](\omega) = f'(x(\omega), \omega)\,h(\omega), \quad \omega \in \Omega.$$

It is a continuous linear operator.

Let us calculate the directional derivative of the function F at x in the direction h. First, we observe that for any $h \in \mathcal{L}_{p'}(\Omega, \mathcal{F}, P; U)$, all $\tau > 0$, and almost all $\omega \in \Omega$,

$$\frac{1}{\tau}\| f(x(\omega)+\tau h(\omega), \omega) - f(x(\omega), \omega) - \tau f'(x(\omega), \omega)\,h(\omega)\| \le 2C(\omega)\|h(\omega)\|,$$

and the function at the right-hand side is p-integrable. Using Lebesgue's dominated convergence theorem, we obtain

$$\begin{aligned}\lim_{\tau\downarrow 0}\frac{1}{\tau}\big\| F(x+\tau h) - F(x) - \tau J(x)\,h\big\| = \\ \bigg(\int \lim_{\tau\to 0}\Big\|\frac{1}{\tau}\big(f(x(\omega)+\tau h(\omega), \omega) - f(x(\omega), \omega) \\ - \tau f'(x(\omega), \omega)\,h(\omega)\big)\Big\|^p \,P(\mathrm{d}\omega)\bigg)^{\frac{1}{p}} = 0.\end{aligned}$$

Therefore, $J(x)$ is the Gâteaux derivative of $F(\cdot)$ at x. □

Unfortunately, in a nonlinear stochastic setting, the Fréchet differentiability of $F(\cdot)$ is very difficult to guarantee, except for very special cases. As an illustration, we provide the following example.

Example 3.26. Let $\Omega = [0, 1]$ and P be the Lebesgue measure. We define the operator $F : \mathcal{L}_1(\Omega, \mathcal{F}, P) \to \mathcal{L}_1(\Omega, \mathcal{F}, P)$ by

$$\big[F(x)\big](\omega) = f(x(\omega), \omega) = \begin{cases}\big(x(\omega)\big)^2 & \text{if } -1 \le x(\omega) \le 1,\\ 2\big|x(\omega)\big| - 1 & \text{otherwise.}\end{cases}$$

Since $\|F(x)\| \leq 2\|x\|$, then $F : \mathcal{L}_1(\Omega, \mathcal{F}, P) \to \mathcal{L}_1(\Omega, \mathcal{F}, P)$. In fact, $F(\cdot)$ is even convex, but it is not Fréchet differentiable at 0. According to Lemma 3.25, the Gâteaux derivative of F at 0 is 0, and thus the Fréchet derivative, if it existed, would be $F'(0) = 0$ as well. Consider the sequence of functions $\{x_n\}_{n\in\mathbb{N}}$ given by

$$x_n(\omega) = \begin{cases} 1 & \text{if } 0 \leq \omega \leq \frac{1}{n}, \\ 0 & \text{otherwise.} \end{cases}$$

We have $\|x_n\| = \frac{1}{n}$ and thus $x_n \to 0$. By construction, $F(x_n) = x_n$, $F(0) = 0$, and then, by the definition of the Fréchet derivative, we would have

$$0 = \lim_{n\to\infty} \frac{F(x_n) - F(0) - F'(0)x_n}{\|x_n\|} = \lim_{n\to\infty} \frac{x_n}{\|x_n\|}.$$

This is a contradiction because all elements on the right-hand side have norm 1. □

Still, we can calculate the Clarke subdifferential of $\varphi(\cdot)$. The following theorem is very similar to Theorem 3.24. Observe that we do not assume $\varrho(\cdot)$ to be nondecreasing.

Theorem 3.27. *Suppose $F : \mathcal{L}_{p'}(\Omega, \mathcal{F}, P; U) \to \mathcal{L}_p(\Omega, \mathcal{F}, P; \mathcal{Y})$ with $1 \leq p \leq p' < \infty$, is an operator given by (3.32) satisfying Assumption 3.22, the functions $f(\cdot, \omega)$ are continuously Fréchet differentiable, and the functional $\varrho : \mathcal{L}_p(\Omega, \mathcal{F}, P; \mathcal{Y}) \to \mathbb{R}$ is convex and continuous. Then the composite functional $\varphi = \varrho \circ F$ is subdifferentially regular, and its Clarke subdifferential has the following form:*

$$\partial\varphi(x) = \bigcup_{\zeta\in\partial\varrho(F(x))} \{g \in \mathcal{L}_q(\Omega, \mathcal{F}, P; U^*) : g(\omega) = \zeta(\omega)\, \nabla f(x(\omega), \omega),\ \text{a.s.}\},$$

where $1/q + 1/p = 1$.

Proof. The estimate (3.38) is still valid. The only difference is that F and ζ are vector-valued. To overcome this hurdle, we use Lemma 3.25. For any fixed $\zeta \in \mathcal{L}_q(\Omega, \mathcal{F}, P; \mathcal{Y}^*)$, by virtue of the mean value theorem,

$$\langle \zeta, F(y + \tau h) - F(y)\rangle = \tau\langle \zeta, F'(\bar{y})h\rangle,$$

where $\bar{y} = y + \theta\tau h$, with some $\theta \in [0, 1]$ (depending on ζ, y, and τ, in general). Thus

$$\frac{1}{\tau}\langle \zeta(y,\tau), \big(F(y + \tau h) - F(y)\big)\rangle = \langle \zeta, F'(x)h\rangle + \Delta(y, \tau),$$

with

$$\Delta(y,\tau) = \big\langle \zeta(y,\tau), (F'(\bar{y}(y,\tau)) - F'(x))h \big\rangle,$$

and $\bar{y}(y,\tau)$ in the segment from y to $y + \tau h$.

Consider the sequences $\{y_k\} \to x$, $\{\tau_k\} \downarrow 0$, as $k \to \infty$, for which the "lim sup" in (3.38) is realized, with some $\zeta_k = \zeta(y_k, \tau_k)$, $k = 1, 2, \ldots$:

$$\varphi^0(x;h) \le \lim \Big\langle \zeta_k, \frac{1}{\tau_k}\big(F(y_k + \tau_k h) - F(y_k)\big)\Big\rangle \le \lim \big\langle \zeta_k, F'(x)h \big\rangle + \Delta_k,$$

where

$$\Delta_k = \big\langle \zeta_k, (F'(\bar{y}(y_k,\tau_k)) - F'(x))\, h \big\rangle,$$

and $\bar{y}(y_k, \tau_k)$ in the segment from y_k to $y_k + \tau_k h$. As in the proof of Theorem 3.24, we establish that $\Delta_k \to 0$ over a subsequence, and we follow the remaining steps of that proof *verbatim*. □

3.1.7 Composition with a Semismooth Function

A practically relevant case is when the mapping $F : \mathcal{X} \to \mathcal{L}_p(\Omega, \mathcal{F}, P)$, $p \in [1, \infty)$, is such that for almost all $\omega \in \Omega$ the relation (3.22) is still satisfied, but the functions $f(\cdot, \omega)$, although Lipschitz continuous, are not subdifferentially regular.

In this case, we need to restrict the decision space $\mathcal{X}$ to the finite-dimensional space $\mathbb{R}^n$.

We start with the definition of the class of functions under consideration.

Definition 3.28. A function $f : \mathbb{R}^n \to \mathbb{R}$ is *differentiable in a generalized sense at a point* $x \in \mathbb{R}^n$, if an open set $U \subset \mathbb{R}^n$ containing x, and a nonempty, convex, compact-valued, and upper semicontinuous multifunction $\hat{\partial} f : U \rightrightarrows \mathbb{R}^n$ exist, such that for all $y \in U$ and all $g \in \hat{\partial} f(y)$ the following equation is true:

$$f(y) = f(x) + \langle g, y - x \rangle + o(x, y, g), \tag{3.40}$$

with

$$\lim_{y \to x} \sup_{g \in \hat{\partial} f(y)} \frac{\big|o(x,y,g)\big|}{\|y - x\|} = 0.$$

The set $\hat{\partial} f(y)$ is the *generalized subdifferential* of f at y. If a function is differentiable in a generalized sense at every $x \in \mathbb{R}^n$ with the same subdifferential mapping $\hat{\partial} f : \mathbb{R}^n \rightrightarrows \mathbb{R}^n$, we call it *differentiable in a generalized sense*.

A function $f : \mathbb{R}^n \to \mathbb{R}^m$ is differentiable in a generalized sense, if each of its component functions, $f_i : \mathbb{R}^n \to \mathbb{R}$, $i = 1, \ldots, m$, has this property.

We also call such functions *Norkin differentiable*, and their generalized subdifferentials - *Norkin subdifferentials*.

The generalized subdifferential is not unique; however, the Clarke subdifferential $\partial f(x)$ is the inclusion-minimal realization of $\hat{\partial} f(x)$. We may remark here that if we use $\hat{\partial} f(x) = \partial f(x)$ in (3.40), and require directional differentiability of $f(\cdot)$, then we obtain the property of *semismoothness* of $f(\cdot)$.

The class of generalized differentiable functions is included in the class of locally Lipschitz continuous functions, but is much broader than the class of subdifferentially regular functions; in particular, it contains all compositions of subdifferentially regular or semismooth functions. The following theorem is the key for the calculus of the generalized subdifferentials.

Theorem 3.29. *If $h : \mathbb{R}^m \to \mathbb{R}$ and $f_i : \mathbb{R}^n \to \mathbb{R}$, $i = 1, \dots, m$, are differentiable in a generalized sense, then the composition $\Psi(x) = h\big(f_1(x), \dots, f_m(x)\big)$ is differentiable in a generalized sense, and at any point $x \in \mathbb{R}^n$ we can define the generalized subdifferential of Ψ as follows:*

$$\begin{aligned} \hat{\partial}\Psi(x) = \operatorname{conv} \big\{ &g \in \mathbb{R}^n : g = \big[g_1 \cdots g_m \big] g_0, \\ &\textit{with } g_0 \in \hat{\partial} h\big(f_1(x), \dots, f_m(x)\big) \textit{ and } g_j \in \hat{\partial} f_j(x),\ j = 1, \dots, m \big\}. \end{aligned} \tag{3.41}$$

Furthermore, if $h(\cdot)$ is nondecreasing, then the convex hull operator in (3.41) *may be omitted.*

Proof. As the multifunctions $\hat{\partial} h(\cdot)$ and $\hat{\partial} f_j(\cdot)$, $j = 1, \dots, m$, are compact-valued, the multifunction $\hat{\partial}\Psi(\cdot)$ is compact-valued as well. Its values are convex by construction. It remains to verify its upper-semicontinuity and the relation (3.40). For brevity, denote $F(x) = (f_1(x), \dots, f_m(x))$. Consider a sequence $\{y^k\}_{k \in \mathbb{N}}$ converging to a point x, and let $g^k \in \hat{\partial}\Psi(y^k)$. By formula (3.41), for each k, coefficients $\alpha_{k,\ell}$ and subgradients $g_j^{k,\ell}$ exist, where $\ell = 1, \dots, n+1$ and $j = 0, \dots, m$, such that

$$g^k = \sum_{\ell=1}^{n+1} \alpha_{k,\ell} g^{k,\ell}, \quad g^{k,\ell} = \Big[g_1^{k,\ell} \cdots g_m^{k,\ell} \Big] g_0^{k,\ell}, \tag{3.42a}$$

$$g_0^{k,\ell} \in \hat{\partial} h\big(F(y^k)\big), \quad g_j^{k,\ell} \in \hat{\partial} f_j(y^k),\ j = 1, \dots, m, \tag{3.42b}$$

$$\sum_{\ell=1}^{n+1} \alpha_{k,\ell} = 1, \quad \alpha_{k,\ell} \geq 0,\ \ell = 1, \dots, n+1. \tag{3.42c}$$

Suppose the sequence $\{g^k\}_{k \in \mathbb{N}}$ is convergent to a limit $\bar{g}$. As all subdifferentials $\hat{\partial} h(\cdot)$ and $\hat{\partial} f_j(\cdot)$, $j = 1, \dots, m$, are upper semicontinuous, and bounded in a neighborhood of x, we can choose an infinite set of indices $\mathcal{K} \subset \mathbb{N}$, such that the subsequences $\{g_0^{k,\ell}\}_{k \in \mathcal{K}}$ and $\{g_j^{k,\ell}\}_{k \in \mathcal{K}}$, $j = 1, \dots, m$, are convergent with limits $\bar{g}_0^\ell \in \hat{\partial} h\big(F(x)\big)$, $\bar{g}_j^\ell \in \hat{\partial} f_j(x)$, $j = 1, \dots, m$, respectively, for each

$\ell = 1, \dots, n+1$. Passing to a sub-subsequence, if necessary, we can also assume that the sequences $\{\alpha_{k,\ell}\}_{k\in\mathcal{K}}$, for each $\ell = 1, \dots, n+1$, are convergent as well. Let $\bar{\alpha}_\ell$, $\ell = 1, \dots, n+1$, be their corresponding limits. Passing to the limit over $k \in \mathcal{K}$ in (3.42), we conclude that $\bar{g} \in \hat{\partial}\Psi(x)$. Consequently, $\hat{\partial}\Psi(\cdot)$ is upper semicontinuous.

It follows from (3.40) that

$$f_j(y^k) = f_j(x) + \langle g_j^{k,\ell}, y^k - x\rangle + o_j(x, y^k, g_j^{k,\ell}),\ j = 1, \dots, m,$$

and

$$\begin{aligned} h(F(y^k)) &= h(F(x)) + \langle g_0^{k,\ell}, F(y^k) - F(x)\rangle + o_0(x, y^k, g_0^{k,\ell}) \\ &= h(F(x)) + \langle g^{k,\ell}, y^k - x\rangle \\ &\quad + \sum_{j=1}^{m} \big(g_0^{k,\ell}\big)_j o_j(x, y^k, g_j^{k,\ell}) + o_0(x, y^k, g_0^{k,\ell}). \end{aligned}$$

Since the subgradients $g_0^{k,\ell}$ are uniformly bounded, and the function $F(\cdot)$ is Lipschitz continuous,

$$\lim_{k\to\infty} \frac{h(F(y^k)) - h(F(x)) - \langle g^{k,\ell}, y^k - x\rangle}{\|y^k - x\|} = 0.$$

Taking the convex combinations, as in (3.42), we also obtain

$$\lim_{k\to\infty} \frac{h(F(y^k)) - h(F(x)) - \langle g^{k}, y^k - x\rangle}{\|y^k - x\|} = 0,$$

which proves the relation (3.40) for the composite function.

Finally, if $h(\cdot)$ is nondecreasing, then the convexity of the set

$$\hat{\partial}\Psi(x) = \big\{g \in \mathbb{R}^n : g = \big[g_1 \cdots g_m\big]\, g_0, \\ \text{with } g_0 \in \hat{\partial}h\big(F(x)\big) \text{ and } g_j \in \hat{\partial}f_j(x),\ j = 1, \dots, m\big\}$$

follows from the nonnegativity of each $g_0 \in \partial h(F(x))$ and the convexity of each $\hat{\partial}f_j(x)$, in the same way as in the proof of Theorem 3.7. □

Even if we take $\hat{\partial}h(\cdot) = \partial h(\cdot)$ and $\hat{\partial}f_j(\cdot) = \partial f_j(\cdot)$, $j = 1, \dots, m$, we may obtain $\hat{\partial}\Psi(\cdot) \neq \partial\Psi(\cdot)$, but $\hat{\partial}\Psi$ defined above satisfies Definition 3.28. The main practical conclusion from the above theorem is that we are allowed to apply standard calculus rules, and the resulting "derivatives" will be legitimate elements of a Norkin subdifferential.

We now pass to the functions $F : \mathbb{R}^n \times \Omega \to \mathbb{R}$ satisfying Assumption 3.18. We make the following additional assumption.

Assumption 3.30. *The functions $f(\cdot,\omega)$, $\omega \in \Omega$, are subdifferentiable in the Norkin sense with subdifferentials $\hat{\partial} f(x,\omega)$, and satisfy Assumption* 3.18.

We can now derive the formula for the Norkin subdifferential of the composite risk function $\varphi(\cdot)$.

Theorem 3.31. *Suppose $F : \mathbb{R}^n \to \mathcal{L}_p(\Omega, \mathcal{F}, P)$, with $p \in [1,\infty)$, is defined by* (3.22) *and satisfies Assumptions* 3.18 *and* 3.30. *Furthermore, assume that it is Hadamard directionally differentiable at every point $x \in \mathbb{R}^n$ and in every direction $d \in \mathbb{R}^n$. Then the composite function $\varphi = \varrho \circ F$ is Norkin subdifferentiable at every point $x \in \mathcal{X}$, with the Norkin subdifferential given by the following formula:*

$$\hat{\partial}\varphi(x) = \bigcup_{\zeta \in \partial\varrho(F(x))} \int_\Omega \zeta(\omega)\, \hat{\partial} f(x,\omega)\; P(\mathrm{d}\omega). \tag{3.43}$$

Proof. We verify Definition 3.28. Consider a sequence $\{y^k\}$ converging to x, and take any $G^k \in \hat{\partial}\varphi(y^k)$, as defined in (3.43). It follows that

$$G^k = \int_\Omega \zeta^k(\omega)\, g^k(\omega)\; P(\mathrm{d}\omega), \quad \zeta^k \in \partial\varrho(F(y^k)), \quad g^k(\omega) = \hat{\partial} f(y^k,\omega).$$

By Definition 3.28, for each function $f(\cdot,\omega)$,

$$f(x,\omega) - f(y^k,\omega) = \langle g^k(\omega), x - y^k\rangle + o(x, y^k, g^k(\omega); \omega), \tag{3.44}$$

with

$$\lim_{k\to\infty} \frac{o(x, y^k, g^k(\omega); \omega)}{\|y^k - x\|} = 0.$$

Moreover, due to Assumptions 3.18 and 3.30,

$$\frac{o(x, y^k, g^k(\omega); \omega)}{\|y^k - x\|} \le 2C(\omega).$$

By the Lebesgue theorem,

$$\lim_{k\to\infty} \frac{1}{\|y^k - x\|} \left(\int_\Omega \big(o(x, y^k, g^k(\omega); \omega)\big)^p \; P(\mathrm{d}\omega)\right)^{1/p} = 0. \tag{3.45}$$

Multiplying (3.44) by $\zeta^k(\omega)$ and integrating, we obtain

$$\begin{aligned}\varphi(x) - \varphi(y^k) &\ge \int_\Omega \big[f(x,\omega) - f(y^k,\omega)\big]\, \zeta^k(\omega)\; P(\mathrm{d}\omega)\\ &= \langle G^k, x - y^k\rangle + \int_\Omega o(x, y^k, g^k(\omega); \omega)\, \zeta^k(\omega)\; P(\mathrm{d}\omega).\end{aligned}$$

Thus,

$$\liminf_{k\to\infty} \frac{\varphi(x) - \varphi(y^k) - \langle G^k, x - y^k\rangle}{\|y^k - x\|} \geq 0.$$

Suppose an infinite set $\mathcal{K} \subset \mathbb{N}$ exists, such that

$$\limsup_{\substack{k\to\infty\\ k\in\mathcal{K}}} \frac{\varphi(x) - \varphi(y^k) - \langle G^k, x - y^k\rangle}{\|y^k - x\|} > 0. \tag{3.46}$$

Without loss of generality, by passing to a subsequence, if needed, we may assume that $\lim_{k\to\infty, k\in\mathcal{K}} \frac{y^k - x}{\|y^k - x\|} = d$. For these k, due to the Hadamard directional differentiability of $F(\cdot)$, we have

$$\lim_{\substack{k\to\infty\\ k\in\mathcal{K}}} \frac{F(y^k) - F(x)}{\|y^k - x\|} = F'(x; d).$$

Since convex continuous functionals are semismooth, for some $\lambda \in \partial\varrho(F(x))$,

$$\lim_{\substack{k\to\infty\\ k\in\mathcal{K}}} \langle \zeta^k, F'(x; d)\rangle = \varrho'\big(F(x); F'(x; d)\big) = \langle \lambda, F'(x, d)\rangle.$$

Then, by the subgradient inequality and the above relations,

$$\begin{aligned}\lim_{\substack{k\to\infty\\ k\in\mathcal{K}}} \frac{\varphi(y^k) - \varphi(x)}{\|y^k - x\|} &\geq \lim_{k\to\infty} \frac{\langle \lambda, F(y^k) - F(x)\rangle}{\|y^k - x\|} \\ &= \langle \lambda, F'(x, d)\rangle = \lim_{k\to\infty} \langle \zeta^k, F'(x; d)\rangle.\end{aligned} \tag{3.47}$$

Using the definition of the generalized differentiability and (3.45), we can continue this chain of relations as follows:

$$\begin{aligned}\lim_{\substack{k\to\infty\\ k\in\mathcal{K}}} \frac{\varphi(y^k) - \varphi(x)}{\|y^k - x\|} &\geq \lim_{\substack{k\to\infty\\ k\in\mathcal{K}}} \int_\Omega f'(x; d, \omega)\, \zeta^k(\omega)\, P(\mathrm{d}\omega) \\ = \lim_{\substack{k\to\infty\\ k\in\mathcal{K}}} \frac{1}{\|y^k - x\|} \int_\Omega \langle g^k(\omega), y^k - x\rangle\, \zeta^k(\omega) P(\mathrm{d}\omega) &= \lim_{\substack{k\to\infty\\ k\in\mathcal{K}}} \frac{\langle G^k, y^k - x\rangle}{\|y^k - x\|}.\end{aligned}$$

This is a contradiction with (3.46). We conclude that

$$\lim_{k\to\infty} \frac{\varphi(x) - \varphi(y^k) - \langle G^k, x - y^k\rangle}{\|y^k - x\|} = 0.$$

This means that $\varphi(\cdot)$ with the multifunction $\hat{\partial}\varphi(\cdot)$ satisfy the expansion (3.40). The convexity of the set (3.43) follows from the nonnegativity of each $\zeta \in \partial\varrho(F(x))$ and the convexity of $\hat{\partial} f(x, \omega)$, in the same way as in the proof of Theorem 3.7. The upper semicontinuity of (3.43) follows from the upper semicontinuity of the multifunctions involved there. □

3.1.8 Conservativeness of the Subdifferential Field

Intricate properties of the subdifferential of the composite functional $\varphi(\cdot)$ are essential to establish the conservativeness of the field $\partial\varphi(\cdot)$, as defined below. For an absolutely continuous function $X : [0, \infty) \to \mathcal{X}$, which we call a *path*, we denote by $\dot{X}(\cdot)$ its weak derivative: a measurable function such that for every $T > 0$ we have

$$X(T) = X(0) + \int_0^T \dot{X}(t)\, \mathrm{d}t. \tag{3.48}$$

It is unique up to a set of Lebesgue measure 0. We call the field $\partial\varphi(\cdot)$ *conservative* if for every path $X : [0, \infty) \to \mathcal{X}$, for any selector $g(\cdot) \in \partial\varphi(\cdot)$ and for all $T > 0$ we have

$$\varphi\big(X(T)\big) - \varphi\big(X(0)\big) = \int_0^T \big\langle g(t), \dot{X}(t)\big\rangle\, \mathrm{d}t. \tag{3.49}$$

It turns out that under our general assumptions the field $\partial\varphi(\cdot)$ is indeed conservative.

Theorem 3.32. *Suppose $\mathcal{X}$ is a Banach space. If the functional $\varphi : \mathcal{X} \to \mathbb{R}$ is subdifferentially regular at every $x \in \mathcal{X}$, then its subdifferential field $\partial\varphi(\cdot)$ is conservative.*

Proof. Consider an arbitrary path $X : [0, \infty) \to \mathcal{X}$ and any $T > 0$. Define the function

$$S(\varepsilon) = \int_0^T \varphi\big(X(t+\varepsilon)\big)\, \mathrm{d}t, \quad \varepsilon \ge 0.$$

We can calculate its right derivative at $\varepsilon = 0$ in two ways.

First,

$$
\begin{aligned}
S'_+(0) &= \lim_{\varepsilon\downarrow 0}\frac{1}{\varepsilon}\big[S(\varepsilon)-S(0)\big] = \lim_{\varepsilon\downarrow 0}\frac{1}{\varepsilon}\Big[\int_0^T \varphi\big(X(t+\varepsilon)\big)\,\mathrm{d}t - \int_0^T \varphi\big(X(t)\big)\,\mathrm{d}t\Big] \\
&= \lim_{\varepsilon\downarrow 0}\frac{1}{\varepsilon}\Big[\int_\varepsilon^{T+\varepsilon} \varphi\big(X(\tau)\big)\,\mathrm{d}\tau - \int_0^T \varphi\big(X(t)\big)\,\mathrm{d}t\Big] \\
&= \lim_{\varepsilon\downarrow 0}\frac{1}{\varepsilon}\Big[\int_T^{T+\varepsilon} \varphi\big(X(t)\big)\,\mathrm{d}t - \int_0^\varepsilon \varphi\big(X(t)\big)\,\mathrm{d}t\Big] = \varphi\big(X(T)\big)-\varphi\big(X(0)\big).
\end{aligned} \tag{3.50}
$$

On the other hand,

$$
S'_+(0) = \lim_{\varepsilon\downarrow 0}\int_0^T \frac{1}{\varepsilon}\big[\varphi\big(X(t+\varepsilon)\big)-\varphi\big(X(t)\big)\big]\,\mathrm{d}t. \tag{3.51}
$$

By (3.48), the differential quotient under the integral can be expanded as follows:

$$
\frac{1}{\varepsilon}\big[\varphi\big(X(t+\varepsilon)\big)-\varphi\big(X(t)\big)\big] = \frac{1}{\varepsilon}\Big[\varphi\Big(X(t)+\int_t^{t+\varepsilon}\dot{X}(\tau)\,\mathrm{d}\tau\Big)-\varphi\big(X(t)\big)\Big].
$$

Define

$$
h(t,\varepsilon) = \frac{1}{\varepsilon}\int_t^{t+\varepsilon}\dot{X}(\tau)\,\mathrm{d}\tau.
$$

For almost all t,

$$
\lim_{\varepsilon\downarrow 0} h(t,\varepsilon) = \dot{X}(t).
$$

Due to the subdifferential regularity, $\varphi(\cdot)$ is Hadamard directionally differentiable, which entails

$$
\lim_{\varepsilon\downarrow 0}\frac{1}{\varepsilon}\big[\varphi\big(X(t)+\varepsilon h(t,\varepsilon)\big)-\varphi\big(X(t)\big)\big] = \varphi'\big(X(t);\dot{X}(t)\big).
$$

It follows that the limit for $\varepsilon\downarrow 0$ of the quotients under the integral on the right-hand side of (3.51) exists for almost all t. They are uniformly bounded, due to the local Lipschitz continuity of $\varphi(\cdot)$. By the Lebesgue theorem,

$$
\begin{aligned}
S'_+(0) &= \int_0^T \frac{1}{\varepsilon}\lim_{\varepsilon\downarrow 0}\big[\varphi\big(X(t+\varepsilon)\big)-\varphi\big(X(t)\big)\big]\,\mathrm{d}t \\
&= \int_0^T \varphi'\big(X(t);\dot{X}(t)\big)\,\mathrm{d}t = \int_0^T \max_{g(t)\in\partial\varphi(X(t))}\big\langle g(t),\dot{X}(t)\big\rangle\,\mathrm{d}t.
\end{aligned}
$$

The comparison with (3.50) yields

$$
\varphi\big(X(T)\big)-\varphi\big(X(0)\big) = \int_0^T \max_{g(t)\in\partial\varphi(X(t))}\big\langle g(t),\dot{X}(t)\big\rangle\,\mathrm{d}t.
$$

By reversing time and considering the path $Y(t) = X(T-t)$, after an absolutely continuous extension of $X(\cdot)$ to small negative arguments, in the same way, we obtain the equation

$$\varphi\big(X(0)\big) - \varphi\big(X(T)\big) = \int_0^T \varphi'\big(X(t); -\dot{X}(t)\big)\, \mathrm{d}t = \\ -\int_0^T \min_{g(t)\in\partial\varphi(X(t))} \big\langle g(t), \dot{X}(t)\big\rangle\, \mathrm{d}t.$$

The last two equations imply that (3.49) is true for any selection $g(\cdot) \in \partial\varphi(\cdot)$. □

In the finite-dimensional case, we have a stronger result for Norkin differentiable functions without subdifferential regularity.

Theorem 3.33. *If the function* $\varphi : \mathbb{R}^n \to \mathbb{R}$ *is Norkin differentiable, then every its generalized subdifferential field* $\hat{\partial}\varphi(\cdot)$ *is conservative.*

Proof. Consider an absolutely continuous path $X : [0, \infty) \to \mathbb{R}^n$ and any time $T > 0$. The composition $\varphi \circ X$ is also absolutely continuous, hence differentiable almost everywhere. Therefore, a full-measure subset $D \subset [0, T]$ exists such that both X and $\varphi \circ X$ are differentiable at every point of D.

Suppose the chain rule is not valid along X, that is, a positive measure set $E \subset D$ exists such that for all $t \in E$ there is $g \in \hat{\partial}\varphi(X(t))$ at which

$$\frac{\mathrm{d}}{\mathrm{d}t}(\varphi \circ X)(t) \neq \langle g, \dot{X}(t)\rangle.$$

This implies that $\dot{X}(t) \neq 0$ for all $t \in E$, since if $\dot{X}(t) = 0$ then $\frac{\mathrm{d}}{\mathrm{d}t}(\varphi \circ X)(t) = 0$ as well.

Reducing E, we may assume without loss of generality that for all $t \in E$ we can find $g(t) \in \hat{\partial}\varphi(X(t))$ such that

$$\frac{\mathrm{d}}{\mathrm{d}t}(\varphi \circ X)(t) < \langle g(t), \dot{X}(t)\rangle. \tag{3.52}$$

If the inequality is reversed on a positive measure set, the proof is identical.

Consider a particular function $g : [0, T] \to \mathbb{R}^n$ satisfying

$$g(t) = \operatorname*{arg\,max}_{v\in\hat{\partial}\varphi(X(t))} \langle v, \dot{X}(t)\rangle \quad \text{if} \quad t \in E,$$

and $g(t) = 0$, otherwise. It exists, because $\hat{\partial}\varphi(X(t))$ is compact, and may be assumed to be measurable, by the measurable selection theorem. By virtue of Lusin's theorem, a closed subset $E_1 \subset E$ of positive measure exists, such that

$g(\cdot)$ is continuous on E_1. Furthermore, due to the Lebesgue density theorem, a point $t_* \in E_1$ exists such that

$$\lim_{\delta \downarrow 0} \frac{1}{2\delta} \operatorname{meas}\big(E_1 \cap [t_* - \delta, t_* + \delta]\big) = 1.$$

This implies that we can find a sequence of points $\{t_k\}_{k \in N} \subset E_1 \setminus \{t_*\}$ convergent to t_*. If $X(t_k) = X(t_*)$ for all but finitely many k, then we obtain $\dot{X}(t_*) = 0$, which is impossible in E. If $X(t_k) \neq X(t_*)$ for k in an infinite set $\mathcal{K} \subset N$, for all these k we may write the equation

$$\begin{aligned} &\frac{\varphi(X(t_k)) - \varphi(X(t_*))}{t_k - t_*} \\ &= \frac{\|X(t_k) - X(t_*)\|}{t_k - t_*} \left(\frac{\varphi(X(t_k)) - \varphi(X(t_*)) - \langle g(t_k), X(t_k) - X(t_*) \rangle}{\|X(t_k) - X(t_*)\|} \right) \\ &\quad + \frac{\langle g(t_k), X(t_k) - X(t_*) \rangle}{t_k - t_*}. \end{aligned} \tag{3.53}$$

As both X and $\varphi \circ X$ are differentiable at t_*, and g is continuous at t_*, we infer that when $k \to \infty$ in the set $\mathcal{K}$, then

$$\begin{aligned} \frac{\varphi(X(t_k)) - \varphi(X(t_*))}{t_k - t_*} &\to \frac{\mathrm{d}}{\mathrm{d}t}(\varphi \circ X)(t_*), \\ \frac{\|X(t_k) - X(t_*)\|}{t_k - t_*} &\to \|\dot{X}(t_*)\|, \\ \frac{\langle g(t_k), X(t_k) - X(t_*) \rangle}{t_k - t_*} &\to \langle g(t_*), \dot{X}(t_*) \rangle. \end{aligned}$$

Furthermore, due to the generalized differentiability of $\varphi(\cdot)$, we have the expansion (3.40):

$$\frac{\varphi(X(t_k)) - \varphi(X(t_*)) - \langle g(t_k), X(t_k) - X(t_*) \rangle}{\|X(t_k) - X(t_*)\|} \to 0.$$

Using these relations and passing to the limit with $k \to \infty$, $k \in \mathcal{K}$, in (3.53), we conclude that $\frac{\mathrm{d}}{\mathrm{d}t}(\varphi \circ X)(t_*) = \langle g(t_*), \dot{X}(t_*) \rangle$, which contradicts (3.52) in this case as well. □

3.2 Optimization of Composite Risk Functions

In this section, we discuss optimization problems of the form (3.1):

$$\min_{x \in \mathcal{X}} \left\{ \varphi(x) \triangleq \varrho\big(F(x)\big) \right\}, \tag{3.54}$$

where $F : \mathfrak{X} \to \mathcal{L}_p(\Omega, \mathcal{F}, P)$, $p \in [1, \infty)$, $\varrho : \mathcal{L}_p(\Omega, \mathcal{F}, P) \to \mathbb{R}$ is a measure of risk, and $\mathcal{X}$ is a subset of a separable Banach space $\mathfrak{X}$. We focus on finite-valued measures of risk, to avoid unnecessary complications.

3.2.1 Composition with a Convex Function

We start from the case of a convex mapping $F(\cdot)$ of the form (3.19) discussed in § 3.1.3. The set $\mathcal{X}$ is convex as well.

The first result follows from the optimality conditions in convex optimization.

Theorem 3.34. *Suppose $\mathfrak{X}$ is a separable Banach space, the mapping $F : \mathfrak{X} \to \mathcal{L}_p(\Omega, \mathcal{F}, P)$ is given by* (3.19)*, $f(\cdot, \omega)$ is convex and continuous for almost all $\omega \in \Omega$, $\mathbb{E}[F(\cdot)]$ is continuous as well, the set $\mathcal{X}$ is convex and closed, and the risk measure $\varrho : \mathcal{L}_p(\Omega, \mathcal{F}, P) \to \mathbb{R}$ satisfies the conditions of convexity and monotonicity. Then $\bar{x}$ is an optimal solution of problem* (3.54) *if and only if $\bar{\zeta} \in \partial\varrho\big(F(\bar{x})\big)$ and subgradients $s(\omega) \in \partial f(\bar{x}, \omega)$, $\omega \in \Omega$, exist, such that*

$$0 \in \int_{\Omega} s(\omega)\, \bar{\zeta}(\omega)\; P(\mathrm{d}\omega) + \mathcal{N}_{\mathcal{X}}(\bar{x}). \tag{3.55}$$

Proof. The result follows immediately from the necessary and sufficient condition of optimality $0 \in \partial\varphi(\bar{x}) + \mathcal{N}_{\mathcal{X}}(\bar{x})$ and the formula for the subdifferential of $\varphi(\cdot)$ derived in Theorem 3.12. □

Example 3.35. Suppose $Y \in \mathcal{L}_p(\Omega, \mathcal{F}, P)$, with $p \in [1, \infty)$, represents random losses. The losses have to be shared among N agents, whose risk is measured by coherent measures of risk $\varrho_j : \mathcal{L}_p(\Omega, \mathcal{F}, P) \to \mathbb{R}$, $j = 1, \ldots, N$. Our objective is to split the losses among the agents in a way that minimizes their cumulative risk.

We define $\mathfrak{X} = \big(\mathcal{L}_p(\Omega, \mathcal{F}, P)\big)^N$, $\mathcal{X} = \big\{X \in \mathfrak{X} : X_1 + \cdots + X_N = Y\big\}$ and consider the problem

$$\min_{X \in \mathcal{X}} \varphi(X) \triangleq \varrho_1(X_1) + \cdots + \varrho_N(X_N).$$

In this case the random variables X_j, $j = 1, \ldots, N$, are decisions themselves, and we do not need to deal with a composition. Since the space $\mathcal{L}_p(\Omega, \mathcal{F}, P)$ is a Banach lattice, each $\varrho_i(\cdot)$ is continuous (Theorem 2.32). The subdifferential of $\varphi(\cdot)$ is the Cartesian product

$$\partial\varphi(X) = \partial\varrho_1(X_1) \times \cdots \times \partial\varrho_N(X_N).$$

The set $\mathcal{X}$ is a hyperplane, its normal is the subspace

$$\mathcal{N}_{\mathcal{X}}(X) = \{(\lambda_1, \ldots, \lambda_N) \in \big(\mathcal{L}_q(\Omega, \mathcal{F}, P)\big)^N : \lambda_1 = \cdots = \lambda_N\},$$

where $1/p + 1/q = 1$. The necessary and sufficient condition of optimality $0 \in \partial\varphi(X) + \mathcal{N}_{\mathcal{X}}(X)$ is equivalent to the system of relations

$$\bigcap_{j=1}^{N} \partial\varrho_j(X_j) \neq \emptyset \quad \text{and} \quad \sum_{j=1}^{N} X_j = Y.$$

□

Let us now consider the case of the mapping $F(\cdot)$ defined by (3.11). For a general convex set $\mathcal{X} \subset \mathcal{L}_{p'}(\Omega, \mathcal{F}, P; U)$, we have the following immediate consequence of the subdifferential representation of Theorem 3.10.

Theorem 3.36. *Suppose $F : \mathcal{L}_{p'}(\Omega, \mathcal{F}, P; U) \to \mathcal{L}_p(\Omega, \mathcal{F}, P)$ is given by* (3.11), *with $1 \leq p \leq p' \leq \infty$, $p < \infty$, $f(\cdot, \omega)$ is convex and continuous at $\bar{x}(\omega)$ for almost all $\omega \in \Omega$, and $\mathbb{E}[F(\cdot)]$ is continuous at $\bar{x}$ as well. Further, let $\varrho : \mathcal{L}_p(\Omega, \mathcal{F}, P) \to \mathbb{R}$ satisfy the conditions of convexity and monotonicity. Then $\bar{x}$ is an optimal solution of problem* (3.54) *if and only if $\bar{\zeta} \in \partial\varrho\big(F(\bar{x})\big)$ and subgradients $s(\omega) \in \partial f(\bar{x}(\omega), \omega)$, $\omega \in \Omega$, exist, such that for almost all $\omega \in \Omega$ we have:*

$$0 \in \bar{\zeta} s + \mathcal{N}_{\mathcal{X}}(\bar{x}). \tag{3.56}$$

If the convex set $\mathcal{X}$ is defined by a measurable convex-valued multifunction $X : \Omega \rightrightarrows U$, so that

$$\mathcal{X} = \big\{x \in \mathcal{L}_{p'}(\Omega, \mathcal{F}, P; U) : x(\omega) \in X(\omega), \text{ a.s.}\big\}, \tag{3.57}$$

then its normal cone is decomposable as well:

$$\mathcal{N}_{\mathcal{X}}(\bar{x}) = \big\{d \in \mathcal{L}_{q'}(\Omega, \mathcal{F}, P; U) : d(\omega) \in \mathcal{N}_{X(\omega)}(\bar{x}(\omega)), \text{ a.s.}\big\},$$

with $1/p' + 1/q' = 1$. In this case, the optimality condition (3.56) decomposes by elementary events:

$$0 \in \bar{\zeta}(\omega)s(\omega) + \mathcal{N}_{X(\omega)}(\bar{x}(\omega)), \quad \omega \in \Omega.$$

If $\zeta(\omega) > 0$, we can divide both sides by it, and conclude that the optimality conditions for the scenario-based problem (associated with the elementary events) are satisfied. The measure of risk is inessential because the problem can be solved separately for each ω.

3.2.2 Composition with a Lipschitz Continuous Function

In this section, we develop optimality conditions and duality relations for problem (3.54) under the conditions of the Lipschitz continuity of the mapping F and the convexity of the set $\mathcal{X}$. The decision space $\mathfrak{X}$ is a separable Banach space.

Theorem 3.37. *Suppose* $F : \mathfrak{X} \to \mathcal{L}_p(\Omega, \mathcal{F}, P)$, *with* $p \in [1, \infty)$, *is defined by* (3.22) *and satisfies Assumption* 3.18. *Furthermore, suppose* $\varrho : \mathcal{L}_p(\Omega, \mathcal{F}, P) \to \mathbb{R}$ *is nondecreasing and convex, and let* $\mathcal{X} \subset \mathfrak{X}$ *be a convex closed set. If* $\bar{x}$ *is a local minimum of* (3.54), *then* $\zeta \in \partial\varrho\big(F(\bar{x})\big)$ *exists, such that*

$$0 \in \int_{\Omega} \zeta(\omega)\, \partial f(\bar{x}, \omega)\; P(\mathrm{d}\omega) + \mathcal{N}_{\mathcal{X}}(\bar{x}). \tag{3.58}$$

Proof. The statement is an immediate consequence of Theorem 3.21 and standard optimality conditions for problems with Lipschitz continuous objective functions. □

If $\varrho(\cdot)$ is coherent, each subgradient ζ is a probability density function.

Again, if the feasible set $\mathcal{X}$ is decomposable and given by (3.57), then the optimality condition decomposes:

$$0 \in \bar{\zeta}(\omega)\, \partial f(\bar{x}(\omega), \omega) + \mathcal{N}_{X(\omega)}(\bar{x}(\omega)), \quad \omega \in \Omega.$$

As $\bar{\zeta} \geq 0$, the optimality conditions for problems associated with each elementary event ω are satisfied as well: the entire problem decomposes by ω and the presence of the measure of risk is inessential.

3.3 Duality Theory

Many approaches to duality theory in stochastic optimization are possible, depending on which constraints are selected to incorporate into the Lagrange function. We focus here on two approaches, which are specific for measures of risk and relevant to solution methods.

3.3.1 Duality Associated with Risk Measures

Owing to Theorem 2.25, we can use the dual representation (2.27) to write problem (3.54) in the following *primal* form:

$$\min_{x \in \mathcal{X}} \sup_{\zeta \in \mathcal{A}} \Phi(x, \zeta), \tag{3.59}$$

where $\mathcal{A} = \text{dom}(\varrho^*)$ and the function $\Phi : \mathcal{X} \times \mathcal{L}_q(\Omega, \mathcal{F}, P) \to \overline{\mathbb{R}}$ is defined by

$$\Phi(x, \zeta) \triangleq \int_{\Omega} f(x, \omega)\, \zeta(\omega)\ P(d\omega) - \varrho^*(\zeta). \tag{3.60}$$

Together with problem (3.59) we consider its *dual problem*, resulting from the interchange of the minimization and maximization operations:

$$\max_{\zeta \in \mathcal{A}} \inf_{x \in \mathcal{X}} \Phi(x, \zeta). \tag{3.61}$$

Since for every $\zeta \in \mathcal{A}$,

$$\inf_{x \in \mathcal{X}} \Phi(x, \zeta) \leq \inf_{x \in \mathcal{X}} \sup_{\zeta \in \mathcal{A}} \Phi(x, \zeta),$$

the optimal value of the dual problem (3.61) is a lower bound for the optimal value of the primal problem (3.59):

$$\sup_{\zeta \in \mathcal{A}} \inf_{x \in \mathcal{X}} \Phi(x, \zeta) \leq \inf_{x \in \mathcal{X}} \sup_{\zeta \in \mathcal{A}} \Phi(x, \zeta).$$

The fundamental questions of the theory of duality are whether both problems have solutions and whether a gap exists between their optimal values. These questions are closely related to the existence of a saddle point of the function $\Phi(\cdot, \cdot)$.

Definition 3.38. A point $(\bar{x}, \bar{\zeta}) \in \mathcal{X} \times \mathcal{A}$ is a *saddle point* of $\Phi(\cdot, \cdot)$ in $\mathcal{X} \times \mathcal{A}$, if

$$\Phi(\bar{x}, \zeta) \leq \Phi(\bar{x}, \bar{\zeta}) \leq \Phi(x, \bar{\zeta}), \quad \forall\, (x, \zeta) \in \mathcal{X} \times \mathcal{A}. \tag{3.62}$$

The following observation follows directly from the definition of a saddle point.

Theorem 3.39. *If the function $\Phi(\cdot, \cdot)$ has a saddle point $(\bar{x}, \bar{\zeta})$ in $\mathcal{X} \times \mathcal{A}$, then $\bar{x}$ is a solution of the primal problem* (3.59), *$\bar{\zeta}$ is a solution of the dual problem* (3.61) *and the optimal values of both problems coincide.*

Proof. The minimization of the left-hand side of (3.62) with respect to x can only strengthen these inequalities, so

$$\inf_{x \in \mathcal{X}} \Phi(x, \zeta) \leq \Phi(\bar{x}, \zeta) \leq \Phi(\bar{x}, \bar{\zeta}) = \min_{x \in \mathcal{X}} \Phi(x, \bar{\zeta}), \quad \forall\, \zeta \in \mathcal{A}.$$

This means that $\bar{\zeta}$ is an optimal solution of the dual problem (3.61). In a similar way, the maximization of the right hand side of (3.62) with respect to ζ yields

$$\max_{\zeta \in \mathcal{A}} \Phi(\bar{x}, \zeta) = \Phi(\bar{x}, \bar{\zeta}) \leq \Phi(x, \bar{\zeta}) \leq \sup_{\zeta \in \mathcal{A}} \Phi(x, \zeta),$$

which proves that $\bar{x}$ solves the primal problem (3.59). Both optimal values are equal to $\Phi(\bar{x}, \bar{\zeta})$. □

The existence of a saddle point of $\Phi(\cdot, \cdot)$ and the optimality conditions of Theorem 3.34 are closely related.

Theorem 3.40. *Suppose $\mathcal{X}$ is a separable Banach space, the mapping $F : \mathcal{X} \to \mathcal{L}_p(\Omega, \mathcal{F}, P)$ is convex and continuous a.s., $\mathbb{E}[F(\cdot)]$ is continuous as well, the risk measure $\varrho : \mathcal{L}_p(\Omega, \mathcal{F}, P) \to \mathbb{R}$ satisfies the conditions of convexity and monotonicity, and $X \subset \mathcal{X}$ is convex and closed. Then $(\bar{x}, \bar{\zeta}) \in \mathcal{X} \times \mathcal{A}$ is a saddle point of $\Phi(x, \zeta)$ if and only if $\bar{\zeta} \in \partial\varrho(F(\bar{x}))$ and subgradients $g(\omega) \in \partial f(\bar{x}, \omega)$, $\omega \in \Omega$, exist, such that relations* (3.55) *are satisfied.*

Proof. Suppose $(\bar{x}, \bar{\zeta})$ is a saddle point. Since X is convex and $\Phi(\cdot, \bar{\zeta})$ is convex and finite-valued, the optimality conditions for the problem

$$\min_{x \in X} \Phi(x, \bar{\zeta}), \tag{3.63}$$

whose solution by virtue of the right relation in (3.62) is $\bar{x}$, read:

$$0 \in \partial\Phi(\bar{x}, \bar{\zeta}) + \mathcal{N}_X(\bar{x}).$$

Applying Theorem 3.13, we obtain relations (3.55) with $s(\omega) \in \partial f(\bar{x}, \omega)$, $\omega \in \Omega$. The left relation in (3.62) implies that $\bar{\zeta} \in \partial\varrho(F(\bar{x}))$, by virtue of Theorem 3.34.

Conversely, suppose relations (3.55) are satisfied with $\bar{\zeta} \in \partial\varrho(F(\bar{x}))$ and $s(\omega) \in \partial f(\bar{x}, \omega)$, $\omega \in \Omega$. This means that optimality conditions for problem (3.63) are satisfied, and thus the right relations in (3.62) are true. The left relation in (3.62) follows from Theorem 3.34 again. □

Corollary 3.41. *Suppose problem* (3.54) *has an optimal solution $\bar{x}$. Then there is no duality gap between problems* (3.59) *and* (3.61)*, and the dual problem* (3.61) *has an optimal solution.*

Example 3.42. We have n assets with random return rates $R_1, \ldots, R_n$. We assume that the n-dimensional random vector $R = \begin{bmatrix} R_1 \ \ldots \ R_n \end{bmatrix}^T$ has K realizations $R^1, \ldots, R^K$, attained with probabilities $p_1, \ldots, p_K$.

The capital to invest is fixed and equals S. For a portfolio $x = \begin{bmatrix} x_1 \ x_2 \ \ldots x_n \end{bmatrix}^T$, the profit is random and has the form $R^T x$. Our problem has the following form:

$$\begin{aligned} &\min \varrho\big(-R^T x\big) \\ &\text{s.t. } x_1 + x_2 + \cdots + x_n = S, \\ &\quad x \in X_0, \end{aligned} \tag{3.64}$$

where $\mathcal{X}_0$ represents additional constraints on the amounts invested (for example, no short-selling, limits on individual investments, etc.). The function $\varrho(\cdot)$ is a coherent measure of risk. We use the negative profit as the argument of $\varrho(\cdot)$, because we are assuming that the argument of the measure of risk represents losses.

For a coherent measure of risk $\varrho(Z)$, in the case of finitely many realizations $z_1, \dots, z_K$ attained with probabilities $p_1, \dots, p_K$, the dual representation takes on the form

$$\varrho(Z) = \max_{\mu \in \mathcal{A}'} \left\{ \sum_{k=1}^{K} \mu_k z_k \right\}. \tag{3.65}$$

The set $\mathcal{A}'$ is a subset of the set

$$\mathcal{P}' = \left\{ \mu \in \mathbb{R}^K : \sum_{k=1}^{K} \mu_k = 1,\ \mu \geq 0 \right\}.$$

The probabilities $\mu_k = \zeta_k p_k$, $k = 1, \dots, K$, where $\zeta \in \mathcal{A}$.

Substituting $Z(x) = -R^T x$, that is, $z_k(x) = -\sum_{j=1}^{n} r_j^k x_j$, and the dual representation (3.65), we obtain a min–max formulation of the portfolio problem (3.64):

$$\min_{x \in \mathcal{X}} \max_{\mu \in \mathcal{A}'} \left\{ \sum_{k=1}^{K} \mu_k z_k(x) \right\} = -\max_{x \in \mathcal{X}} \min_{\mu \in \mathcal{A}'} \mu^T R^T x.$$

Both sets, $\mathcal{X}$ and $\mathcal{A}'$, are convex and compact and the function $\mu^T R^T x$ is bilinear. Therefore, the "min" and "max" operators can be exchanged. With $\hat{x}$ denoting the solution of the problem, we conclude that

$$\varrho\big(Z(\hat{x})\big) = -\max_{x \in \mathcal{X}} \min_{\mu \in \mathcal{A}'} \mu^T R^T x = -\min_{\mu \in \mathcal{A}'} \max_{x \in \mathcal{X}} \mu^T R^T x. \tag{3.66}$$

Due to the positive homogeneity of the risk measure ϱ, with no loss of generality, we may assume that the capital $S = 1$. If $\mathcal{X}_0$ involves the "no shorting" constraints $x \geq 0$, then the problem (3.66) becomes equivalent to a matrix game, in which x is the mixed strategy of the first player (the investor), and μ is the mixed strategy of the second player (the market). The game has a *saddle point* $(\hat{x}, \hat{\mu})$, at which

$$\max_{x \in \mathcal{X}} \hat{\mu}^T R^T x = \hat{\mu}^T R^T \hat{x} = \min_{\mu \in \mathcal{A}'} \mu^T R^T \hat{x}.$$

□

3.3.2 Duality Associated with Nonanticipativity Constraints

We can reformulate problem (3.54) in yet another way. We assume that $\mathcal{X}$ is a separable Banach space, $F : \mathcal{X} \to \mathcal{L}_p(\Omega, \mathcal{F}, P)$, $p \in [1, \infty)$, is convex and continuous a.s, $\mathbb{E}[F(\cdot)]$ is continuous as well, and $\varrho : \mathcal{L}_p(\Omega, \mathcal{F}, P) \to \mathbb{R}$ satisfies the conditions of convexity and monotonicity.

Suppose we replace the decision vector x by a measurable function $\mathbb{x} : \Omega \to \mathcal{X}$. In order to advance our construction, we restrict $\mathbb{x}$ to be an element of the Banach space $\mathbb{X} = \mathcal{L}_{p'}(\Omega, \mathcal{F}, P; \mathcal{X})$, with $p' \in [p, \infty]$. We already explored this idea in the proof of Theorem 3.11; now, we will derive duality relations associated with this transformation.

Define $\big[\mathbb{F}(\mathbb{x})\big](\omega) \triangleq f(\mathbb{x}(\omega), \omega)$ and let

$$\widetilde{\mathcal{X}} = \{\mathbb{x} \in \mathbb{X} : \mathbb{x} \in \mathcal{X} \text{ a.s.}\}. \tag{3.67}$$

We assume that the space $\mathbb{X}$ and the set $\mathcal{X}$ are such that $\mathbb{F}(\mathbb{x}) \in \mathcal{Z}$ for every $\mathbb{x} \in \widetilde{\mathcal{X}}$; for example, $\mathcal{X}$ is a bounded set, or $p' = \infty$. In the space $\mathbb{X}$ we consider the subspace of constant functions:

$$\mathcal{S} = \{\mathbb{x} \in \mathbb{X} : \exists (x \in \mathcal{X})\ \mathbb{x} = x \text{ a.s.}\}.$$

We can represent it by a linear operator equation,

$$\mathbb{x} = \mathbb{E}[\mathbb{x}] \quad \text{a.s.}, \tag{3.68}$$

which we call the *nonanticipativity constraint*. The expected value in (3.68) is understood in the weak* sense: for all $y \in \mathcal{X}^*$ we have

$$\langle y, \mathbb{E}[\mathbb{x}] \rangle = \int_\Omega \langle y, \mathbb{x}(\omega) \rangle \ P(d\omega).$$

We can now equivalently reformulate problem (3.54) as follows:

$$\begin{aligned} &\min_{\mathbb{x}} \varrho\big(\mathbb{F}(\mathbb{x})\big) \\ &\text{subject to } \mathbb{x} = \mathbb{E}[\mathbb{x}] \quad \text{a.s.}, \\ &\qquad \mathbb{x} \in \widetilde{\mathcal{X}}. \end{aligned} \tag{3.69}$$

The virtues of this reformulation become evident when we relax the constraint (3.68). We associate with it a Lagrange multiplier $\pi \in \mathcal{L}_{q'}(\Omega, \mathcal{F}, P; \mathcal{X}^*)$, $1/p' + 1/q' = 1$, and formulate the Lagrangian $L : \widetilde{\mathcal{X}} \times \mathcal{L}_{q'}(\Omega, \mathcal{F}, P; \mathcal{X}^*) \to \mathbb{R}$,

$$L(\mathbb{x}, \pi) = \varrho\big(\mathbb{F}(\mathbb{x})\big) + \langle \pi, \mathbb{E}[\mathbb{x}] - \mathbb{x} \rangle.$$

When $p' = \infty$, the dual space $\mathbb{X}^*$ is larger than $\mathcal{L}_1(\Omega, \mathcal{F}, P; \mathfrak{X}^*)$, but we shall show that restricting the multipliers to this subspace is sufficient.

The value of the Lagrangian does not change if π is shifted by a constant vector $c\mathbb{1}$, because the definition of the integral $\mathbb{E}[x]$ implies that

$$\langle c\mathbb{1}, \mathbb{E}[x] - x \rangle = \int_\Omega \langle c, \mathbb{E}[x] - x(\omega)\rangle \; P(\mathrm{d}\omega) = 0.$$

Therefore, with no loss of generality, we can restrict the range of the multiplier π to the set

$$\mathcal{S}^\perp = \big\{\pi \in \mathcal{L}_{q'}(\Omega, \mathcal{F}, P; \mathfrak{X}^*) : \mathbb{E}[\pi] = 0\big\},$$

and write the Lagrangian $L : \widetilde{\mathcal{X}} \times \mathcal{S}^\perp \to \mathbb{R}$ in a simplified form:

$$L(x, \pi) = \varrho\big(\mathbb{F}(x)\big) - \langle \pi, x \rangle. \tag{3.70}$$

The dual function associated with the Lagrangian is

$$L_D(\pi) = \inf_{x \in \widetilde{\mathcal{X}}} L(x, \pi), \quad \pi \in \mathcal{S}^\perp.$$

Since choosing $x \in \mathcal{S}$ annihilates the Lagrangian term $\langle \pi, x \rangle$, the function $L(x, \pi)$ coincides with $\varrho(F(x))$ on constant functions $x = x$ a.s.. Therefore, $L_D(\pi)$ provides a lower bound for the optimal value of problems (3.69) and (3.54).

The dual problem associated with the Lagrangian (3.70) has the form:

$$\max_{\pi \in \mathcal{S}^\perp} L_D(\pi). \tag{3.71}$$

A question arises whether the dual problem has a solution and whether the optimal values of the original (primal) and the dual problems are equal.

Theorem 3.43. *Suppose $\mathfrak{X}$ is a separable Banach space, the mapping $F : \mathfrak{X} \to \mathcal{L}_p(\Omega, \mathcal{F}, P)$, with $p \in [1, \infty)$, is convex and continuous, $\mathbb{E}[F(\cdot)]$ is continuous as well, the set $\mathcal{X}$ is convex and closed, and the risk measure $\varrho : \mathcal{L}_p(\Omega, \mathcal{F}, P) \to \mathbb{R}$ satisfies the conditions of convexity and monotonicity. If the primal problem* (3.54) *has an optimal solution, then the dual problem* (3.71) *has an optimal solution, its optimal value is equal to the optimal value of problem* (3.54)*, and for every optimal solution $\hat{\pi}$ of the dual problem, a constant function $x \equiv \hat{x}$ is a minimizer of $L(\cdot, \hat{\pi})$ if and only if $\hat{x}$ is an optimal solution of problem* (3.54).

Proof. By virtue of Lemma 3.5, the composite mapping $\varphi = \varrho \circ F$ is convex. According to Theorem 2.32, it is continuous.

Let $\hat{x}$ be a solution of problem (3.54). By virtue of Theorem 3.34, $\hat{\zeta} \in \partial\varrho\big(F(\hat{x})\big)$ and subgradients $s(\omega) \in \partial f(\hat{x}, \omega)$, $\omega \in \Omega$, exist, such that equation (3.55) is satisfied. Define

$$\hat{\pi} = \hat{\zeta}s - \mathbb{E}[\hat{\zeta}s].$$

The equation $\mathbb{E}[\hat{\pi}] = 0$ is satisfied by construction, and thus $\hat{\pi} \in \mathcal{S}^{\perp}$.

Let $\hat{\mathbb{x}} \in \mathbb{X}$ be such that $\hat{\mathbb{x}}(\cdot) = \hat{x}$ a.s.. Since $\hat{\zeta} \in \partial\varrho(F(\hat{x}))$, for any function $\mathbb{x} \in \widetilde{\mathcal{X}}$ we have

$$\begin{aligned}\varrho(F(\mathbb{x})) &\geq \varrho(F(\hat{\mathbb{x}})) + \langle\hat{\zeta}, F(\mathbb{x}) - F(\hat{\mathbb{x}})\rangle \\ &\geq \varrho(F(\hat{\mathbb{x}})) + \int_{\Omega} \hat{\zeta}(\omega)\langle s(\omega), \mathbb{x}(\omega) - \hat{x}\rangle \; P(\mathrm{d}\omega).\end{aligned}$$

In the last inequality we used the relations $s(\omega) \in \partial f(\hat{x}, \omega)$ and $\hat{\zeta} \geq 0$. Substitution to the Lagrangian yields the following chain of relations:

$$\begin{aligned}&L(\mathbb{x}, \hat{\pi}) - L(\hat{\mathbb{x}}, \hat{\pi}) \\ &\quad\geq \int_{\Omega} \langle\hat{\zeta}(\omega)\, s(\omega), \mathbb{x}(\omega) - \hat{x}\rangle \; P(\mathrm{d}\omega) - \int_{\Omega} \langle\hat{\pi}(\omega), \mathbb{x}(\omega) - \hat{x}\rangle \; P(\mathrm{d}\omega) \\ &\quad= \int_{\Omega} \big\langle\mathbb{E}[\hat{\zeta}s], \mathbb{x}(\omega) - \hat{x}\big\rangle \; P(\mathrm{d}\omega).\end{aligned}$$

Due to (3.55),

$$\mathbb{E}[\hat{\zeta}s] \in -\mathcal{N}_{\mathcal{X}}(\hat{x}).$$

Since $\mathbb{x}(\omega) \in \mathcal{X}$, the last chain of inequalities can be completed as follows:

$$L(\mathbb{x}, \hat{\pi}) - L(\hat{\mathbb{x}}, \hat{\pi}) \geq 0.$$

This proves that the constant function $\hat{\mathbb{x}} \equiv \hat{x}$ is the minimizer of $L(\cdot, \hat{\pi})$ in $\widetilde{\mathcal{X}}$. Consequently,

$$\begin{aligned}L_D(\hat{\pi}) = L(\hat{\mathbb{x}}, \hat{\pi}) &= \varrho(F(\hat{\mathbb{x}})) - \langle\hat{\pi}, \hat{\mathbb{x}}\rangle \\ &= \varrho(F(\hat{x})) - \int_{\Omega} \langle\hat{\pi}(\omega), \hat{x}\rangle \; P(\mathrm{d}\omega) = \varrho(F(\hat{x})).\end{aligned}$$

This means that the optimal value of the dual problem is equal to the optimal value of the primal problem, and that the vector $\hat{\pi}$ is an optimal solution of the dual problem.

If $\bar{\pi}$ is another solution to the dual problem, then the constant function $\hat{\mathbb{x}}$ still yields

$$L(\hat{\mathbb{x}}, \bar{\pi}) = \varrho(F(\hat{x})) = L_D(\bar{\pi}),$$

and is a minimizer of $L(\cdot, \bar{\pi})$ in $\widetilde{\mathcal{X}}$. Suppose another constant function $\bar{X} \equiv \bar{x} \in \mathcal{X}$ minimizes $L(\cdot, \bar{\pi})$. The last displayed equation is still true for $\bar{X}$ and implies that $\varrho(F(\bar{x})) = \varrho(F(\hat{x}))$. □

We can now integrate the results of the last two sections. Consider the problem of minimizing the Lagrangian (3.70), and use the dual representation (2.27) of the risk measure $\varrho(\cdot)$:

$$\begin{aligned} L_D(\pi) &= \min_{x\in\mathcal{X}} \max_{\zeta\in\mathcal{A}} \big\{\langle\zeta, F(x)\rangle - \varrho^*(\zeta) - \langle\pi, x\rangle\big\} \\ &= \max_{\zeta\in\mathcal{A}} \min_{x\in\mathcal{X}} \big\{\langle\zeta, F(x)\rangle - \varrho^*(\zeta) - \langle\pi, x\rangle\big\}. \end{aligned}$$

The interchange on the "min" and "max" operators is possible due to the weak* compactness of the set $\mathcal{A}$. It follows that the dual problem can be equivalently written as a max–min problem involving *both* dual vectors: the nonanticipativity multiplier π and the risk subgradient ζ. We obtain the following dual problem:

$$\max_{\substack{\zeta\in\mathcal{A}\\ \pi\in\mathcal{S}^\perp}} \Big[-\varrho^*(\zeta) + \min_{x\in\mathcal{X}} \big\{\langle\zeta, F(x)\rangle - \langle\pi, F(x)\rangle\big\}\Big].$$

By virtue of the interchangeability principle, the inner problem of minimizing with respect to x can be transformed as follows:

$$\begin{aligned} &\min_{x\in\mathcal{X}} \int_\Omega \big[\zeta(\omega) f(x(\omega), \omega) - \langle\pi(\omega), x(\omega)\rangle\big]\, P(\mathrm{d}\omega) \\ &\qquad = \int_\Omega \min_{x\in\mathcal{X}} \big[\zeta(\omega) f(x, \omega) - \langle\pi(\omega), x\rangle\big]\, P(\mathrm{d}\omega). \end{aligned}$$

Consequently, its solution $\hat{x}(\pi, \zeta)$ can be obtained by solving scenario subproblems, one for each ω.

3.4 Stochastic Subgradient Methods

In this section, we are concerned with stochastic methods for solving a special case of the general optimization problem (3.1):

$$\min_{x\in\mathcal{X}} \Big\{\varphi(x) \triangleq \varrho\big(F(x)\big)\Big\}, \tag{3.72}$$

where $\varrho : \mathcal{L}_p(\Omega, \mathcal{F}.P) \to \mathbb{R}$ is a measure of risk, $p \in [2, \infty)$, $F : \mathbb{R}^n \to \mathcal{L}_p(\Omega, \mathcal{F}.P)$, and $\mathcal{X} \subset \mathbb{R}^n$ is a convex closed set. As before, we write $f(x, \omega)$ for $[F(x)](\omega)$. We restrict our considerations to a finite-dimensional decision space $\mathcal{X} = \mathbb{R}^n$, and to random outcomes with a finite variance ($p \ge 2$). We also require that Assumption 3.18 be satisfied with some $p \ge 2$.

As discussed in § 3.2, in many applications, the mapping F results from the composition of a cost function $f : \mathbb{R}^n \times \mathfrak{D} \to \mathbb{R}$ and a random data vector $D : \Omega \to \mathfrak{D}$, where $\mathfrak{D}$ is a Polish space (usually, a finite-dimensional space). Then $\big[F(x)\big](\omega) \triangleq f(x, D(\omega))$, $\omega \in \Omega$. We shall make use of this representation.†

Three difficulties are associated with problem (3.72). First, the values of the operator $F(\cdot)$ are elements of an infinite-dimensional space and can be calculated and represented in the simplest cases only. Second, the risk measure $\varrho(\cdot)$ is nondifferentiable. In many applications, the function $f(\cdot, D)$ is nondifferentiable as well. Third, the distribution of the data may be accessible only by sampling: either via real-time observations, or through a simulation procedure.

Example 3.44. A typical statistical learning problem can be formulated as an expected-value stochastic optimization problem:

$$\min_{x \in \mathcal{X}} \mathbb{E}\left[f(x, D)\right], \tag{3.73}$$

where $f : \mathbb{R}^n \times \mathfrak{D} \to \mathbb{R}$ is the loss function of the predictor x on the random data D with a distribution P_D. In a typical setting, the data D comprise the "features" Φ and the "labels" L, and the loss function measures the error between L and a parametric model $M(x, \Phi)$, such as a neural network.‡

A central issue in this context is to make decisions that *generalize* well (work well on unseen data) and are robust to perturbations in the underlying data distribution. This motivates the following *distributionally robust* version of problem (3.73):

$$\min_{x \in \mathcal{X}} \max_{Q \in \mathcal{M}(P_D)} \mathbb{E}_{D \sim Q}\left[f(x, D)\right], \tag{3.74}$$

where the *ambiguity set* $\mathcal{M}(P_D)$ is a closed convex set of probability measures that models perturbations to the law P; the predictor x is chosen to accommodate the worst-case perturbations.

In problem (3.72), with $F(x) = f(x, D)$, we achieve an implicit robust formulation. The set $\mathcal{A} = \partial\varrho(0)$ plays the role of the ambiguity set $\mathcal{M}(P_D)$. All probability measures in it are absolutely continuous with respect to the original measure P, thus excluding perturbations that are structurally impossible in the problem. □

We assume that problem (3.72) has a solution, and we consider two stochastic algorithms. The first one is the basic stochastic subgradient method with projection:

---

† Previously, we used the notation $f(x, \omega)$ in this context, but the current more specific representation is more convenient for us because we shall be using samples of D.

‡ In the statistical learning literature, the features are usually denoted by X, the labels by Y, and the model parameters by β.

$$x^{k+1} = \Pi_{\mathcal{X}}\big(x^k - \tau_k \tilde{g}^k\big), \quad k = 0, 1, 2, \ldots. \tag{3.75}$$

In the formula above, the directions $\tilde{g}^k \in \mathbb{R}^n$ are *stochastic subgradients* of $\varphi(\cdot)$ at x^k, that is, random vectors such that

$$\mathbb{E}\big[\tilde{g}^k \,\big|\, x^0, \ldots, x^k\big] \in \partial\varphi(x^k), \quad k = 0, 1, \ldots, \tag{3.76}$$

Depending on the application, the subdifferential $\partial\varphi(\cdot)$ is understood in the convex analysis sense, if $\varphi(\cdot)$ is convex, in the Clarke sense, if $\varphi(\cdot)$ is subdifferentially regular, and in the Norkin sense, if $\varphi(\cdot)$ is Norkin subdifferentiable.

The coefficients τ_k are positive scalars, which may be random, and the operator $\Pi_{\mathcal{X}}(\cdot)$ is the orthogonal projection on the convex closed set $\mathcal{X}$. Thus, the successive iterates $\{x^k\}$ are random as well.

In the following subsections, we consider two versions of the method (3.75): the convex case, where $\varphi(\cdot)$ is convex and $\partial\varphi(\cdot)$ is the usual subdifferential, and the Lipschitz continuous case, where $\partial\varphi(\cdot)$ is understood as the Norkin subdifferential $\hat{\partial}\varphi(\cdot)$ of a generalized differentiable function $\varphi(\cdot)$.

The second algorithm is the method with subgradient averaging (momentum), which will be presented for the nonconvex and nonsmooth case, with a Norkin subdifferentiable function $\varphi(\cdot)$. We defer its description to subsection 3.4.3.

3.4.1 Stochastic Subgradients of Mini-Batch Risk Functions

Stochastic subgradients satisfying (3.76) can be constructed when the measure of risk $\varrho(\cdot)$ is a mini-batch risk measure, as defined in (2.88). We adapt that definition to our setting by effectively replacing the abstract probability space $(\Omega, \mathcal{F}, P)$ by a more explicit model on the Polish data space $\mathfrak{D}$, with its Borel σ-algebra $\mathcal{B}(\mathfrak{D})$, and the probability measure $P_D \triangleq P \circ D^{-1}$. Then we can interpret the measure of risk ϱ as a functional on a space of measurable functions on $\mathfrak{D}$.

Suppose we have a coherent risk form $\varrho : \mathbb{B}(\mathfrak{D}) \times \mathcal{P}(\mathfrak{D}) \to \mathbb{R}$ having the support property. For a bounded measurable function $Z : \mathfrak{D} \to \mathbb{R}$ we can define the mini-batch risk measure as in (2.88):

$$\varrho^{(N)}[Z] \triangleq \mathbb{E}\big\{\varrho[Z, P_D^{(N)}]\big\}. \tag{3.77}$$

Here, $P_D^{(N)}$ is a random empirical measure associated with a sample $D^{1:N} \triangleq (D^1, \ldots, D^N)$ of the data. From now on, we skip the dependence of $\varrho^{(N)}[\cdot]$ on the probability measure P_D, because we are concerned with the case of a fixed P_D.

As discussed in § 2.7.5, the risk measure (3.77) is well-defined on the space of integrable random variables $\mathcal{L}_1(\mathfrak{D}, \mathcal{B}(\mathfrak{D}), P_D)$. Particularly relevant to our purposes is formula (2.97), which now reads:

$$\varrho^{(N)}[Z] = \mathbb{E}\left\{ \max_{\xi \in \mathcal{A}^{(N)}} \sum_{j=1}^{N} \xi_j Z(D^j) \right\}, \tag{3.78}$$

where the expectation is over $D^{1:N} \sim P_D^N$. The set $\mathcal{A}^{(N)}$ is a closed convex subset of the simplex (2.95). Examples in § 2.3.1 provide explicit representations of the set $\mathcal{A}^{(N)}$ for some popular measures of risk.[†]

For a sample $D^{1:N}$ and a function $Z(\cdot)$, we construct a random measure $\mu(Z; D^{1:N})$ in the following way. We solve the maximization problem inside (3.78) obtaining a vector $\hat{\xi}(Z; D^{1:N})$. Then we set

$$\mu(Z; D^{1:N}) = \sum_{j=1}^{N} \hat{\xi}_j(Z; D^{1:N})\, \delta_{D^j}. \tag{3.79}$$

By construction, for any $W \in \mathbb{B}(\mathfrak{D})$,

$$\begin{aligned}
&\max_{\xi \in \mathcal{A}^{(N)}} \sum_{j=1}^{N} \xi_j W(D^j) \\
&\quad \geq \sum_{j=1}^{N} \hat{\xi}_j(Z; D^{1:N}) Z(D^j) + \sum_{j=1}^{N} \hat{\xi}_j(Z; D^{1:N}) \big(W(D^j) - Z(D^j)\big) \\
&\quad = \sum_{j=1}^{N} \hat{\xi}_j(Z; D^{1:N}) Z(D^j) + \int_{\mathfrak{D}} \big(W(v) - Z(v)\big) \big[\mu(Z; D^{1:N})\big](dv).
\end{aligned}$$

Taking the expected value of both sides with respect to the sample $D^{1:N}$, we conclude that

$$\varrho^{(N)}[W] \geq \varrho^{(N)}[Z] + \int_{\mathfrak{D}} \big(W(v) - Z(v)\big) \big[\bar{\mu}(Z)\big](dv), \tag{3.80}$$

where

$$\bar{\mu}(Z) = \mathbb{E}_{D^{1:N} \sim P_D^N}\big[\mu(Z; D^{1:N})\big]. \tag{3.81}$$

The expected value is understood in the weak* sense here.

[†] In § 2.3.1, the dual representations are expressed in terms of the densities with respect to the reference measure. Now, the reference measure is the empirical measure with all probabilities equal to $\frac{1}{N}$. We skip the normalization by N here, and thus $\mathcal{A}^{(N)} = \frac{1}{N}\mathcal{A}$, with the set $\mathcal{A}$ defined in (2.28) and calculated in the examples in that section.

It follows from (3.80) that the measure $\bar{\mu}(Z)$ is a subgradient of $\varrho^{(N)}[\cdot]$ at Z, and the random measure $\mu(Z; D^{1:N})$ can be interpreted as an unbiased stochastic subgradient at Z.

Using (3.81) in the weak* sense, for any bounded measurable function $W(\cdot)$ we obtain

$$\begin{aligned}
&\int_{\mathfrak{D}} \big(W(v) - Z(v)\big) \, \big[\bar{\mu}(Z)\big](\mathrm{d}v) \\
&\quad = \mathbb{E}_{D^{1:N} \sim P_D^N}\Big[\int_{\mathfrak{D}} \big(W(v) - Z(v)\big) \, \big[\mu(Z; D^{1:N})\big](\mathrm{d}v)\Big] \\
&\quad = \mathbb{E}_{D^{1:N} \sim P_D^N}\Big[\sum_{j=1}^{N} \hat{\xi}_j(Z; D^{1:N})\big(W(D^j) - Z(D^j)\big)\Big].
\end{aligned} \tag{3.82}$$

As in (2.99) we define

$$\big[\Pi_j^{(N)}(\hat{\xi})\big](v) = \mathbb{E}\big[\hat{\xi}_j(Z; D^{1:N}) \,\big|\, D^j = v\big], \quad j = 1, \dots, N.$$

By using the fact that all D^j are distributed according to P_D, we obtain from from (3.80) and (3.82) the inequality

$$\begin{aligned}
\varrho^{(N)}[W] - \varrho^{(N)}[Z] &\geq \mathbb{E}_{D^{1:N} \sim P_D^N}\Big\{ \sum_{j=1}^{N} \big[\Pi_j^{(N)}(\hat{\xi})\big](D^j)\big(W(D^j) - Z(D^j)\big)\Big\} \\
&= \int_{\mathfrak{D}} \sum_{j=1}^{N} \big[\Pi_j^{(N)}(\hat{\xi})\big](v)\big(W(v) - Z(v)\big) \; P_D(\mathrm{d}v).
\end{aligned}$$

This means that the subgradient $\bar{\mu}(Z)$ is absolutely continuous with respect to P_D, with the density

$$\zeta = \sum_{j=1}^{N} \Pi_j^{(N)}(\hat{\xi}).$$

If we extend the domain of $\varrho^{(N)}[\cdot]$ to $\mathcal{L}_1(\mathfrak{D}, \mathcal{B}(\mathfrak{D}), P_D)$, we get $\zeta \in \partial\varrho^{(N)}[Z]$.

Consider now the case when $Z = F(x)$ is a random "cost" dependent on the decision vector $x \in \mathbb{R}^n$ and random data D. We discussed this situation in § 3.1. We assume that $F : \mathbb{R}^n \to \mathcal{L}_2(\mathfrak{D}, \mathcal{B}(\mathfrak{D}), P_D)$ and we use the same notation $f(x, D) = [F(x)](D)$.

First, we consider the case of a *convex* operator F, as in subsection 3.1.1. Setting $V = F(y)$ and $Z = F(x)$, we obtain from (3.82) the inequality:

$$\begin{aligned}
\varrho^{(N)}[F(y)] - \varrho^{(N)}[F(x)] &\geq \int_{\mathfrak{D}} \big(f(y, D) - f(x, D)\big)\, \mathrm{d}\bar{\mu}(Z) \\
&= \mathbb{E}\Bigg[\sum_{j=1}^{N} \hat{\xi}_j(F(x); D^{1:N})\big(f(y, D^j) - f(x, D^j)\big)\Bigg] \\
&\geq \mathbb{E}\Bigg[\sum_{j=1}^{N} \hat{\xi}_j(F(x); D^{1:N})\big\langle g(x, D^j), y - x\big\rangle\Bigg],
\end{aligned}$$

where $g(x, D^j) \in \partial f(x, D^j)$. It follows that the vector

$$\tilde{g}(x; D^{1:N}) = \sum_{j=1}^{N} \hat{\xi}_j(F(x); D^{1:N})\, g(x, D^j) \tag{3.83}$$

is a stochastic subgradient of the composite function $\varphi = \varrho^{(N)} \circ F$. Indeed,

$$\begin{aligned}
\mathbb{E}\big[\tilde{g}(x; D^{1:N})\big] &= \mathbb{E}\Bigg\{\sum_{j=1}^{N} \mathbb{E}\big[\hat{\xi}_j(F(x); D^{1:N}) \,\big|\, D^j = v\big]\, g(x, v)\Bigg\} \\
&= \int_{\mathfrak{D}} \sum_{j=1}^{N} \bar{\xi}_j(F(x); v)\, g(x, v)\ P(\mathrm{d}v) \\
&= \int_{\mathfrak{D}} \zeta(F(x); v)\, g(x, v)\ P(\mathrm{d}v),
\end{aligned} \tag{3.84}$$

which is an element of the subdifferential (3.21).

Since Assumption 3.18 with $p = 2$ is satisfied, then $\mathbb{E}\Big[\big\|\tilde{g}(x; D^{1:N})\big\|^2\Big]$ is uniformly bounded.

The second case is when the mapping $F : \mathbb{R}^n \to \mathcal{L}_2(\mathfrak{D}, \mathcal{B}(\mathfrak{D}), P_D)$ is Lipschitz continuous, but not necessarily convex. Under the conditions of Theorem 3.31, the composite function $\varphi = \varrho^{(N)} \circ F$ is Lipschitz continuous and Norkin subdifferentiable, and its Norkin subdifferential is given by the formula

$$\hat{\partial}\varphi(x) = \bigcup_{\zeta \in \partial\varrho^{(N)}(F(x))} \int_{\mathfrak{D}} \zeta(v)\, \hat{\partial} f(x, v)\ P_D(\mathrm{d}v). \tag{3.85}$$

Our objective is to generate a stochastic Norkin subgradient: a random vector whose expected value is an element of the set (3.85). Consider a particular sample $D^{1:N}$, the associated random risk measure

$$\tilde{\varrho}(Z; D^{1:N}) = \max_{\xi \in \mathcal{A}^{(N)}} \sum_{j=1}^{N} \xi_j Z(D^j), \tag{3.86}$$

and the composite risk function

$$\tilde{\varphi}(x; D^{1:N}) = \tilde{\varrho}\big(F(x); D^{1:N}\big).$$

By construction

$$\varphi(x) = \varrho^{(N)}[F(x)] = \mathbb{E}\big[\tilde{\varphi}(x; D^{1:N})\big]. \tag{3.87}$$

The function $\tilde{\varphi}(\cdot; D^{1:N})$ is a special case of the case considered in Theorem 3.31, just with the measure P replaced by the empirical measure. Under the assumptions of that theorem, $\tilde{\varphi}(\cdot, D^{1:N})$ is Norkin subdifferentiable with the subdifferential is given by the formula

$$\hat{\partial}\tilde{\varphi}(x; D^{1:N}) = \bigcup_{\xi \in \hat{\mathcal{A}}(x; D^{1:N})} \sum_{j=1}^{n} \xi_j \hat{\partial} f(x, D^j),$$

where $\hat{\mathcal{A}}(x; D^{1:N})$ is the set of maximizers in (3.86) at $Z = F(x)$. It follows that for an arbitrary $\hat{\xi} \in \hat{\mathcal{A}}(x; D^{1:N})$, the vector

$$\tilde{g}(x; D^{1:N}) = \sum_{j=1}^{N} \hat{\xi}_j(F(x); D^{1:N})\, g(x, D^j), \tag{3.88}$$

where $g(x, D^j) \in \hat{\partial} f(x, D^j)$, $j = 1, \ldots, N$, is a Norkin subgradient of $\tilde{\varphi}(\cdot; D^{1:N})$ at x. Due to (3.87), $\tilde{g}(x; D^{1:N})$ is a stochastic subgradient of $\varphi(\cdot)$. The verification that $\mathbb{E}\big[\tilde{g}(x; D^{1:N})\big]$ is indeed an element of the set (3.85) is identical to the calculation in (3.84). As in the preceding case, since Assumption 3.18 is satisfied with $p = 2$ everywhere, then $\mathbb{E}\big[\|\tilde{g}(x; D^{1:N})\|^2\big]$ is uniformly bounded.

Example 3.45. Consider the mini-batch mean–semideviation risk measure derived from the risk form discussed in Example 2.72:

$$\varrho^{(N)}[F(x)] = \mathbb{E}_{D^{1:N} \sim P_D^N}\left\{ \frac{1}{N} \sum_{j=1}^{N} f(x, D^j) \right.$$
$$\left. + \frac{\varkappa}{N} \sum_{j=1}^{N} \Big(f(x, D^j) - \frac{1}{N} \sum_{k=1}^{N} f(x, D^k) \Big)_+ \right\}. \tag{3.89}$$

To apply the formula (3.88), we use the calculations of Example 2.27. The set (2.32) was formulated there in terms of the densities with respect to the reference measure $(p_1, \ldots, p_N)$. Using $p_k = \frac{1}{N}$, $k = 1, \ldots, N$, and switching to probability vectors $\xi_k = \zeta_k p_k$, $k = 1, \ldots, N$, we obtain

$$\mathcal{A}_{\varkappa}^{(N)} = \left\{ \xi \in \mathbb{R}^n : \xi_k = \frac{1}{N}\Big(1+\delta_k - \frac{1}{N}\sum_{j=1}^{N}\delta_j\Big),\ 0 \le \delta_j \le \varkappa,\ j = 1,\dots,N \right\}.$$

Thus, a stochastic subgradient $\tilde{g}$ of $\varrho^{(N)}[F(x)]$ can be obtained by sampling $D^{1:N}$ from P_D^N, solving the problem

$$\max_{\xi \in \mathcal{A}_{\varkappa}^{(N)}} \sum_{j=1}^{N} \xi_j f(x, D^j),$$

and using (3.88). This method requires N observations.

The number of observations can be reduced for small $\varkappa$ by the following randomized procedure. The formula (3.89) can be equivalently written as:

$$\begin{aligned}&\varrho^{(N)}[F(x)] = (1-\varkappa)\mathbb{E}_{D\sim P_D}[f(x,D)]\\ &+\varkappa\mathbb{E}_{D^{1:N}\sim P_D^N}\left\{\frac{1}{N}\sum_{j=1}^{N} f(x,D^j) + \frac{1}{N}\sum_{j=1}^{N}\Big(f(x,D^j) - \frac{1}{N}\sum_{k=1}^{N} f(x,D^k)\Big)_+\right\}.\end{aligned}$$

Thus, we may sample a Bernoulli random variable β such that $P[\beta = 1] = \varkappa$, independently of all other samples. If $\beta = 0$, we need only one sample of D and set $\tilde{g} \in \partial f(x, D)$. If $\beta = 1$, we sample $D^{1:N}$ from P_D^N, solve the problem

$$\max_{\xi \in \mathcal{A}_{1}^{(N)}} \sum_{j=1}^{N} \xi_j f(x, D^j),$$

and use (3.88) to calculate the stochastic subgradient. This method uses $1 + \kappa(N-1)$ observations per one subgradient, on average.

In the special case of $N = 2$ further simplifications are possible. The definition of the risk measure can be manipulated to the following form:

$$\begin{aligned}\varrho^{(2)}[F(x)] &= \mathbb{E}[f(x,D)] + \frac{\varkappa}{4}\mathbb{E}_{D^{1:2}\sim P_D^2}\Big[\big|f(x,D^1) - f(x,D^2)\big|\Big]\\ &= \Big(1-\frac{\varkappa}{2}\Big)\mathbb{E}\big[f(x,D)\big] + \frac{\varkappa}{2}\mathbb{E}_{D^{1:2}\sim P_D^2}\Big[\max\big(f(x,D^1), f(x,D^2)\big)\Big].\end{aligned}$$

A stochastic subgradient $\tilde{g}$ of $\varrho^{(2)}[F(x)]$ can be obtained in the following way. We sample a Bernoulli random variable β such that $P[\beta = 1] = \varkappa/2$. If $\beta = 0$, we sample $D \sim P_D$ and set $\tilde{g} \in \hat{\partial} f(x, D)$. If $\beta = 1$, we sample $D^{1:2} \sim P_D^2$, choose $D^* = \arg\max\{f(x, D^1), f(x, D^2)\}$, and set $\tilde{g} \in \hat{\partial} f(x, D^*)$. In this way, on average, $1 + \varkappa/2$ data observations are needed. □

3.4.2 The Basic Subgradient Method

In this section, we consider the basic stochastic subgradient method for solving problem (3.72). The method is described by the iterative formula (3.75). We use $(\Omega, \mathcal{F}, P)$ to denote the probability space on which the sequence $\{x^k\}$ is defined, and we let

$$\mathcal{F}_k = \sigma(x^0, \tilde{g}^0, x^1, \dots, \tilde{g}^{k-1}, x^k)$$

to denote the σ-algebra generated by the history of the method up to the point x^k. We make the following assumptions.

Assumption 3.46. *The stepsizes τ_k are $\mathcal{F}_k$-measurable, and such that*

(i) $\tau_k > 0, \quad k = 0, 1, 2, \dots;$

(ii) $\displaystyle\sum_{k=0}^{\infty} \tau_k = \infty \quad a.s.;$

(iii) $\displaystyle\mathbb{E} \sum_{k=0}^{\infty} \tau_k^2 < \infty.$

Assumption 3.47. *The random vectors $\tilde{g}^k$ satisfy for $k = 0, 1, \dots$ the conditions:*

(i) $\tilde{g}^k = g^k + e^k$, *with* $g^k \in \partial\varphi(x^k)$;

(ii) $\mathbb{E}\big[e^k \,\big|\, \mathcal{F}_k\big] = 0$;

(iii) $\mathbb{E}\big[\|e^k\|^2 \,\big|\, \mathcal{F}_k\big] \le \sigma^2$, *with some constant* σ^2.

As mentioned before, the meaning of the subdifferential $\partial\varphi(\cdot)$ will be appropriate for the class of functions considered.

Remark 3.48. We could make a slightly weaker assumption that $\tilde{g}^k = g^k + e^k + b^k$, with some bias b^k such that $\sum_{k=0}^{\infty} \tau_k b^k$ is convergent. It will be evident, how the proofs can be modified to accommodate this version. □

3.4.2.1 The Convex Case

First, we analyze the case of a *convex* operator F, as in subsection 3.1.1.

The formula (3.83) for generating the stochastic subgradients of a mini-batch risk function satisfies Assumption 3.47 with the usual subdifferential of a convex function, if a new sample $D^{1:N}$ is generated independently at each iteration k and if Assumption 3.18 with $p = 2$ is satisfied. Moreover, the latter assumption implies that a constant C exists, such that $\|g\| \le C$ for all $g \in \partial\varphi(x)$ and all x.

Before formulating the convergence theorem, we note that these assumptions and our results can be generalized in many ways. We chose this setting to show the main ideas most transparently.

Theorem 3.49. *If Assumptions* 3.46 *and* 3.47 *are satisfied, then, with probability 1, the sequence* $\{x^k\}$ *generated by* (3.75) *converges to an optimal solution of* (3.72).

Proof. Suppose x^* is a solution of (3.72). To estimate the (squared) distance of the next iterate x^{k+1} to x^*, we use the nonexpansiveness of the projection operator:

$$\begin{aligned}\|x^{k+1} - x^*\|^2 &= \left\|\Pi_X(x^k - \tau_k \tilde{g}^k) - x^*\right\|^2 \le \left\|x^k - \tau_k \tilde{g}^k - x^*\right\|^2\\ &= \|x^k - x^*\|^2 - 2\tau_k \langle \tilde{g}^k, x^k - x^*\rangle + \tau_k^2 \|\tilde{g}^k\|^2\\ &= \|x^k - x^*\|^2 - 2\tau_k \langle g^k, x^k - x^*\rangle - 2\tau_k \langle e^k, x^k - x^*\rangle + \tau_k^2 \|\tilde{g}^k\|^2.\end{aligned}$$

Since $g^k \in \partial\varphi(x^k)$ and x^* is optimal,

$$\langle g^k, x^* - x^k\rangle \le \varphi(x^*) - \varphi(x^k) \le 0, \tag{3.90}$$

and thus

$$\|x^{k+1} - x^*\|^2 \le \|x^k - x^*\|^2 - 2\tau_k \big(\varphi(x^k) - \varphi(x^*)\big) - 2\tau_k \langle e^k, x^k - x^*\rangle + \tau_k^2 \|\tilde{g}^k\|^2. \tag{3.91}$$

By Assumption 3.47,

$$\mathbb{E}\big[\|\tilde{g}^k\|^2 \,|\, \mathcal{F}_k\big] = \mathbb{E}\big[\|g^k\|^2 + 2\langle g^k, e^k\rangle + \|e^k\|^2 \,|\, \mathcal{F}_k\big] \le C^2 + \sigma^2.$$

Denote $\tilde{\sigma}^2 = C^2 + \sigma^2$. We take the conditional expectation of both sides of (3.91) with respect to $\mathcal{F}_k$ and use Assumption 3.47 to obtain

$$\mathbb{E}\big[\|x^{k+1} - x^*\|^2 \,\big|\, \mathcal{F}_k\big] \le \|x^k - x^*\|^2 - 2\tau_k \big(\varphi(x^k) - \varphi(x^*)\big) + \tilde{\sigma}^2 \tau_k^2. \tag{3.92}$$

Consider the sequence

$$U_k = \|x^k - x^*\|^2 + \tilde{\sigma}^2 \mathbb{E}\Big[\sum_{j=k}^{\infty} \tau_j^2 \,\Big|\, \mathcal{F}_k\Big], \quad k = 0, 1, 2, \ldots.$$

Due to Assumption 3.46(iii), the series $\sum_{j=1}^{\infty} \tau_j^2$ is a convergent submartingale. Therefore, the quantities $\{U_k\}$ are well-defined. It follows from (3.92) and (3.90) that

$$\begin{aligned}\mathbb{E}\big[U_{k+1}\,\big|\,\mathcal{F}_k\big] &= \mathbb{E}\big[\|x^{k+1}-x^*\|^2\,\big|\,\mathcal{F}_k\big]+\tilde{\sigma}^2\mathbb{E}\Big[\sum_{j=k+1}^{\infty}\tau_j^2\,\Big|\,\mathcal{F}_k\Big]\\ &\le \|x^k-x^*\|^2+\tilde{\sigma}^2\tau_k^2+\tilde{\sigma}^2\mathbb{E}\Big[\sum_{j=k+1}^{\infty}\tau_j^2\,\Big|\,\mathcal{F}_k\Big]=U_k.\end{aligned}$$

This means that the sequence $\{U_k\}$ is a nonnegative supermartingale; therefore, it is convergent with probability 1 to some random variable U_∞. We obtain

$$\lim_{k\to\infty}\|x^k-x^*\|^2=\lim_{k\to\infty}U_k-\tilde{\sigma}^2\lim_{k\to\infty}\mathbb{E}\Big[\sum_{j=k}^{\infty}\tau_j^2\,\Big|\,\mathcal{F}_k\Big]=U_\infty,\quad \text{a.s.}$$

Moreover, $\mathbb{E}\big[\|x^k-x^*\|^2\big]\le\mathbb{E}[U_k]\le\mathbb{E}[U_0]$.

Consider an arbitrary $m>0$. We obtain from (3.91) that with probability 1

$$\begin{aligned}&\|x^{k+m}-x^*\|^2\le\|x^k-x^*\|^2\\ &-2\sum_{j=k}^{k+m-1}\tau_j\big(\varphi(x^j)-\varphi(x^*)\big)+2\sum_{j=k}^{k+m-1}\tau_j\langle \mathrm{e}^j,x^j-x^*\rangle+\tilde{\sigma}^2\sum_{j=k}^{k+m-1}\tau_j^2.\end{aligned}\tag{3.93}$$

It follows from Assumption 3.47(ii) that the sequence

$$L_m=\sum_{j=1}^{m-1}\tau_j\langle \mathrm{e}^j,x^j-x^*\rangle,\quad m=1,2,\ldots$$

is a martingale, and

$$\begin{aligned}\mathbb{E}\big[L_{m+1}^2\,\big|\,\mathcal{F}_m\big]&=L_m^2+\tau_m^2\|x^m-x^*\|\,\mathbb{E}\big[\|\mathrm{e}^m\|\,\big|\,\mathcal{F}_m\big]\\ &\le L_m^2+\tau_m^2\Big(\|x^m-x^*\|^2+\mathbb{E}\big[\|\mathrm{e}^m\|^2\,\big|\,\mathcal{F}_m\big]\Big).\end{aligned}$$

Since both $\mathbb{E}\,\|x^m-x^*\|^2$ and $\mathbb{E}\big[\|\mathrm{e}^m\|^2\,\big|\,\mathcal{F}_m\big]$ are uniformly bounded, Assumption 3.46(iii) implies that $\mathbb{E}\,[L_m^2]$ is convergent. Consequently, the martingale $\{L_m\}$ is convergent with probability 1. The last sum in (3.93) is a convergent submartingale directly from Assumption 3.47(ii). Therefore, passing to the limit with $m\to\infty$ in (3.93), we conclude that

$$\sum_{j=0}^{\infty}\tau_j\big(\varphi(x^j)-\varphi(x^*)\big)<\infty\quad\text{a.s..}$$

However, Assumption 3.46(ii) requires that the series of the steplengths is divergent. It follows that a subsequence $\mathcal{K}$ (depending on ω) must exist, such that

$$\lim_{k\in\mathcal{K}} \varphi(x^k) = \varphi(x^*).$$

Let $\hat{x}(\omega)$ be its accumulation point. Obviously, it is an optimal solution of problem (3.72).

Let us fix ω and consider a particular path of the method. Substituting $\hat{x}(\omega)$ for x^* in inequality (3.93), we observe that $\|x^{k+m}(\omega) - \hat{x}(\omega)\|$ can be made arbitrarily close to 0 for all $m \geq 0$, if $k \in \mathcal{K}$ is large enough. This implies that the entire sequence $\{x^j(\omega)\}$ is convergent to $\hat{x}(\omega)$. This is true for all $\omega \in \Omega$, except, possibly, for a set of zero probability. □

We can also obtain objective function error bounds after finitely many iterations.

Suppose we use deterministic stepsizes $\tau_k > 0$, $k = 1, 2, \ldots$. Taking the expected value of both sides of inequality (3.92), we obtain

$$\tau_k \mathbb{E}\big[\varphi(x^k) - \varphi(x^*)\big] \leq \mathbb{E}\big[\|x^k - x^*\|^2\big] - \mathbb{E}\big[\|x^{k+1} - x^*\|^2\big] + \tilde{\sigma}^2 \mathbb{E}\big[\tau_k^2\big].$$

The summation of these inequalities for $k = 0, 1, \ldots, N-1$ yields

$$\mathbb{E} \sum_{k=0}^{N-1} \tau_k \big[\varphi(x^k) - \varphi(x^*)\big] \leq \|x^0 - x^*\|^2 + \tilde{\sigma}^2 \sum_{k=0}^{N-1} \tau_k^2. \tag{3.94}$$

Define the average points

$$\bar{x}^n = \frac{\sum_{k=0}^n \tau_k x^k}{\sum_{k=0}^n \tau_k}, \quad n = 1, 2, \ldots$$

As the function $\varphi(\cdot)$ is convex,

$$\varphi(\bar{x}^n) \leq \frac{\sum_{k=0}^n \tau_k \varphi(x^k)}{\sum_{k=0}^n \tau_k}.$$

We divide (3.94) by $\sum_{k=0}^{N-1} \tau_k$ and use the above inequality to obtain

$$\mathbb{E}\big[\varphi(\bar{x}^{N-1})\big] - \varphi(x^*) \leq \frac{\|x^0 - x^*\|^2 + \tilde{\sigma}^2 \sum_{k=0}^{N-1} \tau_k^2}{\sum_{k=0}^{N-1} \tau_k}.$$

Setting $\tau_k = \alpha/\sqrt{N}$ we get

$$\mathbb{E}\big[\varphi(\bar{x}^{N-1})\big] - \varphi(x^*) \leq \frac{\|x^0 - x^*\|^2 + \tilde{\sigma}^2 \alpha^2}{\alpha\sqrt{N}}.$$

We can thus choose N and α to obtain a desired quality guarantee.

3.4.2.2 The Nonsmooth Nonconvex Case

We now consider the case of problem (3.72) when the mapping $F : \mathbb{R}^n \to \mathcal{L}_p(\mathfrak{D}, \mathcal{B}(\mathfrak{D}), P_D)$, with $p \geq 2$, is Lipschitz continuous, but not necessarily convex. In this case, the composite function $\varphi = \varrho^{(N)} \circ F$ is neither differentiable nor convex. However, under the assumptions of Theorem 3.31, it is Lipschitz continuous and Norkin subdifferentiable, and its generalized subdifferential is given by formula (3.43).

We consider the basic stochastic subgradient method (3.75), with stochastic generalized subgradients $\tilde{g}^k$. We make the same Assumptions 3.46 and 3.47 as in the preceding subsection, but with $\partial\varphi(\cdot)$ now understood as the Norkin subdifferential $\hat{\partial}\varphi(\cdot)$. Observe that the method (3.88) for generating the stochastic subgradients of a mini-batch risk function satisfies Assumption 3.47, if an independent sample $D^{1:N}$ is generated at each iteration k, and Assumption 3.18 is satisfied for all x. As before, a constant C exists, such that $\|g\| \leq C$ for all $g \in \hat{\partial}\varphi(x)$ and all x.

We cannot, in general, expect the convergence of the method (3.75) to the global minimum of problem (3.72). Instead, we establish convergence to the set of critical points,

$$\mathcal{X}^* = \big\{x \in \mathcal{X} : 0 \in \hat{\partial}\varphi(x) + \mathcal{N}_{\mathcal{X}}(x)\big\}. \tag{3.95}$$

We also make an additional assumption about the problem.

Assumption 3.50.

(i) *The set $\mathcal{X}$ is bounded;*

(ii) *The set $\{\varphi(x) : x \in \mathcal{X}^*\}$ does not contain an interval of nonzero length.*[†]

Our proof of convergence of the method (3.75) contains several steps. We focus on a specific realization $\{x^k\}$ of the random sequence generated by the method. To simplify notation, we still write x^k for the kth iterate on this path, instead of $x^k(\omega)$.

First, we construct continuous-time trajectories by linear interpolation. We introduce the accumulated stepsizes

$$t_k = \sum_{j=0}^{k-1} \tau_j, \quad k = 0, 1, 2 \dots,$$

and we construct the interpolated trajectory

$$X_0(t) = x^k + \frac{t - t_k}{\tau_k}(x^{k+1} - x^k), \quad t_k \leq t \leq t_{k+1}, \quad k = 0, 1, 2, \dots.$$

[†] Condition (ii) is known as the *Sard condition*.

Then, for an increasing sequence of positive numbers $\{s_m\}$ diverging to ∞, we define the shifted trajectories

$$X_m(t) = X_0(t + s_m), \quad m = 0, 1, \ldots, \quad t \geq 0. \tag{3.96}$$

Lemma 3.51. *The sequence of functions $\{X_m\}$ is bounded and equicontinuous at every $t \geq 0$.*

Proof. The sequence of functions $\{X_m\}$ is bounded, because the sequence of points $\{x^k\}$ is bounded by Assumption 3.50(i). It is convenient to write the algorithm (3.75) as follows:

$$x^{k+1} = x^k - \tau_k g^k - \tau_k e^k - \tau_k q^k, \quad k = 0, 1, 2, \ldots,$$

with $\tau_k q^k = x^k - \tau_k \tilde{g}^k - \Pi_{\mathcal{X}}(x^k - \tau_k \tilde{g}^k)$. By construction, $q^k \in \mathcal{N}_{\mathcal{X}}(x^{k+1})$.

We define piecewise constant functions $\hat{g}(\cdot)$, $\hat{e}(\cdot)$, and $\hat{q}(\cdot)$ on $[0, \infty)$ as follows

$$\hat{g}(t) = \sum_{k=0}^{\infty} g^k \mathbb{1}_{[t_k, t_{k+1})}, \quad \hat{e}(t) = \sum_{k=0}^{\infty} e^k \mathbb{1}_{[t_k, t_{k+1})}, \quad \hat{q}(t) = \sum_{k=0}^{\infty} q^k \mathbb{1}_{[t_k, t_{k+1})}.$$

Denote for $t \geq 0$

$$G_0(t) = \int_0^t \hat{g}(s)\, \mathrm{d}s, \quad E_0(t) = \int_0^t \hat{e}(s)\, \mathrm{d}s, \quad Q_0(t) = \int_0^t \hat{q}(s)\, \mathrm{d}s.$$

We also define for $m = 0, 1, \ldots$ the shifted functions

$$G_m(t) = G_0(s_m + t), \quad E_m(t) = E_0(s_m + t), \quad Q_m(t) = Q_0(s_m + t), \quad t \geq 0.$$

With this notation, we can decompose $X_m(\cdot)$ as follows:

$$X_m(t) = x^0 - G_m(t) - E_m(t) - Q_m(t), \quad t \geq 0. \tag{3.97}$$

As the sequence $\{g^k\}$ is uniformly bounded by Assumption 3.18, the functions $G_m(\cdot)$ are Lipschitz continuous.

We need to verify the equicontinuity of the functions $E_m(\cdot)$ at the point t. For an arbitrary $\nu > 0$, we define

$$b_{m,\nu} = \sup_{|s-t|<1/\nu} \| E_m(s) - E_m(t) \|$$

and

$$a_\nu = \sup_{m \geq 1} b_{m,\nu}, \quad \nu = 1, 2, \ldots.$$

Since the sequence $\{a_\nu\}$ is nonnegative and nonincreasing by construction, it is convergent to some limit. We will show that the limit is zero, which will imply the equicontinuity at t.

By virtue of Assumptions 3.47 and 3.46, the series $\sum_{k=0}^{\infty} \tau_k e^k$ is a convergent martingale. Therefore, for every $\varepsilon \geq 0$ there exists $M > 0$ such that

$$\sup_{L\geq \ell \geq M} \left\| \sum_{i=\ell}^{L} \gamma_i E^i \right\| \leq \varepsilon.$$

If m is such that $s_m + t \geq t_M + 1$, then the above inequality implies that $\|E_m(s) - E_m(t)\| \leq \varepsilon$ for all s such that $|s - t| < 1$. The set $\mathcal{K}_M = \{m : s_m + t < t_M + 1\}$ is finite, because $s_m \to \infty$ as $m \to \infty$. We can thus choose ν' such that for all $\nu \geq \nu'$ and all $n \in \mathcal{K}_M$, if $|s - t| < 1/\nu$ then $\|E_m(s) - E_m(t)\| \leq \varepsilon$ again. It follows that $a_\nu \leq \varepsilon$ for all $\nu \geq \nu'$. As $\varepsilon > 0$ was arbitrary, $\lim_{\nu\to\infty} a_\nu = 0$.

Due to the nonexpansiveness of the projection, $\|q^k\| \leq \|g^k + e^k\|$, and thus the sequence $\{Q_m(\cdot)\}$ is equicontinuous as well. □

Our next step is the derivation of a continuous-time model of the method. For an absolutely continuous function $X : [0, \infty) \to \mathbb{R}^n$ we denote by $\dot{X}(\cdot)$ its weak derivative: a measurable function such that for every $T > 0$ we have

$$X(T) = X(0) + \int_0^T \dot{X}(t)\, \mathrm{d}t.$$

It is unique up to a set of Lebesgue measure 0.

We can now prove a fundamental property of our sequence of piecewise linear interpolations.

Theorem 3.52. *The sequence of functions $\{X_m\}_{m\in\mathbb{N}}$ has a subsequence convergent a.s. to an absolutely continuous function $X_\infty : [0, \infty) \to \mathcal{X}$ which satisfies the differential inclusion:*

$$\dot{X}(t) \in -\hat{\partial}\varphi(X(t)) - \mathcal{N}_{\mathcal{X}}(X(t)), \quad t \geq 0. \tag{3.98}$$

Proof. By the Arzela–Ascoli theorem, in view of Lemma 3.51, a subsequence $\{X_{m_j}\}_{j\in\mathbb{N}}$ of $\{X_m\}$ and a function $X_\infty : [0, \infty) \to \mathbb{R}^n$ exist such that for every $T > 0$

$$\lim_{j\to\infty} \sup_{0\leq t\leq T} \|X_{m_j}(t) - X_\infty(t)\| = 0.$$

To simplify notation, by redefining the sequence $\{s_m\}$, we may assume that the entire sequence $\{X_m\}$ is convergent to X_∞ uniformly on bounded intervals. Because the set $\mathcal{X}$ is convex and closed, $X_\infty(t) \in \mathcal{X}$ for all $t \geq 0$.

Each function $X_m(\cdot)$ is absolutely continuous by construction. Consider their weak derivatives $\dot{X}_m(\cdot)$. From (3.97) we obtain the relations

$$\dot{X}_m(t) = -\dot{G}_m(t) - \dot{E}_m(t) - \dot{Q}_m(t), \quad t \geq 0.$$

As the sequence $\sum_{k=0}^{\infty} \tau_k e^k$ is a convergent martingale, the sequence of functions $E_m(\cdot)$ converges to a function equal to zero, uniformly on compact intervals. The functions $G_m(\cdot)$ are uniformly Lipschitz continuous. Furthermore, $\|\dot{Q}_m(t)\| \leq \|\dot{G}_m(t)\| + \|\dot{E}_m(t)\|$.

It follows that for every interval $[a, b]$ the functions $\mathbb{1}_{[a,b]}\dot{X}_m(t)$ are uniformly integrable. By the Dunford–Pettis theorem, a weakly convergent subsequence $\{\mathbb{1}_{[a,b]}\dot{X}_{n_j}\}_{j\in N}$ exists. Let $\mathbb{1}_{[a,b]}H(\cdot)$ be its weak limit. Thus, for every $c \in [a, b]$ we have

$$\begin{aligned}\int_a^c H(t)\,\mathrm{d}t &= \lim_{j\to\infty} \int_a^c \dot{X}_{n_j}(t)\,\mathrm{d}t \\ &= \lim_{j\to\infty} \left[X_{n_j}(c) - X_{n_j}(a)\right] = X_\infty(c) - X_\infty(a),\end{aligned}$$

which means that $H(\cdot) = \dot{X}_\infty(\cdot)$ in the interval $[a, b]$. As every subsequence of $\{\dot{X}_m\}$ has a sub-subsequence convergent weakly to $\dot{X}_\infty(\cdot)$ on $[a, b]$, the entire sequence $\{\dot{X}_m\}$ converges weakly to $\dot{X}_\infty(\cdot)$ on any compact interval $[a, b]$. By Mazur's lemma, coefficients $w_{m,i} \geq 0$, $i = m, \ldots, N(n)$, exist, with $\sum_{i=m}^{N(n)} w_{m,i} = 1$, such that the convex combination sequence

$$\dot{X}^w_m(\cdot) = \sum_{i=m}^{N(n)} w_{m,i}\dot{X}_i(\cdot), \quad m = 1, 2, \ldots,$$

is convergent strongly to $\dot{X}_\infty(\cdot)$ on $[a, b]$. Then a subsequence $\{\dot{X}^w_{n_j}(\cdot)\}$ exists which is convergent almost surely to $\dot{X}_\infty(\cdot)$ on $[a, b]$. Since the half-line $[0, \infty)$ can be covered by countably many intervals, by diagonalization, we can choose a convex combination subsequence $\{\dot{X}^w_{n_j}(\cdot)\}$ convergent almost surely to $\dot{X}_\infty(\cdot)$ on $[0, \infty)$. Due to the convexity and closedness of the set $\mathcal{X}$, the function $X_\infty(\cdot)$ has values in $\mathcal{X}$.

Summing up, after an appropriate redefinition of the sequence $\{s_m\}$, for almost all $t \in [0, \infty)$, we have almost surely the relations

$$\lim_{m\to\infty} X_m(t) = X_\infty(t) \quad \text{and} \quad \lim_{m\to\infty} \dot{X}^w_m(t) = \dot{X}_\infty(t). \tag{3.99}$$

Let us fix $t \geq 0$. Define $k(m, t) = \max\{k : t_k \leq s_m + t\}$. Since $k(m, t) \to \infty$ as $n \to \infty$, we obtain

$$\lim_{m\to\infty} \left\| X_m(t) - x^{k(m,t)} \right\| = 0, \quad \dot{X}_m(t) = -g^{k(m,t)} - q^{k(m,t)} - \dot{E}_m(t),$$

where $g^{k(m,t)} \in \hat{\partial}\varphi\big(x^{k(m,t)}\big)$ and $q^{k(m,t)} \in \mathcal{N}_{\mathcal{X}}\big(x^{k(m,t)+1}\big)$ for all n. As discussed before, $\lim_{m\to\infty} \dot{E}_m(t) = 0$, and thus (3.99) implies that

$$\lim_{m\to\infty} x^{k(m,t)} = X_\infty(t), \quad \text{and} \quad -\lim_{m\to\infty} \sum_{i=m}^{N(n)} w_{m,i}\big[g^{k(i,t)} + q^{k(i,t)}\big] = \dot{X}_\infty(t).$$

Since the multifunctions $\hat{\partial}\varphi(\cdot)$ and $\mathcal{N}_{\mathcal{X}}(\cdot)$ are upper semicontinuous and convex-valued, then for almost all $t \ge 0$ we obtain

$$\dot{X}_\infty(t) = -\hat{g}(t) - \hat{q}(t).$$

with some $\hat{g}(t) \in \hat{\partial}\varphi(X_\infty(t))$ and $\hat{q}(t) \in \mathcal{N}_{\mathcal{X}_\infty}(X(t))$, which is equivalent to (3.98). □

We could write the differential inclusion (3.98) in an equivalent way

$$\dot{X}(t) \in \Pi_{\mathcal{T}_{\mathcal{X}}(X(t))}\big(-\partial\varphi(X(t))\big), \quad t \ge 0,$$

which underlines the nature of the limiting subgradient process.

We are now ready to prove the main convergence theorem.

Theorem 3.53. *Suppose Assumptions* 3.18 *(with* $p = 2$*),* 3.19, 3.46, 3.47, *and* 3.50 *are satisfied. Then, with probability 1, the sequence* $\{\varphi(x^k)\}_{k\in N}$ *is convergent and every accumulation point of the sequence* $\{x^k\}_{k\in N}$ *is an element of* $\mathcal{X}^*$.

Proof. Define

$$\underline{\varphi} = \liminf_{k\to\infty} \varphi(x^k), \qquad \overline{\varphi} = \limsup_{k\to\infty} \varphi(x^k).$$

Suppose $\underline{\varphi} < \overline{\varphi}$. By Assumption 3.50(ii), a value $v \in (\underline{\varphi}, \overline{\varphi})$ exists, such that $v \notin \varphi(\mathcal{X}^*)$. Take any $w \in (v, \overline{\varphi})$. Define $\nu_1 = \min\big\{k : \varphi(x^k) \le v\big\}$ and for $j = 1, 2, \dots$

$$\begin{aligned} \ell_j &= \min\big\{k \ge \nu_j : \varphi(x^k) > w\big\}, \\ \nu_{j+1} &= \min\big\{k \ge \ell_j : \varphi(x^k) \le v\big\}. \end{aligned}$$

Then we define

$$k_j = \max\big\{k \in [\nu_j, \ell_j] : \varphi(x^k) \le v\big\}, \quad j = 1, 2, \dots.$$

Both $k_j \to \infty$ and $\ell_j \to \infty$, as $j \to \infty$.

Consider the sums $T_j = \sum_{k=k_j}^{\ell_j-1} \tau_k$. Denoting by C the Lipschitz constant of $\varphi(\cdot)$, we obtain

$$w - v \le \varphi(x^{\ell_j}) - \varphi(x^{k_j}) \le C\|x^{\ell_j} - x^{k_j}\|. \tag{3.100}$$

For $k = k_j, \dots, \ell_j - 1$ we have

$$\begin{aligned}\|x^{k+1} - x^{k_j}\|^2 &= \big\|\Pi_{\mathcal{X}}(x^k - \tau_k \tilde{g}^k) - x^{k_j}\big\|^2 \le \big\|x^k - \tau_k \tilde{g}^k - x^{k_j}\big\|^2 \\ &= \|x^k - x^{k_j}\|^2 - 2\tau_k\langle g^k, x^k - x^{k_j}\rangle - 2\tau_k\langle e^k, x^k - x^{k_j}\rangle + \tau_k^2\|\tilde{g}^k\|^2.\end{aligned}$$

Thus

$$\begin{aligned}\|x^{\ell_j} - x^{k_j}\|^2 \le 2\sum_{k=k_j}^{\ell_j-1} \tau_k\|g^k\|\|x^k - x^{k_j}\| \\ + \sum_{k=k_j}^{\ell_j-1}\big[-\tau_k\langle e^k, x^k - x^{k_j}\rangle + \tau_k^2\|\tilde{g}^k\|^2\big].\end{aligned}$$

Since $\sum_{k=1}^{\infty} \tau_k e^k$ is a convergent martingale, and $\sum_{k=1}^{\infty} \tau_k^2$ is a convergent submartingale, the last sum is negligible, when $j \to \infty$. We conclude that for all sufficiently large j,

$$\|x^{\ell_j} - x^{k_j}\|^2 \le 3CDT_j,$$

where D is an upper bound on $\|x^k - x^{k_j}\|$. With a view on (3.100),

$$T_j \ge \frac{(w-v)^2}{3C^3D}.$$

Set $s_j = t_{k_j}$ in (3.96), and consider the interval $[0, T]$ with $T = \frac{(w-v)^2}{3C^3D}$. By Theorem 3.52, a subsequence of the sequence $\{X_j\}$ is convergent to a solution X_∞ of the differential inclusion (3.98). For simplicity, we still denote it by $\{X_j\}$. Due to Theorem 3.33, the generalized Newton–Leibniz formula (3.49) is true:

$$\varphi(X_\infty(T)) = \varphi(X_\infty(0)) + \int_0^T \langle g(t), \dot{X}(t)\rangle\, \mathrm{d}t, \tag{3.101}$$

for any selection $g(t) \in \hat{\partial}\varphi(X_\infty(t))$, $t \in [0, T]$. From (3.98), we deduce that

$$\dot{X}_\infty(t) = -\hat{g}(t) - \hat{q}(t).$$

with some $\hat{g}(t) \in \hat{\partial}\varphi(X_\infty(t))$ and $\hat{q}(t) \in \mathcal{N}_{\mathcal{X}}(X_\infty(t))$. From the inclusion $X_\infty(t) \in \mathcal{X}$ for all t, we conclude that $\dot{X}_\infty(t) \perp \hat{q}(t)$ for almost all t. Setting $g(t) = \hat{g}(t)$ in (3.101) we obtain

$$
\begin{aligned}
\varphi(X_\infty(T)) &= \varphi(X_\infty(0)) - \int_0^T \langle \hat{g}(t), \hat{g}(t) + \hat{q}(t)\rangle \, dt \\
&= \varphi(X_\infty(0)) - \int_0^T \|\hat{g}(t) + \hat{q}(t)\|^2 \, dt.
\end{aligned} \tag{3.102}
$$

As $v \notin \varphi(\mathcal{X}^*)$, we have $X_\infty(0) = \lim_{j\to\infty} x^{k_j} \notin \mathcal{X}^*$, and thus

$$
0 \notin \hat{\partial}\varphi(X_\infty(0)) + \mathcal{N}_{\mathcal{X}}(X_\infty(0)).
$$

The Norkin subdifferential $\hat{\partial}\varphi(\cdot)$ and the normal cone $\mathcal{N}_{\mathcal{X}}(\cdot)$ are upper semicontinuous and closed valued. This implies that

$$
\operatorname{dist}\big(0, \hat{\partial}\varphi(X_\infty(t)) + \mathcal{N}_{\mathcal{X}}(X_\infty(t))\big) > 0
$$

in some interval $t \in [0, \bar{t}\,]$ of positive length. It follows from (3.102) that $\varphi(X_\infty(T)) < \varphi(X_\infty(0)) = v$. However, $X_\infty(T)$ is the limit of the sequence $\{x^{m_j}\}$, where

$$
m_j = \min\Bigg\{ m \ge k_j : \sum_{k=k_j}^{m_j-1} \tau_k \ge T \Bigg\}, \quad j = 1, 2, \ldots. \tag{3.103}
$$

For all sufficiently large j we have $k_j \le m_j \le \ell_j$ and $\varphi(x^{m_j}) > v$, which is a contradiction. Therefore, $\underline{\varphi} = \overline{\varphi}$ and the sequence $\big\{\varphi(x^k)\big\}$ has a limit.

Suppose a subsequence $\{x^{k_j}\}$ is convergent to a point $\bar{x} \notin \mathcal{X}^*$. Again, we set $s_j = t_{k_j}$ in (3.96), consider the interval $[0, T]$ with $T > 0$, and conclude that a sub-subsequence of $\{X_j\}$ exists, which is uniformly convergent to a solution $X_\infty(\cdot)$ of the differential inclusion (3.98). Then the points $\{x^{m_j}\}$ are convergent to $X_\infty(T)$ on this sub-subsequence, where m_j are defined as in (3.103). As $X_\infty(0) = \bar{x} \notin \mathcal{X}^*$, we obtain $X_\infty(T) < X_\infty(0)$. This would mean that a sub-subsequence of $\big\{\varphi(x^{m_j})\big\}$ has the limit $\varphi(X_\infty(T)) < \varphi(\bar{x})$, which contradicts the convergence of the sequence $\big\{\varphi(x^k)\big\}$. □

3.4.3 The Method with Dual Averaging

We now focus on a popular stochastic subgradient method involving an auxiliary filter that averages the stochastic subgradients observed in the course of its operation. Its versions are known under various names: a heavy ball method, a method with momentum, or a method with averaging. Particularly useful in our context is the single time-scale variant, where the stepsizes in the main iterations and in the filter are proportional.

The method generates two random sequences: approximate solutions $\{x^k\}$ and path-averaged stochastic subgradients $\{z^k\}$, defined on a certain probability space $(\Omega, \mathcal{F}, P)$. We let $\mathcal{F}_k$ to be the σ-algebra generated by $\{x^0, \ldots, x^k, z^0, \ldots, z^k\}$. We assume that for each k, we can observe an $\mathcal{F}_k$-measurable random vector $\tilde{g}^k \in \mathbb{R}^n$, such that, for some $\mathcal{F}_k$-measurable vector e^k, we have $\tilde{g}^k = g^k + e^k$, with $g^k \in \partial\varphi(x^k)$, and Assumption 3.47 satisfied. Depending on the application, the subdifferential $\partial\varphi(\cdot)$ is understood in the convex analysis sense, if $\varphi(\cdot)$ is convex, in the Clarke sense, if $\varphi(\cdot)$ is subdifferentially regular, and in the Norkin sense, if $\varphi(\cdot)$ is Norkin subdifferentiable.

The method proceeds for $k = 0, 1, 2\ldots$ as follows ($a > 0$ and $\beta > 0$ are fixed parameters). We compute

$$y^k = \operatorname*{arg\,min}_{y\in\mathcal{X}} \left\{\langle z^k, y - x^k\rangle + \frac{\beta}{2}\|y - x^k\|^2\right\}, \tag{3.104}$$

and, with an $\mathcal{F}_k$-measurable stepsize $\tau_k \in \big(0, \min(1, 1/a)\big]$, we set

$$x^{k+1} = x^k + \tau_k(y^k - x^k). \tag{3.105}$$

Then we observe g^{k+1} at x^{k+1}, and update the averaged stochastic subgradient as

$$z^{k+1} = (1 - a\tau_k)z^k + a\tau_k g^{k+1}. \tag{3.106}$$

Recall that the set $\mathcal{X}^*$ of critical points of problem (3.72) is defined in (3.95).

Before proceeding to the convergence analysis in various cases, we establish useful properties of the gap function $\eta : \mathcal{X} \times \mathbb{R}^n \to (-\infty, 0]$,

$$\eta(x, z) = \min_{y\in\mathcal{X}} \left\{ \langle z, y - x\rangle + \frac{\beta}{2}\|y - x\|^2\right\}. \tag{3.107}$$

We denote the minimizer in (3.107) by $\bar{y}(x, z)$.

Lemma 3.54. *For all $x \in \mathcal{X}$ and $z \in \mathbb{R}^n$*

(i) $\eta(\cdot,\cdot)$ *is continuously differentiable, with*

$$\nabla_x\eta(x, z) = -z + \beta(x - \bar{y}(x, z)) \quad \textit{and} \quad \nabla_z\eta(x, z) = \bar{y}(x, z) - x;$$

furthermore, the gradients are Lipschitz continuous with respect to (x, z);

(ii) $\langle z, \bar{y}(x, z) - x\rangle + \beta\|\bar{y}(x, z) - x\|^2 \le 0$;

(iii) $x^* \in \mathcal{X}^*$ *if and only if $g^* \in \hat{\partial}\varphi(x^*)$ exists such that $\eta(x^*, g^*) = 0$.*

Proof. (i) Due to the convexity and closedness of the set $\mathcal{X}$, the minimizer $\bar{y}(x, z)$ always exists and is unique. In fact, it is the orthogonal projection of $x - z/\beta$ on $\mathcal{X}$. The differentiability of the gap function and the formulae for the gradient follow from the Danskin theorem. The Lipschitz continuity of the

gradient follows from the nonexpansiveness of the projection operator.
(ii) As $\bar{y}(x,z) = \Pi_{\mathcal{X}}(x - z/\beta)$ and $x \in \mathcal{X}$, we have

$$\big\langle x - z/\beta - \bar{y}(x,z), x - \bar{y}(x,z)\big\rangle \le 0,$$

which is the postulated inequality.
(iii) $x^* \in \mathcal{X}^*$ if and only if $g^* \in \hat{\partial}\varphi(x^*)$ exists such that $-g^* \in \mathcal{N}_{\mathcal{X}}(x^*)$. The latter property is equivalent to $x^* = \Pi_{\mathcal{X}}(x^* - g^*/\beta) = \bar{y}(x^*, g^*)$. □

We analyze the most general case and establish asymptotic convergence of the algorithm (3.104)–(3.106) under the Assumptions 3.46, 3.47, and 3.50 (with $\partial\varphi$ understood as the Norkin subdifferential $\hat{\partial}\varphi$).

We start from the following elementary property of the sequence $\{z^k\}$.

Lemma 3.55. *If the sequence $\{x^k\}$ is included in a set $A \subset \mathbb{R}^n$, then*

$$\lim_{k\to\infty} \operatorname{dist}(z^k, B) = 0, \quad \textit{where} \quad B = \operatorname{conv}\Big(\bigcup_{x\in A} \hat{\partial} f(x)\Big).$$

Proof. Consider the quantities $\tilde{z}^k = z^k + a\sum_{j=k}^{\infty} \tau_j e^j$, $k \in \mathbb{N}$. They are well-defined, because the Assumptions 3.46 and 3.47 imply that $\sum_{j=0}^{\infty} \tau_j e^j$ is a convergent martingale. This allows us to establish the recursive relation

$$\tilde{z}^{k+1} = (1 - a\tau_k)\tilde{z}^k + a\tau_k(g^k + \Delta_k), \quad k = 0, 1, 2, \ldots,$$

where $g^k \in B$ and $\Delta_k = \sum_{j=k}^{\infty} \tau_j e^j \to 0$ a.s.. By the convexity of the distance function,

$$\operatorname{dist}(\tilde{z}^{k+1}, B) \le (1 - a\tau_k)\operatorname{dist}(\tilde{z}^k, B) + a\tau_k\|\Delta_k\|, \quad k = 0, 1, 2, \ldots.$$

Since $\|\Delta_k\| \to 0$ and $\sum_{k=0}^{\infty} \tau_k = \infty$, we conclude that $\operatorname{dist}(\tilde{z}^k, B) \to 0$ and $\operatorname{dist}(z^k, B) \to 0$ as well. □

Theorem 3.56. *If Assumptions* 3.46, 3.47, *and* 3.50 *are satisfied, then, with probability 1, every accumulation point $\hat{x}$ of the sequence $\{x^k\}$ is stationary, and the sequence $\{\varphi(x^k)\}$ is convergent.*

Proof. By virtue of Lemma 3.55, the sequence $\{z^k\}$ is bounded. We divide the proof into three standard steps.

Step 1: The Limiting Dynamical System. We define $p^k = (x^k, z^k)$, accumulated stepsizes $t_k = \sum_{j=0}^{k-1} \tau_j$, $k = 0, 1, 2\ldots$, and we construct the interpolated trajectory

$$P_0(t) = p^k + \frac{t - t_k}{\tau_k}(p^{k+1} - p^k), \quad t_k \le t \le t_{k+1}, \quad k = 0, 1, 2, \ldots.$$

For an increasing sequence of positive numbers $\{s_m\}$ diverging to ∞, we define shifted trajectories $P_m(t) = P_0(t + s_m)$. We write $P_m(t) = \big(X_m(t), Z_m(t)\big)$.

In the same way as in the proof of Theorem 3.52, there exist a subsequence $\{m_j\}_{j\in\mathbb{N}}$ and absolutely continuous functions $X_\infty : [0, +\infty) \to X$ and $Z_\infty : [0, +\infty) \to \mathbb{R}^n$ such that for any $T > 0$

$$\lim_{j\to\infty} \sup_{t\in[0,T]} \big(\|X_{m_j}(t) - X_\infty(t)\| + \|Z_{m_j}(t) - Z_\infty(t)\|\big) = 0,$$

and $(X_\infty(\cdot), Z_\infty(\cdot))$ is a solution of the system of differential equations and inclusions:

$$\dot{X}(t) = \bar{y}\big(X(t), Z(t)\big) - X(t), \tag{3.108}$$

$$\dot{Z}(t) \in a\big(\hat{\partial}\varphi(X(t)) - Z(t)\big). \tag{3.109}$$

Moreover, for any $t \ge 0$, the pair $(X_\infty(t), Z_\infty(t))$ is an accumulation point of the sequence $\{(x^k, z^k)\}$.

Step 2: Descent Along a Path. We use the Lyapunov function

$$W(x, z) = a\varphi(x) - \eta(x, z).$$

For any solution $(X(\cdot), Z(\cdot))$ of the system (3.108)–(3.109), and for any $T > 0$, we estimate the difference $W(X(T), Z(T)) - W(X(0), Z(0))$. We split $W(X(\cdot), Z(\cdot))$ into a generalized differentiable composition $f(X(\cdot))$ and the continuously differentiable part $\eta(X(\cdot), Z(\cdot))$.

Since the path $X(\cdot)$ satisfies (3.108) and $\bar{y}(\cdot, \cdot)$ is continuous, $X(\cdot)$ is continuously differentiable. Thus, we can use Theorems 3.32 or 3.33 to conclude that for any $g(X(\cdot)) \in \hat{\partial}\varphi(X(\cdot))$

$$\begin{aligned}\varphi(X(T)) - \varphi(X(0)) &= \int_0^T \big\langle g(X(t)), \dot{X}(t)\big\rangle\, dt \\ &= \int_0^T \big\langle g(X(t)), \bar{y}(X(t), Z(t)) - X(t)\big\rangle\, dt.\end{aligned} \tag{3.110}$$

Due to Lemma 3.54(i), the function $\eta(\cdot, \cdot)$ is continuously differentiable, and the chain formula holds for it as well:

$$\begin{aligned}&\eta(X(T), Z(T)) - \eta(X(0), Z(0)) \\ &= \int_0^T \big\langle \nabla_x \eta(X(t), Z(t)), \dot{X}(t)\big\rangle\, dt + \int_0^T \big\langle \nabla_z \eta(X(t), Z(t)), \dot{Z}(t)\big\rangle\, dt.\end{aligned}$$

Substituting $\nabla_x \eta(x,z) = -z + \beta(x - \bar{y}(x,z))$, $\nabla_z \eta(x,z) = \bar{y}(x,z) - x$ and $\dot{Z}(t) = a\big(\hat{g}(X(t)) - Z(t)\big)$ with some $\hat{g}(X(\cdot)) \in \hat{\partial} f(X(\cdot))$, and using Lemma 3.54(ii) we obtain

$$\begin{aligned}
&\eta(X(T), Z(T)) - \eta(X(0), Z(0)) \\
&= \int_0^T \big\langle -Z(t) + \beta(X(t) - \bar{y}(X(t), Z(t))), \, \bar{y}(X(t), Z(t)) - X(t) \big\rangle \, \mathrm{d}t \\
&\quad + a \int_0^T \big\langle \bar{y}(X(t), Z(t)) - X(t), \, \hat{g}(X(t)) - Z(t) \big\rangle \, \mathrm{d}t \\
&\geq a \int_0^T \big\langle \bar{y}(X(t), Z(t)) - X(t), \, \hat{g}(X(t)) - Z(t) \big\rangle \, \mathrm{d}t \\
&\geq a \int_0^T \big\langle \bar{y}(X(t), Z(t)) - X(t), \, \hat{g}(X(t)) \big\rangle \, \mathrm{d}t \\
&\quad + a\beta \int_0^T \big\| \bar{y}(X(t), Z(t)) - X(t) \big\|^2 \, \mathrm{d}t.
\end{aligned}$$

We substitute the subgradient selector $g(X(t)) = \hat{g}(X(t))$ into (3.110) and combine it with the last inequality, concluding that

$$\begin{aligned}
&W(X(T), Z(T)) - W(X(0), Z(0)) \\
&\leq -a\beta \int_0^T \big\| \bar{y}(X(t), Z(t)) - X(t) \big\|^2 \, \mathrm{d}t = -a\beta \int_0^T \big\| \dot{X}(t) \big\|^2 \, \mathrm{d}t.
\end{aligned} \tag{3.111}$$

Step 3: Analysis of the Limit Points. Define the set $\mathcal{S} = \big\{(x,z) \in \mathcal{X}^* \times \mathbb{R}^n : \eta(x,z) = 0\big\}$. Suppose $(\bar{x}, \bar{z})$ is an accumulation point of the sequence $\{(x^k, z^k)\}$. If $\eta(\bar{x}, \bar{z}) < 0$, then every solution $(X(t), Z(t))$ of the system (3.108)–(3.109), starting from $(X(0), Z(0)) = (\bar{x}, \bar{z})$ has $\dot{X}(0) \neq 0$. Using (3.111) and arguing as in the proof of Theorem 3.53, we obtain a contradiction. Therefore, we must have $\eta(\bar{x}, \bar{z}) = 0$. Suppose $\bar{x} \notin \mathcal{X}^*$. Then

$$\operatorname{dist}\big(0, \hat{\partial}\varphi(\bar{x}) + \mathcal{N}_{\mathcal{X}}(\bar{x})\big) > 0. \tag{3.112}$$

Suppose $X(t) = \bar{x}$ for all $t \geq 0$. The inclusion (3.109) simplifies: $\dot{Z}(t) \in a\big(\hat{\partial}\varphi(\bar{x}) - Z(t)\big)$. By using the convex Lyapunov function $V(z) = \operatorname{dist}\big(z, \hat{\partial}\varphi(\bar{x})\big)$ and applying the classical chain formula on the path $Z(\cdot)$, we deduce that

$$\lim_{t\to\infty} \operatorname{dist}\big(Z(t), \hat{\partial}\varphi(\bar{x})\big) = 0. \tag{3.113}$$

It follows from (3.112)–(3.113) that $T > 0$ exists, such that $-Z(T) \notin \mathcal{N}_{\mathcal{X}}(\bar{x})$, which yields $\dot{X}(T) \neq 0$. Consequently, the path $X(t)$ starting from $\bar{x}$ cannot be constant. But then again $T > 0$ exists, such that $\dot{X}(T) \neq 0$. By Step 1, the pair $(X(T), Z(T))$ would have to be an accumulation point of of the sequence $\{(x^k, z^k)\}$, a case already excluded. We conclude that every accumulation point $(\bar{x}, \bar{z})$ of the sequence $\{(x^k, z^k)\}$ is in $\mathcal{S}$. The convergence of the sequence $\{W(x^k, z^k)\}$ then follows in the same way as in the proof of Theorem 3.53. As $\eta(x^k, z^k) \to 0$, the convergence of $\{\varphi(x^k)\}$ follows as well. □

Directly from Lemma 3.55 we obtain the convergence of the averaged stochastic subgradients.

Corollary 3.57. *If the sequence $\{x^k\}$ is convergent to a single point $\bar{x}$, then every accumulation point of $\{z^k\}$ is an element of $\hat{\partial}\varphi(\bar{x})$.*

Chapter 4
Dynamic Risk Optimization

In this chapter, we present methods to quantify the risk of a random sequence $Z_1,\dots,Z_T$. When the sequence depends on our decisions made at stages $1,\dots,T$, the risk evaluation and decision-making should form a cohesive approach. This implies that at each stage t, we need to be able to evaluate the risk of the current and future outcomes $Z_t,\dots,Z_T$. The theory of dynamic risk is concerned with consistent ways of obtaining such evaluations and using them in optimization models.

Even in the expected value case, dynamic optimization models pose serious theoretical and computational difficulties. In our setting, they are compounded with the use of dynamic risk measures.

4.1 Dynamic Measures of Risk

4.1.1 Time Consistency

Consider a probability space $(\Omega, \mathcal{F}, P)$, a filtration $\mathcal{F}_1 \subset \cdots \subset \mathcal{F}_T \subset \mathcal{F}$, and an adapted sequence of random variables $\{Z_t\}_{t=1,\dots,T}$. We assume that $\mathcal{F}_1 = \{\Omega, \emptyset\}$, and thus Z_1 is in fact deterministic. In our all considerations, we interpret the variables Z_t as stage-wise costs. Define the spaces $\mathcal{Z}_t = \mathcal{L}_p(\Omega, \mathcal{F}_t, P)$, $p \in [1,\infty)$, $t = 1,\dots,T$, and let $\mathcal{Z}_{t:s} \triangleq \mathcal{Z}_t \times \cdots \times \mathcal{Z}_s$ for $1 \le t \le s \le T$.

We need to evaluate the risk of the tail subsequences $Z_t,\dots,Z_T$ from the perspective of each stage t. This motivates the following definition.

Definition 4.1. A *conditional risk measure* is a mapping $\varrho_{t,T} : \mathcal{Z}_{t:T} \to \mathcal{Z}_t$, where $1 \le t \le T$. A *dynamic risk measure* is a sequence of conditional risk measures $\varrho_{t,T} : \mathcal{Z}_{t:T} \to \mathcal{Z}_t$, $t = 1,\dots,T$.

D. Dentcheva, A. Ruszczyński, *Risk-Averse Optimization and Control*, Springer Series in Operations Research and Financial Engineering,
https://doi.org/10.1007/978-3-031-57988-2_4

The value of the conditional risk measure $\varrho_{t,T}(Z_t,\dots,Z_T)$ can be interpreted as a fair one-time $\mathcal{F}_t$-measurable charge we would be willing to incur at time t, instead of the sequence of random costs $Z_t,\dots,Z_T$.

In the following definition and later, the equation "$=$" and the inequality "$\leq$" between random sequences are understood almost surely and at each time.

Definition 4.2. A conditional risk measure $\varrho_{t,T}$, where $1 \leq t \leq T$,

(i) is *normalized*, if $\varrho_{t,T}(0,\dots,0) = 0$;

(ii) is *monotonic*, if the inequality $(Z_t,\dots,Z_T) \leq (V_t,\dots,V_T)$ implies that

$$\varrho_{t,T}(Z_t,\dots,Z_T) \leq \varrho_{t,T}(V_t,\dots,V_T);$$

(iii) has the *translation property*, if

$$\varrho_{t,T}(Z_t,\dots,Z_T) = Z_t + \varrho_{t,T}(0, Z_{t+1},\dots,Z_T).$$

A dynamic risk measure $\{\varrho_{t,T}\}_{1\leq t\leq T}$ is normalized, monotonic, or has the translation property, if each of its conditional risk measures has the corresponding property.

The key issue associated with dynamic preferences is the question of their consistency over time.

Definition 4.3. A dynamic risk measure $\{\varrho_{t,T}\}_{1\leq t\leq T}$ is *time-consistent* if for all $1 \leq t < T$ and all sequences $Z, W \in \mathcal{Z}_{t:T}$ the conditions

$$Z_t = W_t \quad \text{and} \quad \varrho_{t+1,T}(Z_{t+1},\dots,Z_T) \leq \varrho_{t+1,T}(W_{t+1},\dots,W_T) \tag{4.1}$$

imply that

$$\varrho_{t,T}(Z_t,\dots,Z_T) \leq \varrho_{t,T}(W_t,\dots,W_T). \tag{4.2}$$

In other words, if Z will be at least as good as W from the perspective of the next stage $t+1$, and they are identical now, then Z should not be worse than W from the present perspective.

Remark 4.4. Definition 4.3 implies that if $Z_\tau = W_\tau$ for $\tau = t,\dots,\theta-1$, and $\varrho_{\theta,T}(Z_\theta,\dots,Z_T) \leq \varrho_{\theta,T}(W_\theta,\dots,W_T)$, then (4.2) is true. This can be established by applying time consistency recursively from $\theta-1$ back to t. □

For a dynamic risk measure $\{\varrho_{t,T}\}_{1\leq t\leq T}$ we define a broader family of conditional risk measures, by setting

$$\varrho_{t,\theta}(Z_t,\dots,Z_\theta) = \varrho_{t,T}(Z_t,\dots,Z_\theta,0,\dots,0), \quad 1 \leq t \leq \theta \leq T. \tag{4.3}$$

The following structure of a time-consistent dynamic risk measure can be established.

Theorem 4.5. *Suppose a dynamic risk measure $\{\varrho_{t,T}\}_{1\le t\le T}$ is normalized and satisfies the translation property. Then it is time-consistent if and only if it is monotonic and for all $1 \le t \le \theta \le T$ and all $Z \in \mathcal{Z}_{t:T}$ the following identity is true:*

$$\varrho_{t,T}\big(Z_t,\dots,Z_T\big) = \varrho_{t,\theta}\big(Z_t,\dots,Z_{\theta-1},\varrho_{\theta,T}(Z_\theta,\dots,Z_T)\big). \tag{4.4}$$

Proof. We consider two sequences in $\mathcal{Z}_{t:T}$:

$$\begin{aligned} Z &= (Z_t,\dots,Z_{\theta-1},Z_\theta,Z_{\theta+1},\dots,Z_T),\\ W &= \big(Z_t,\dots,Z_{\theta-1},\varrho_{\theta,T}(Z_\theta,\dots,Z_T),0,\dots,0\big). \end{aligned}$$

By the translation property and normalization,

$$\begin{aligned} \varrho_{\theta,T}(W_\theta,\dots,W_T) &= \varrho_{\theta,T}(\varrho_{\theta,T}(Z_\theta,\dots,Z_T),0,\dots,0)\\ &= \varrho_{\theta,T}(Z_\theta,\dots,Z_T) + \varrho_{\theta,T}(0,\dots,0) = \varrho_{\theta,T}(Z_\theta,\dots,Z_T). \end{aligned}$$

Suppose the measure $\{\varrho_{t,T}\}_{1\le t\le T}$ is time-consistent. Using the above equation, we get $\varrho_{t,T}(Z) = \varrho_{t,T}(W)$. By (4.3), we conclude that

$$\varrho_{t,T}(Z_t,\dots,Z_T) = \varrho_{t,T}(W_t,\dots,W_\theta,0,\dots,0) = \varrho_{t,\theta}(W_t,\dots,W_\theta).$$

Substitution of W yields (4.4). We verify the monotonicity by induction backward in time. The measure $\rho_{T,T}(\cdot)$ is monotonic, because translation and normalization imply that $\rho_{T,T}(Z_T) = Z_T$. Suppose $\rho_{t+1,T}(\cdot)$ is monotonic for some $t < T$. If $Z_\theta \le W_\theta$ for $\theta = t,\dots,T$ then we have

$$\rho_{t+1,T}(Z_{t+1},\dots,Z_T) \le \rho_{t+1,T}(W_{t+1},\dots,W_T).$$

Therefore, the translation and the time consistency imply that

$$\begin{aligned} \rho_{t,T}(Z_t,\dots,Z_T) &= Z_t + \rho_{t,T}(0,Z_{t+1},\dots,Z_T)\\ &\le W_t + \rho_{t,T}(0,W_{t+1},\dots,W_T) = \rho_{t,T}(W_t,\dots,W_T). \end{aligned}$$

By induction, all $\rho_{t,T}$, $t = 1,\dots,T$, are monotonic.

Conversely, suppose the identity (4.4) is true and all $\rho_{t,T}$, $t = 1,\dots,T$, are monotonic. Consider Z and W satisfying conditions (4.1). Then, by monotonicity, we have

$$\begin{aligned} &\varrho_{t,T}\big(Z_t,\dots,Z_{\theta-1},\varrho_{\theta,T}(Z_\theta,\dots,Z_T),0,\dots,0\big)\\ &\qquad \le \varrho_{t,T}\big(W_t,\dots,W_{\theta-1},\varrho_{\theta,T}(W_\theta,\dots,W_T),0,\dots,0\big). \end{aligned}$$

Using equation (4.4) we obtain (4.2). □

Observe that (4.4) is equivalent to

$$\varrho_{t,s}\big(Z_t,\ldots,Z_s\big) = \varrho_{t,\theta}\big(Z_t,\ldots,Z_{\theta-1},\varrho_{\theta,s}(Z_\theta,\ldots,Z_s)\big) \tag{4.5}$$

for all $1 \le t \le \theta \le s \le T$. This can be obtained by augmenting Z with zeros at times $s+1,\ldots,T$ in (4.4).

Theorem 4.5 allows us to derive a telescopic structure of time-consistent dynamic measures of risk. Define *one-step conditional risk mappings* ρ_t : $\mathcal{Z}_{t+1} \to \mathcal{Z}_t$, $t = 1,\ldots,T-1$ as follows:

$$\rho_t(Z_{t+1}) = \varrho_{t,t+1}(0, Z_{t+1}).$$

Corollary 4.6. *Suppose a dynamic risk measure* $\{\varrho_{t,T}\}_{1\le t\le T}$ *is normalized and satisfies the translation property. Then it is time-consistent if and only if for all* $1 \le t \le T$ *the one-step measures* ρ_t *are monotonic and for all* $Z \in \mathcal{Z}_{1:T}$ *the following identity is true:*

$$\varrho_{t,T}(Z_t,\ldots,Z_T) = \\ Z_t + \rho_t\Big(Z_{t+1} + \rho_{t+1}\Big(Z_{t+2} + \cdots + \rho_{T-2}\big(Z_{T-1} + \rho_{T-1}(Z_T)\big)\cdots\Big)\Big). \tag{4.6}$$

Proof. Suppose $\{\varrho_{t,T}\}_{1\le t\le T}$ is time-consistent. By Theorem 4.5, it is monotonic, and thus the one-step measures ρ_t are monotonic by construction. Furthermore, from (4.4), we obtain the chain of equations:

$$\begin{aligned}\varrho_{t,T}\big(Z_t,\ldots,Z_{T-1},Z_T\big) &= \varrho_{t,T-1}\big(Z_t,\ldots,\varrho_{T-1,T}(Z_{T-1},Z_T)\big)\\ &= \varrho_{t,T-1}\big(Z_t,\ldots,Z_{T-1}+\varrho_{T-1,T}(0,Z_T)\big)\\ &= \varrho_{t,T-1}\big(Z_t,\ldots,Z_{T-1}+\rho_{T-1}(Z_T)\big).\end{aligned}$$

In the second equation, we used the translation property. We can now use (4.5) with $s = T-1,\ldots,t+1$ in the same way, to obtain (4.6).

Conversely, if (4.6) is true and the one-step measures ρ_t are monotonic, the time consistency follows from the identity

$$\varrho_{t,T}(Z_t,\ldots,Z_T) = Z_t + \rho_t\big(\varrho_{t+1,T}(Z_{t+1},\ldots,Z_T)\big).$$

□

It follows that a time-consistent dynamic risk measure is completely defined by one-step conditional risk mappings ρ_t, $t = 1,\ldots,T-1$. For $t = 1$ formula (4.6) defines a risk measure of the *entire* sequence $Z \in \mathcal{Z}_{1:T}$ (with a deterministic Z_1).

Another important property of a dynamic measure of risk is the following.

Definition 4.7. A conditional risk measure $\varrho_{t,T}$, $1 \le t \le T$, has the *local property* if for every $Z \in \mathcal{Z}_{t:T}$ and every event $A \in \mathcal{F}_t$ the following equation is true:

$$\varrho_{t,T}\big(\mathbb{1}_A Z_t, \dots, \mathbb{1}_A Z_T\big) = \mathbb{1}_A\, \varrho_{t,T}\big(Z_t, \dots, Z_T\big). \tag{4.7}$$

A dynamic risk measure has the local property if all its conditional risk measures have it.

The local property expresses the intuitively understandable requirement that at time t the value of the conditional risk measure must not depend on the realizations of $Z_{t+1}, \dots, Z_T$ that cannot occur.

For time-consistent dynamic risk measures satisfying the normalization and translation properties, their local property is equivalent to the local property of all its one-step conditional risk mappings:

$$\rho_t(\mathbb{1}_A Z_{t+1}) = \mathbb{1}_A\, \rho_t(Z_{t+1}), \quad \forall\, A \in \mathcal{F}_t, \quad \forall\, 1 \le t \le T-1. \tag{4.8}$$

This can be verified directly from the definition, with the use of the telescopic representation (4.6).

Example 4.8. A simple example of a dynamic risk measure is the collection of conditional expectations:

$$\varrho_{t,T}(Z_t, \dots, Z_T) = \mathbb{E}\big[Z_t + \dots + Z_T \,\big|\, \mathcal{F}_t\big], \quad t = 1, \dots, T.$$

By construction, it is normalized, monotonic, and satisfies the translation property. It is also time-consistent, because of the identity:

$$\mathbb{E}\big[Z_t + Z_{t+1} + \dots + Z_T \,\big|\, \mathcal{F}_t\big] = Z_t + \mathbb{E}\big[\mathbb{E}\big[Z_{t+1} + \dots + Z_T \,\big|\, \mathcal{F}_{t+1}\big] \,\big|\, \mathcal{F}_t\big].$$

Formula (4.6) corresponds to the tower property:

$$\begin{aligned} &\mathbb{E}\big[Z_t + Z_{t+1} + \dots + Z_T \,\big|\, \mathcal{F}_t\big] \\ &\quad = Z_t + \mathbb{E}\big[Z_{t+1} + \mathbb{E}\big[Z_{t+2} + \dots + \mathbb{E}\big[Z_T \,\big|\, \mathcal{F}_{T-1}\big] \cdots \,\big|\, \mathcal{F}_{t+1}\big] \,\big|\, \mathcal{F}_t\big]. \end{aligned}$$

The one-step conditional risk mappings are $\rho_t(\cdot) = \mathbb{E}[\cdot \,|\, \mathcal{F}_t]$, $t = 1, \dots, T-1$. They have the local property: for all $A \in \mathcal{F}_t$ the function $\mathbb{1}_A$ is $\mathcal{F}_t$-measurable, and thus $\mathbb{E}[\mathbb{1}_A Z_{t+1} \,|\, \mathcal{F}_t] = \mathbb{1}_A \mathbb{E}[Z_{t+1} \,|\, \mathcal{F}_t]$ □

Example 4.9. The entropic risk measure of Example 2.6 has the following dynamic counterpart:

$$\varrho_{t,T}(Z_t, \dots, Z_T) \triangleq \frac{1}{\kappa} \ln\Big(\mathbb{E}\big[e^{\kappa(Z_t + \dots + Z_T)} \,\big|\, \mathcal{F}_t\big]\Big), \quad t = 1, \dots, T,$$

where $\kappa > 0$. It is normalized and monotonic by construction. The translation property and time consistency follow from the following transformations:

$$\begin{aligned}\varrho_{t,T}(Z_t,\dots,Z_T) &= \frac{1}{\kappa}\ln\Big(\mathrm{e}^{\kappa Z_t}\,\mathbb{E}\big[\mathrm{e}^{\kappa(Z_{t+1}+\dots+Z_T)}\,\big|\,\mathcal{F}_t\big]\Big)\\ &= Z_t + \frac{1}{\kappa}\ln\Big(\mathbb{E}\big[\mathbb{E}\big[\mathrm{e}^{\kappa(Z_{t+1}+\dots+Z_T)}\,\big|\,\mathcal{F}_{t+1}\big]\,\big|\,\mathcal{F}_t\big]\Big)\\ &= Z_t + \frac{1}{\kappa}\ln\Big(\mathbb{E}\big[\mathrm{e}^{\kappa\varrho_{t+1,T}(Z_{t+1},\dots,Z_T)}\,\big|\,\mathcal{F}_t\big]\Big).\end{aligned}$$

The corresponding one-step conditional risk mappings are thus

$$\rho_t(Z_{t+1}) = \frac{1}{\kappa}\ln\Big(\mathbb{E}\big[\mathrm{e}^{\kappa Z_{t+1}}\,\big|\,\mathcal{F}_t\big]\Big). \tag{4.9}$$

They have the local property:

$$\begin{aligned}\rho_t(\mathbb{1}_A Z_{t+1}) &= \frac{1}{\kappa}\ln\Big(\mathbb{E}\big[\mathrm{e}^{\kappa \mathbb{1}_A Z_{t+1}}\,\big|\,\mathcal{F}_t\big]\Big) = \frac{1}{\kappa}\ln\Big(\mathbb{E}\big[\mathbb{1}_A \mathrm{e}^{\kappa Z_{t+1}} + 1 - \mathbb{1}_A\,\big|\,\mathcal{F}_t\big]\Big)\\ &= \frac{1}{\kappa}\ln\Big(\mathbb{1}_A\mathbb{E}\big[\mathrm{e}^{\kappa Z_{t+1}}\,\big|\,\mathcal{F}_t\big] + 1 - \mathbb{1}_A\Big) = \mathbb{1}_A\frac{1}{\kappa}\ln\Big(\mathbb{E}\big[\mathrm{e}^{\kappa Z_{t+1}}\,\big|\,\mathcal{F}_t\big]\Big).\end{aligned}$$

□

In the examples above, we specified a dynamic risk measure and derived the one-step conditional measures from it. If the postulated risk measure is not time-consistent, such derivation is not possible. Therefore, a better modeling approach is to specify one-step conditional risk measures first and to derive the overall measure from them.

We elaborate on this issue in the next section.

4.1.2 One-Step Conditional Risk Mappings

Consider the spaces $\mathcal{Z}_1 = \mathcal{L}_p(\Omega, \mathcal{F}_1, P)$ and $\mathcal{Z}_2 = \mathcal{L}_p(\Omega, \mathcal{F}_2, P)$, with $p \in [1, \infty]$ and $\mathcal{F}_1 \subset \mathcal{F}_2$. Clearly, $\mathcal{Z}_1 \subset \mathcal{Z}_2$.

A mapping $\rho : \mathcal{Z}_2 \to \mathcal{Z}_1$ is a *coherent one-step conditional risk mapping* if for P-almost all $\omega \in \Omega$ it satisfies the following conditions:

Convexity: $\rho(\alpha Z + (1-\alpha)V) \le \alpha\rho(Z) + (1-\alpha)\rho(V)$, $\forall \alpha \in [0,1]$;
Monotonicity: If $Z \le V$, then $\rho(Z) \le \rho(V)$;
Translation equivariance: If $V \in \mathcal{Z}_1$, $Z \in \mathcal{Z}_2$, then $\rho(V + Z) = V + \rho(Z)$;
Positive homogeneity: $\rho(\gamma Z) = \gamma\rho(Z)$, $\forall \gamma \ge 0$.

A *convex* one-step conditional risk mapping is characterized by the first three conditions.

If the sigma algebra $\mathcal{F}_1$ is trivial, *i.e.*, $\mathcal{F}_1 = \{\emptyset, \Omega\}$, then the space $\mathcal{Z}_1$ can be identified with $\mathbb{R}$, and the above conditions define a coherent or convex risk measure.

Examples of law-invariant coherent or convex risk measures from § 2.1.1 have their conditional counterparts. The simplest is the conditional expectation operator $\rho(\cdot) = \mathbb{E}[\cdot \,|\, \mathcal{F}_1]$ of Example 4.8; it is a coherent one-step conditional risk mapping.

The one-step entropic risk mapping (4.9) is convex.

Example 4.10 (Conditional AVaR**).** Using the formula (2.9) of the Average Value at Risk, we can define the conditional risk mapping $\mathrm{AVaR}^+_\alpha(\cdot \,|\, \mathcal{F}_1) : \mathcal{Z}_2 \to \mathcal{Z}_1$ at level $\alpha \in (0, 1)$ as follows:

$$[\mathrm{AVaR}^+_\alpha(Z \,|\, \mathcal{F}_1)](\omega) \triangleq \inf_{\eta \in \mathbb{R}} \left\{ \eta + \frac{1}{\alpha} \mathbb{E}\big[(Z-\eta)_+ \,|\, \mathcal{F}_1\big](\omega) \right\}, \quad \omega \in \Omega. \tag{4.10}$$

The minimum on the right-hand side is attained at $\bar{\eta} = \mathrm{VaR}^+_\alpha(Z \,|\, \mathcal{F}_1)(\omega)$: the $(1-\alpha)$-quantile of the conditional distribution of Z. Exactly as in Example 2.5, we can verify that $\mathrm{AVaR}^+_\alpha(\cdot \,|\, \mathcal{F}_1)$ is a coherent conditional risk mapping. More generally, we may consider the conditional counterpart of the dual risk measure (2.2),

$$\rho(Z \,|\, \mathcal{F}_1)(\omega) \triangleq \inf_{\eta \in \mathbb{R}} \left\{ \eta + \frac{1}{\alpha} \Big(\mathbb{E}\big[(Z-\eta)^p_+ \,|\, \mathcal{F}_1\big](\omega) \Big)^{1/p} \right\}, \quad \omega \in \Omega,$$

where $\alpha \in (0, 1)$. In an even more general setting, we may make the probability level α random, but $\mathcal{F}_1$-measurable. □

Example 4.11 (Conditional mean-upper-semideviation). An analog of the mean-upper-semideviation risk measure of order $p \in [1, \infty)$, presented in Example 2.4, can be constructed as follows. For $\varkappa \in [0, 1]$ define

$$\rho(Z \,|\, \mathcal{F}_1) \triangleq \mathbb{E}[Z \,|\, \mathcal{F}_1] + \varkappa \Big(\mathbb{E}\Big[\big(Z - \mathbb{E}[Z \,|\, \mathcal{F}_1]\big)^p_+ \,|\, \mathcal{F}_1 \Big] \Big)^{1/p}. \tag{4.11}$$

Coherence follows by the same argument as in Example 2.4. In a more general setting, we may make the coefficient $\varkappa$ random, but $\mathcal{F}_1$-measurable. □

The following theorem relates coherence to the local property.

Theorem 4.12. *If a conditional risk mapping $\rho : \mathcal{Z}_2 \to \mathcal{Z}_1$ is coherent then it has the local property.*

Proof. Take any set $A \in \mathcal{F}_1$ and a nonnegative function $\xi \in \mathcal{L}_q(\Omega, \mathcal{F}_2, P)$, with $1/p + 1/q = 1$, such that $\mathbb{E}[\xi] = 1$, and which is supported on A. We associate with it the functional

$$\rho_\xi(Z) \triangleq \mathbb{E}[\xi\,\rho(Z)], \quad Z \in \mathcal{Z}_2. \tag{4.12}$$

Since $\rho(\cdot)$ is coherent, we can verify directly from the definition that $\rho_\xi(\cdot)$ is a coherent real-valued risk measure. By virtue of Theorem 2.25, $\rho_\xi(\cdot)$ can be represented in the following form:

$$\rho_\xi(Z) = \sup_{\zeta \in \mathcal{A}(\xi)} \int_\Omega \zeta(\omega) Z(\omega)\, P(\mathrm{d}\omega), \tag{4.13}$$

for some convex closed set $\mathcal{A}(\xi) \subset \mathcal{L}_q(\Omega, \mathcal{F}_2, P)$ of probability density functions. In particular, for any $B \in \mathcal{F}_1$,

$$\rho_\xi(\mathbb{1}_B) = \mathbb{E}[\xi \mathbb{1}_B] = \sup_{\zeta \in \mathcal{A}(\xi)} \int_B \zeta(\omega)\, P(\mathrm{d}\omega).$$

This implies that each density $\zeta \in \mathcal{A}(\xi)$ is supported on the set A as well; otherwise, for $B \cap A = \emptyset$ the last displayed equation leads to a contradiction. Therefore, the integration in (4.13) can be restricted to A:

$$\rho_\xi(Z) = \sup_{\zeta \in \mathcal{A}(\xi)} \int_A \zeta(\omega) Z(\omega)\, \mathrm{d}P(\omega) = \rho_\xi(\mathbb{1}_A Z).$$

Substitution of (4.12) leads to the conclusion that

$$\int_A \xi\, \rho(Z)\, \mathrm{d}P = \int_A \xi\, \rho(\mathbb{1}_A Z)\, \mathrm{d}P,$$

for every $Z \in \mathcal{Z}_2$ and for every density function $\xi \in \mathcal{L}_q(\Omega, \mathcal{F}_1, P)$ supported on A. This is equivalent to

$$\mathbb{1}_A\, \rho(Z) = \mathbb{1}_A\, \rho(\mathbb{1}_A Z), \quad \text{a.s.}, \tag{4.14}$$

for every $Z \in Z_2$. Since the set A was arbitrary, the equation (4.14) holds true for all $A \in \mathcal{F}_1$. Replacing A with A^c and Z with $\mathbb{1}_A Z$ in (4.14), we obtain

$$\mathbb{1}_{A^c}\, \rho(\mathbb{1}_A Z) = \mathbb{1}_{A^c}\, \rho(\mathbb{1}_{A^c} \mathbb{1}_A Z) = 0.$$

We can thus add $\mathbb{1}_{A^c}\, \rho(\mathbb{1}_A Z)$ to the right hand side of (4.14) to conclude that

$$\mathbb{1}_A\, \rho(Z) = \mathbb{1}_A\, \rho(\mathbb{1}_A Z) + \mathbb{1}_{A^c}\, \rho(\mathbb{1}_A Z) = \rho(\mathbb{1}_A Z),$$

as postulated. □

Remark 4.13. The local property can be also derived from continuity, monotonicity, translation and normalization ($\rho(0) = 0$), without the use of the convexity axiom.

Suppose Z is bounded. Then

$$\mathbb{1}_A Z - \mathbb{1}_{A^c}\|Z\|_\infty \le Z \le \mathbb{1}_A Z + \mathbb{1}_{A^c}\|Z\|_\infty.$$

By monotonicity and translation,

$$\rho(\mathbb{1}_A Z) - \mathbb{1}_{A^c}\|Z\|_\infty \le \rho(Z) \le \rho(\mathbb{1}_A Z) - \mathbb{1}_{A^c}\|Z\|_\infty.$$

Multiplying by $\mathbb{1}_A$ we obtain the equation:

$$\mathbb{1}_A\,\rho(\mathbb{1}_A Z) = \mathbb{1}_A\rho(Z) \tag{4.15}$$

On the other hand, by monotonicity, translation, and normalization

$$-\,\mathbb{1}_A\|Z\|_\infty \le \rho(\mathbb{1}_A Z) \le \mathbb{1}_A\|Z\|_\infty.$$

Multiplication by $\mathbb{1}_{A^c}$ yields

$$\mathbb{1}_{A^c}\,\rho(\mathbb{1}_A Z) = 0.$$

We add this equation to (4.15) and obtain the local property for all bounded Z.

Now, for any $Z \in \mathcal{L}_p(\Omega, \mathcal{F}_2, P)$, we can construct a sequence of bounded $Z_k \to Z$. By continuity,

$$\rho(\mathbb{1}_A Z) = \lim_{k\to\infty} \rho(\mathbb{1}_A Z_k) = \lim_{k\to\infty} \mathbb{1}_A\,\rho(Z_k) = \mathbb{1}_A\,\rho(Z),$$

which is the local property of ρ. □

Consequently, the conditional risk mappings of Examples 4.10 and 4.11 have the local property automatically.

The local property allows us to strengthen the convexity and positive homogeneity properties to allow for random coefficients.

Theorem 4.14. *If $\rho : \mathcal{Z}_2 \to \mathcal{Z}_1$ is coherent, then*

(i) *For all $Z \in \mathcal{Z}_2$ and all bounded nonnegative $\gamma \in \mathcal{Z}_1$,*

$$\rho(\gamma Z) = \gamma\rho(Z); \tag{4.16}$$

(ii) *For all $Z, V \in \mathcal{Z}_2$ and all $\gamma \in \mathcal{Z}_1$ such that $0 \le \gamma \le 1$,*

$$\rho\big(\gamma Z + (1-\gamma)V\big) \le \gamma\rho(Z) + (1-\gamma)\rho(V). \tag{4.17}$$

Proof. (i) Consider a nonnegative bounded $\gamma \in \mathcal{Z}_1$. At first, assume that it is a simple function,

$$\gamma = \sum_{j=1}^{n} \alpha_j \mathbb{1}_{A_j}, \tag{4.18}$$

where $A_j \in \mathcal{F}_1$ are disjoint sets, $\bigcup_{j=1}^n A_j = \Omega$, and $\alpha_j \geq 0$. It follows from the local property and the positive homogeneity of $\rho(\cdot)$ that

$$\begin{aligned}\mathbb{1}_{A_i}\rho(\gamma Z) &= \mathbb{1}_{A_i}\rho\Big(\sum_{j=1}^n \alpha_j \mathbb{1}_{A_j} Z\Big)\\ &= \rho\Big(\mathbb{1}_{A_i}\sum_{j=1}^n \alpha_j \mathbb{1}_{A_j} Z\Big) = \rho(\alpha_i \mathbb{1}_{A_i} Z) = \alpha_i \mathbb{1}_{A_i}\rho(Z).\end{aligned}$$

Adding these equations for $i = 1, \dots, n$, we conclude that (4.16) is true for simple functions γ. For a general nonnegative bounded function $\gamma \in \mathcal{Z}_1$, and for every $\varepsilon_k > 0$, we can find a nonnegative simple function γ_k such that

$$\gamma_k \leq \gamma \leq \gamma_k + \varepsilon_k \mathbb{1}.$$

Define $\underline{\gamma}_k = \gamma_k + \varepsilon_k \mathbb{1}_{\{Z<0\}}$ and $\overline{\gamma}_k = \gamma_k + \varepsilon_k \mathbb{1}_{\{Z \geq 0\}}$. Then

$$\underline{\gamma}_k Z \leq \gamma Z \leq \overline{\gamma}_k Z.$$

It follows from the monotonicity and translation properties of $\rho(\cdot)$ that

$$\rho(\underline{\gamma}_k Z) \leq \rho(\gamma Z) \leq \rho(\overline{\gamma}_k Z).$$

Since (4.16) is true for simple functions, we conclude that

$$\underline{\gamma}_k \rho(Z) \leq \rho(\gamma Z) \leq \overline{\gamma}_k \rho(Z).$$

Letting $\varepsilon_k \to 0$ we obtain $\underline{\gamma}_k \to \gamma$ and $\overline{\gamma}_k \to \gamma$ in $\mathcal{Z}_1$, and the last displayed inequalities yield (4.16) in the limit.

(ii) Let γ be a simple function (4.18) such that $0 \leq \gamma \leq 1$. This implies that all $\alpha_j \in [0, 1]$ in (4.18). Using the local property and the convexity of $\rho(\cdot)$, we obtain

$$\begin{aligned}\mathbb{1}_{A_i}\rho(\gamma Z + (1-\gamma)V) &= \mathbb{1}_{A_i}\rho\Big(\sum_{j=1}^n \mathbb{1}_{A_j}[\alpha_j Z + (1-\alpha_j)V]\Big)\\ &= \rho\big(\mathbb{1}_{A_i}[\alpha_i Z + (1-\alpha_i)V]\big) = \mathbb{1}_{A_i}\rho\big(\alpha_i Z + (1-\alpha_i)V\big)\\ &\leq \mathbb{1}_{A_i}\big[\alpha_i \rho(Z) + (1-\alpha_i)\rho(V)\big].\end{aligned}$$

Adding these inequalities for $i = 1, \dots, n$, we conclude that (4.17) is true for simple functions γ.

The case of general functions $\gamma \in \mathcal{Z}_1$, with values in [0, 1], can be dealt with by approximations by simple functions, similar to case (i). For $\varepsilon_k > 0$, we construct a simple function $\overline{\gamma}_k \in \mathcal{Z}_1$, with values in [0, 1], such that $|\overline{\gamma}_k - \gamma| \leq \varepsilon_k$, and

$$\gamma Z + (1-\gamma)V \le \bar{\gamma}_k Z + (1-\bar{\gamma}_k)V.$$

Then the monotonicity of $\rho(\cdot)$ and (4.17) for simple functions yield

$$\rho\big(\gamma Z + (1-\gamma)V\big) \le \rho\big(\bar{\gamma}_k Z + (1-\bar{\gamma}_k)V\big) \le \bar{\gamma}_k \rho(Z) + (1-\bar{\gamma}_k)\rho(V).$$

In the limit, when $\varepsilon_k \to 0$, we obtain (4.17) for γ. □

4.2 Dual Representation

4.2.1 Subdifferentiation of Conditional Risk Mappings

As before, we have $\mathcal{Z}_1 = \mathcal{L}_p(\Omega, \mathcal{F}_1, P)$, $\mathcal{Z}_2 = \mathcal{L}_p(\Omega, \mathcal{F}_2, P)$, with $\mathcal{F}_1 \subset \mathcal{F}_2$, and $p \in [1, \infty)$. Suppose $\rho : \mathcal{Z}_2 \to \mathcal{Z}_1$ is a convex conditional risk mapping having the local property. It is a special case of a convex mapping discussed in § 3.1.1, and our considerations here closely mirror that setting. All inequalities and equations in $\mathcal{Z}_1$ and $\mathcal{Z}_2$ are understood in the almost sure sense.

For any $Z \in \mathcal{Z}_2$ we can define a subgradient of $\rho(\cdot)$ at Z as a linear continuous operator $L : \mathcal{Z}_2 \to \mathcal{Z}_1$ such that

$$Lh \le F(Z+h) - F(Z), \quad \forall\, h \in \mathcal{Z}_2.$$

The set of all such operators is the subdifferential $\partial\rho(Z)$ of $\rho(\cdot)$ at Z.

For every $h \in \mathcal{Z}_2$, the directional derivative

$$\rho'(Z; h) = \lim_{t \downarrow 0} \frac{1}{t}\big[\rho(Z+th) - \rho(Z)\big]$$

is well-defined, because the differential quotients are nonincreasing in $\mathcal{Z}_1$, when $t \downarrow 0$. By construction, $\rho'(Z; \cdot)$ is convex and positively homogeneous. Furthermore, $L \in \partial\rho(Z)$ if and only if for all $h \in \mathcal{Z}_2$

$$Lh \le \rho'(Z; h). \tag{4.19}$$

We can prove the following extension of Theorem 2.32.

Theorem 4.15. *Suppose $\rho : \mathcal{L}_p(\Omega, \mathcal{F}, P) \to \mathcal{L}_r(\Omega, \mathcal{G}, P)$, with $1 \le r \le p \le \infty$ and $\mathcal{G} \subset \mathcal{F}$, is a convex and nondecreasing mapping. Then $\rho(\cdot)$ is continuous, $\partial\rho(Z) \neq \emptyset$, for all $Z \in \mathcal{L}_p(\Omega, \mathcal{F}, P)$, and every $L \in \partial\rho(Z)$ is nondecreasing.*

Proof. The directional derivative $\rho'(Z; \cdot)$ is convex and positively homogeneous. By choosing a one-dimensional subspace $M = \{t\mathbb{1}_\Omega : t \in \mathbb{R}\}$, we can observe that the linear operator $L_M(t\mathbb{1}_\Omega) \triangleq t\rho'(Z;\mathbb{1}_\Omega)$ is a minorant of $\rho'(Z;\cdot)$ on M. By the generalization of the Hahn–Banach theorem to lattice-valued operators, the operator L_M may be extended to an operator $L : \mathcal{L}_p(\Omega, \mathcal{F}, P) \to \mathcal{L}_r(\Omega, \mathcal{G}, P)$, so that $L = L_M$ on M and (4.19) is satisfied everywhere. However, L is not guaranteed to be continuous.

The monotonicity of $\rho(\cdot)$ implies that for any $h \ge 0$, we have $\rho'(Z;-h) \le 0$, and thus $Lh \ge -\rho'(Z;-h) \ge 0$. Thus, L is nondecreasing. To show that L is continuous, we follow the classical line of argument.

Suppose L is not continuous. Then a sequence $\{h_n\}_{n\in N} \subset \mathcal{L}_p(\Omega, \mathcal{F}, P)$ exists, such that $\|h_n\| \le 1$, but $\|Lh_n\| \ge 2^{n+1}$. We can decompose each $h_n = h_n^+ - h_n^-$, where $h_n^+ = \max(0, h_n)$ and $h_n^- = \max(0, -h_n)$. Then

$$\|Lh_n^+\| + \|Lh_n^-\| \ge 2^{n+1}.$$

Thus, for at least one of the two elements, h_n^+ or h_n^-, the value of L has its norm greater than or equal to 2^n: we denote this element by a_n. We conclude that a sequence $\{a_n\}_{n\in N}$ exists, such that for all $n \in N$

$$a_n \ge 0, \quad \|a_n\| \le 1, \quad La_n \ge 0, \quad \|La_n\| \ge 2^n.$$

As the space $\mathcal{L}_p(\Omega, \mathcal{F}, P)$ is complete, the element

$$a = \sum_{n=1}^{\infty} \frac{a_n}{2^n}$$

is well-defined. By construction, $\|a\| \le 1$. Due to the monotonicity of $L(\cdot)$, for all $N \ge 1$

$$La \ge \sum_{n=1}^{N} \frac{La_n}{2^n}.$$

All components of the sum are nonnegative and have the norm of at least 1, therefore $\|La\|^r \ge N$ for all N, which is a contradiction. □

Directly from the definition, we conclude that the set $\partial\rho(x)$ is convex and compact in the weak operator topology. We can also provide a result on the local property of the subgradients, parallel to Lemma 3.8.

Lemma 4.16. *Suppose $\rho : \mathcal{Z}_2 \to \mathcal{Z}_1$ is convex and local. Then every $L \in \partial\rho(Z)$ is also local in the following sense: for all $B \in \mathcal{F}_1$ and all $h \in \mathcal{Z}_2$ we have*

$$L(\mathbb{1}_B h) = \mathbb{1}_B Lh.$$

Proof. If $\rho(\cdot)$ is local, then $\rho'(Z;\cdot)$ is local as well. The rest of the proof is identical to the proof of Lemma 3.8; the only difference is that only the events $B \in \mathcal{F}_1$ need to be considered. □

We can now characterize the subgradients of convex conditional risk mappings.

Theorem 4.17. *Suppose $\rho : \mathcal{Z}_2 \to \mathcal{Z}_1$ is a convex conditional risk mapping having the local property. An operator $L \in \partial\rho(Z)$ if and only if $\lambda \in \Lambda^\rho(Z)$ exists, such that for all $h \in \mathcal{Z}_2$*

$$Lh = \mathbb{E}[\lambda h \,|\, \mathcal{F}_1], \tag{4.20}$$

where $\Lambda^\rho(Z)$ is a nonempty closed convex set defined as follows:

$$\Lambda^\rho(Z) = \big\{\lambda \in \mathcal{Z}_2^* : \lambda \geq 0,\ \mathbb{E}[\lambda \,|\, \mathcal{F}_1] = 1,\\ \rho(Z+h) - \rho(Z) \geq \mathbb{E}[\lambda h \,|\, \mathcal{F}_1],\ \forall\, h \in \mathcal{Z}_2\big\}. \tag{4.21}$$

Proof. It is evident that every operator of the form (4.20) is a subgradient. To prove the opposite, we extend the idea of the proof of Theorem 3.9.

Suppose $L \in \partial\rho(Z)$. Consider the functional $\ell : \mathcal{Z}_2 \to \mathbb{R}$ defined as

$$\ell(h) = \mathbb{E}[Lh], \quad h \in \mathcal{Z}_2.$$

As L is linear and continuous, so is $\ell(\cdot)$. By the local property of L established in Lemma 4.16,

$$\ell(\mathbb{1}_B h) = \mathbb{E}\big[L(\mathbb{1}_B h)\big] = \mathbb{E}\big[\mathbb{1}_B Lh\big], \quad \forall B \in \mathcal{F}_1.$$

Since $p \in [1, \infty)$, by the Riesz representation theorem, we can identify $\ell(\cdot)$ with an element $\lambda \in \mathcal{L}_q(\Omega, \mathcal{F}_2, P)$, where $1/p + 1/q = 1$, such that

$$\ell(h) = \mathbb{E}[\lambda h], \quad \forall\, h \in \mathcal{Z}_2.$$

Combining the last two relations we conclude that

$$\mathbb{E}\big[\mathbb{1}_B Lh\big] = \mathbb{E}[\mathbb{1}_B \lambda h], \quad \forall B \in \mathcal{F}_1.$$

Since Lh is $\mathcal{F}_1$-measurable by definition, the last equation means that (4.20) is true.

Due to the monotonicity of $\rho(\cdot)$, for every $h \geq 0$, we have $Lh \geq \rho(Z) - \rho(Z - h) \geq 0$ and thus $\lambda \geq 0$. Finally, the translation property implies that $L\mathbb{1} = \mathbb{1}$. Hence, $\mathbb{E}[\lambda \,|\, \mathcal{F}_1] = \mathbb{1}$. As $\partial\rho(Z) \neq \emptyset$, due to Theorem 4.15, the set $\Lambda^\rho(Z)$ is nonempty as well. □

This result allows us to identify the subgradients of $\rho(\cdot)$ with the elements of $\Lambda^\rho(\cdot)$. We can simply write "$\lambda \in \Lambda^\rho(Z)$" instead of "$L \in \partial\rho(Z)$ for the operator L associated with λ by the formula (4.20)."

Corollary 4.18. *The set $\Lambda^\rho(Z)$ has the following generalized convexity property: if $\lambda_1, \lambda_2 \in \Lambda^\rho(Z)$ then for all $\mathcal{F}_1$-measurable $\alpha \in [0,1]$, we have $\alpha\lambda_1 + (1-\alpha)\lambda_2 \in \Lambda^\rho(Z)$ as well.*

4.2.2 Dual Representation of Conditional Risk Mappings

We can develop a result parallel to the dual representation of risk measures of Theorem 2.25. We refine the ideas of § 3.1.1.

Definition 4.19. Suppose $\rho : \mathcal{Z}_2 \to \mathcal{Z}_1$ is a convex conditional risk mapping having the local property. The *conjugate mapping* to $\rho(\cdot)$ is defined as

$$\rho^*(\lambda) = \operatorname*{ess\,sup}_{Z \in \mathcal{Z}_2} \big\{ \mathbb{E}[\lambda Z \,|\, \mathcal{F}_1] - \rho(Z) \big\}, \quad \lambda \in \mathcal{Z}_2^*.$$

We also define

$$\operatorname{dom}(\rho^*) = \big\{ \lambda \in \mathcal{Z}_2^* : \rho^*(\lambda) \in \mathcal{Z}_1 \big\}.$$

Further, we define the *biconjugate mapping* to $\rho(\cdot)$ as

$$\rho^{**}(Z) = \operatorname*{ess\,sup}_{\lambda \in \operatorname{dom}(\rho^*)} \big\{ \mathbb{E}[\lambda Z \,|\, \mathcal{F}_1] - \rho^*(\lambda) \big\}, \quad Z \in \mathcal{Z}_2.$$

Unlike in the general convex lattice-valued mapping case, we define $\rho^*(\cdot)$ as a function of $\lambda \in \mathcal{Z}_2^*$, as if $\rho(\cdot)$ was real-valued[†]. Furthermore, $\rho^{**}(\cdot)$ is also not the conjugate to $\rho^*(\cdot)$ in the classical sense, because we restrict its domain to $\mathcal{Z}_2$. These simplifications are possible thanks to the local property, as we shall see below.

Lemma 4.20. *If $\rho : \mathcal{Z}_2 \to \mathcal{Z}_1$ is a convex conditional risk mapping having the local property, then the following statements are equivalent*

(i) $\rho(Z) + \rho^*(\lambda) = \mathbb{E}[\lambda Z \,|\, \mathcal{F}_1]$;

(ii) $\lambda \in \Lambda^\rho(Z)$.

Proof. The statement (i) is true if and only if for all $V \in \mathcal{Z}_2$

$$\mathbb{E}[\lambda Z \,|\, \mathcal{F}_1] - \rho(Z) \ge \mathbb{E}[\lambda V \,|\, \mathcal{F}_1] - \rho(V).$$

This is equivalent to (ii). □

We can now prove the following duality theorem.

[†] In general, we would have to work with linear continuous operators from $\mathcal{Z}_2$ to $\mathcal{Z}_1$.

Theorem 4.21. *Suppose* $\rho : \mathcal{Z}_2 \to \mathcal{Z}_1$ *is a convex conditional risk mapping. Then* $\text{dom}(\rho^*)$ *is a nonempty convex closed set included in* $\{\lambda \in \mathcal{Z}_2^* : \lambda \geq 0,\ \mathbb{E}[\lambda \,|\, \mathcal{F}_1] = 1\}$ *and for every* $Z \in \mathcal{Z}_2$, *we have*

$$\rho(Z) = \rho^{**}(Z). \tag{4.22}$$

Proof. Suppose $\lambda \in \text{dom}(\rho^*)$ exists such that $P[\lambda < 0] > 0$. Then

$$\mathbb{E}[\mathbb{1}_{\{\lambda<0\}}\lambda \,|\, \mathcal{F}_1] < 0$$

with positive probability. For any Z_0 we consider the ray $Z_t = Z_0 - t\mathbb{1}_{\{\lambda<0\}}$. By construction, $Z_0 \geq Z_t$ for all $t \geq 0$, and the monotonicity condition implies that $\varrho(Z_0) \geq \varrho(Z_t)$ for all $t \geq 0$. Consequently,

$$\begin{aligned}\rho^*(\lambda) &\geq \operatorname*{ess\,sup}_{t\geq 0} \big\{\mathbb{E}[Z_t\lambda \,|\, \mathcal{F}_1] - \rho(Z_t)\big\}\\ &\geq \mathbb{E}[Z_0\lambda \,|\, \mathcal{F}_1] - \rho(Z_0) + \operatorname*{ess\,sup}_{t\geq 0}\big\{-t\,\mathbb{E}[\mathbb{1}_{\{\lambda<0\}}\lambda \,|\, \mathcal{F}_1]\big\}.\end{aligned}$$

The last expression is not in $\mathcal{L}_p(\Omega, \mathcal{F}_1, P)\}$ because it is equal to ∞ with positive probability. Therefore, $\lambda \geq 0$ for all $\lambda \in \text{dom}(\rho^*)$.

By virtue of the translation equivariance, for every Z, every $B \in \mathcal{F}_1$, and every $\lambda \in \mathcal{Z}^*$ we have

$$\begin{aligned}\rho^*(\lambda) &\geq \sup_{c\in\mathbb{R}} \big\{\mathbb{E}[(Z + c\mathbb{1}_B)\lambda \,|\, \mathcal{F}_1] - \rho(Z + c\mathbb{1}_B)\big\}\\ &= \sup_{c\in\mathbb{R}} \big\{c\mathbb{1}_B\big(\mathbb{E}[\lambda \,|\, \mathcal{F}_1] - 1\big)\big\} + \mathbb{E}[Z\lambda \,|\, \mathcal{F}_1] - \rho(Z).\end{aligned}$$

For $\rho^*(\lambda)$ to be finite, it is necessary that $\mathbb{E}[\lambda \,|\, \mathcal{F}_1] = 1$ a.s..

It remains to show the relation (4.22). By the definition of $\rho^*(\cdot)$, we have

$$\rho(Z) \geq \mathbb{E}[\lambda Z \,|\, \mathcal{F}_1] - \rho^*(\lambda), \quad \forall\, \lambda \in \text{dom}(\rho^*).$$

Therefore, $\rho(Z) \geq \rho^{**}(Z)$ for all $Z \in \mathcal{Z}_2$. Substituting $\lambda \in \Lambda^\rho(Z) \neq \emptyset$, and invoking Lemma 4.20, we obtain the equality. □

We now focus on coherent measures, because they play a key role in the control problems we will be dealing with later in the book.

Theorem 4.22. *If* $\rho : \mathcal{Z}_2 \to \mathcal{Z}_1$ *is a coherent conditional risk mapping, then for all* $Z \in \mathcal{Z}_2$,

$$\rho(Z) = \operatorname*{ess\,sup}_{\lambda\in\Lambda} \mathbb{E}[\lambda Z \,|\, \mathcal{F}_1], \tag{4.23}$$

where $\Lambda = \Lambda^{\rho}(0)$ is a nonempty closed convex set defined as follows:

$$\Lambda = \{\lambda \in \mathcal{Z}_2^* : \lambda \geq 0,\ \mathbb{E}[\lambda \mid \mathcal{F}_1] = 1,\ \rho(Z) \geq \mathbb{E}[\lambda Z \mid \mathcal{F}_1], \forall Z \in \mathcal{Z}_2\}. \tag{4.24}$$

Proof. If $\rho(\cdot)$ is positively homogeneous, then its conjugate function satisfies for every $\gamma > 0$ the chain of relations:

$$\begin{aligned}\rho^*(\lambda) &= \operatorname*{ess\,sup}_{\gamma Z \in \mathcal{Z}} \{\mathbb{E}[\gamma Z \lambda \mid \mathcal{F}] - \rho(\gamma Z)\} \\ &= \gamma \operatorname*{ess\,sup}_{Z \in \mathcal{Z}} \{\mathbb{E}[Z\lambda \mid \mathcal{F}] - \varrho(Z)\} = \gamma \varrho^*(\zeta).\end{aligned}$$

Therefore $\rho^*(\lambda)$ is either 0 or $+\infty$. For it to be a member of $\mathcal{L}_p(\Omega, \mathcal{F}_1, P)$ it must be equal to 0 a.s. on $\operatorname{dom}(\rho^*)$. Then the relation (4.22) implies the representation

$$\rho(Z) = \operatorname*{ess\,sup}_{\lambda \in \operatorname{dom}(\rho^*)} \mathbb{E}[\lambda Z \mid \mathcal{F}_1].$$

The characterization of $\operatorname{dom}(\rho^*)$ by (4.24) is evident from the definition. □

The dual representation has a useful form in the case of product spaces of elementary events. Suppose $\Omega = \Omega_1 \times \Omega_2$, where Ω_1 and Ω_2 are two measure spaces, with σ-algebras $\mathcal{G}_1$ and $\mathcal{G}_2$, respectively. Let $\mathcal{F}_2 = \mathcal{G}_1 \otimes \mathcal{G}_2$ be the product σ-algebra generated by $\mathcal{G}_1 \times \mathcal{G}_2$, and let $\mathcal{F}_1 = \mathcal{G}_1 \otimes \Omega_2$. If Ω_2 is standard, then the probability measure P on $(\Omega, \mathcal{F}_2)$ can be disintegrated into its marginal measure P_1 on $(\Omega_1, \mathcal{G}_1)$ and a stochastic kernel $P_{2|1} : \Omega_1 \to \mathcal{P}(\Omega_2, \mathcal{G}_2)$. We write it symbolically as

$$P = P_1 \otimes P_{2|1}.$$

Then the dual representation (4.23) takes on the form:

$$\rho(Z)(\omega_1) = \sup_{\lambda \in \Lambda} \int_{\Omega_2} \lambda(\omega_1, \omega_2)\, Z(\omega_1, \omega_2)\, P_{2|1}(d\omega_2 \mid \omega_1). \tag{4.25}$$

Due to (4.24),

$$\int_{\Omega_2} \lambda(\omega_1, \omega_2)\, P_{2|1}(d\omega_2 \mid \omega_1) = 1, \quad a.s.,$$

and thus $\lambda(\omega_1, \cdot)$ is a density function with respect to the conditional probability measure $P_{2|1}(\cdot \mid \omega_1)$. It follows that the set Λ in (4.24) is the graph of a multifunction $\Lambda : \Omega_1 \rightrightarrows \mathcal{L}_q(\Omega_2, \mathcal{G}_2, P_{2|1}(\omega_1))$, whose values are convex sets of probability density functions on $(\Omega_2, \mathcal{G}_2, P_{2|1}(\omega_1))$:

$$\Lambda(\omega_1) = \Big\{ \ell \in \mathcal{L}_q\big(\Omega_2, \mathcal{G}_2, P_{2|1}(\omega_1)\big) :$$
$$\rho(Z)(\omega_1) \geq \int_{\Omega_2} \ell(\omega_2) Z(\omega_1, \omega_2)\; P_{2|1}(\mathrm{d}\omega_2 \,|\, \omega_1) \Big\}, \qquad \omega_1 \in \Omega_1. \tag{4.26}$$

We can, therefore, rewrite (4.25) in a more explicit way:

$$\rho(Z)(\omega_1) = \sup_{\mu \in \Lambda(\omega_1)} \int_{\Omega_2} \mu(\omega_2) Z(\omega_1, \omega_2)\; P_{2|1}(\mathrm{d}\omega_2 \,|\, \omega_1). \tag{4.27}$$

Every measurable selector of the multifunction (4.26) defines a probability kernel, which is absolutely continuous with respect to the conditional probability $P_{2|1}$. Therefore, we can also identify Λ with the convex set of kernels

$$\mathcal{Q} = \Big\{ Q : \Omega_1 \to \mathcal{P}(\Omega_2) : \frac{\mathrm{d}Q}{\mathrm{d}P_{2|1}} \in \Lambda \Big\}.$$

We call it the *risk multikernel* of $\rho(\cdot)$. Such multikernels can be calculated for particular examples of conditional measures of risk directly from the dual representations of their static counterparts.

Example 4.23. For the conditional risk measure of Example 4.10, we can calculate the sets (4.26) by modifying the derivations of Example 2.29:

$$\Lambda(\omega_1) = \Big\{ \mu \in \mathcal{L}_\infty\big(\Omega_2, \mathcal{G}_2, P_{2|1}(\omega_1)\big) : \mu \text{ is a density},\ 0 \leq \mu \leq \frac{1}{\alpha} \Big\}.$$

□

Example 4.24. The conditional risk measure of Example 4.11 has the following risk multikernel:

$$\Lambda(\omega_1) = \big\{ \mu = \mathbb{1} + \xi - \mathbb{E}_{2|1}[\xi \,|\, \omega_1] : \xi \in \mathcal{L}_q\big(\Omega_2, \mathcal{G}_2, P_{2|1}(\omega_1)\big),$$
$$\mathbb{E}_{2|1}\big[|\xi|^q \,|\, \omega_1\big] \leq \varkappa^q(\omega_1),\ \xi \geq 0 \big\}.$$

Here, $\mathbb{E}_{2|1}[\xi \,|\, \omega_1] = \int_{\Omega_2} \xi(\omega_2)\, P_{2|1}(\mathrm{d}\omega_2 \,|\, \omega_1)$. □

4.2.3 Dual Representation of Dynamic Risk Measures

We consider a dynamic risk measure $\{\varrho_{t,T}\}_{1 \leq t \leq T}$, where each $\varrho_{t,T} : \mathcal{Z}_{t:T} \to \mathcal{Z}_t$. If it is time-consistent, normalized, and satisfies the translation property, then Corollary 4.6 implies that it has the telescopic structure (4.6): for all $1 \leq t < T$ we have

$$\varrho_{t,T}(Z_t, \dots, Z_T) = \rho_t\big(\rho_{t+1}(\cdots \rho_{T-1}(Z_t + Z_{t+1} + \cdots + Z_T) \cdots)\big).$$

Consider the composition

$$\varrho_{t,T} = \rho_t \circ \rho_{t+1} \circ \cdots \circ \rho_{T-1}. \tag{4.28}$$

The following fact follows directly from the definition.

Lemma 4.25. *If the conditional risk mappings $\rho_t : \mathcal{Z}_{t+1} \to \mathcal{Z}_t$, $t = 1, \ldots, T-1$, are coherent, then for each $t = 1, \ldots, T$ the mapping $\rho_{t,T} : \mathcal{Z}_T \to \mathcal{Z}_t$ given by* (4.28) *is a coherent conditional risk measure.*

Proof. We verify the definition by induction. For $t = T - 1$ there is nothing to prove. Supposing the result is true for $t + 1$, we verify it for t. By the convexity of $\varrho_{t+1,T}(\cdot)$,

$$\varrho_{t+1,T}\big(\alpha V_T + (1-\alpha)W_T\big) \le \alpha\varrho_{t+1,T}(V_T) + (1-\alpha)\varrho_{t+1,T}(W_T).$$

Applying ρ_t to both sides and using its monotonicity and convexity, we obtain

$$\begin{aligned}\varrho_{t,T}\big(\alpha V_T + (1-\alpha)W_T\big) &\le \rho_t\Big(\alpha\varrho_{t+1,T}(V_T) + (1-\alpha)\varrho_{t+1,T}(W_T)\Big)\\ &\le \alpha\varrho_{t,T}(V_T) + (1-\alpha)\varrho_{t,T}(W_T),\end{aligned}$$

which proves the convexity of $\rho_{t,T}$. The monotonicity, translation property, and positive homogeneity are obvious. □

If all $\rho_t : \mathcal{Z}_{t+1} \to \mathcal{Z}_t$ are coherent, then by virtue of Theorem 4.22 they have the dual representation:

$$\rho_t(V_{t+1}) = \operatorname*{ess\,sup}_{\lambda_t \in \Lambda_t} \mathbb{E}\,[\lambda_t V_{t+1} \,|\, \mathcal{F}_t], \tag{4.29}$$

with

$$\begin{aligned}\Lambda_t = \big\{\lambda_t \in \mathcal{Z}^*_{t+1} : \lambda_t \ge 0,\ \mathbb{E}\,[\lambda_t \,|\, \mathcal{F}_t] = 1,&\\ \rho_t(V) \ge \mathbb{E}\,[\lambda_t V \,|\, \mathcal{F}_t], \forall V \in \mathcal{Z}_{t+1}\big\}.&\end{aligned} \tag{4.30}$$

This equation allows us to derive the dual representation of the composite risk measure (4.28).

Theorem 4.26. *If the conditional risk mappings $\rho_t : \mathcal{Z}_{t+1} \to \mathcal{Z}_t$, $t = 1, \ldots, T-1$, are coherent, then for each $t = 1, \ldots, T$ the mapping $\varrho_{t,T} : \mathcal{Z}_T \to \mathcal{Z}_t$ defined in* (4.28) *has the following representation*

$$\varrho_{t,T}(V_T) = \operatorname*{ess\,sup}_{\lambda \in \Lambda_{t,T}} \mathbb{E}\,[\lambda V_T \,|\, \mathcal{F}_t], \tag{4.31}$$

where

$$\begin{aligned}\Lambda_{t,T} &= \Lambda_t \circ \Lambda_{t+1} \circ \cdots \circ \Lambda_{T-1} \\ &\triangleq \big\{\lambda_t\lambda_{t+1}\cdots\lambda_{T-1} : \lambda_s \in \Lambda_s,\ s = t,\ldots,T-1\big\}. \quad (4.32)\end{aligned}$$

Proof. We prove the result by induction. For $t = T-1$ our assertion is identical with (4.29). Supposing the formula (4.32) is true for $t+1$, we verify it for t using the dual representation (4.29) again:

$$\begin{aligned}\varrho_{t,T}(V_T) = \rho_t\big(\rho_{t+1,T}(V_T)\big) &= \operatorname*{ess\,sup}_{\lambda_t\in\Lambda_t} \mathbb{E}\Big[\lambda_t \operatorname*{ess\,sup}_{\xi\in\Lambda_{t+1,T}} \mathbb{E}\left[\xi V_T \,|\, \mathcal{F}_{t+1}\right] \Big|\, \mathcal{F}_t\Big] \\ &= \operatorname*{ess\,sup}_{\lambda_t\in\Lambda_t} \mathbb{E}\Big[\operatorname*{ess\,sup}_{\xi\in\Lambda_{t+1,T}} \mathbb{E}\left[\lambda_t\xi V_T \,|\, \mathcal{F}_{t+1}\right] \Big|\, \mathcal{F}_t\Big] \\ &= \operatorname*{ess\,sup}_{\lambda_t\in\Lambda_t} \operatorname*{ess\,sup}_{\xi\in\Lambda_{t+1,T}} \mathbb{E}\big[\lambda_t\xi V_T \,\big|\, \mathcal{F}_t\big].\end{aligned}$$

In the last equation, we could exchange the optimization and the conditional expectation operations, because we allow ξ to be $\mathcal{F}_{t+1}$-measurable. This verifies (4.31) for t. □

The dual sets $\Lambda_{t,T}$ enjoy a generalized convexity property.

Corollary 4.27. *For every $t = 1,\ldots,T-1$, the set $\Lambda_{t,T}$ is convex in the following generalized sense: for all $\mathcal{F}_t$-measurable $\alpha : \Omega \to [0,1]$ and for all $\lambda'_{t,T}, \lambda''_{t,T} \in \Lambda_{t,T}$, we have $\alpha\lambda'_{t,T} + (1-\alpha)\lambda''_{t,T} \in \Lambda_{t,T}$ as well.*

Proof. For $t = T-1$ we have $\Lambda_{T-1,T} = \Lambda_{T-1}$ and the assertion follows from Corollary 4.18. Suppose the assertion is true for $t+1$; we shall verify it for t. We have

$$\Lambda_{t,T} = \Lambda_t \circ \Lambda_{t+1,T} = \big\{\lambda_t\xi : \lambda_t \in \Lambda_t,\ \xi \in \Lambda_{t+1,T}\big\}.$$

Suppose $\lambda'_{t,T} = \lambda'_t\xi'$, $\lambda''_{t,T} = \lambda''_t\xi''$, with $\lambda'_t, \lambda''_t \in \Lambda_t$ and $\xi', \xi'' \in \Lambda_{t+1,T}$. Let α be $\mathcal{F}_t$-measurable with values in $[0,1]$ and consider the element

$$\begin{aligned}\lambda_{t,T} &= \alpha\lambda'_{t,T} + (1-\alpha)\lambda''_{t,T} = \alpha\lambda'_t\xi' + (1-\alpha)\lambda''_t\xi'' \\ &= \big(\alpha\lambda'_t + (1-\alpha)\lambda''_t\big)\big(\theta\xi' + (1-\theta)\xi''\big),\end{aligned}$$

with

$$\theta = \frac{\alpha\lambda'_t}{\alpha\lambda'_t + (1-\alpha)\lambda''_t}.$$

As λ'_t and λ''_t are nonnegative, we have $\theta \in [0,1]$. Since α is $\mathcal{F}_t$-measurable and θ is $\mathcal{F}_{t+1}$-measurable, we have $\alpha\lambda'_t + (1-\alpha)\lambda''_t \in \Lambda_t$ and $\theta\xi' + (1-\theta)\xi'' \in \Lambda_{t+1,T}$, and thus $\lambda_{t,T} \in \Lambda_{t,T}$. □

As in the case of conditional mappings, the dual representation has a useful form on product spaces. Suppose $\Omega = \Omega_1 \times \cdots \times \Omega_T$, where $(\Omega_t, \mathcal{G}_t)$ are standard measure spaces, with σ-algebras $\mathcal{G}_t$, $t = 1, \dots, T$, respectively. We may assume Ω_1 to be just a single point representing the initial data, but a more general setting with a prior distribution of the data is covered by our approach as well.

Define $\Omega_{1:t} = \Omega_1 \times \cdots \times \Omega_t$, $t = 1, \dots, T$, and let

$$\mathcal{F}_t = \mathcal{G}_1 \otimes \cdots \otimes \mathcal{G}_t \otimes \Omega_{t+1} \otimes \cdots \otimes \Omega_T, \quad t = 1, \dots, T,$$

be the product σ-algebra of events observable up to time t. We use $\omega_{1:t}$ to denote an element of $\Omega_{1:t}$.

Since the spaces Ω_t are standard, the probability measure P on $(\Omega, \mathcal{F}_T)$ can be disintegrated into its marginal measure P_1 on Ω_1 and a composition of stochastic kernels $P_{t+1|t} : \Omega_{1:t} \to \mathcal{P}(\Omega_{t+1}, \mathcal{G}_{t+1})$, $t = 1, \dots, T-1$. We write it symbolically as

$$P = P_1 \otimes P_{2|1} \otimes \cdots \otimes P_{T|T-1}.$$

Exactly as in (4.25) each $\lambda_t \in \Lambda_t$ is a density function with respect to the conditional probability measure $P_{t+1|t}\big(\cdot \,\big|\, \omega_{1:t}\big)$. It follows that each set (4.30) can be viewed as the graph of a multifunction

$$\Lambda_t : \Omega_{1:t} \rightrightarrows \mathcal{L}_q\big(\Omega_{t+1}, \mathcal{G}_{t+1}, P_{t+1|t}(\omega_{1:t})\big),$$

whose values are convex sets of probability density functions on the probability space $\big(\Omega_{t+1}, \mathcal{G}_{t+1}, P_{t+1|t}(\omega_{1:t})\big)$. Every measurable selector of this multifunction defines a probability kernel, which is absolutely continuous with respect to the conditional probability $P_{t+1|t}$. Therefore, we can also identify each Λ_t with the convex set of kernels

$$\mathcal{A}_t = \left\{ A_t : \Omega_{1:t} \to \mathcal{P}(\Omega_{t+1}) : \frac{\mathrm{d}A_t}{\mathrm{d}P_{t+1|t}} \in \Lambda_t \right\}, \quad t = 1, \dots, T-1.$$

In the special case, when the process $\{Z_t\}$ is Markov, further simplifications are possible. We discuss them in Chapter 8.

4.2.4 Scenario Tree Models

We assume now that all sigma-algebras are finite and thus all vector spaces $\mathcal{Z}_t$ are finite-dimensional. In this setting, the realizations of data form a *scenario tree*. The nodes of the tree are organized in levels associated with stages $1, \dots, T$; they correspond to the nested partitions of Ω given by the filtration $\mathcal{F}_1 \subset \mathcal{F}_2 \subset \cdots \subset \mathcal{F}_T$.

Level $t = 1$ consists of only one node, called the root node. We index the nodes by the symbol ν and assign $\nu = 1$ to the root node. The nodes located at further levels $t = 2, \ldots, T$ correspond to elementary events in $\mathcal{F}_t$. Every node ν at level $t = 2, \ldots, T$ is connected to its unique ancestor node $a(\nu)$ at level $t-1$. The ancestor node corresponds to the elementary event in $\mathcal{F}_{t-1}$ containing the event associated with ν. Every node ν at levels $t = 1, \ldots, T-1$ is connected to a collection $C(\nu)$ of children nodes at level $t+1$. The children nodes correspond to the elementary events in $\mathcal{F}_{t+1}$ included in the event associated with ν. The set of all nodes at stage t is denoted by $\mathcal{N}_t$, $t = 1, \ldots, T$. A *scenario* is a path s from the root to a node at the last stage T. We use the symbol $\mathcal{S}(\nu)$ to denote the set of scenarios passing through node ν. In this setting, the measure P can be specified by conditional probabilities $p_{\nu\eta} = P[\eta\,|\,\nu]$, where $\nu \in \mathcal{N}_t$, $\eta \in C(\nu)$, and $t = 1, \ldots, T-1$. Every node ν at level t has a *history*: the path $(\nu_1, \ldots, \nu_{t-1}, \nu)$ from the root to ν. The probability of ν is the product of the corresponding conditional probabilities in its history: $p_\nu = p_{\nu_1\nu_2}p_{\nu_2\nu_3}\cdots p_{\nu_{t-1}\nu}$. When $t = T$, this formula describes the probability of a scenario $s \in \mathcal{N}_T$.

For every node $\nu \in \mathcal{N}_t$, an $\mathcal{F}_t$-measurable random variable Z has identical values on all scenarios $s \in \mathcal{S}(\nu)$. It can, therefore, be equivalently represented as a function of a node in $\mathcal{N}_t$; we write this function as $Z^{\mathcal{N}_t}$.

The value of a coherent one-step conditional measure of risk $\rho_t(\cdot)$ is $\mathcal{F}_t$-measurable, and is also a function of a node ν at level t. Due to Theorem 4.12, $\rho_t(\cdot)$ has the local property. In the scenario tree model it means that the value of $\rho_t(Z_{t+1})$ at node ν depends only on the values of $Z_{t+1}^{\mathcal{N}_{t+1}}$ at nodes $\eta \in C(\nu)$. We denote the vector of these values by $Z_{t+1}^{C(\nu)}$, and we write the conditional risk measure equivalently as $\rho_t^\nu\big(Z_{t+1}^{C(\nu)}\big)$.

Since the random variables

$$V_t = \rho_t\Big(Z_{t+1} + \rho_{t+1}\big(Z_{t+2} + \cdots + \rho_{T-1}(Z_T)\ldots\big)\Big) \tag{4.33}$$

are $\mathcal{F}_t$-measurable, we only need to consider their values V_t^ν at nodes $\nu \in \mathcal{N}_t$, $t = 1, \ldots, T$. Consequently, the value of a time-consistent measure of risk (4.6) can be written recursively as follows:

$$\begin{aligned} &V_T^\nu = Z_T^\nu, \quad \nu \in \mathcal{N}_T, \\ &V_t^\nu = Z_t^\nu + \rho_t^\nu\big(V_{t+1}^{C(\nu)}\big), \quad \nu \in \mathcal{N}_t, \quad t = T-1, \ldots, 1. \end{aligned} \tag{4.34}$$

We denote by $\mathcal{P}(C)$ the set of probability distributions on a collection of nodes $C \subset \mathcal{N}_t$. By (4.27), for every $t = 1, \ldots, T-1$ and every node $\nu \in \mathcal{N}_t$ a convex closed set

$$\mathcal{A}_t(\nu) \subset \mathcal{P}(C(\nu)) \tag{4.35}$$

exists such that

$$\rho_t^\nu\big(V_{t+1}^{C(\nu)}\big) = \max_{\mu\in\mathcal{A}_t(\nu)} \big\langle \mu, V_{t+1}^{C(\nu)}\big\rangle. \tag{4.36}$$

In fact, $\mathcal{A}_t(\nu) = \partial\rho_t^\nu\big(0\big)$ (the subdifferential of $\rho^\nu(\cdot)$ at zero).

The mappings $\mathcal{A}_t : \mathcal{N}_t \rightrightarrows \mathcal{P}(\mathcal{N}_{t+1})$ are the risk multikernels associated with the conditional risk measures $\rho_t(\cdot)$. Their values are convex and closed sets of probability measures.

If a kernel μ_t satisfies the relation $\mu_t(\nu) \in \mathcal{A}_t(\nu)$ for all $\nu \in \mathcal{N}_t$, we call it a selector of $\mathcal{A}_t$ and write $\mu_t \lessdot \mathcal{A}_t$. The value of $\mu(\nu)$ at a node $\eta \in C(\nu)$ is written as $\mu(\nu, \eta)$.

Much of our analysis uses compositions of multikernels. For a probability measure $q_t \in \mathcal{P}(\mathcal{N}_t)$ and a kernel $\mu_t \lessdot \mathcal{A}_t$, their composition is a probability measure on $\mathcal{N}_{t+1}$ satisfying the following equations:

$$(q_t \otimes \mu_t)(\eta) = q_t(a(\eta))\, \mu_t(a(\eta), \eta), \;\; \eta \in \mathcal{N}\mathcal{N}_{t+1}.$$

The composition of a set of probability distributions $\mathcal{Q}_t \subset \mathcal{P}(\mathcal{N}_t)$ with a multikernel $\mathcal{A}_t : \mathcal{N}_t \rightrightarrows \mathcal{P}(\mathcal{N}_{t+1})$ satisfying (4.35), is the following set of probability measures on $\mathcal{N}_{t+1}$:

$$Q_t \otimes \mathcal{A}_t = \big\{q_t \otimes \mu_t : q_t \in Q_t,\, \mu_t \lessdot \mathcal{A}_t\big\}.$$

If Q_t and $\mathcal{A}_t$ are convex and compact, then their composition $\mathcal{Q}_t \otimes A_t$ is convex and compact as well.

We can now formulate a corollary of Theorem 4.26.

Corollary 4.28. *In a scenario tree model, suppose a dynamic risk measure* $\varrho(\cdot)$ *is given by* (4.6) *with coherent conditional risk measures* $\rho_t(\cdot)$. *Then for every* $t = 1, \ldots, T$ *and every* $V_T \in \mathcal{Z}_T$

$$\varrho_{t,T}(V_T) = \max_{\mu\in\mathcal{A}_{t,T}} \big\langle \mu, V_T\big\rangle, \tag{4.37}$$

where $\mathcal{A}_{t,T} : \mathcal{N}_t \to \mathcal{P}(\mathcal{N}_T)$ *is a convex closed multikernel defined as follows:*

$$\mathcal{A}_{t,T} = \mathcal{A}_t \otimes \mathcal{A}_{t+1} \otimes \ldots \otimes \mathcal{A}_{T-1}. \tag{4.38}$$

The maximization in (4.37) *is understood node-wise, for every* $\nu \in \mathcal{N}_t$.

4.3 Multistage Risk Optimization Problems

4.3.1 Problem Formulation

In a multistage optimization problem, decisions $x_1, \ldots, x_T$ are made in periods $1, \ldots, T$. They may depend on uncertain data $D_1, \ldots, D_T$, which are revealed gradually over time. The decision process has the form:

$$\text{observation } (D_1) \rightsquigarrow \text{decision } (x_1) \rightsquigarrow \text{observation } (D_2) \rightsquigarrow \text{decision } (x_2) \rightsquigarrow \cdots$$
$$\cdots \rightsquigarrow \text{observation } (D_T) \rightsquigarrow \text{decision } (x_T).$$

The data sequence $D_t \in \mathbb{R}^d, t = 1, \ldots, T$, is a stochastic process on a probability space $(\Omega, \mathcal{F}, P)$ and its distribution does not depend on decisions. We write $D_{1:t} \triangleq (D_1, \ldots, D_t)$ to denote the history of the process up to time t. Since D_1 is observed before any decisions are made, we may consider it known.

The decision vector x_t (with values in $\mathbb{R}^n$) at stage t may depend on the data $D_{1:t}$ observed up to time t, but not on the results of future observations. This is the requirement of nonanticipativity. To formalize this condition, for $t = 1, \ldots, T$, we denote by $\mathcal{F}_t = \sigma\big(D_{1:t}\big)$ the σ-subfield generated by the observations up to time t. With no loss of generality, we may assume that $\mathcal{F}_1 = \{\emptyset, \Omega\}$ and $\mathcal{F} = \mathcal{F}_T$. The nonanticipativity condition means that each x_t must be an $\mathcal{F}_t$-measurable random vector, for $t = 1, \ldots, T$. We call the sequence $x \triangleq (x_1, \ldots, x_T)$ a *policy*: it chooses a decision at each stage using the information available up to this stage. We write $x_{1:t} \triangleq (x_1, \ldots, x_t)$.

To facilitate our analysis, we require that x_t is an element of the space $\mathcal{X}_t = \mathcal{L}_p(\Omega, \mathcal{F}_t, P; \mathbb{R}^n)$ with $p \in [1, \infty)$. We define the space $\mathcal{X} \triangleq \mathcal{X}_1 \times \cdots \times \mathcal{X}_T$ and thus $x \in \mathcal{X}$.

Fundamental to the multistage model are the relations between the decision vectors in successive stages: they describe the evolution of the system in time. We assume that at each stage t a function $\psi : \mathbb{R}^{nt} \times \mathbb{R}^{nd} \to \mathbb{R}^m$ is defined such that for almost all $\omega \in \Omega$ we must satisfy the relation

$$\psi_t(x_{1:t}, D_{1:t}) \in \mathcal{Y}_t, \quad t = 1, \ldots, T, \tag{4.39}$$

where $\mathcal{Y}_t$ is a random closed convex set, measurable with respect to $\mathcal{F}_t$. To formalize the resulting constraint, at each stage $t = 1, \ldots, T$ we consider the spaces $\mathfrak{Y}_t = \mathcal{L}_p(\Omega, \mathcal{F}_t, P; \mathbb{R}^m)$, in which the values of our dynamics operators will lie, and we define $\mathfrak{Y} \triangleq \mathfrak{Y}_1 \times \cdots \times \mathfrak{Y}_T$. Then we define the operators $G_t : \mathcal{X} \times \Omega \to \mathfrak{Y}_t$ as follows:

$$G_t(x, \omega) = \psi_t(x_{1:t}(\omega), D_{1:t}(\omega)), \quad t = 1, \ldots, T, \quad \omega \in \Omega.$$

Evidently, we need some assumptions on ψ_t to ensure that the values of the operators G_t are indeed elements of the spaces $\mathfrak{Y}_t$; we shall discuss them in due course. But the most important feature of these operators, if they are properly defined, is *causality* as described in the following definition.

Definition 4.29. An operator $G : \mathcal{X} \to \mathfrak{Y}$ is *causal* if functions $g_t : \mathbb{R}^{nt} \times \Omega \to \mathbb{R}^m$ exist such that for all $t = 1, \ldots, T$

$$G_t(x)(\omega) = g_t(x_{1:t}(\omega), \omega), \quad \omega \in \Omega, \tag{4.40}$$

and each $g_t(\cdot, \cdot)$ is superpositionally measurable.

Our operators evidently have the causality property by construction, with

$$g_t(x_{1:t}(\omega), \omega) = \psi_t(x_{1:t}(\omega), D_{1:t}(\omega)). \tag{4.41}$$

Superpositional measurability is guaranteed for functions that satisfy the assumption below.

Assumption 4.30. *For all* $t = 1, \dots, T$*:*

(i) $g_t(\xi, \cdot)$ *is an element of* $\mathfrak{Y}_t$ *for all* $\xi \in \mathbb{R}^{nt}$*;*

(ii) *For almost all* $\omega \in \Omega$*,* $g_t(\cdot, \omega)$ *is continuously differentiable with respect to its first argument, with the Jacobian* $g_t'(\cdot, \omega)$*;*

(iii) *A constant* C_g *exists, such that* $\|g_t'(\cdot, \omega)\| \le C_g$*, almost surely.*

Under Assumption 4.30, each G_t given by (4.40) indeed maps the product space $\mathfrak{X}_{1:t}$ into a subset of $\mathfrak{Y}_t$.

The dynamics of the system is thus abstractly represented by the relation

$$G(x) \in \mathcal{Y}, \tag{4.42}$$

where $G : \mathfrak{X} \to \mathfrak{Y}$ is a causal operator, and $\mathcal{Y} = \mathcal{Y}_1 \times \cdots \times \mathcal{Y}_T$, with each $\mathcal{Y}_t : \Omega \rightrightarrows \mathbb{R}^m$, $t = 1, \dots, T$, being an $\mathcal{F}_t$-measurable multifunction with convex and closed values. In a more explicit way, the relation (4.42) has the form:

$$G_t(x_{1:t}) \in \mathcal{Y}_t, \quad t = 1, \dots, T, \tag{4.43}$$

and, due to the causality of $G(\cdot)$ and the decomposability of $\mathcal{Y}$,

$$g_t(x_{1:t}(\omega), \omega) \in \mathcal{Y}_t(\omega), \quad t = 1, \dots, T, \quad \omega \in \Omega.$$

Additionally, $\mathcal{F}_t$-measurable multifunctions with closed convex images $\mathcal{X}_t : \Omega \rightrightarrows \mathbb{R}^n$, $t = 1, \dots, T$, may be defined. They describe the stage-wise feasible sets for the decisions x_t.

The cost incurred at stage $t = 1, \dots, T$ is represented by a function $f_t : \mathbb{R}^n \times \Omega \to \mathbb{R}$. In applications, the function f_t results from a composition of a cost function $c_t : \mathfrak{X}_t \times \mathfrak{D} \to \mathbb{R}$ and an $\mathcal{F}_t$-measurable random data vector $D_t : \Omega \to \mathbb{R}^d$. Then $f_t(x_t, \omega) \triangleq c_t(x_t(\omega), D_t(\omega))$, $\omega \in \Omega$.

We assume that it satisfies Assumption 4.30, with the spaces $\mathfrak{Y}_t$ replaced by $\mathcal{Z}_t = \mathcal{L}_p(\Omega, \mathcal{F}_t, P)$, $t = 1, \dots, T$, in which the random costs lie. Similar to § 3.1, we define the operators $F_t : \mathfrak{X}_t \to \mathcal{Z}_t$, as

$$F_t(x_t)(\omega) = f_t(x_t(\omega), \omega), \quad \omega \in \Omega, \quad t = 1, \dots, T. \tag{4.44}$$

Evidently, the operator $F = (F_1, \dots, F_T)$ is causal.

Our objective is to find the decision policy that minimizes the risk of the sequence

$$Z_t = F_t(x_t), \quad t = 1, \dots, T, \tag{4.45}$$

where risk is measured by a dynamic risk measure $\{\varrho_{t,T}\}_{t=1,\dots,T}$.

The multistage stochastic optimization problem can be now formulated as follows:

$$\min \varrho_{1,T}\big(F(x)\big) \tag{4.46}$$

$$\text{s.t. } G_t(x_{1:t}) \in \mathcal{Y}_t \quad \text{a.s.}, \quad t = 1, \dots, T, \tag{4.47}$$

$$x_t \in \mathcal{X}_t \quad \text{a.s.}, \quad t = 1, \dots, T. \tag{4.48}$$

Evidently, we could have aggregated the relations (4.47) and (4.48) into one inclusion, but it is convenient to distinguish between the causal relations describing the dynamics of the system, and the stage-wise constraints.

4.3.2 Convexity and Subdifferentiation of Composite Dynamic Risk Measures

In order to develop optimality conditions and solution methods for problem (4.46)–(4.48), we need to study the properties of the composite function

$$\varphi(x_1, \dots, x_T) \triangleq \varrho_{1,T}\big[F_1(x_1), \dots, F_T(x_T)\big]. \tag{4.49}$$

We focus on the case of a *coherent* dynamic risk measure ϱ, that is, a measure of the form (4.6) with coherent one-step conditional risk mappings $\rho_t : \mathcal{Z}_{t+1} \to \mathcal{Z}_t$, $t = 1, \dots, T-1$.

The following statement follows from the direct application of Lemma 3.5 to our case. Its proof is valid for vector-valued F and f.

Lemma 4.31. *If the functions $f_t(\cdot, \omega)$ are convex for all $\omega \in \Omega$ and all $t = 1, \dots, T$, and ϱ has the properties of convexity and monotonicity, then the composite function $\varphi = \varrho_{1,T} \circ F$ is convex.*

Let us discuss the differentiability properties of the composite function $\varphi = \varrho \circ F$ at a point $\bar{x} \in \mathcal{X}$. If F is convex, then $f(\cdot, \omega)$ is a convex $\mathbb{R}^T$-valued function; hence, it is continuous and has finite directional derivatives at $\bar{x}$:

$$f'(\bar{x}, \omega; h) \triangleq \lim_{t \downarrow 0} \frac{1}{t}\big[f(\bar{x} + th, \omega) - f(\bar{x}, \omega)\big], \quad h \in \mathcal{X}.$$

These properties are inherited by the mapping F, under the following local growth condition.

Assumption 4.32. *For each $t = 1, \ldots, T$ a function $C_t \in \mathcal{L}_p(\Omega, \mathcal{F}_t, P)$ and $\varepsilon > 0$ exist, such that*

$$\left| f_t(y, \omega) - f_t(\bar{y}, \omega) \right| \le C_t(\omega) \| y - \bar{y} \|, \quad \omega \in \Omega, \tag{4.50}$$

whenever $\| x - \bar{x} \| \le \varepsilon$.

The following property is evident.

Lemma 4.33. *Suppose for all $t = 1, \ldots, T$ the functions $f_t(\cdot, \omega)$ are convex and satisfy Assumption 4.32. Then $F(\cdot)$ is continuous and directionally differentiable at $\bar{x}$, and its directional derivative is given by*

$$\left[F_t'(\bar{x}; h) \right](\omega) = f_t'(\bar{x}_t(\omega), \omega; h_t(\omega)), \quad t = 1, \ldots, T, \quad \omega \in \Omega, \quad h \in \mathcal{X}. \tag{4.51}$$

We can now derive the subdifferential of the composite function.

Theorem 4.34. *Suppose for all $t = 1, \ldots, T$ the functions $f_t(\cdot, \omega)$ are convex and satisfy Assumption 4.32. Suppose $\varrho_{1,T} : \mathcal{Z}_{1,T} \to \mathbb{R}$ satisfies the conditions of convexity and monotonicity. Then the composite function $\varphi = \varrho_{1,T} \circ F$ is subdifferentiable at $\bar{x}$ and*

$$\partial \varphi(\bar{x}) = \bigcup_{\zeta \in \partial \varrho_{1,T}(F(\bar{x}))} \begin{bmatrix} \partial f_1(\bar{x}_1) \\ \partial f_2(\bar{x}_2(\omega), \omega)\, \zeta_1(\omega) \\ \vdots \\ \partial f_T(\bar{x}_T(\omega), \omega)\, \zeta_{T-1}(\omega) \end{bmatrix}. \tag{4.52}$$

Proof. We proceed as in the proof of Theorem 3.12 to obtain the following expression for the directional derivative of $\varphi(\cdot)$ at $\bar{x}$ in a direction $h \in \mathcal{X}$:

$$\begin{aligned} \varphi'(\bar{x}; h) &= \varrho_{1,T}'\big(F(\bar{x}); F'(\bar{x}; h)\big) \\ &= \max_{\zeta \in \partial \varrho_{1,T}(F(\bar{x}))} \int_{\Omega} \big\langle \zeta(\omega), [F'(\bar{x}; h)](\omega) \big\rangle \, P(d\omega). \end{aligned}$$

The only difference is that the subgradients $\zeta(\omega)$ are vector-valued. Substituting (4.51), we obtain the formula for the directional derivative of $\varphi(\cdot)$:

$$\begin{aligned} \varphi'(\bar{x}; h) &= \max_{\zeta \in \partial \varrho_{1,T}(F(\bar{x}))} \int_{\Omega} \sum_{t=1}^{T} \big\langle \zeta_t(\omega), f_t'(\bar{x}_t(\omega), \omega; h_t(\omega)) \big\rangle \, P(d\omega) \\ &= \max_{\zeta \in \partial \varrho(F(\bar{x}))} \int_{\Omega} \sum_{t=1}^{T} \Big\langle \zeta_t(\omega), \max_{\mu_t(\omega) \in \partial f_t(\bar{x}_t(\omega), \omega)} \langle \mu_t(\omega), h_t(\omega) \rangle \Big\rangle \, P(d\omega). \end{aligned} \tag{4.53}$$

For monotonic $\varrho_{1,T}(\cdot)$ every $\zeta \in \partial\varrho_{1,T}(\bar{Z})$ is nonnegative. By interchangeability, measurable selectors $\mu_t(\cdot)$ of $\partial f_t(\bar{x}(\cdot), \cdot)$ exist, which achieve the pointwise maximum under the integral P-almost surely. Therefore,

$$\varphi'(\bar{x}; h) = \max_{\zeta \in \partial\varrho(F(\bar{x}))} \max_{\mu(\cdot) \lessdot \partial f(\bar{x}(\cdot), \cdot)} \int_{\Omega} \sum_{t=1}^{T} \zeta_t(\omega) \langle \mu_t(\omega), h_t \rangle \; P(d\omega).$$

It follows from the last equation that the subdifferential of the function $\varphi'(\bar{x}; \cdot)$ at 0 is equal to the set on the right-hand side of (4.52). Since $\partial\varphi(\bar{x}) = \partial\varphi'(\bar{x}; 0)$, the formula (4.52) is true. □

If $\varrho_{1,T}(\cdot)$ is coherent, the subgradients $\zeta \in \partial\varrho(F(\bar{x}))$ have a special form and the formula (4.52) can be written in a more explicit way.

By virtue of Theorem 4.26,

$$\varrho_{1,T}(Z_1, \ldots, Z_T) = \sup_{\lambda \in \Lambda_{1,T}} \mathbb{E}\big[\lambda(Z_1 + \cdots + Z_T)\big]$$

where $\Lambda_{1,T} = \{\lambda_1\lambda_2 \cdots \lambda_{T-1} : \lambda_s \in \Lambda_s,\ s = 1, \ldots, T-1\}$, with each set Λ_s is is given by the formula (4.30). As the sequence $\{Z_t\}$ is adapted and $\mathbb{E}[\lambda_s | \mathcal{F}_t] = 1$ for all $s = 1, \ldots, T$, we obtain

$$\zeta_t = \lambda_1 \cdots \lambda_{t-1}, \quad t = 1, \ldots, T,$$

where the sequences $\lambda = (\lambda_1, \ldots, \lambda_{T-1})$ are elements of the set

$$\Lambda(\bar{x}) = \operatorname*{Arg\,max}_{\substack{\lambda_s \in \Lambda_s \\ s=1,\ldots,T-1}} \mathbb{E}\bigg[\sum_{t=1}^{T} \Big(\prod_{s=1}^{t-1} \lambda_s\Big) F_t(\bar{x}_t)\bigg]. \tag{4.54}$$

Therefore

$$\partial\varphi(\bar{x}) = \bigcup_{\lambda \in \Lambda(\bar{x})} \begin{bmatrix} \partial f_1(\bar{x}_1) \\ \partial f_2(\bar{x}_2(\omega), \omega)\, \lambda_1(\omega) \\ \partial f_3(\bar{x}_3(\omega), \omega)\, \lambda_1(\omega)\lambda_2(\omega) \\ \vdots \\ \partial f_T(\bar{x}_T(\omega), \omega) \left(\prod_{s=1}^{T-1} \lambda_s(\omega)\right) \end{bmatrix}. \tag{4.55}$$

We may remark here that the problem appearing in (4.54) can be solved in a recursive way because all λ's are nonnegative.

4.3.3 Differentiation of Causal Operators

In multistage problems, the decisions x_t are elements of the Lebesgue spaces $\mathcal{X}_t = \mathcal{L}_p(\Omega, \mathcal{F}_t, P; \mathbb{R}^n)$, the values of the costs are elements of the spaces $\mathcal{Z}_t = \mathcal{L}_p(\Omega, \mathcal{F}_t, P)$, and the values of the constraints are elements of the spaces

$\mathcal{Y}_t = \mathcal{L}_p(\Omega, \mathcal{F}_t, P; \mathbb{R}^n)$. Because of that, as we shall see soon, nonlinear and nonconvex problems are very difficult to analyze, because of the inherent absence of the Fréchet differentiability in the functional spaces in this setting.

We focus on the case when the functions $f_t(\cdot, \omega)$ and $g_t(\cdot, \omega)$, $t = 1, \dots, T$, are Fréchet differentiable for almost all elementary events $\omega \in \Omega$.

We write

$$g'(x(\omega), \omega) = \big\{g_t'(x_{1:t}(\omega), \omega)\big\}_{t=1,\dots,T}$$

to represent the Jacobian of $[G(x)](\omega)$ with respect to $x(\omega)$.

In the next result, we provide the Clarke subdifferential of the function

$$\Phi(\cdot) = \varrho\big(F(\cdot)\big), \tag{4.56}$$

with $F(\cdot)$ being only Gâteaux differentiable, and $\varrho(\cdot)$ convex and subdifferentiable. It is a direct reproduction of Theorem 3.24.

Theorem 4.35. *Suppose $F : \mathcal{X} \to \mathcal{Y}$ is a causal operator satisfying Assumption 4.30, and $\varrho : \mathcal{Y} \to \mathbb{R}$ is a convex functional which is subdifferentiable at $F(x)$. Then the Clarke subdifferential of the composite functional has the following form:*

$$\partial\Phi(x) = \big[F'(x)\big]^* \, \partial\varrho(F(x)),$$

where $\big[F'(x)\big]^*$ is the adjoint operator to the Gâteaux derivative $F'(x)$.

The proof is identical to the proof of Theorem 3.24.

We shall use this result in two ways. First, we calculate the Clarke subdifferential of the objective (risk) functional

$$\varphi(\cdot) = \varrho_{1,T}\big(F(\cdot)\big). \tag{4.57}$$

We consider a dynamic risk measure $\big\{\varrho_{t,T}\big\}_{1\le t\le T}$, where each $\varrho_{t,T} : \mathcal{Z}_{t:T} \to \mathcal{Z}_t$. If it is time-consistent, normalized, and satisfies the translation property, then Corollary 4.6 implies that it has the telescopic structure (4.6). The corollary below follows directly from Theorems 4.35 and 4.26.

Corollary 4.36. *Suppose $F : \mathcal{X} \to \mathcal{Y}$ is a causal operator satisfying Assumption 4.30 and all one-step conditional risk mappings $\rho_t(\cdot)$ are convex and monotonic, $t = 1, \dots, T-1$. Then the Clarke subdifferential of the composite function* (4.57) *has the following form:*

$$\partial\varphi(\bar{x}) = \bigcup_{\zeta\in\partial\varrho_{1,T}(F(\bar{x}))} \begin{bmatrix} \nabla f_1(\bar{x}_1) \\ \nabla f_2(\bar{x}_2(\omega), \omega)\, \zeta_2(\omega) \\ \nabla f_3(\bar{x}_3(\omega), \omega)\, \zeta_3(\omega) \\ \vdots \\ \nabla f_T(\bar{x}_T(\omega), \omega)\, \zeta_T(\omega) \end{bmatrix}. \tag{4.58}$$

The proof follows the proof of Theorem 4.34.

We can also calculate the subdifferential of the penalty function

$$\Phi(\cdot) = \operatorname{dist}\big(G(\cdot), \mathcal{Y}\big). \tag{4.59}$$

Corollary 4.37. *Suppose* $G : \mathfrak{X} \to \mathfrak{Y}$ *is a causal operator satisfying Assumption* 4.30. *Then*

$$\partial\Phi(x) = \big[G'(x)\big]^* \left(\mathcal{N}_{\mathcal{Y}}(G(x)) \cap \mathbb{B}_{\mathfrak{Y}^*}\right),$$

where $\big[G'(x)\big]^*$ *is the adjoint operator to the Gâteaux derivative* $G'(x)$*,* $\mathcal{N}_{\mathcal{Y}}(\cdot)$ *is the normal cone to* $\mathcal{Y}$*, and* $\mathbb{B}_{\mathfrak{Y}^*}$ *is the closed unit ball in* $\mathfrak{Y}^*$*.*

To derive a more explicit form of the subdifferential, we denote:

$$G_t'(\hat{x}_{1:t}) = A_t = \big(A_{t,1}, \ldots, A_{t,t}\big), \quad t = 1, \ldots, T,$$

with partial Jacobians $A_{t,\ell} : \mathfrak{X}_\ell \to \mathfrak{Y}_t$,

$$A_{t,\ell} = \frac{\partial G_t(\hat{x}_{1:t})}{\partial x_\ell}, \quad \ell = 1, \ldots, t, \quad t = 1, \ldots, T. \tag{4.60}$$

These linear operators are defined pointwise:

$$A_{t,\ell}(\omega) = \frac{\partial g_t(\hat{x}_{1:t}(\omega), \omega)}{\partial x_\ell(\omega)}, \quad \ell = 1, \ldots, t, \quad t = 1, \ldots, T, \quad \omega \in \Omega. \tag{4.61}$$

Due to Assumption 4.30, all $A_{t,\ell}$ are continuous linear operators. We can derive a more explicit form of the vector $\big[G'(\hat{x})\big]^* \hat{\psi}$ for $\hat{\psi} \in \mathcal{N}_{\mathcal{Y}}(G(x)) \cap \mathbb{B}_{\mathfrak{Y}^*}$. Due to the decomposability of $\mathcal{Y}_t$ the normal cone $\mathcal{N}_{\mathcal{Y}_t}(G_t(x))$ is composed of elements which are selectors of $\mathcal{N}_{\mathcal{Y}_t(\cdot)}\big(G_t(x(\cdot), \cdot)\big)$; we have $\mathcal{N}_{\mathcal{Y}_t}(G_t(x))(\omega) = \mathcal{N}_{\mathcal{Y}_t(\omega)}\big(g_t(x(\omega), \omega)\big)$ a.s.. Using the causality of G_t, we obtain

$$\psi_t(\omega) \in \mathcal{N}_{\mathcal{Y}_t(\omega)}\big(g_t(\hat{x}_{1:t}(\omega), \omega)\big) \quad t = 1, \ldots, T, \quad \text{for almost all } \omega \in \Omega.$$

Now, using the block-triangular form of $A = G'(\hat{x})$, for any $h \in \mathcal{X}$ we can write

$$\begin{aligned} \langle A^*\hat{\psi}, h\rangle &= \langle \hat{\psi}, Ah\rangle = \sum_{t=1}^{T} \langle \hat{\psi}_t, A_t h\rangle = \sum_{t=1}^{T}\sum_{\ell=1}^{t} \langle \hat{\psi}_t, A_{t,\ell} h_\ell\rangle \\ &= \sum_{\ell=1}^{T}\sum_{t=\ell}^{T} \langle A_{t,\ell}^* \hat{\psi}_t, h_\ell\rangle = \sum_{\ell=1}^{T} \Big\langle \mathbb{E}\Big[\sum_{t=\ell}^{T} A_{t,\ell}^* \hat{\psi}_t \,\Big|\, \mathcal{F}_\ell\Big], h_\ell \Big\rangle. \end{aligned} \tag{4.62}$$

It follows that every subgradient of the function $\Phi(\cdot)$ has block-coordinates

$$\mathbb{E}\left[\sum_{t=\ell}^{T} A_{t,\ell}^{*}\hat{\psi}_t \,\middle|\, \mathcal{F}_\ell\right], \quad \ell = 1, \ldots, T. \tag{4.63}$$

4.4 Optimality Conditions

The multistage problems can be formulated in two different ways regarding the way implementability of the solution is reflected in the model. One possibility is to formulate the model in such a way that the definition of the decision spaces includes the $\mathcal{F}_t$-measurability of the decisions at time t, $t = 1, \ldots, T$. This is the model (4.46)–(4.48). In another formulation, we consider decision spaces of $\mathcal{F}$-measurable decisions at each stage, but we add additional linear constraints enforcing $\mathcal{F}_t$-measurability.

We start from the basic model.

4.4.1 Subregular Recourse

Our key idea is to use uniform parametric subregularity of deterministic finite-dimensional set-constrained systems associated with each stage $t = 1, \ldots, T$ and each elementary event $\omega \in \Omega$:

$$g_t(\zeta_{1:t-1}, \xi, \omega) \in \mathcal{Y}_t(\omega), \tag{4.64}$$

$$\xi \in \mathcal{X}_t(\omega). \tag{4.65}$$

Here, $\zeta_{1:t-1} \in \mathbb{R}^{n(t-1)}$ represents the history of decisions at the particular elementary event, and the elementary event $\omega \in \Omega$ itself are parameters of the system. For uniformity of notation, for $t = 1$ the parameter $\zeta_{1:t-1}$ is non-existent.

We introduce the following concept.

Definition 4.38. The system (4.64)–(4.65) *admits complete subregular recourse*, if a constant C exist, such that for almost all $\omega \in \Omega$, every $\zeta_{1:t-1} \in \mathcal{X}_{1:t-1}(\omega)$ and every $\eta \in \mathbb{R}^n$, a solution ξ of (4.64)–(4.65) exists, satisfying the inequality

$$\|\xi - \eta\| \le C\big(\mathrm{d}(g_t(\zeta_{1:t-1}, \eta, \omega), \mathcal{Y}_t(\omega)) + \mathrm{d}(\eta, \mathcal{X}_t(\omega))\big).$$

In two-stage stochastic linear programming the concept of *relatively complete recourse* is well-established. It means the solvability of the system (4.64)–(4.65) for almost all $\omega \in \Omega$ and every $\zeta_1 \in \mathcal{X}_1(\omega)$. But even in this case, the uniform subregularity of this system is not guaranteed.

We shall prove the subregularity of the entire infinite-dimensional system of constraints (4.47)–(4.48) when complete subregular recourse is admitted.

Theorem 4.39. *If the system* (4.64)–(4.65) *admits complete subregular recourse, then the system* (4.47)–(4.48) *is subregular at any feasible point* $\hat{x}_{1:T}$.

Proof. Let $u_{1:T} \in \mathfrak{X}$ be chosen from a sufficiently small neighborhood of $\hat{x}_{1:T}$. We shall construct a solution $\bar{x}$ of (4.47)–(4.48) which is close to u, with an appropriate error bound.

For $t = 1, \dots, T$ we consider the system in the space $\mathfrak{X}_t$:

$$\begin{aligned} G_t(\bar{x}_{1:t-1}, x_t) &\in \mathcal{Y}_t, \\ x_t &\in \mathcal{X}_t. \end{aligned}$$

Our intention is to find a solution $\bar{x}_t$ to this system, which is sufficiently close to u_t. By the Lipschitz continuity of $G_t(\cdot, \cdot)$,

$$\big\|G_t\big(\bar{x}_{1:t-1}, u_t\big)\big\| \le \big\|G_t\big(u_{1:t}\big)\big\| + L\big\|\bar{x}_{1:t-1} - u_{1:t-1}\big\|. \tag{4.66}$$

We define a multifunction $\mathfrak{G} : \Omega \rightrightarrows \mathbb{R}^n$ by the relations

$$\mathfrak{G}(\omega) = \Big\{\xi \in \mathbb{R}^n : g_t(\bar{x}_{1:t-1}(\omega), \xi, \omega) \in \mathcal{Y}_t(\omega),\ \xi \in \mathcal{X}_t(\omega),$$
$$\big\|\xi - u_t(\omega)\big\| \le C\Big(\mathrm{d}\big(g_t(\bar{x}_{1:t-1}(\omega), u_t(\omega), \omega), \mathcal{Y}_t(\omega)\big) + \mathrm{d}\big(u_t(\omega), \mathcal{X}_t(\omega)\big)\Big)\Big\}.$$

We observe that both distance functions in the definition of $\mathfrak{G}(\cdot)$ are $\mathcal{F}_t$-measurable, by virtue of the Berge theorem. Therefore, the multifunction $\mathfrak{G}$ is $\mathcal{F}_t$-measurable. It has non-empty images due to Definition 4.38 applied with $\eta = u_t(\omega)$ and $\zeta_{1:t-1} = \bar{x}_{1:t-1}(\omega)$. Hence, a measurable selection $\bar{x}_t$ of $\mathfrak{G}$ exists. From the construction of the multifunction $\mathfrak{G}$,

$$\big\|\bar{x}_t(\omega) - u_t(\omega)\big\| \le C\Big(\mathrm{d}\big(g_t(\bar{x}_{1:t-1}(\omega), u_t(\omega), \omega), \mathcal{Y}_t(\omega)\big) + \mathrm{d}\big(u_t(\omega), \mathcal{X}_t(\omega)\big)\Big).$$

Therefore, with the norms and distances in the spaces $\mathfrak{X}_t$ and $\mathfrak{Y}_t$,

$$\big\|\bar{x}_t - u_t\big\| \le C\Big(\operatorname{dist}\big(G_t(\bar{x}_{1:t-1}, u_t), \mathcal{Y}_t\big) + \operatorname{dist}\big(u_t, \mathcal{X}_t\big)\Big). \tag{4.67}$$

Combining inequalities (4.67) and (4.66), we infer that

$$\big\|\bar{x}_t - u_t\big\| \le C\Big(\operatorname{dist}\big(G_t(u_{1:t}), \mathcal{Y}_t\big) + L\big(\big\|\bar{x}_{1:t-1} - u_{1:t-1}\big\|\big) + \operatorname{dist}\big(u_t, \mathcal{X}_t\big)\Big). \tag{4.68}$$

We can now prove by induction that constants $\bar{C}_t$ exist such that

$$\|\bar{x}_t - u_t\| \le \bar{C}_t \sum_{\ell=1}^{t} \Big(\operatorname{dist}\big(G_\ell(u_{1:\ell}), \mathcal{Y}_\ell\big) + \operatorname{dist}\big(u_\ell, \mathcal{X}_\ell\big)\Big).$$

For $t = 1$, the result is provided by (4.68), because the term $\|\bar{x}_{1:t-1} - u_{1:t-1}\|$ is not present there. Supposing it is true for $t-1$, we verify it for t using (4.68). The last relation for $t = T$ establishes the subregularity of the system (4.47)–(4.48). □

4.4.2 Optimality Conditions for Problems with Built-In Nonanticipativity

Now, we establish the necessary conditions of optimality for problem (4.46)–(4.48). Recall the notation

$$A_{t,\ell}(\omega) = \frac{\partial g_t(\hat{x}_{1:t}(\omega), \omega)}{\partial x_\ell(\omega)}, \quad \ell = 1, \dots, t, \quad t = 1, \dots, T, \quad \omega \in \Omega.$$

Theorem 4.40. *Suppose the system* (4.64)–(4.65) *admits complete subregular recourse and the policy* $\hat{x}$ *is a local minimum of problem* (4.46)–(4.48). *Then a subgradient* $\hat{s} \in \partial\varphi(\hat{x})$, *multipliers* $\hat{\psi}_t \in \mathcal{N}_{\mathcal{Y}_t}(G_t(\hat{x}_{1:t}))$, $t = 1, \dots, T$, *and normal elements* $\hat{n}_t \in \mathcal{N}_{\mathcal{X}_t}(\hat{x}_t)$, $t = 1, \dots, T$, *exist, such that for* P*-almost all* $\omega \in \Omega$ *we have:*

$$\hat{s}_t + A_{t,t}^\top \hat{\psi}_t + \mathbb{E}_t\Big[\sum_{\ell=t+1}^{T} A_{\ell,t}^\top \hat{\psi}_\ell\Big] + \hat{n}_t = 0, \quad t = 1, \dots, T. \tag{4.69}$$

Proof. Since $\varphi(\cdot)$ is Lipschitz continuous about $\hat{x}$ with some constant L_φ, then for every $K > L_\varphi$ the point $\hat{x}$ is a local minimum of the function

$$\varphi(x) + K \operatorname{dist}(x, \mathcal{X} \cap G^{-1}(\mathcal{Y})).$$

The system (4.47)–(4.48) is subregular with some constant $\bar{C}$ by virtue of Theorem 4.39. Consequently, $\hat{x}$ is a local minimum of the function

$$\varphi(x) + K\bar{C}\big(\operatorname{dist}(G(x), \mathcal{Y}) + \operatorname{dist}(x, \mathcal{X})\big).$$

We use Clarke's necessary conditions of optimality for Lipschitz continuous functions:

$$0 \in \partial\varphi(\hat{x}) + K\bar{C}\,\partial\big[\operatorname{dist}(G(\cdot), \mathcal{Y})\big](\hat{x}) + K\bar{C}\,\partial\big[\operatorname{dist}(\cdot, \mathcal{X})\big](\hat{x}).$$

The Clarke subdifferential of the function $\operatorname{dist}(G(\cdot), \mathcal{Y})$ is calculated in Corollary 4.37:

$$\partial\big[\operatorname{dist}(G(\cdot), \mathcal{Y})\big](\hat{x}) = \big[G'(\hat{x})\big]^* \big(\mathcal{N}_{\mathcal{Y}}(G(\hat{x})) \cap \mathbb{B}_{\mathfrak{Y}^*}\big).$$

The subdifferential of $\operatorname{dist}(\hat{x}, \mathcal{X})$ is $\mathcal{N}_{\mathcal{X}}(\hat{x}) \cap \mathbb{B}_{\mathfrak{X}^*}$. We infer that a subgradient $\hat{s} \in \partial\varphi(\hat{x})$, an element $\hat{\psi} \in \mathcal{N}_{\mathcal{Y}}(G(\hat{x}))$, and a normal vector $\hat{n} \in \mathcal{N}_{\mathcal{X}}(\hat{x})$ exist, such that

$$\hat{s} + \big[G'(\hat{x})\big]^* \hat{\psi} + \hat{n} = 0.$$

Equation (4.63) yields the equations (4.69). □

We can also use Corollary 4.36 to obtain a more explicit representation of the subgradient $\hat{s}$ in (4.69): a subgradient $\hat{\zeta} \in \partial\varrho_{1,T}(F(\hat{x}))$ exists, such that

$$\hat{s}(\omega) = \begin{bmatrix} \nabla f_1(\hat{x}_1) \\ \nabla f_2(\hat{x}_2(\omega), \omega)\, \hat{\zeta}_2(\omega) \\ \nabla f_3(\hat{x}_3(\omega), \omega)\, \hat{\zeta}_3(\omega) \\ \vdots \\ \nabla f_T(\hat{x}_T(\omega), \omega)\, \hat{\zeta}_T(\omega) \end{bmatrix}, \quad \omega \in \Omega.$$

4.4.3 Optimality Conditions for Problems with Nonanticipativity Constraints

A different situation arises with the use of *nonanticipativity constraints*. The fundamental idea reflected in this formulation is to consider extended spaces $\mathbb{X}_t = \mathcal{L}_p(\Omega, \mathcal{F}, P; \mathbb{R}^n)$, $t = 1, \dots, T$ and a relaxed policy

$$x = (x_1, \dots, x_T) \in \mathbb{X}_1 \times \cdots \times \mathbb{X}_T = \mathbb{X}.$$

In order to enforce that the relaxed policy can be identified with an element of the space $\mathfrak{X}$, we impose the following requirement known as the *nonaticipativity constraint*:

$$x_t = \mathbb{E}[x_t \,|\, \mathcal{F}_t], \quad t = 1, \dots, T. \tag{4.70}$$

The equations (4.70) define a closed subspace $\mathcal{S}$ in $\mathbb{X}$. This subspace can be identified with the space $\mathfrak{X}$ in the original problem.

Several important theoretical and practical advantages are associated with this reformulation. It allows to study individual scenario models, for each $\omega \in \Omega$, and to analyze the effect of the information constraint. It may also serve as the theoretical foundation for a variety of decomposition methods, similar to the case of linear dynamics and integral functionals.

To formally define the nonlinear problem in the space $\mathbb{X}$ we need to extend the domains and the ranges of the operators $F(\cdot)$ and $G(\cdot)$, and to re-define the composition $\varphi(\cdot)$.

An extension of the causal operator $G(\cdot)$ is natural from its definition; it is still given by (4.40). Its value space is $\mathbb{Y} = \mathbb{Y}_1 \times \cdots \times \mathbb{Y}_T$ with $\mathbb{Y}_t = \mathcal{L}_p(\Omega, \mathcal{F}, P; \mathbb{R}^m)$, $t = 1, \ldots, T$. The sets $\mathcal{X}_t$ and $\mathcal{Y}_t$ can still be viewed as subsets $\widetilde{\mathcal{X}}_t$ of $\mathbb{X}_t$ and $\widetilde{\mathcal{Y}}_t$ of $\mathbb{Y}_t$:

$$\begin{aligned} \widetilde{\mathcal{X}}_t &= \{x_t \in \mathbb{X}_t : x_t(\omega) \in \mathcal{X}_t(\omega) \text{ a.s.}\}, \\ \widetilde{\mathcal{Y}}_t &= \{y_t \in \mathbb{Y}_t : y_t(\omega) \in \mathcal{Y}_t(\omega) \text{ a.s.}\}, \quad t = 1, \ldots, T. \end{aligned}$$

Notice that the sets $\widetilde{\mathcal{X}}_t$ and $\widetilde{\mathcal{Y}}_t$ contain more elements than their counterparts in the previous formulation, because they allow for a broader class of measurable selections from $\mathcal{X}_t(\cdot)$ and $\mathcal{Y}_t(\cdot)$, respectively.

Similarly, we extend $F(\cdot)$ by allowing x_t in (4.44) to be an element of the space $\mathbb{X}_t$. Then $\widetilde{Z}_t \triangleq F_t(x_t)$ is an element of the space $\mathbb{Z}_t = \mathcal{L}_p(\Omega, \mathcal{F}, P)$, $t = 1, \ldots, T$. Evidently, we cannot evaluate the dynamic risk measure $\varrho_{1,T}$ on the sequence $(\widetilde{Z}_1, \ldots, \widetilde{Z}_T)$. However, by using the time consistency, monotonicity, translation, and normalization, we can invoke Corollary 4.6 to obtain the formula (4.6). If the one-step conditional risk mappings have the translation property, for an adapted sequence $Z_1, \ldots, Z_t$, we obtain

$$\begin{aligned} &\varrho_{1,T}(Z_1, \ldots, Z_T) = \\ &Z_1 + \rho_1\Big(Z_2 + \rho_2\Big(Z_3 + \cdots + \rho_{T-2}\big(Z_{T-1} + \rho_{T-1}(Z_T)\big) \cdots\Big)\Big) \\ &= \rho_1\Big(\rho_2\Big(\cdots\Big(\rho_{T-2}\big(\rho_{T-1}(Z_1 + Z_2 + Z_3 + \cdots + Z_{T-1} + Z_T)\big)\Big)\cdots\Big)\Big) \end{aligned}$$

Using this identity, we can formally define

$$\tilde{\varrho}_{1,T}\big(\widetilde{Z}_1, \ldots, \widetilde{Z}_T\big) = \rho_1 \circ \rho_2 \circ \cdots \circ \rho_{T_1}\big(\widetilde{Z}_1 + \cdots + \widetilde{Z}_T\big). \tag{4.71}$$

It is indeed, an extension of $\varrho_{1,T}$ to the space $\mathbb{Z}_{1,T} = \mathbb{Z}_1 \times \cdots \times \mathbb{Z}_T$.

Now, we can define extension $\tilde{\varphi}(\cdot)$ of $\varphi(\cdot)$ as follows:

$$\tilde{\varphi}(x_1, x_2, \ldots, x_T) = \tilde{\varrho}_{1,T}\big(F_1(x_1), \ldots, F_T(x_t)\big). \tag{4.72}$$

The problem is reformulated as follows:

$$\min \tilde{\varphi}(x_1, \ldots, x_T) \tag{4.73}$$

$$\text{s.t. } x_t - \mathbb{E}_t x_t = 0 \quad \text{a.s.}, \quad t = 1, \ldots, T, \tag{4.74}$$

$$G_t(x_{1:t}) \in \widetilde{\mathcal{Y}}_t \quad \text{a.s.}, \quad t = 1, \ldots, T, \tag{4.75}$$

$$x_t \in \widetilde{\mathcal{X}}_t \quad \text{a.s.}, \quad t = 1, \ldots, T. \tag{4.76}$$

First, we prove the subregularity of the constraints present in the problem formulation with explicit nonanticipativity constraints.

Theorem 4.41. *If the system* (4.64)–(4.65) *admits complete subregular recourse, then the system* (4.74)–(4.76) *is subregular at any feasible point* $\hat{x}_{1:T}$.

Proof. Let $u_{1:T} \in \mathcal{X}$ be fixed. We shall construct a solution $\bar{x}$ of (4.74)–(4.76) which is close to u, with an appropriate error bound.

For $t = 1, \dots, T$, we consider the following system in the space $\mathcal{X}_t$:

$$
\begin{aligned}
G_t(\bar{x}_{1:t-1}, x_t) &\in \widetilde{\mathcal{Y}}_t, \\
x_t - \mathbb{E}_t[x_t] &= 0, \\
x_t &\in \widetilde{\mathcal{X}}_t.
\end{aligned}
$$

Our intention is to find a solution $\bar{x}_t$ to this system, which is sufficiently close to $\mathbb{E}_t[u_t]$. Using the Lipschitz continuity of $G_t(\cdot)$, we obtain

$$
\big\|G_t\big(\bar{x}_{1:t-1}, \mathbb{E}_t[u_t]\big)\big\| \le \big\|G_t\big(u_{1:t}\big)\big\| + L\big(\big\|\bar{x}_{1:t-1} - u_{1:t-1}\big\| + \big\|u_t - \mathbb{E}_t[u_t]\big\|\big). \tag{4.77}
$$

We define a multifunction $\mathfrak{G} : \Omega \rightrightarrows \mathbb{R}^n$ by the relations

$$
\begin{aligned}
\mathfrak{G}(\omega) = \Big\{\xi :\ & g_t(\bar{x}_{1:t-1}(\omega), \xi, \omega) \in \mathcal{Y}_t(\omega), \quad \xi \in \mathcal{X}_t(\omega), \\
& \|\xi - \mathbb{E}_t[u_t](\omega)\| \\
& \le C\Big(\mathrm{d}\big(g_t\big(\bar{x}_{1:t-1}(\omega), \mathbb{E}_t[u_t](\omega), \omega\big), \mathcal{Y}_t(\omega)\big) + \mathrm{d}\big(\mathbb{E}_t[u_t](\omega), \mathcal{X}_t(\omega)\big)\Big)\Big\}.
\end{aligned}
$$

We observe that both distance terms on the right-hand side are $\mathcal{F}_t$-measurable. Therefore, the multifunction $\mathfrak{G}$ is $\mathcal{F}_t$-measurable. It has non-empty images due to Definition 4.38 applied with $\eta = \mathbb{E}_t[u_t](\omega)$ and $\zeta_{1:t-1} = \bar{x}_{1:t-1}(\omega)$. Hence, an $\mathcal{F}_t$-measurable selection $\bar{x}_t$ of $\mathfrak{G}$ exists. From the construction of $\mathfrak{G}$,

$$
\begin{aligned}
&\big\|\bar{x}_t(\omega) - \mathbb{E}_t[u_t](\omega)\big\| \\
&\quad \le C\Big(\mathrm{d}\big(g_t\big(\bar{x}_{1:t-1}(\omega), \mathbb{E}_t[u_t](\omega), \omega\big), \mathcal{Y}_t(\omega)\big) + \mathrm{d}\big(\mathbb{E}_t[u_t](\omega), \mathcal{X}_t(\omega)\big)\Big).
\end{aligned} \tag{4.78}
$$

We view both sides of this inequality as nonnegative elements of the space $\mathcal{L}_p(\Omega, \mathcal{F}_t, P)$. Since it is a Banach lattice, the functional norm of the element on the left-hand side does not exceed the functional norm of the element on right-hand side. The triangle inequality yields:

$$
\big\|\bar{x}_t - \mathbb{E}_t[u_t]\big\| \le C\Big(\operatorname{dist}\big(G_t(\bar{x}_{1:t-1}, \mathbb{E}_t[u_t]), \mathcal{Y}_t\big) + \operatorname{dist}\big(\mathbb{E}_t[u_t], \mathcal{X}_t\big)\Big). \tag{4.79}
$$

For every $\tilde{x}_t \in \mathcal{X}_t$, Jensen inequality implies that

$$\big\|\mathbb{E}_t[u_t] - \mathbb{E}_t[\tilde{x}_t]\big\| \leq \big\|\mathbb{E}_t[u_t] - \tilde{x}_t\big\|$$

and $\mathbb{E}_t[\tilde{x}_t] \in \mathcal{X}_t$ by convexity. Therefore, $\operatorname{dist}\big(\mathbb{E}_t[u_t], \mathcal{X}_t\big) = \operatorname{dist}\big(\mathbb{E}_t[u_t], \widetilde{\mathcal{X}}_t\big)$. Using a similar argument, we have

$$\operatorname{dist}\big(G_t(\bar{x}_{1:t-1}, \mathbb{E}_t[u_t]), \mathcal{Y}_t\big) = \operatorname{dist}\big(G_t(\bar{x}_{1:t-1}, \mathbb{E}_t[u_t]), \widetilde{\mathcal{Y}}_t\big).$$

Both distances above are finite because $\big\|g_t(\bar{x}_{1:t-1}(\cdot), \mathbb{E}_t[u_t](\cdot), \cdot)\big\|$ has a finite $\mathcal{L}_p$-norm by virtue of (4.77) and the term $\mathrm{d}\big(\mathbb{E}_t[u_t](\cdot), \mathcal{X}_t(\cdot)\big)$ is bounded from above by $\|\mathbb{E}_t[u_t](\cdot) - \hat{x}_t(\cdot)\|$, which has a finite $\mathcal{L}_p$-norm by assumption.

Combining these observations with inequalities (4.79) and (4.77), we infer that

$$\begin{aligned}\|\bar{x}_t - \mathbb{E}_t[u_t]\| \leq C\Big(&\operatorname{dist}\big(G_t(u_{1:t}), \widetilde{\mathcal{Y}}_t\big) + L\big(\big\|\bar{x}_{1:t-1} - u_{1:t-1}\big\| \\ &+ \big\|u_t - \mathbb{E}_t[u_t]\big\|\big) + \operatorname{dist}\big(\mathbb{E}_t[u_t], \widetilde{\mathcal{X}}_t\big)\Big).\end{aligned}$$

Since $\operatorname{dist}\big(\mathbb{E}_t[u_t], \widetilde{\mathcal{X}}_t\big) \leq \operatorname{dist}\big(u_t, \widetilde{\mathcal{X}}_t\big) + \big\|u_t - \mathbb{E}_t[u_t]\big\|$, we conclude that

$$\begin{aligned}&\|\bar{x}_t - u_t\| \leq (1 + C + CL)\big\|u_t - \mathbb{E}_t[u_t]\big\| \\ &+ C\Big(\operatorname{dist}\big(G_t(u_{1:t}), \widetilde{\mathcal{Y}}_t\big) + L\big\|\bar{x}_{1:t-1} - u_{1:t-1}\big\| + \operatorname{dist}\big(u_t, \widetilde{\mathcal{X}}_t\big)\Big). \quad (4.80)\end{aligned}$$

We can now prove by induction that constants $\bar{C}_t$ exist such that

$$\|\bar{x}_t - u_t\| \leq \bar{C}_t \sum_{\ell=1}^{t}\Big(\big\|u_\ell - \mathbb{E}_\ell[u_\ell]\big\| + \operatorname{dist}\big(F_\ell(u_{1:\ell}), \widetilde{\mathcal{Y}}_\ell\big) + \operatorname{dist}\big(u_\ell, \widetilde{\mathcal{X}}_\ell\big)\Big).$$

For $t = 1$, the result follows from (4.80), because the term $\big\|\bar{x}_{1:t-1} - u_{1:t-1}\big\|$ is not present. Supposing it is true for $t-1$, we verify it for t using (4.80). The last relation for $t = T$ establishes the subregularity of the system (4.74)–(4.76). □

Abusing notation, we shall use the same notation for the operators

$$G_t'(\hat{x}_{1:t}) = A_t = \big(A_{t,1}, \ldots, A_{t,t}\big), \quad t = 1, \ldots, T,$$

referring to the partial Jacobians $A_{t,\ell} : \mathbb{X}_\ell \to \mathbb{Y}_t$, which are defined by (4.60)–(4.61), but are acting as linear operators between larger spaces.

Now, we can formulate the main result of this section.

Theorem 4.42. *Suppose the system* (4.64)–(4.65) *admits complete subregular recourse. If a policy $\hat{x}$ is a local minimum of problem* (4.73)–(4.76) *then a subgradient $\tilde{s} \in \partial\tilde{\varphi}(\hat{x})$, multipliers $\pi_t \in \mathcal{X}_t^*$, $\tilde{\psi}_t \in N_{\widetilde{\mathcal{Y}}_t}(G_t(\hat{x}_{1:t}))$, $t = 1, \ldots, T$,* *and normal elements* $\tilde{n}_t \in N_{\widetilde{\mathcal{X}}_t}(\hat{x}_t)$*,* $t = 1, \ldots, T$*, exist, such that for* P*-almost all* $\omega \in \Omega$ *we have:*

$$\tilde{s}_t + \pi_t + \sum_{\ell=t}^{T} A_{\ell,t}^\top \tilde{\psi}_\ell + \tilde{n}_t = 0, \quad t = 1, \ldots, T, \tag{4.81}$$

$$\mathbb{E}_t[\pi_t] = 0, \quad t = 1, \ldots, T. \tag{4.82}$$

Proof. We follow a similar line of argument as in Theorem 4.40. Using the Lipschitz continuity of $\tilde{\varphi}(\cdot)$ about $\hat{x}$ with some Lipschitz constant L_φ, we infer that, for every $K > L_\varphi$, the point $\hat{x}$ is a local minimum of the function

$$\tilde{\varphi}(x) + K \operatorname{dist}(x, \widetilde{\mathcal{X}} \cap G^{-1}(\widetilde{\mathcal{Y}}) \cap \mathcal{S}).$$

We define the linear operator $\Pi : \mathbb{X} \to \mathbb{X}$, by

$$\Pi(x_1, \ldots, x_T) = (\mathbb{E}_1[x_1], \ldots, \mathbb{E}_T[x_T]). \tag{4.83}$$

Theorem 4.41 implies that the system (4.74)–(4.76) is metrically subregular with some constant $\bar{C}$. Consequently, $\hat{x}$ is a local minimum of the function

$$\tilde{\varphi}(x) + K\bar{C}\big(\operatorname{dist}(G(x), \widetilde{\mathcal{Y}}) + \operatorname{dist}(x, \widetilde{\mathcal{X}}) + \|x - \Pi x\|\big).$$

We use the necessary conditions of optimality for Lipschitz continuous functions:

$$0 \in \partial\varphi(\hat{x}) + K\bar{C}\,\partial_x\big[\operatorname{dist}(G(\hat{x}), \widetilde{\mathcal{Y}})\big] + K\bar{C}\,\partial\big[\operatorname{dist}(\hat{x}, \widetilde{\mathcal{X}})\big] + K\bar{C}\,\partial\|\hat{x} - \Pi\hat{x}\|.$$

By virtue of Theorem 4.35, the subdifferential of the function $\operatorname{dist}(G(\cdot), \widetilde{\mathcal{Y}})$ it is equal to $\big[G'(\hat{x})\big]^* \big(\mathcal{N}_{\widetilde{\mathcal{Y}}}(G(\hat{x})) \cap \mathbb{B}_{\mathbb{Y}^*}\big)$. The subdifferential of $\operatorname{dist}(\hat{x}, \widetilde{\mathcal{X}})$ is $\mathcal{N}_{\widetilde{\mathcal{X}}}(\hat{x}) \cap \mathbb{B}_{\mathbb{X}^*}$. The subdifferential of the last term is $(I - \Pi^*)\mathbb{B}_{\mathbb{X}^*}$. Using the tower property of conditional expectations, we see that

$$\Pi^*(v_1, \ldots, v_T) = (\mathbb{E}_1[v_1], \ldots, \mathbb{E}_T[v_T]).$$

Therefore,

$$\partial\|\hat{x} - \Pi\hat{x}\| = (I - \Pi^*)\mathbb{B}_{\mathbb{X}^*} = [\ker(\Pi^*)] \cap \mathbb{B}_{\mathbb{X}^*}.$$

Summing up, it follows that a subgradient $\tilde{s} \in \partial\tilde{\varphi}(\hat{x})$, an element $\tilde{\psi} \in \mathcal{N}_{\widetilde{\mathcal{Y}}}(G(\hat{x}))$, a normal vector $\tilde{n} \in \mathcal{N}_{\widetilde{\mathcal{X}}}(\hat{x})$, and a multiplier $\pi \in \ker(\Pi^*)$ exist, such that

$$\tilde{s} + \pi + \big[G'(\hat{x})\big]^* \tilde{\psi} + \tilde{n} = 0.$$

The condition $\pi \in \ker(\Pi^*)$ is equivalent to (4.82). Equations (4.81) can now be derived as in the proof of Theorem 4.40, using the block-triangular form of

$A = G'(\hat{x})$, and equation (4.62) for any $h \in \mathbb{X}$. Since both spaces, $\mathbb{Y}_t^*$ and $\mathbb{X}_\ell^*$, are defined with the use of the full σ-algebra $\mathcal{F}$, we simply have $A^*_{t,\ell} = A^\top_{t,\ell}$. That is why no conditional expectation appears in (4.81). □

Again, we can use Corollary 4.36 to obtain a more explicit representation of the subgradient $\tilde{s}$ in (4.81) for the specific form of the functional (4.72): a subgradient

$$\hat{\zeta} \in \partial \varrho_{1,T}\left(0, \dots, 0, \sum_{t=1}^{T} F_t(\hat{x}_t)\right)$$

exists, such that

$$\tilde{s}(\omega) = \begin{bmatrix} \nabla f_1(\hat{x}_1(\omega), \omega)\, \hat{\zeta}_T(\omega) \\ \nabla f_2(\hat{x}_2(\omega), \omega)\, \hat{\zeta}_T(\omega) \\ \vdots \\ \nabla f_T(\hat{x}_T(\omega), \omega)\, \hat{\zeta}_T(\omega) \end{bmatrix}, \quad \omega \in \Omega. \tag{4.84}$$

It may be of interest to explore the relations of two sets of optimality conditions of Theorems 4.40 and 4.42.

Corollary 4.43. *The subgradient $\hat{s} \in \partial\varphi(\hat{x})$ given by $\hat{s}_t = \mathbb{E}_t[\tilde{s}_t]$, $t = 1, \dots, T$, together with the multipliers $\hat{\psi}_t = \mathbb{E}_t[\tilde{\psi}_t]$, $t = 2, \dots, T$, and normal vectors $\hat{n}_t = \mathbb{E}_t[\tilde{n}_t]$ satisfy the optimality conditions* (4.69).

Proof. We take the conditional expectation of both sides of a typical relation in (4.81), first with respect to with respect to $\mathcal{F}_t$. Since $\mathbb{E}_t[\pi_t] = 0$, using the tower property and $\mathcal{F}_\ell$-measurability of $A_{\ell,t}$, we obtain

$$\begin{aligned} 0 &= \mathbb{E}_t\big[\tilde{s}_t\big] + \mathbb{E}_t\Big[\sum_{\ell=t}^{T} A^\top_{\ell,t} \tilde{\psi}_\ell\Big] + \mathbb{E}_t\big[\tilde{n}_t\big] \\ &= \mathbb{E}_t\big[\tilde{s}_t\big] + \mathbb{E}_t\Big[\sum_{\ell=t}^{T} A^\top_{\ell,t} \mathbb{E}_\ell[\tilde{\psi}_\ell]\Big] + \mathbb{E}_t\big[\tilde{n}_t\big]. \end{aligned}$$

We shall verify that $\hat{s}$ is a subgradient of $\varphi(\cdot)$ at $\hat{x}$. Having in mind that $\tilde{s} \in \partial\tilde{\varphi}(\hat{x})$, for any $x \in \mathcal{X}$, we have

$$\varphi(x) - \varphi(\hat{x}) \geq \sum_{t=1}^{T} \langle \tilde{s}_t, x_t - \hat{x}_t \rangle = \sum_{t=1}^{T} \big\langle \tilde{s}_t, \mathbb{E}_t[x_t - \hat{x}_t] \big\rangle = \sum_{t=1}^{T} \big\langle \mathbb{E}_t[\tilde{s}_t], x_t - \hat{x}_t \big\rangle,$$

and, thus, $\hat{s} \in \partial\varphi(\hat{x})$.

In a similar way, if $\tilde{n}_t \in N_{\widetilde{\mathcal{X}}_t}(\hat{x}_t)$, then, for every $x_t \in \mathcal{X}_t$, we have

$$0 \geq \langle \tilde{n}_t, x_t - \hat{x}_t \rangle = \big\langle \tilde{n}_t, \mathbb{E}_t[x_t - \hat{x}_t] \big\rangle = \big\langle \mathbb{E}_t[\tilde{n}_t], x_t - \hat{x}_t \big\rangle.$$

This proves that $\hat{n}_t \in N_{\mathcal{X}_t}(\hat{x}_t)$, $t = 1, \dots, T$. Likewise, we obtain $\hat{\psi}_t \in N_{\mathcal{Y}_t}(G_t(\hat{x}_{1:t}))$ for $t = 1, \dots, T$. □

We may also remark here that the equation $\hat{s}_t = \mathbb{E}_t[\tilde{s}_t]$, $t = 1, \dots, T$, follows directly from (4.84) and the relations $\hat{\zeta}_t = \mathbb{E}_t[\hat{\zeta}_{t+1}]$, $t = 1, \dots, T-1$.

4.4.4 Duality Associated with Nonanticipativity Constraints

In the convex case, the formulation (4.73)–(4.76) with explicit nonanticipativity constraints allows for the development of a duality theory, which generalizes the results of § 3.3.2.

The virtues of this reformulation become evident when we relax the constraint (4.74). We associate with it a Lagrange multiplier $\pi \in \mathcal{X}^*$, and formulate the Lagrangian $L : \widetilde{\mathcal{X}} \times \mathcal{X}^* \to \mathbb{R}$,

$$L(x, \pi) = \tilde{\varrho}_{1,T}(F_1(x_1), \dots, F_T(x_T)) + \sum_{t=1}^{T} \langle \pi_t, x_t - \mathbb{E}_t[x_t] \rangle.$$

Recall that

$$\langle \pi_t, x_t \rangle = \int_{\Omega} \langle \pi_t(\omega), x_t(\omega) \rangle \, P(d\omega).$$

The value of the Lagrangian does not change if π is shifted by an adapted vector $\nu \in \mathcal{X}^*$, because

$$\langle \nu_t, x_t - \mathbb{E}_t[x_t] \rangle = \langle \nu_t - \mathbb{E}_t[\nu_t], x_t \rangle = 0, \quad t = 1, \dots, T.$$

Therefore, with no loss of generality, we can restrict the range of the multiplier π to the set

$$\mathcal{S}^{\perp} = \big\{ \pi \in \mathcal{X}^* : \mathbb{E}_t[\pi_t] = 0, \; t = 1, \dots, T \big\},$$

and write the Lagrangian $L : \widetilde{\mathcal{X}} \times \mathcal{S}^{\perp} \to \mathbb{R}$ in a simplified form:

$$L(x, \pi) = \tilde{\varrho}_{1,T}(F_1(x_1), \dots, F_T(x_T)) + \sum_{t=1}^{T} \langle \pi_t, x_t \rangle. \tag{4.85}$$

The dual function $L_D(\pi)$ associated with the Lagrangian is the optimal value of the problem

$$\min L(x,\pi) \tag{4.86}$$
$$\text{s.t. } G_t(x_{1:t}) \in \widetilde{\mathcal{Y}}_t \quad \text{a.s.,} \quad t=1,\dots,T, \tag{4.87}$$
$$x_t \in \widetilde{\mathcal{X}}_t \quad \text{a.s.,} \quad t=1,\dots,T. \tag{4.88}$$

Since choosing $x \in \mathcal{S}$ annihilates the Lagrangian term $\langle \pi, x\rangle$, the function $L(x,\pi)$ coincides with $\varrho_{1,T}(F(x))$ on adapted policies x. Therefore, $L_D(\pi)$ provides a lower bound for the optimal value of the problem (4.73)–(4.76).

The dual problem associated with the Lagrangian (4.85) has the form:

$$\max_{\pi \in \mathcal{S}^{\perp}} L_D(\pi). \tag{4.89}$$

A question arises whether the dual problem has a solution and whether the optimal values of the original (primal) and dual problems are equal.

Theorem 4.44. *Suppose for all $t = 1,\dots,T$ the functions $f_t(\cdot,\omega)$ are convex and satisfy Assumption* 4.32 *at any $\bar{x} \in \mathcal{X}$, the operators G_t are linear, the sets $\mathcal{X}$ and $\mathcal{Y}$ are convex and closed, and the risk measure $\varrho_{1,T} : \mathcal{Z}_{1,T} \to \mathbb{R}$ is coherent and time-consistent. Suppose the system* (4.64)–(4.65) *admits complete subregular recourse.*

If the primal problem (4.73)–(4.76) *has an optimal solution, then the dual problem* (4.89) *has an optimal solution, its optimal value is equal to the optimal value of problem* (4.73)–(4.76)*, and for every optimal solution $\hat{\pi}$ of the dual problem, an adapted policy $\hat{x}$ is a minimizer of $L(\cdot,\hat{\pi})$ if and only if $\hat{x}$ is an optimal solution of problem* (4.73)–(4.76).

Proof. By virtue of Lemma 4.31, the composite mapping $\tilde{\varphi} = \tilde{\varrho}_{1,T} \circ F$ is convex. According to Theorem 4.34, it is continuous and subdifferentiable.

Let $\hat{x}$ be a solution of problem (4.73)–(4.76). By Theorem 4.42, a subgradient $\tilde{s} \in \partial\tilde{\varphi}(\hat{x})$, multipliers $\hat{\pi}_t \in \mathcal{X}_t^*$, $\tilde{\psi}_t \in N_{\widetilde{\mathcal{Y}}_t}(G_t(\hat{x}_{1:t}))$, $t = 1,\dots,T$, and normal elements $\tilde{n}_t \in N_{\widetilde{\mathcal{X}}_t}(\hat{x}_t)$, $t = 1,\dots,T$, exist, such that for P-almost all $\omega \in \Omega$ the equations (4.81) are satisfied (with $\pi = \hat{\pi}$). Due to (4.82), $\hat{\pi} \in \mathcal{S}^{\perp}$. It follows from (4.81) that the optimality conditions for problem (4.86)–(4.88) are satisfied. It is elementary to verify that these conditions are sufficient optimality conditions if this problem is convex. Therefore, the dual function $L_D(\hat{\pi})$ is equal to the optimal value of the original problem (4.73)–(4.76), which proves that $\hat{\pi}$ is the optimal solution of the dual problem.

Suppose $\bar{\pi}$ is an optimal solution of the dual problem. Every subgradient of the dual function is equal to a minimizer $\bar{x}$ of problem (4.86)–(4.88) with $\pi = \bar{\pi}$. Since $\bar{\pi}$ maximizes the dual function over $\mathcal{S}^{\perp}$, a minimizer $\bar{x} \in \mathcal{S}$ of problem (4.86)–(4.88) must exists. It follows that $\bar{x}$ is also a minimizer of problem (4.73)–(4.76). ◻

We can now integrate the results of this section with the dual representation of the dynamic risk measure from Theorem 4.26:

$$\tilde{\varrho}_{1,T}(F_1(x_1),\dots,F_T(x_T)) = \max_{\lambda\in\Lambda_{1,T}} \mathbb{E}\big[\lambda(F_1(x_1)+\dots+F_T(x_T))\big].$$

Consider the problem (4.86)–(4.88) and denote the set defined by the pointwise constraints (4.87)–(4.88) by $\widetilde{\mathcal{D}}$. We can express the dual functional as follows:

$$\begin{aligned} L_D(\pi) &= \min_{x\in\widetilde{\mathcal{D}}}\ \max_{\lambda\in\Lambda_{1,T}} \Big\{\langle\lambda, F_1(x_1)+\dots+F_T(x_T)\rangle + \langle\pi, x\rangle\Big\} \\ &= \max_{\lambda\in\Lambda_{1,T}}\ \min_{x\in\widetilde{\mathcal{D}}} \Big\{\langle\lambda, F_1(x_1)+\dots+F_T(x_T)\rangle + \langle\pi, x\rangle\Big\}. \end{aligned}$$

We could exchange the "min" and "max" operator due to the weak* compactness of the set $\Lambda_{1,T}$ and its convexity (see Corollary 4.27).

It follows that the dual problem can be equivalently written as a max–min problem involving *both* dual vectors: the nonanticipativity multiplier π and the risk subgradient λ. We obtain the following dual problem:

$$\max_{\substack{\lambda\in\Lambda_{1,T}\\ \pi\in\mathcal{S}^\perp}}\ \min_{x\in\widetilde{\mathcal{D}}} \sum_{t=1}^{T} \big\{\langle\lambda, F_t(x_t)\rangle + \langle\pi_t, x_t\rangle\big\}. \tag{4.90}$$

By virtue of the interchangeability principle, the inner problem of minimizing with respect to $x\in\widetilde{\mathcal{D}}$ decomposes into scenario subproblems, one for each ω. The outer maximization problem is convex, because the inner minimum, as the minimum over a family of linear functions, is concave, the set $\Lambda_{1,T}$ is convex (by Corollary 4.27), and $\mathcal{S}^\perp$ is a subspace. We explore these features in section 4.6.

4.4.5 Dynamic Programming Relations

For the scenario tree model discussed in § 4.2.4, we can re-formulate the problem (4.46)–(4.48) in a dynamic programming form. Recall that we denoted by $\mathcal{N}_t$ the set of nodes of the tree at level $t = 1,\dots,T$. The set of all subsets of $\mathcal{N}_t$ is the σ-algebra $\mathcal{F}_t$, and thus we define the decision spaces $\mathcal{X}_t$ as the spaces of functions $x_t : \mathcal{N}_t \to \mathbb{R}^n$, $t = 1,\dots,T$.

Define for $t = 2,\dots,T$ the sets:

$$\mathcal{D}_{1:t-1} = \big\{x_{1:t-1}\in\mathcal{X}_{1:t-1} : G_s(x_{1:s})\in\mathcal{Y}_s,\ s=1,\dots,t-1\big\}, \tag{4.91}$$

$$\mathcal{D}_{t:T}(x_{1:t-1}) = \big\{x_{t:T}\in\mathcal{X}_{t:T} : G_s(x_{1:s})\in\mathcal{Y}_s,\ s=t,\dots,T\big\}, \tag{4.92}$$

and

$$\mathcal{D}_t(x_{1:t-1}) = \{x_t \in \mathcal{X}_t : G_t(x_{1:t-1}, x_t) \in \mathcal{Y}_t\}. \tag{4.93}$$

For every $t = 2, \dots, T$, the set $\mathcal{D}_{1:t-1}$ of policies in the periods $1, \dots, t-1$ is decomposable: for every $\nu \in \mathcal{N}_{t-1}$ we have

$$\mathcal{D}_{1:t-1}(\nu) = \{x_{1:t-1}(\nu) \in \mathcal{X}_{1:t-1}(\nu) : g_s(x_{1:s}(\nu), \nu) \in \mathcal{Y}_s(\nu),\ s=1, \dots, t-1\}.$$

Furthermore, the sets $\mathcal{D}_{t:T}(x_{1:t-1})$ and $\mathcal{D}_t(x_{1:t-1})$ are random: they dependent on $\nu \in \mathcal{N}_{t-1}$ and represent the sets of feasible policies (actions) following ν.

We need the following assumption of *relatively complete recourse*.

Assumption 4.45. For all $t = 1, \dots, T$, all $\nu \in \mathcal{N}_{t-1}$ and for all $x_{1:t-1}(\nu) \in \mathcal{D}_{1:t-1}(\nu)$ the set $\mathcal{D}_t(x_{1:t-1})(\nu)$ is nonempty.

By induction forward in time, for $t = 1, \dots, T$, we can verify that Assumption 4.45 is satisfied then also For all $t = 1, \dots, T$, all $\nu \in \mathcal{N}_{t-1}$ and for all $x_{1:t-1}(\nu) \in \mathcal{D}_{1:t-1}(\nu)$ the set $\mathcal{D}_{t:T}(x_{1:t-1})(\nu)$ is nonempty as well.

Suppose problem (4.46)–(4.48) has an optimal solution. We convert it into a two-stage minimization problem, in which "stage 1" corresponds to periods $1, \dots, t-1$, and "stage 2" to periods $t, \dots, T$:

$$\min_{x_{1:t-1} \in \mathcal{D}_{1:t-1}} \inf_{x_{t:T} \in \mathcal{D}_{t:T}(x_{1:t-1})} \rho_1 \circ \dots \circ \rho_{T-1}\big(F_1(x_1) + \dots + F_T(x_T)\big). \tag{4.94}$$

Due to the monotonicity of the one-step conditional measures $\rho_1, \dots, \rho_{t-1}$, we can use the interchangeability principle and move the minimization with respect to $x_{t:T}$ under $\rho_1 \circ \dots \circ \rho_{t-1}$. We obtain the problem:

$$\begin{aligned} \min_{x_{1:t-1} \in \mathcal{D}_{1:t-1}} & \rho_1 \circ \dots \circ \rho_{t-1}\Big[F_1(x_1) + \dots + F_{t-1}(x_{t-1}) \\ & + \inf_{x_{t:T} \in \mathcal{D}_{t:T}(x_{1:t-1})} \rho_t \circ \dots \circ \rho_{T-1}\big(F_t(x_t) + \dots + F_T(x_T)\big)\Big]. \end{aligned} \tag{4.95}$$

The inner minimization with respect to $x_{t:T}$ can be carried out for every $\nu \in \mathcal{N}_t$ separately. We can thus define the stage t *value function*

$$Q_t(x_{1:t-1}) = \inf_{x_{t:T} \in \mathcal{D}_{t:T}(x_{1:t-1})} \rho_t \circ \dots \circ \rho_{T-1}\big(F_t(x_t) + \dots + F_T(x_T)\big), \tag{4.96}$$

which depends on the decision history $x_{1:t-1}$ and the node $\nu \in \mathcal{N}_t$. Problem (4.95) takes on the form of a $(t-1)$-stage problem:

$$\min_{x_{1:t-1} \in \mathcal{D}_{1:t-1}} \rho_1 \circ \dots \circ \rho_{t-1}\Big(F_1(x_1) + \dots + F_{t-1}(x_{t-1}) + Q_t(x_{1:t-1})\Big). \tag{4.97}$$

We can also represent the infimum in (4.96) in a two-stage format, with stage 1 corresponding to period t, and stage 2 corresponding to periods $t + 1, \dots, T$. We obtain

$$\begin{aligned} &Q_t(x_{1:t-1}) \\ &= \inf_{x_t \in \mathcal{D}_t(x_{1:t-1})} \inf_{x_{t+1:T} \in \mathcal{D}_{t+1:T}(x_{1:t})} \rho_t \circ \cdots \circ \rho_{T-1}\big(F_t(x_t) + \cdots + F_T(x_T)\big) \\ &= \inf_{x_t \in \mathcal{D}_t(x_{1:t-1})} \Big\{ F_t(x_t) + \\ &\quad \rho_t\Big(\inf_{x_{t+1:T} \in \mathcal{D}_{t+1:T}(x_{1:t})} \rho_{t+1} \circ \cdots \circ \rho_{T-1}\big(F_{t+1}(x_{t+1}) + \cdots + F_T(x_T)\big)\Big)\Big\}. \end{aligned}$$

We observe that the inner infimum is (4.97) for $t + 1$. Therefore, the following *dynamic programming equation* is true:

$$Q_t(x_{1:t-1}) = \inf_{x_t \in \mathcal{D}_t(x_{1:t-1})} \Big\{ F_t(x_t) + \rho_t\big(Q_{t+1}(x_{1:t})\big)\Big\}, \quad t = 1, \dots, T. \tag{4.98}$$

For $t = T$ the second term (with ρ_T) is void, and for $t = 1$ the history $x_{1:t-1}$ is void and $\mathcal{N}_1$ is just the root node, and thus Q_1 is a number: the optimal value of the problem.

The dynamic programming equation (4.98) should be understood node-wise. For every $\nu \in \mathcal{N}_t$,

$$Q_t^\nu(x_{1:t-1}) = \inf_{x_t(\nu) \in \mathcal{D}_t(x_{1:t-1};\nu)} \Big\{ f_t(x_t(\nu), \nu) + \rho_t^\nu\Big(Q_{t+1}^{C(\nu)}\big(x_{1:t}(\nu)\big)\Big)\Big\}, \quad t = 1, \dots, T. \tag{4.99}$$

Here, according to the notation of § 4.2.4, $Q_{t+1}^{C(\nu)}(x_{1:t}(\nu))$ represents a random variable with realizations $Q_{t+1}^{\nu'}(x_{1:t}(\nu)))$, $\nu' \in C(\nu)$. Observe that Assumption 4.45 and the existence of an optimal solution guarantee that these random variables are always well-defined.

4.5 Primal Decomposition Methods

Two- and multistage stochastic optimization problems have very specific structures which can be exploited by decomposition.

To illustrate the main concepts, let us consider the simplified two-stage version of problem (4.46)–(4.48):

$$\min \varrho_{1,2}\big(F_1(x_1), F_2(x_2)\big) \tag{4.100}$$

$$\text{s.t. } G(x_1, x_2) = 0 \quad \text{a.s.,} \tag{4.101}$$

$$x_1 \in \mathcal{X}_1, \quad x_2 \in \mathcal{X}_2 \quad \text{a.s.,} \tag{4.102}$$

in which we assume that $\mathcal{Y}_2 \equiv \{0\}$ (this can always be achieved by introducing "slack variables" to the second stage decisions). Suppose the probability space is finite: $\Omega = \{\omega_1, \ldots, \omega_N\}$, with the elementary events occurring with probabilities $p_\nu = P[\omega_\nu]$, $\nu = 1, \ldots, N$. To simplify notation, we write $x_2(\nu)$ for $x_2(\omega_\nu)$, $\mathcal{X}_2(\nu)$ for $\mathcal{X}_2(\omega_\nu)$, $f_2(x_2(\nu), \nu)$ for $f_2(x_2(\omega_\nu), \omega_\nu)$, and $g(x_1, x_2(\nu); \nu)$ for $\big[G(x_1, x_2)\big](\omega_\nu)$. As in the general case, $F_1 : \mathbb{R}^n \to \mathbb{R}$, $f_2 : \mathbb{R}^n \times \Omega \to \mathbb{R}$, and $g : \mathbb{R}^n \times \mathbb{R}^n \times \Omega \to \mathbb{R}$. We may have the dimensions of the first and the second stage vectors x_1 and x_2 different, say n_1 and n_2. To simplify notation, we keep them the same, but we never use this fact in our derivations.

We focus on the case where the risk measure $\varrho_{1,2}(\cdot, \cdot)$ has the form (4.6) for $T = 2$, with a coherent one-step conditional risk mapping $\rho : \mathcal{Z}_2 \to \mathbb{R}$.

Then, problem (4.100)–(4.102) can be written in a more explicit form as follows

$$\min F_1(x_1) + \rho\Big[\big\{f_2(x_2(\nu), \nu)\big\}_{\nu=1,\ldots,N}\Big] \tag{4.103}$$

$$\text{s.t. } g(x_1, x_2(\nu); \nu) = 0, \quad \nu = 1, \ldots, N, \tag{4.104}$$

$$x_1 \in \mathcal{X}_1, \quad x_2(\nu) \in \mathcal{X}_2(\nu), \quad \nu = 1, \ldots, N. \tag{4.105}$$

In (4.103), $f_2(x_2(\nu), \nu)$, $\nu = 1, \ldots, N$, are understood as the realizations of a discrete random variable which are attained with probabilities $\big\{p_\nu\big\}_{\nu=1,\ldots,N}$.

Problem (4.103)–(4.105) is a very large-scale nonsmooth optimization problem, which can hardly be solved by the standard methods. We need to exploit its special structure to develop efficient solution approaches.

Throughout this section, we make the following general assumptions.

Assumption 4.46. *The functions $F_1(\cdot)$ and $f_2(\cdot, \nu)$, $\nu = 1, \ldots, N$, are convex, the functions $g(\cdot, \cdot; \nu)$, $\nu = 1, \ldots, N$, are affine, and the sets $\mathcal{X}_1$ and $\mathcal{X}_2(\nu)$, $\nu = 1, \ldots, N$, and convex and compact.*

From now on, we shall use an explicit form of an affine function $g(\cdot, \cdot; \nu)$:

$$g(x_1, x_2(\nu); \nu) = A(\nu)x_1 + B(\nu)x_2(\nu) - h(\nu),$$

where $A(\nu)$ and $B(\nu)$ are $m \times n$ matrices, and $h(\nu) \in \mathbb{R}^m$.

We also assume a simplified version of the complete subregular recourse property of Definition 4.38, which we specialize to the case under consideration. We consider deterministic set-constrained systems, parametrized by the first-stage decision $\zeta \in \mathcal{X}_1$ and the elementary event $\nu = 1, \ldots, N$:

$$A(\nu)\zeta + B(\nu)\xi = h(\nu), \tag{4.106}$$

$$\xi \in \mathcal{X}_2(\nu). \tag{4.107}$$

Assumption 4.47. *The system* (4.106)–(4.107) *admits* complete subregular recourse*: a constant C exist, such that for all $\nu = 1, \dots, N$, every $\zeta \in \mathcal{X}_1$, and every $\eta \in \mathbb{R}^n$, a solution ξ of* (4.106)–(4.107) *exists, satisfying the inequality*

$$\|\xi - \eta\| \le C\big(\|A(\nu)\zeta + B(\nu)\eta - h(\nu)\| + \mathrm{d}(\eta, \mathcal{X}_2(\nu))\big). \tag{4.108}$$

4.5.1 The Value Function

The primal decomposition methods are based on the dynamic programming equation (4.99), which can be now simplified as follows. We consider the second stage problems for scenarios $\nu = 1, \dots, N$:

$$\min_{x_2(\nu)} f_2(x_2(\nu), \nu) \tag{4.109}$$

$$\text{s.t. } A(\nu)x_1 + B(\nu)x_2(\nu) = h(\nu), \tag{4.110}$$

$$x_2(\nu) \in \mathcal{X}_2(\nu). \tag{4.111}$$

The optimal value of problem (4.109)–(4.111) will be denoted by $Q(x_1, \nu)$ (it corresponds to the value function $Q_2(x_1, \nu)$ from the previous section). Owing to the Assumptions 4.46 and 4.47, the feasible set $S(x_1)$ defined by (4.110)–(4.111) is nonempty and compact, and thus $Q(x_1, \nu)$ is well-defined for all $x_1 \in \mathcal{X}_1$ and all $\nu = 1, \dots, N$.

It is easy to prove that each function $Q(\cdot, \nu)$ is convex on $\mathcal{X}_1$. Suppose x_1 and y_1 are in $\mathcal{X}_1$ and consider the point $z_1 = \alpha x_1 + (1-\alpha)y_1$, where $\alpha \in [0, 1]$. If $x_2(\nu)$ and $y_2(\nu)$ are the optimal solutions of problem (4.109)–(4.111) for x_1 and y_1, respectively, then $z_2 = \alpha x_2 + (1-\alpha)y_2$ is feasible for the second stage problem at z_1, because (4.110) is affine and $\mathcal{X}_2(\nu)$ convex. By the convexity of $f_2(\cdot, \nu)$,

$$\begin{aligned} Q(z_1, \nu) \le f_2(z_2, \nu) \le \alpha f_2(x_2, \nu) + (1-\alpha) f_2(y_2, \nu) \\ = \alpha Q(x_1, \nu) + (1-\alpha) Q(y_1, \nu). \end{aligned}$$

We now study the subdifferentiability of $Q(\cdot, \nu)$ and calculate its subgradients. For this purpose, we consider the Lagrangian of problem (4.109)–(4.111):

$$L(x_2(\nu), \psi; x_1, \nu) \triangleq f_2(x_2(\nu), \nu) + \big\langle \psi, A(\nu)x_1 + B(\nu)x_2(\nu) - h(\nu)\big\rangle, \tag{4.112}$$

with multipliers $\psi \in \mathbb{R}^m$. The corresponding dual function is

$$L_D(\psi; x_1, \nu) = \min_{x_2(\nu) \in \mathcal{X}_2(\nu)} L(x_2(\nu), \psi; x_1, \nu), \tag{4.113}$$

and the dual problem has the form:

$$\max_{\psi \in \mathbb{R}^m} L_D(\psi; x_1, \nu). \tag{4.114}$$

Lemma 4.48. *Suppose Assumptions* 4.46 *and* 4.47 *are satisfied. Then for every* $x_1 \in \mathcal{X}_1$ *the dual problem* (4.114) *has an optimal solution and the optimal values of the primal problem* (4.109)–(4.111) *and the dual problem* (4.114) *coincide.*

Proof. Let us fix $\bar{x}_1 \in \mathcal{X}_1$. Suppose $\bar{x}_2(\nu)$ is an optimal solution of problem (4.109)–(4.111). Let K be the Lipschitz constant of $f_2(\cdot, \nu)$. Owing to Assumption 4.47, the point $\bar{x}_2(\nu)$ is also the unconstrained minimum of the function

$$\begin{aligned}\Psi(x_2(\nu)) = f_2(x_2(\nu), \nu) + KC\big(&\|A(\nu)x_1 + B(\nu)x_2(\nu) - h(\nu)\| \\ &+ \operatorname{dist}(x_2(\nu), \mathcal{X}_2(\nu))\big), \quad x_2(\nu) \in \mathbb{R}^n.\end{aligned}$$

Indeed, if $\Psi(\eta) < \Psi(x_2(\nu))$ for some η, then we can find ξ feasible for (4.110)–(4.111) and such that the bound (4.108) is satisfied. We obtain the chain of relations

$$\begin{aligned}f_2(\xi, \nu) &\le f_2(\eta, \nu) + K\|\xi - \eta\| \\ &\le f_2(\eta, \nu) + KC\big(\|A(\nu)x_1 + B(\nu)\eta - h(\nu)\| + \mathrm{d}(\eta, \mathcal{X}_2(\nu))\big) \\ &= \Psi(\eta) < \Psi(\bar{x}_2(\nu)) = f_2(\bar{x}_2(\nu), \nu),\end{aligned}$$

which contradicts the optimality of $\bar{x}_2(\nu)$ in (4.109)–(4.111).

The function $\Psi(\cdot)$ is convex and its subdifferential can be calculated by applying the Rockafellar–Moreau theorem:

$$\partial\Psi(\bar{x}_2(\nu)) = \partial f_2(\bar{x}_2(\nu), \nu) + KCB(\nu)^\top \mathbb{B}_n + \mathcal{N}_{\mathcal{X}_2(\nu)}(\bar{x}_2(\nu)),$$

where $\mathbb{B}_n$ is the unit ball in $\mathbb{R}^n$ and $\mathcal{N}_{\mathcal{X}_2(\nu)}(\bar{x}_2(\nu))$ is the normal cone to $\mathcal{X}_2(\nu)$ at $\bar{x}_2(\nu)$. We infer that a subgradient $\hat{s}(\nu) \in \partial f_2(\bar{x}_2(\nu), \nu)$, an element $\hat{\psi}(\nu) \in KC\,\mathbb{B}_n$, and a normal vector $\hat{n}(\nu) \in \mathcal{N}_{\mathcal{X}_2(\nu)}(\bar{x}_2(\nu))$ exist, such that

$$\hat{s}(\nu) + B(\nu)^\top \hat{\psi} + \hat{n}(\nu) = 0. \tag{4.115}$$

This is the necessary and sufficient condition for the optimality of the point $\bar{x}_2(\nu)$ in the problem on the right-hand side of (4.113) at $\psi = \hat{\psi}$. Hence, $Q(\bar{x}_1, \nu) = L_D(\hat{\psi}; \bar{x}_1, \nu)$. As $L_D(\psi; \bar{x}_1, \nu) \leq Q(\bar{x}_1, \nu)$ for all $\psi \in \mathbb{R}^m$, the point $\hat{\psi}$ is indeed an optimal solution of (4.114). □

The duality allows us to completely describe the subdifferential of the optimal value function $Q(\cdot, \nu)$.

Theorem 4.49. *Suppose Assumptions* 4.46 *and* 4.47 *are satisfied. Then for every* $\bar{x}_1 \in \mathcal{X}_1$

$$\partial Q(\bar{x}_1, \nu) = \Big\{ A(\nu)^\top \bar{\psi} : L_D(\bar{\psi}, \nu) = \max_{\psi \in \mathbb{R}^m} L_D(\psi; \bar{x}_1, \nu) \Big\}. \tag{4.116}$$

Proof. Suppose $x_1 = \bar{x}_1 \in \mathcal{X}_1$ is fixed. Consider the function

$$v(b) \triangleq \inf_{\substack{x_2(\nu) \in \mathcal{X}_2(\nu) \\ B(\nu) x_2(\nu) = b}} f_2(x_2(\nu), \nu)$$

and let $\bar{b} = h(\nu) - A(\nu)\bar{x}_1$. Evidently, $Q(x_1, \nu) = v(h(\nu) - A(\nu)x_1)$. We shall describe the subdifferential of $v(\cdot)$ at $\bar{b}$.

On the one hand, we can rewrite the dual function (4.113) as follows:

$$\begin{aligned}
&L_D(\psi; \bar{x}_1, \nu) \\
&= \min_{x_2(\nu) \in \mathcal{X}_2(\nu)} \Big\{ f_2(x_2(\nu), \nu) + \big\langle \psi, A(\nu)\bar{x}_1 + B(\nu)x_2(\nu) - h(\nu) \big\rangle \Big\} \\
&= \min_{x_2(\nu) \in \mathcal{X}_2(\nu)} \Big\{ f_2(x_2(\nu), \nu) + \big\langle \psi, B(\nu)x_2(\nu) - \bar{b} \big\rangle \Big\} \\
&= \inf_{b \in \mathbb{R}^m} \inf_{\substack{x_2(\nu) \in \mathcal{X}_2(\nu) \\ B(\nu) x_2(\nu) = b}} \Big\{ f_2(x_2(\nu), \nu) + \big\langle \psi, b - \bar{b} \big\rangle \Big\} \\
&= \inf_{b \in \mathbb{R}^m} \Big\{ v(b) + \big\langle \psi, b - \bar{b} \big\rangle \Big\} = -v^*(-\psi) - \langle \psi, \bar{b} \rangle.
\end{aligned}$$

On the other hand, a vector $\bar{\psi}$ is a solution of the dual problem (4.114) at $x_1 = \bar{x}_1$ if and only if

$$L_D(\bar{\psi}; \bar{x}_1, \nu) = Q(\bar{x}_1, \nu) = v(\bar{b}).$$

Combining the last two displayed equations we infer for all dual solutions $\bar{\psi}$ the equation:

$$v(\bar{b}) + v^*(-\bar{\psi}) + \langle \bar{\psi}, \bar{b} \rangle = 0.$$

By conjugate duality, a vector $s \in \partial v(\bar{b})$ if and only if

$$v(\bar{b}) + v^*(s) = \langle s, \bar{b} \rangle.$$

Consequently,

$$\partial v(\bar{b}) = \Big\{ -\bar{\psi} : L_D(\bar{\psi}; \bar{x}_1, \nu) = \max_{\psi \in \mathbb{R}^m} L_D(\psi; \bar{x}_1, \nu) \Big\}.$$

The substitution $b = h(\nu) - A(\nu)x_1$, together with the formula for the subdifferential of a pre-composition of a convex function with an affine operator, yield the assertion of the theorem.† □

While Theorem 4.49 provides a complete characterization of the subdifferential of the optimal value function, we shall use only specific subgradients in our methods. In this context, essential is the identification of an optimal solution $\hat{\psi}$ of the dual problem (4.114) in the proof of Lemma 4.48.

It follows from Theorem 4.49 that for every $x_1 \in \mathcal{X}_1$ we have

$$Q(x_1, \nu) \geq Q(\bar{x}_1, \nu) + \big\langle A(\nu)^\top \hat{\psi}, x_1 - \bar{x}_1 \big\rangle.$$

Observe that $\|\hat{\psi}\| \leq KC$. Therefore

$$Q(x_1, \nu) \geq Q(\bar{x}_1, \nu) - KC \|A(\nu)\| \, \|x_1 - \bar{x}_1\|.$$

Reversing the roles of x_1 and $\bar{x}_1$ and using a new $\hat{\psi}$ obtained at x_1, we conclude that the function $Q(\cdot, \nu)$ is Lipschitz continuous on $\mathcal{X}_1$ with the Lipschitz constant $KC\|A(\nu)\|$. In our methods, we always assume that we generate subgradients that are uniformly bounded.

4.5.2 The Risk-Averse Multi-Cut Method

Due to the convexity and subdifferentiability of the optimal value functions $Q(\cdot, \nu)$ we can develop a specialized cutting plane method for solving problem (4.103)–(4.105). First, we convert the problem to a compact dynamic programming formulation, which is a special case of (4.99):

$$\min_{x_1 \in \mathcal{X}_1} \Big\{ \varphi(x_1) \triangleq F_1(x_1) + \rho\Big[\big\{Q(x_1, \nu)\big\}_{\nu=1,\ldots,N}\Big]\Big\}. \tag{4.117}$$

Due to Assumption 4.46, it is a convex optimization problem.

The method operates iteratively and uses at each iteration $k = 1, 2, \ldots$ lower polyhedral approximations $F_1^k(\cdot)$ of the first stage cost $F_1(\cdot)$, lower polyhedral approximations $Q^k(\cdot, \nu)$ of the value functions $Q(\cdot, \nu)$, $\nu = 1, \ldots, N$, and lower polyhedral approximations $\rho^k[\cdot]$ of the risk measure $\rho[\cdot]$.

† A minor technical issue is the qualification condition $\operatorname{ri} \operatorname{dom}(v) \cap \{h(\nu) - A(\nu)x_1 : x_1 \in \mathbb{R}^n\} \neq \emptyset$, which is implied by Assumption 4.46.

We assume that at every point x_1^k generated by the method we can calculate a subgradient $s^k \in \partial F_1(x^k)$. Then the functions

$$F_1^k(x_1) = \max_{1\le j\le k} \{F_1(x_1^j) + \langle s^j, x_1 - x_1^j\rangle\}, \quad x_1 \in \mathcal{X}_1, \quad k = 1, 2, \ldots,$$

are indeed lower polyhedral approximations of $F_1(\cdot)$.

Furthermore, we assume that at each point x_1^k we can solve the second stage problems (4.109)–(4.111) and obtain, for all $\nu = 1, \ldots, N$, their solutions $x_2^k(\nu)$ and Lagrange multiplier vectors $\psi^k(\nu)$ which are the solutions of the dual problems (4.114). We assume that the norms of these multiplier vectors are bounded over $k = 1, 2, \ldots$. This is always possible, as discussed in the proof of Lemma 4.48. By virtue of Theorem 4.49, the vectors $A(\nu)^\top \psi^k(\nu) \in \partial Q(x_1^k, \nu)$. Therefore, the polyhedral functions

$$Q^k(x_1, \nu) = \max_{1\le j\le k} \Big\{ Q(x_1^j, \nu) + \big\langle A(\nu)^\top \psi^j(\nu), x_1 - x_1^j \big\rangle \Big\}, \quad x \in \mathcal{X}_1, \tag{4.118}$$

are lower bounds for $Q(\cdot, \nu)$ in $\mathcal{X}_1$, for $\nu = 1, \ldots, N$ and $k = 1, 2, \ldots$.

Finally, at each point x_1^k, we evaluate the risk measure $\rho[\cdot]$ at the discrete random variable $Q(x_1^k)$ attaining the realizations $Q(x_1^k, \nu)$ with probabilities p_ν, $\nu = 1, \ldots, N$. We also calculate $\zeta^k \in \partial\rho[Q(x_1^k)]$. This allows us to construct a lower-bounding risk measure

$$\rho^k[Z] = \max_{1\le j\le k} \langle \zeta^j, Z\rangle.$$

As we are dealing with discrete random variables Z having N realizations, we assume that all ζ^j are probability vectors, rather than the densities with respect to p, and thus the scalar products above are equal to $\sum_{\nu=1}^{n} \zeta^j(\nu) Z(\nu)$.

These approximations allow us to construct a lower bound of the objective function in (4.117):

$$\varphi^k(x_1) = F_1^k(x_1) + \rho^k\Big[\big\{Q^k(x_1, \nu)\big\}_{\nu=1,\ldots,N}\Big], \tag{4.119}$$

and to construct the *master problem* at iteration k:

$$\min_{x_1 \in \mathcal{X}_1} \varphi^k(x_1). \tag{4.120}$$

The solution of this problem is our next iterate x^{k+1}, and the iteration continues. The fact that $\varphi^k(x_1)$ is indeed a lower bound on $\varphi(\cdot)$ follows from the fact that $\rho[\cdot]$ and each $\rho^k[\cdot]$ are nondecreasing.

Remark 4.50. The master problem (4.120) can be equivalently modeled as a constrained optimization problem as follows:

$$\begin{aligned}
\min_{x_1,Z,w,r} \quad & w + r \\
& w \geq F_1(x_1^j) + \langle s^j, x_1 - x_1^j \rangle, \quad j = 1,\dots,k, \\
& r \geq \langle \zeta^j, Z \rangle, \quad j = 1,\dots,k, \\
& Z(\nu) \geq Q(x_1^j, \nu) + \big\langle A(\nu)^\top \psi^j(\nu), x_1 - x_1^j \big\rangle, \\
& \qquad\qquad\qquad j = 1,\dots,k, \quad \nu = 1,\dots,N, \\
& x_1 \in \mathcal{X}_1.
\end{aligned}$$

If $\mathcal{X}_1$ is a convex polyhedral set, it is a linear programming problem. This is the only reason for replacing $\rho[\cdot]$ by $\rho^k[\cdot]$ in (4.119); our considerations remain valid in both versions. □

The method stops when $\varphi(x_1^{k+1}) = \varphi^k(x_1^{k+1})$.

Our method is a special version of the classical cutting plane method of convex optimization and its convergence can be proved in the standard way.

Theorem 4.51. *Suppose Assumptions* 4.46 *and* 4.47 *are satisfied. Then every accumulation point* x_1^* *of the sequence* $\{x_1^k\}$ *generated by the risk-averse multicut method is an optimal solution of problem* (4.117).

Proof. By construction, the optimal value $\varphi^k(x_1^{k+1})$ of the master problem (4.120) is a lower bound for the optimal value φ^* of problem (4.117). Therefore the stopping test is correct; if $\varphi(x_1^{k+1}) = \varphi^k(x_1^{k+1})$, then $\varphi(x_1^{k+1}) = \varphi^*$.

It remains to analyze the case of infinitely many steps. As the set of cuts in the master problem grows, $\theta^{k+1} \geq \theta^k$ for all k. For $\varepsilon > 0$ we define the set

$$\mathcal{K}_\varepsilon = \{k : \varphi^* + \varepsilon < \varphi(x_1^k) < +\infty\}.$$

Let $k_1, k_2 \in \mathcal{K}_\varepsilon$ with $k_1 < k_2$. Since $\varphi(x^{k_1}) > \varphi^* + \varepsilon$ and $\varphi^* \geq \varphi^{k_1-1}(x^{k_1})$, there will be new cuts generated at x^{k_1}. Define

$$g^{k_1} = s^{k_1} + \sum_{\nu=1}^{N} \zeta^{k_1}(\nu) A(\nu)^\top \psi^{k_1}(\nu)$$

the resulting subgradient of $\varphi(\cdot)$ at x^{k_1}. The cuts will stay in the master problem from k_1 on, so they have to be satisfied at x^{k_2}:

$$\varphi(x^{k_1}) + \langle g^{k_1}, x^{k_2} - x^{k_1} \rangle \leq \varphi^{k_2-1}(x^{k_2}) \leq \varphi^*.$$

On the other hand, $\varepsilon < \varphi(x^{k_2}) - \varphi^*$, which combined with the last inequality yields

$$\varepsilon < \varphi(x^{k_2}) - \varphi(x^{k_1}) - \langle g^{k_1}, x^{k_2} - x^{k_1} \rangle.$$

The function $\varphi(\cdot)$ is Lipschitz continuous on $\mathcal{X}_1$: there exists a constant C such that $\varphi(x_1) - \varphi(x_2) \le C\|x_1 - x_2\|$, for all $x_1, x_2 \in \mathcal{X}_1$. The subgradients used by the method are bounded, and thus we can choose C large enough so that $\|g^{k_1}\| \le C$ for all $k_1 \in \mathcal{K}_\varepsilon$. It follows from the last displayed inequality that

$$\varepsilon < 2C\|x^{k_1} - x^{k_2}\| \quad \text{for all} \quad k_1, k_2 \in \mathcal{K}_\varepsilon,\ k_1 \ne k_2.$$

As the set $\mathcal{X}$ is compact, there can exist only finitely many points in $\mathcal{X}$ having the distance of at least $\varepsilon/(2C)$ from each other. Thus the last inequality implies that the set $\mathcal{K}_\varepsilon$ is finite for each $\varepsilon > 0$. This means that for every $\varepsilon > 0$ all but finitely many points x_1^k have the objective function values in the interval $[\varphi^*, \varphi^* + \varepsilon]$. Therefore, the sequence $\{\varphi(x_1^k)\}$ is convergent to φ^*. □

Similarly, we can construct a risk-averse version of the regularized decomposition method, which is in itself a special form of the bundle method of convex optimization.

4.5.3 Primal Multistage Decomposition

The ideas discussed in the preceding sections can be extended to the scenario tree model of the multistage problem. According to (4.99), with each node ν of the scenario tree we can associate the value function

$$Q_t^\nu(x_{1:t-1}) = \inf_{x_t(\nu)\in\mathcal{D}_t(x_{1:t-1};\nu)} \left\{ f_t(x_t(\nu),\nu) + \rho_t^\nu\left[Q_{t+1}^{C(\nu)}\big(x_{1:t}(\nu)\big)\right]\right\},$$
$$\nu \in \mathcal{N}_t, \quad t = 1,\dots,T. \tag{4.121}$$

To develop the method, we make analogous assumptions to those for the two-stage case.

Assumption 4.52. *For every $t = 1,\dots,N$ and for every node $\nu \in \mathcal{N}_t$, the functions $f_t(\cdot;\nu)$ are convex, the functions $g_t(\cdot;\nu)$ are affine, the sets $\mathcal{X}_t(\nu)$ are convex and compact, and the sets $\mathcal{Y}_t(\nu) = \{0\}$.*

From now on, we shall use the following explicit form of the affine functions $g_t(\cdot;\nu)$:

$$g_t(x_{1:t}(\nu);\nu) = A(\nu)x(a(\nu)) + B(\nu)x(\nu) - h(\nu), \tag{4.122}$$

where $A(\nu)$ and $B(\nu)$ are $m \times n$ matrices, and $h(\nu) \in \mathbb{R}^m$. We suppress the stage index t, because its value is implied by the node ν.

We also assume a simplified version of the complete subregular recourse property of Definition 4.38, which we specialize to the case under consideration. We consider deterministic set-constrained systems, parametrized by the ancestor node decision ζ and the node ν:

$$A(\nu)\zeta + B(\nu)\xi = h(\nu), \tag{4.123}$$
$$\xi \in \mathcal{X}(\nu). \tag{4.124}$$

Assumption 4.53. *The system* (4.123)–(4.124) *admits* complete subregular recourse*: a constant C exist, such that for all $\nu \in \mathcal{N}$, every $\zeta \in \mathcal{X}(a(\nu))$, and every $\eta \in \mathbb{R}^n$, a solution ξ of* (4.123)–(4.124) *exists, satisfying the inequality*

$$\|\xi - \eta\| \le C\big(\|A(\nu)\zeta + B(\nu)\eta - h(\nu)\| + \mathrm{d}(\eta, \mathcal{X}(\nu))\big). \tag{4.125}$$

Due to (4.122), the value function (4.121) simplifies to a function of the decision vector at stage t. For every node $\nu \in \mathcal{N}_t$, the optimal value $Q_t^\nu(x_{1:t-1})$ depends on $x(a(\nu))$ only and is the optimal value of the following problem:

$$\min_{x(\nu)} f(x(\nu), \nu) + \rho^\nu\Big[\big\{Q\big(x(\nu), \nu'\big)\big\}_{\nu' \in C(\nu)}\Big] \tag{4.126}$$
$$\text{subject to } A(\nu)x(a(\nu)) + B(\nu)x(\nu) = h(\nu), \tag{4.127}$$
$$x(\nu) \in \mathcal{X}(\nu). \tag{4.128}$$

With a slight abuse of notation, we denote this optimal value by $Q(x(a(\nu)), \nu)$. For $\nu = 1$ (the root node), the constraint (4.127) is not present, and for $\nu \in \mathcal{N}_T$ (the leaves), the risk part in the objective function is void.

Solving the multistage risk-averse problem is equivalent to solving all problems (4.126)–(4.128) arranged in the nodes of the scenario tree.

The difficulty is that the optimal value functions $Q(\cdot, \nu')$ are not known. To overcome it, we replace them with lower polyhedral approximations, in a way similar to (4.118). For each node ν, we construct an approximate problem which passes its solution to the children nodes $\nu' \in C(\nu)$, constructs a cut for its optimal value function $Q(\cdot, \nu)$, and passes it to the ancestor node $a(\nu)$.

Let us focus on a generic node ν of the scenario tree, not a leaf node, and let us use $k_\nu = 0, 1, 2, \ldots$ to denote the number the iterations of the method when the subproblem associated with this node has been solved (k_ν is a "local clock" at ν). The subproblem has the following form:

$$\min_{x(\nu)} f(x(\nu), \nu) + \rho^\nu\Big[\big\{Q^{k_\nu}\big(x(\nu), \nu'\big)\big\}_{\nu' \in C(\nu)}\Big] \tag{4.129}$$
$$\text{subject to } A(\nu)x^{k_\nu}(a(\nu)) + B(\nu)x(\nu) = h(\nu), \tag{4.130}$$
$$x(\nu) \in \mathcal{X}(\nu). \tag{4.131}$$

The solution of this problem is denoted by $x^{k_\nu+1}(\nu)$.

In (4.129), the functions $Q^{k_\nu}\big(\cdot, \nu'\big)$, $\nu' \in C(\nu)$, are lower polyhedral approximations of $Q\big(\cdot, \nu'\big)$. They are formed as follows:

$$Q^{k_\nu}(x(\nu), \nu') = \max_{j \in J(k_\nu, \nu')} \Big\{ Q^j(x^j(\nu), \nu') + \big\langle A(\nu')^\top \psi^j(\nu'), x(\nu) - x^j(\nu) \big\rangle \Big\}. \tag{4.132}$$

In the last formula, $J(k_\nu, \nu')$ is the subset of iterations $\{1, \dots, k_\nu)$ when new cuts were obtained from a node $\nu' \in C(\nu)$, $Q^j(\cdot, \nu')$ is a lower approximation of the value function $Q(\cdot, \nu')$ at that node, and $A(\nu')^\top \psi^j(\nu')$ is a subgradient of that approximation.

Theorem 4.54. *Suppose Assumptions* 4.52 *and* 4.53 *are satisfied. Then every accumulation point* $\big(x^*(a(\nu)), x^*(\nu)\big)$ *of the sequence* $\big(x^{k_\nu}(a(\nu)), x^{k_\nu}(\nu)\big)$ *is such that* $x^*(\nu)$ *is an optimal solution of the problem* (4.126)–(4.128) *with* $x(a(\nu)) = x^*(a(\nu))$.

Proof. By construction, the optimal value $\varphi^k(x_1^{k+1})$ of the master problem (4.120) is a lower bound for the optimal value φ^* of problem (4.117). Therefore the stopping test is correct; if $\varphi(x_1^{k+1}) = \varphi^k(x_1^{k+1})$, then $\varphi(x_1^{k+1}) = \varphi^*$.

It remains to analyze the case of infinitely many steps. As the set of cuts in the master problem grows, $\theta^{k+1} \ge \theta^k$ for all k. For $\varepsilon > 0$ we define the set

$$\mathcal{K}_\varepsilon = \{k : \varphi^* + \varepsilon < \varphi(x_1^k) < +\infty\}.$$

Let $k_1, k_2 \in \mathcal{K}_\varepsilon$ with $k_1 < k_2$. Since $\varphi(x^{k_1}) > \varphi^* + \varepsilon$ and $\varphi^* \ge \varphi^{k_1-1}(x^{k_1})$, there will be new cuts generated at x^{k_1}. Define

$$g^{k_1} = s^{k_1} + \sum_{\nu=1}^{N} \zeta^{k_1}(\nu) A(\nu)^\top \psi^{k_1}(\nu)$$

the resulting subgradient of $\varphi(\cdot)$ at x^{k_1}. The cuts will stay in the master problem from k_1 on, so they have to be satisfied at x^{k_2}:

$$\varphi(x^{k_1}) + \langle g^{k_1}, x^{k_2} - x^{k_1} \rangle \le \varphi^{k_2-1}(x^{k_2}) \le \varphi^*.$$

On the other hand, $\varepsilon < \varphi(x^{k_2}) - \varphi^*$, which combined with the last inequality yields

$$\varepsilon < \varphi(x^{k_2}) - \varphi(x^{k_1}) - \langle g^{k_1}, x^{k_2} - x^{k_1} \rangle.$$

The function $\varphi(\cdot)$ is Lipschitz continuous on $\mathcal{X}_1$: there exists a constant C such that $\varphi(x_1) - \varphi(x_2) \le C\|x_1 - x_2\|$, for all $x_1, x_2 \in \mathcal{X}_1$. The subgradients used by the method are bounded, and thus we can choose C large enough so that $\|g^{k_1}\| \le C$ for all $k_1 \in \mathcal{K}_\varepsilon$. It follows from the last displayed inequality that

$$\varepsilon < 2C\|x^{k_1} - x^{k_2}\| \quad \text{for all} \quad k_1, k_2 \in \mathcal{K}_\varepsilon,\ k_1 \ne k_2.$$

As the set $\mathcal{X}$ is compact, there can exist only finitely many points in $\mathcal{X}$ having the distance of at least $\varepsilon/(2C)$ from each other. Thus the last inequality implies that the set $\mathcal{K}_\varepsilon$ is finite for each $\varepsilon > 0$. This means that for every $\varepsilon > 0$ all but finitely many points x_1^k have the objective function values in the interval $[\varphi^*, \varphi^* + \varepsilon]$. Therefore, the sequence $\{\varphi(x_1^k)\}$ is convergent to φ^*. □

4.6 Scenario Decomposition

Scenario decomposition methods solve the dual problem (4.90). For the scenario tree model introduced in § 4.2.4, we can rewrite this problem in a more specific form. Recall that we denoted by $\mathcal{N}_T$ the set of nodes of the tree at the level T. This set can be identified with the space Ω. For every $\omega \in \mathcal{N}_T$ there is a unique path $\{\nu_1, \nu_2(\omega), \ldots, \nu_{T-1}(\omega), \omega\}$ from the root to ω, which we call a *scenario* (for simplicity, we frequently speak of "scenario ω"). Recall that $p(\omega)$ is the probability of the scenario ω.

Under the assumptions of Theorem 4.44, for a coherent and time-consistent dynamic risk measure $\rho_{1,T}(\cdot)$, the dual problem (4.90) can written in a more direct form:

$$\max_{\substack{\lambda \in \Lambda_{1,T} \\ \pi \in \mathcal{S}^\perp}} \min_{x \in \widetilde{\mathcal{D}}} \sum_{\omega \in \Omega} p(\omega) \sum_{t=1}^{T} \big\{ \lambda(\omega) f_t(x_t(\omega), \omega) + \langle \pi_t(\omega), x_t(\omega) \rangle \big\}. \tag{4.133}$$

The set $\widetilde{\mathcal{D}}$ is the Cartesian product:

$$\widetilde{\mathcal{D}} = \bigtimes_{\omega \in \Omega} \widetilde{\mathcal{D}}(\omega),$$

with each scenario-specific set $\widetilde{\mathcal{D}}(\omega)$ defined by the relations:

$$\begin{aligned} g_t(\tilde{x}_{1:t}(\omega), \omega) &\in \mathcal{Y}_t(\omega), \\ \tilde{x}_t(\omega) \in \mathcal{X}_t(\omega), &\quad t = 1, \ldots, T. \end{aligned} \tag{4.134}$$

Therefore, the calculation of the minimum inside (4.133) decomposes into scenario subproblems

$$\min_{x(\omega) \in \widetilde{\mathcal{D}}(\omega)} \sum_{t=1}^{T} \big\{ \lambda(\omega) f_t(x_t(\omega), \omega) + \langle \pi_t(\omega), x_t(\omega) \rangle \big\}, \quad \omega \in \Omega; \tag{4.135}$$

we denote their optimal values by $W(\lambda, \pi; \omega)$, $\omega \in \Omega$. Observe that each function $W(\cdot, \cdot, \omega)$, as a minimum of a family of linear functions, is concave. In general, it may take the value $-\infty$; to eliminate this case, we make the following assumption.

Assumption 4.55. *We know a convex compact set $\Pi \subset \mathcal{S}^{\perp}$, such that an optimal solution of problem* (4.133) *is an element of $\Lambda_{1,T} \times \Pi$, and for all $(\lambda, \pi) \in \Lambda_{1,T} \times \Pi$ the problems* (4.135) *have optimal solutions $\hat{x}(\omega)(\lambda(\omega), \pi(\omega))$, $\omega \in \Omega$.*

Under this assumption, the dual problem

$$\max_{\substack{\lambda \in \Lambda_{1,T} \\ \pi \in \mathcal{S}^{\perp}}} \left\{ W(\lambda, \pi) = \sum_{\omega \in \Omega} p(\omega)\, W(\lambda, \pi; \omega) \right\}. \tag{4.136}$$

is a concave nonsmooth optimization problem. The subgradients of the objective function are readily available:

$$\sum_{\omega \in \Omega} p(\omega) \begin{bmatrix} \sum_{t=1}^{T} F_t\big(\hat{x}_t(\lambda(\omega), \pi(\omega), \omega), \omega\big) \\ \big\{\hat{x}_t(\lambda(\omega), \pi(\omega), \omega)\big\}_{t=1,\dots,T} \end{bmatrix} \in \partial W(\lambda, \pi).$$

Furthermore, the set $\mathcal{S}^{\perp}$ is easy to represent algebraically. Let

$$\mathfrak{B}_t(\omega) = \big\{ s \in \Omega : \nu_t(s) = \nu_t(\omega) \big\},$$

be the set of scenarios sharing the path $\{\nu_1, \nu_2(s), \dots, \nu_t(s)\}$ with the scenario $\omega \in \Omega$. At each stage t, only $|\mathcal{N}_t|$ such sets exist, we denote them (with a slight abuse of notation) by $\mathfrak{B}_t^{\nu}$, $\nu \in \mathcal{N}_t$. Then

$$\mathcal{S}^{\perp} = \left\{ \pi \in \mathbb{R}^{nT} : \sum_{s \in \mathfrak{B}_t^{\nu}} \pi(s) = 0, \ \forall\, \nu \in \mathcal{N}_t, \ \forall\, t \in \{1, \dots, T\} \right\}.$$

At the last stage, every set $\mathfrak{B}_T^{\nu}$ has one element, and thus $\pi_T = 0$.

There are two difficulties associated with this approach. First, the algebraic description of the set $\Lambda_{1,T}$ is hard to develop from the description of the stage-wise sets Λ_t, $t = 1, \dots, T-1$. Secondly, the dual problem (4.136) has a very large dimension.

Chapter 5
Optimization with Stochastic Dominance Constraints

5.1 Stochastic Orders

Preference between random variables is frequently expressed by comparing their distributions via a *stochastic order*. A variety of stochastic-order relations exist (also called stochastic orderings), which express preferences among univariate random outcomes, random vectors, sequences, or processes. A large body of literature is dedicated to their properties and relations among them. We focus on the most popular orders, which are suitable to model risk-averse preferences. The presentation of their properties is limited to those that are relevant to the analysis and numerical solution of the related stochastic optimization problems.

5.1.1 Definitions Based on Distribution Functions

Consider a random variable $X \in \mathcal{L}_p(\Omega, \mathcal{F}, P)$ with $p \in [1, \infty]$ and its distribution function $F_X(\eta) = P[X \le \eta]$ for $\eta \in \mathbb{R}$. We define the higher-order integrated distribution functions as follows:

$$F_X^{(2)}(\eta) = \int_{-\infty}^{\eta} F_X(\alpha)\, \mathrm{d}\alpha \quad \text{for } \eta \in \mathbb{R}. \tag{5.1}$$

For an integer k such that $2 \le k \le p + 1$, we can define recursively the functions

$$F_X^{(k)}(\eta) = \int_{-\infty}^{\eta} F_X^{(k-1)}(\alpha)\, \mathrm{d}\alpha \quad \text{for } \eta \in \mathbb{R}, \tag{5.2}$$

with the convention that $F_X^{(1)}(\cdot) = F_X(\cdot)$.

D. Dentcheva, A. Ruszczyński, *Risk-Averse Optimization and Control*, Springer Series in Operations Research and Financial Engineering,
https://doi.org/10.1007/978-3-031-57988-2_5

The functions $F_X^{(k)}(\cdot)$ for $k \in \mathbb{N}, k \geq 2$, are convex because each of them is an integral of a nondecreasing function.

Changing the order of integration in (5.1), we obtain the following representation

$$F_X^{(2)}(\eta) = \int_{-\infty}^{\eta} (\eta - t)\; P_X(\mathrm{d}t) = \mathbb{E}\big[\max(\eta - X, 0)\big], \tag{5.3}$$

with $P_X = P \circ X^{-1}$ being the induced probability measure on $\mathbb{R}$. The representation (5.3) allows us to generalize the definition of the integrated distribution function for any order $p \geq 1$. We shall use the Gamma function, which is defined as follows:

$$\Gamma(p) = \int_0^{\infty} z^{p-1} \mathrm{e}^{-z}\, \mathrm{d}z.$$

Definition 5.1. For a random variable $X \in \mathcal{L}_{p-1}(\Omega, \mathcal{F}, P)$ with $p \in [1, \infty)$, we define the function

$$\begin{aligned} F_X^{(p)}(\eta) &= \frac{1}{\Gamma(p)} \int_{-\infty}^{\eta} (\eta - t)^{p-1}\; P_X(\mathrm{d}t) \\ &= \frac{1}{\Gamma(p)} \mathbb{E}\big[\max(0, \eta - X)^{p-1}\big] \end{aligned} \tag{5.4}$$

with the convention that $\max(0, \eta - x)^0 = \mathbb{1}_{(-\infty, \eta]}(x)$.

Definition 5.2. A random variable $X \in \mathcal{L}_{p-1}(\Omega, \mathcal{F}, P)$ with $p \in [1, \infty)$ *dominates in the p-th order* another random variable $Y \in \mathcal{L}_{p-1}(\Omega, \mathcal{F}, P)$ on the interval $I = [a, b] \subseteq \mathbb{R}$ if

$$F_X^{(p)}(\eta) \leq F_Y^{(p)}(\eta) \quad \text{for all } \eta \in I. \tag{5.5}$$

If $I = \mathbb{R}$, we denote the relation (5.5) by $X \succeq_{(p)} Y$; otherwise, we use the notation $X \succeq_{(p)}^{I} Y$. For every $p \in [1, \infty)$, the stochastic dominance relation $\succeq_{(p)}$ introduces a partial order among the random variables in $\mathcal{L}_{p-1}(\Omega, \mathcal{F}, P)$. More precisely, the order is defined on the corresponding space of distributions. We adopt the convention that $\mathcal{L}_0(\Omega, \mathcal{F}, P)$ stands for the space of all $\mathcal{F}$-measurable mappings from Ω to $\mathbb{R}$, which are indistinguishable on sets of P-measure zero. For $p \in (0, 1)$, the set $\mathcal{L}_p(\Omega, \mathcal{F}, P)$ contains those random variables Z, for which $\mathbb{E}[|Z|^p]$ is finite although $\big(\mathbb{E}[|Z|^p]\big)^{\frac{1}{p}}$ is not a norm.

The first-order stochastic dominance is also known as the usual stochastic order.

It follows from (5.3) that an equivalent representation of the second-order stochastic dominance relation $\succeq_{(2)}$ is given by the relation of the shortfall functions $S(X, \eta) = \mathbb{E}[\max(0, \eta - X)]$, *i.e.*,

$$\mathbb{E}[\max(0, \eta - X)] \leq \mathbb{E}[\max(0, \eta - Y)] \quad \text{for all } \eta \in \mathbb{R}. \tag{5.6}$$

We shall show that stochastic dominance of order p implies stochastic dominance of higher order $q > p$ if the compared random variables have finite moments of order $q - 1$.

Theorem 5.3. *If* $X \succeq_{(p)} Y$, *then* $X \succeq_{(q)} Y$ *hold for any* $q > p \geq 1$, *provided* $X, Y \in \mathcal{L}_{q-1}(\Omega, \mathcal{F}, P)$.

Proof. We show first that for $r > 0$ the following holds

$$F_X^{(p+r)}(\eta) = \frac{1}{\Gamma(r)} \int_{-\infty}^{\eta} (\eta - t)^{r-1} F_X^{(p)}(t) \, \mathrm{d}t. \tag{5.7}$$

We consider the right-hand side of (5.7), and transform it as follows:

$$\begin{aligned}
&\frac{1}{\Gamma(r)} \int_{-\infty}^{\eta} (\eta - t)^{r-1} F_X^{(p)}(t) \, \mathrm{d}t \\
&\quad = \frac{1}{\Gamma(r)} \int_{-\infty}^{\eta} (\eta - t)^{r-1} \Big(\frac{1}{\Gamma(p)} \int_{-\infty}^{t} (t - s)^{p-1} \, P_X(\mathrm{d}s) \Big) \, \mathrm{d}t \\
&\quad = \frac{1}{\Gamma(r)\Gamma(p)} \int_{-\infty}^{\eta} \int_{-\infty}^{t} (\eta - t)^{r-1} (t - s)^{p-1} \, P_X(\mathrm{d}s) \, \mathrm{d}t \\
&\quad = \frac{1}{\Gamma(r)\Gamma(p)} \int_{-\infty}^{\eta} \int_{s}^{\eta} (\eta - t)^{r-1} (t - s)^{p-1} \, \mathrm{d}t \, P_X(\mathrm{d}s).
\end{aligned} \tag{5.8}$$

The order of integration in the last equation is changed by using Fubini's theorem. We introduce the substitution $t = s + (\eta - s)z$ and rewrite the inner integral as follows:

$$\begin{aligned}
\int_{s}^{\eta} (\eta - t)^{r-1} (t - s)^{p-1} \, \mathrm{d}t &= \int_{0}^{1} (\eta - s)^{p+r-1} z^{p-1} (1 - z)^{r-1} \, \mathrm{d}z \\
&= (\eta - s)^{p+r-1} \int_{0}^{1} z^{p-1} (1 - z)^{r-1} \, \mathrm{d}z = (\eta - s)^{p+r-1} B(p, r) \\
&= (\eta - s)^{p+r-1} \frac{\Gamma(p)\Gamma(r)}{\Gamma(p + r)}.
\end{aligned} \tag{5.9}$$

Here, $B(p, r)$ is the Beta function and we have used its relation to the Gamma function. Substituting (5.9) back into (5.8), we obtain

$$\begin{aligned}
\frac{1}{\Gamma(r)} \int_{-\infty}^{\eta} (\eta - t)^{r-1} F_X^{(p)}(t) \, \mathrm{d}t &= \frac{1}{\Gamma(p + r)} \int_{-\infty}^{\eta} (\eta - s)^{p+r-1} \, P_X(\mathrm{d}s) \\
&= F_X^{(p+r)}(\eta),
\end{aligned}$$

which proves (5.7). Thus, $F_X^{(p)}(\eta) \leq F_Y^{(p)}(\eta)$ for all $\eta \in \mathbb{R}$ implies

$$F_X^{(q)}(\eta) = \frac{1}{\Gamma(q-p)} \int_{-\infty}^{\eta} (\eta - t)^{q-p-1} F_X^{(p)}(t)\, \mathrm{d}t$$
$$\leq \frac{1}{\Gamma(q-p)} \int_{-\infty}^{\eta} (\eta - t)^{q-p-1} F_Y^{(p)}(t)\, \mathrm{d}t = F_Y^{(q)}(\eta).$$

Hence, $X \succeq_{(q)} Y$ as claimed. □

Lemma 5.4. *Consider $\mathcal{L}_p(\Omega, \mathcal{F}, P)$ with $p \geq 1$.*

(i) *For all $\eta \in \mathbb{R}$, the mapping $X \mapsto F_X^{(p+1)}(\eta)$ is convex. Additionally, for all $\eta \in \mathbb{R}$, and for all $t \in [0, 1]$, the following holds*

$$F_{tX+(1-t)Y}^{(p+1)}(\eta) \leq \Big(t\big(F_X^{(p+1)}(\eta)\big)^{\frac{1}{p}} + (1-t)\big(F_Y^{(p+1)}(\eta)\big)^{\frac{1}{p}}\Big)^p; \tag{5.10}$$

(ii) *The mapping $\eta \mapsto F_X^{(p+1)}(\eta)$ is convex. Furthermore, for all $\eta, \zeta \in \mathbb{R}$ and for all $t \in [0, 1]$,*

$$F_X^{(p+1)}(t\eta + (1-t)\zeta) \leq \Big(t\big(F_X^{(p+1)}(\eta)\big)^{\frac{1}{p}} + (1-t)\big(F_X^{(p+1)}(\zeta)\big)^{\frac{1}{p}}\Big)^p. \tag{5.11}$$

Proof. To show claim (i), we define the random variables $Z = \max(0, \eta - X)$, $W = \max(0, \eta - Y)$, and $U = \max(0, \eta - tX - (1-t)Y)$. Using the convexity of the function $z \mapsto \max(0, \eta - z)$, we infer that

$$0 \leq U \leq tZ + (1-t)W \quad \text{a.s.}$$

Hence,

$$\|U\|_p \leq \|tZ + (1-t)W\|_p \leq t\|Z\|_p + (1-t)\|W\|_p.$$

Definition 5.1 implies that $\Gamma(p+1)F_X^{(p+1)}(\eta) = \|\max(0, \eta - X)\|_p^p$. Hence, inequality (5.10) follows from the last displayed inequality after substitution and cancellation of the common multiplier. On the other hand, raising both sides of the inequality $\|U\|_p \leq \|tZ + (1-t)W\|_p$ to power p and using the convexity and monotonicity of the mapping $x \to x^p$ for positive x and $p \geq 1$, we obtain

$$\|U\|_p^p \leq t\|Z\|_p^p + (1-t)\|W\|_p^p.$$

Hence, the convexity in statement (i) follows from the last displayed inequality using Definition 5.1 and cancelling the common multiplier.

Statement (ii) follows by the same arguments applied to the variables Z, $W' = \max(0, \zeta - X)$, and $U' = \max(0, t\eta + (1-t)\zeta - X)$. □

The stochastic dominance relation between two random variables implies relations between their moments. For a random variable $X \in \mathcal{L}_m(\Omega, \mathcal{F}, P)$ let us denote its integral moment of order k, $k = 1, \ldots, m$, by μ_X^k.

Theorem 5.5. *For two random variables $X, Y \in \mathcal{L}_m(\Omega, \mathcal{F}, P)$, the functions $F_X^{(m+1)}$ and $F_Y^{(m+1)}$ are finite-valued. If $X \succeq_{(p)} Y$ for some $p \in [1, m+1]$, then one of the following two moment relations holds:*

(a) *All moments $\mu_X^k = \mu_Y^k$, for $k \in \mathbb{N}$, $k \leq m$.*

(b) *An integer k with $1 \leq k \leq m$ exists such that $\mu_X^i = \mu_Y^i$, for all $i = 1, \dots, k-1$ and $(-1)^k \mu_X^k < (-1)^k \mu_Y^k$.*

Proof. Notice that if $X \succeq_{(p)} Y$ for $1 \leq p < m+1$, then also $X \succeq_{(m+1)} Y$. Using the binomial expansion, we represent

$$\begin{aligned} F_X^{(m+1)}(\eta) &= \frac{1}{m!} \int_{-\infty}^{\eta} (\eta - t)^m \, \mathrm{d}F_X(t) \\ &= \frac{1}{m!} \sum_{i=0}^{m} (-1)^i \binom{m}{i} \eta^{m-i} \int_{-\infty}^{\eta} t^i \, \mathrm{d}F_X(t) \\ &= \frac{1}{m!} \sum_{i=0}^{m} (-1)^i \binom{m}{i} \eta^{m-i} \Big(\mu_X^i - \int_{\eta}^{\infty} t^i \, \mathrm{d}F_X(t) \Big). \end{aligned}$$

Observe that for every $\eta > 0$

$$0 \leq \eta^m (1 - F_X(\eta)) \leq \int_{\eta}^{\infty} t^m \, \mathrm{d}F_X(t),$$

and the right-hand side of the last inequality converges to zero when $\eta \to \infty$. Hence, for each $i \leq m$, we have

$$\int_{\eta}^{\infty} t^i \, \mathrm{d}F_X(t) \leq \eta^{i-m} \int_{\eta}^{\infty} t^m \, \mathrm{d}F_X(t) = o(\eta^{i-m}).$$

Therefore, we obtain that

$$F_X^{(m+1)}(\eta) = \frac{1}{m!} \sum_{i=0}^{m} (-1)^i \binom{m}{i} \eta^{m-i} \mu_X^i + o(1).$$

This proves that the function $F_X^{(m+1)}(\cdot)$ is finite for all values of the argument. Expressing the corresponding function for Y in the same way, we infer

$$\begin{aligned} 0 \leq F_Y^{(m+1)}(\eta) - F_X^{(m+1)}(\eta) &= \frac{1}{m!} \sum_{i=1}^{m} (-1)^i \binom{m}{i} \eta^{m-i} (\mu_Y^i - \mu_X^i) + o_1(1) \\ &= \frac{1}{m!} \sum_{i=1}^{m} \binom{m}{i} \Delta_i \eta^{m-i} + o_1(1) \end{aligned} \tag{5.12}$$

with $\Delta_i = (-1)^i(\mu_Y^i - \mu_X^i)$. For all sufficiently large η, the polynomial on the right-hand side in (5.12) is dominated by the term with the smallest i for which $\mu_Y^i \neq \mu_X^i$. This proves the claim. □

Notice that the statement of the theorem implies that $X \succeq_{(p)} Y$ for any $p \geq 1$ entails $\mathbb{E}[X] \geq \mathbb{E}[Y]$.

Corollary 5.6. *For two random variables $X, Y \in \mathcal{L}_m(\Omega, \mathcal{F}, P)$, $m \geq 1$, the relations $X \succeq_{(m)} Y$ implies that either $F_X \equiv F_Y$ or for some k with $1 \leq k \leq m$, $\mu_X^i = \mu_Y^i$, for $i = 1, \ldots, k-1$ and $(-1)^k \mu_X^k < (-1)^k \mu_Y^k$.*

Proof. If $F_X \not\equiv F_Y$, then also $F_X^{(m)} \not\equiv F_Y^{(m)}$. Having in mind that X dominates Y in order m, we infer that $F_X^{(m)}(\eta) < F_Y^{(m)}(\eta)$ for some η. Due to the continuity of both functions, the inequality is strict on an interval of positive length. Hence, for some $\varepsilon > 0$, we have $F_Y^{(m+1)}(\eta) - F_X^{(m+1)}(\eta) > \varepsilon$ for all sufficiently large η. Therefore, the result follows from Theorem 5.5. □

5.1.2 Preference to Small Outcomes

For a random variable X, denote $\bar{F}_X(\eta) = P[X > \eta]$. We see that the first-order stochastic dominance relation is symmetric in the sense that

$$X \succeq_{(1)} Y \iff \bar{F}_X(\eta) \geq \bar{F}_Y(\eta) \text{ for all } \eta \in \mathbb{R} \iff Y \preceq_{(1)} X. \tag{5.13}$$

Hence both statements "X is stochastically larger than Y" and "Y is stochastically smaller than X" are intuitive and justified. However, the stochastic order relation $\succeq_{(p)}$ for $p \geq 2$ expresses preference for larger outcomes over smaller ones; we shall bring more evidence to this later in our exposition. When small outcomes are preferred, we compare the survival functions of the variables in question. For a random variable $X \in \mathcal{L}_{p-1}(\Omega, \mathcal{F}, P)$ with $p \in [1, \infty)$, we define the function

$$\begin{aligned} \bar{F}_X^{(p)}(\eta) &= \frac{1}{\Gamma(p)} \int_{-\infty}^{\eta} (t-\eta)^{p-1} P_X(dt) \\ &= \frac{1}{\Gamma(p)} \mathbb{E}\big[\max(0, X-\eta)^{p-1}\big], \end{aligned} \tag{5.14}$$

where $\Gamma(p)$ stands for the Gamma function and $\max(0, x-\eta)^0 = \mathbb{1}_{(\eta,\infty)}(x)$. We have $\bar{F}_X^{(1)} \equiv \bar{F}_X(\eta)$.

A natural counterpart for the comparison of two random outcomes, when we prefer small outcomes, is provided in the following definition.

Definition 5.7. A random variable $X \in \mathcal{L}_{p-1}(\Omega, \mathcal{F}, P)$, $p \in [1, \infty)$, is *stochastically smaller* than another random variable $Y \in \mathcal{L}_{p-1}(\Omega, \mathcal{F}, P)$ with respect to the p-th order restricted to the interval $I \subseteq \mathbb{R}$ iff

$$\bar{F}_X^{(p)}(\eta) \leq \bar{F}_Y^{(p)}(\eta) \quad \text{for all } \eta \in I. \tag{5.15}$$

We denote this relation by $X \preceq_{(p)}^{I} Y$ for $I \subset \mathbb{R}$ and we omit the superscript if $I = \mathbb{R}$. To simplify notation, we write

$$\mathbb{E}\big[Z_+^p\big] \triangleq \mathbb{E}\big[\max(0, Z)^p\big].$$

The integrated survival function of an integrable random variable X:

$$\bar{F}_X^{(2)}(\eta) = \int_\eta^\infty P(X > t)\, dt = \mathbb{E}[(X - \eta)_+],$$

is called sometimes *stop-loss-transform* in actuarial sciences. The relation $\preceq_{(2)}$ is known as the *increasing convex order*. For it, we also use the notation $X \preceq_{(2)} Y$ for $I \subset \mathbb{R}$ omitting the superscript if $I = \mathbb{R}$.

The function $\bar{F}_X^{(p)}(\cdot)$ for $p \geq 2$ is non-increasing and convex. Using the equivalent representation of $\bar{F}_X^{(p)}(\cdot)$ in (5.14), we obtain that an integrable random variable X is stochastically smaller than $Y \in \mathcal{L}_{p-1}(\Omega, \mathcal{F}, P)$ with respect to the order $\preceq_{(p)}$ on the set I if

$$\mathbb{E}[(X - \eta)_+^{p-1}] \leq \mathbb{E}[(Y - \eta)_+^{p-1}], \quad \text{for all } \eta \in I. \tag{5.16}$$

When $I = \mathbb{R}$, we obtain the equivalence of the relation $\preceq_{(p)}$ to the relation (5.3) for the variable $-X$ and $-Y$ by changing signs in (5.16), i.e.,

$$X \preceq_{(p)} Y \iff -X \succeq_{(p)} -Y. \tag{5.17}$$

Notice that Lemma 5.4 remains valid also for the functions $\bar{F}_X^{(p)}(\cdot)$, while the other statements of § 5.1.1 have an obvious adaptation, which we summarize in the following corollary.

Corollary 5.8. (i) *If $X \preceq_{(p)} Y$, then also $X \preceq_{(q)} Y$ holds for any $q > p \geq 1$, provided $X, Y \in \mathcal{L}_{q-1}(\Omega, \mathcal{F}, P)$.*

(ii) *For two random variables $X, Y \in \mathcal{L}_m(\Omega, \mathcal{F}, P)$, $m \geq 1$, the relation $X \preceq_{(m)} Y$ implies that either $F_X \equiv F_Y$ or for some k with $1 \leq k \leq m$, $\mu_X^j = \mu_Y^j$, for $j = 1, \ldots, k-1$ and $\mu_X^k < \mu_Y^k$.*

5.1.3 Inverse Characterizations

Recall that the left-continuous inverse of the cumulative distribution function $F_X(\cdot)$ is defined as follows:

$$F_X^{(-1)}(p) = \inf \{\eta : F_X(\eta) \geq p\} \quad \text{for all } 0 < p < 1.$$

Given $p \in [0, 1]$, the number $q = q_X(p)$ is called a *p-quantile* of the random variable X if

$$P(X < q) \leq p \leq P(X \leq q).$$

For $p \in (0, 1)$ the set of p-quantiles is a closed interval and $F_X^{(-1)}(p)$ represents its left end. We adopt the convention that $F_X^{(-1)}(1) = +\infty$, if the 1-quantile does not exist.

The definition of the first-order stochastic dominance implies that $X \succeq_{(1)} Y$ if and only if

$$F_X^{(-1)}(p) \geq F_Y^{(-1)}(p) \quad \text{for all } 0 < p < 1. \tag{5.18}$$

For a random variable $X \in \mathcal{L}_1(\Omega, \mathcal{F}, P)$, we consider the integrated quantile function $F_X^{(-2)} : \mathbb{R} \to \mathbb{R}$, defined as

$$F_X^{(-2)}(p) = \int_0^p F_X^{(-1)}(\alpha)\, d\alpha \quad \text{for } 0 < p \leq 1, \tag{5.19}$$

For completeness, we also define $F_X^{(-2)}(0) = 0$ and $F_X^{(-2)}(p) = +\infty$ for all real numbers $p \notin [0, 1]$. The function $F_X^{(-2)}(\cdot)$ is the absolute Lorenz function introduced in § 2.2.2.

The Lorenz curves are used in economics for ordering positive random variables relative to their expectations, i.e., the relative Lorenz function is given by $p \mapsto \frac{1}{\mathbb{E}(X)} F_X^{(-2)}(p)$. We encountered the function $F_X^{(-2)}(\cdot)$ in Section 2.2.2 and we shall use some of the results established there.

Similar to $F_X^{(2)}(\cdot)$, the function $F_X^{(-2)}(\cdot)$ is well defined for any random variable $X \in \mathcal{L}_1(\Omega, \mathcal{F}, P)$. It is convex as an integral of a non-decreasing function. We also define the *upper Lorenz function* in the following way:

$$\bar{F}_X^{(-2)}(p) = \int_p^1 F_X^{(-1)}(t)\, dt \quad \text{for } 0 \leq p < 1. \tag{5.20}$$

The function $\bar{F}_X^{(-2)}(\cdot)$ can be extended beyond the interval $(0, 1)$ by setting $\bar{F}_X^{(-2)}(1) = 0$ and $\bar{F}_X^{(-2)}(p) = -\infty$ for all real numbers $p \notin [0, 1]$. It is a concave function because its derivative is non-increasing.

Using the right-continuity of $F_X(\cdot)$ and relation (5.3), we obtain the following form of the convex subdifferential of $F_X^{(2)}(\eta)$:

$$\partial F_X^{(2)}(\eta) = [P(X < \eta), P(X \leq \eta)], \tag{5.21}$$

The Fenchel duality relation between the integrated quantile function $F_X^{(-2)}(\cdot)$ and the integrated distribution function $F_X^{(2)}(\cdot)$ has been established in Theorem 2.15. Additionally, we identify the Fenchel conjugate function to the excess function $\bar{F}_X^{(2)}(\cdot)$.

Theorem 5.9. *For every integrable random variable X, we have*

(i) $F_X^{(-2)} = \big[F_X^{(2)}\big]^*$ *and* $F_X^{(2)} = \big[F_X^{(-2)}\big]^*$.

(ii) $\big[\bar{F}_X^{(2)}\big]^*(q) = -\bar{F}_X^{(-2)}(q+1)$ *and* $\big[-\bar{F}_X^{(-2)}(p)\big]^* = \bar{F}_X^{(2)}(p-1)$.

Proof. The statement (i) is proved in Theorem 2.15. To prove statement (ii), we represent the excess function as follows:

$$\bar{F}_X^{(2)}(\eta) = \mathbb{E}[(X-\eta)_+] = \mathbb{E}[X] - \eta + \mathbb{E}[(\eta - X)_+].$$

Using the statement (i), we obtain

$$\begin{aligned}\big[\bar{F}_X^{(2)}\big]^*(q) &= \sup_\eta \big\{\eta q + \eta - \mathbb{E}[X] - \mathbb{E}[(\eta - X)_+]\big\}\\ &= \sup_\eta \big\{\eta(q+1) - \mathbb{E}[(\eta - X)_+]\big\} - \mathbb{E}[X] = F_X^{(-2)}(q+1) - \mathbb{E}[X]\end{aligned}$$

for $0 < q + 1 \leq 1$. Representing $\mathbb{E}[X] = \int_0^1 F_X^{(-1)}(t)\,\mathrm{d}t$, we obtain

$$\big[\bar{F}_X^{(2)}\big]^*(q) = -\int_{q+1}^1 F_X^{(-1)}(t)\,\mathrm{d}t \quad \text{for } q \in (-1, 0).$$

For $q = 0$, we obtain

$$[\bar{F}_X^{(2)}]^*(0) = \sup_\eta \big\{\eta - \mathbb{E}[(\eta - X)_+]\big\} - \mathbb{E}[X] = 0 = \bar{F}_X^{(-2)}(1).$$

For $q = -1$, we get

$$\big[\bar{F}_X^{(2)}\big]^*(-1) = \sup_\eta \big\{-\mathbb{E}[(\eta - X)_+]\big\} - \mathbb{E}[X] = -\mathbb{E}[X] = -\bar{F}_X^{(-2)}(0).$$

For any real number $q \notin [-1, 0]$, we obtain $[\bar{F}_X^{(2)}]^*(q) = +\infty$, which completes the proof of the first equation in the statement (ii). The second equation in (ii) follows analogously. □

Theorem 5.9 enables us to characterize the second-order stochastic-dominance relation and its counterpart, the increasing convex order, by using the Lorenz functions of the respective random variables.

Corollary 5.10. *The following holds:*

$$\begin{aligned} X \succeq_{(2)} Y &\iff F_X^{(-2)}(p) \geq F_Y^{(-2)}(p) \quad \forall p \in [0,1]; \\ X \preceq_{(2)} Y &\iff \bar{F}_X^{(-2)}(p) \leq \bar{F}_Y^{(-2)}(p) \quad \forall p \in [0,1]. \end{aligned} \tag{5.22}$$

Additionally, we obtain the following relations.

Corollary 5.11. *The following statements are equivalent:*

(i) *η is a p-quantile of X;*

(ii) $F_X^{(-2)}(p) + F_X^{(2)}(\eta) = p\eta$*;*

(iii) $\bar{F}_X^{(-2)}(p) = \bar{F}_X^{(2)}(\eta) - \eta(p-1) = \mathbb{E}[X] - F_X^{(-2)}(p)$;

(iv) *If $P(X \leq \eta) = p \in (0,1)$, then $F_X^{(-2)}(p) = p\,\mathbb{E}[X|X \leq \eta]$.*

Proof. Directly from the definitions (5.3) and (5.19), we observe that assertion (i) is equivalent to $p \in \partial F_X^{(2)}(\eta)$, and $\eta \in \partial F_X^{(-2)}(p)$. The equivalence follows from Theorem 5.9. Relation (iii) is the counterpart of (ii) for the upper Lorenz function.

Finally, let $p \in (0,1)$ and η be such that $P\{X \leq \eta\} = p$. Then the following relations hold:

$$\begin{aligned} F_X^{(-2)}(p) &= p\eta - F_X^{(2)}(\eta) \\ &= p\eta + p\,\mathbb{E}\{X - \eta | X \leq \eta\} = p\,\mathbb{E}\{X|X \leq \eta\}. \end{aligned} \tag{5.23}$$

This proves (iv). □

The last equation provides insight into the nature of the integrated quantile function and extends the observation in Remark 2.14.

5.1.4 Relations to the Expected Utility Theory

In Section 1.2, we discussed the preference relations among probability measures and among random variables. In particular, Theorem 1.24 provides a numerical representation of preferences in terms of expected utility functionals of the form (1.18). We now use this insight to characterize the preferences encoded in the stochastic dominance relations by collections of such functionals.

For a collection $\mathcal{U}$ of functions $u : \mathbb{R} \to \mathbb{R}$, we define a preference relation $\succeq_{\mathcal{U}}$ as follows:

$$X \succeq_{\mathcal{U}} Y \quad \Longleftrightarrow \quad \mathbb{E}\big[u(X)\big] \geq \mathbb{E}\big[u(Y)\big] \quad \text{for all } u(\cdot) \in \mathcal{U}. \tag{5.24}$$

The collection of functions $\mathcal{U}$ is called the *generator* of the order $\succeq_{\mathcal{U}}$. With no loss of generality, the set $\mathcal{U}$ may be assumed a convex cone because for any number of functions $u_i \in \mathcal{U}$, $i = 1, \dots m$, and any positive numbers α_i, we have

$$\mathbb{E}\Big[\sum_{i=1}^{m} \alpha_i u_i(X)\Big] = \sum_{i=1}^{m} \alpha_i \, \mathbb{E}\big[u_i(X)\big] \geq \sum_{i=1}^{m} \alpha_i \, \mathbb{E}\big[u_i(Y)\big] = \mathbb{E}\Big[\sum_{i=1}^{m} \alpha_i u_i(Y)\Big],$$

implying that the conic extension of $\mathcal{U}$, the set $\widetilde{\mathcal{U}} = \text{cone}(\mathcal{U})$, is also a generator of the stochastic order $\succeq_{\mathcal{U}}$. We may also add constants to all functions.

The relation $X \succeq_{(p)} Y$ with $p > 1$ can be expressed in the following way:

$$\int_{-\infty}^{\infty} -\big(\max(0, \eta - z)\big)^{p-1} \; P_X(\mathrm{d}z) \geq \\ \int_{-\infty}^{\infty} -\big(\max(0, \eta - z)\big)^{p-1} \; P_Y(\mathrm{d}z) \quad \text{for all} \quad \eta \in \mathbb{R}. \tag{5.25}$$

Recall that we adopted the convention that $\max(0, \eta - x)^0 = \mathbb{1}_{(-\infty,\eta]}(x)$. This shows that a generator of the p-th order stochastic dominance relation, for $p > 1$, is formed by the functions $x \mapsto -[\max(0, \eta - x)]^{p-1}$, $\eta \in \mathbb{R}$ and the generator of the first-order stochastic dominance relation consists of all characteristic functions $-\mathbb{1}_{(-\infty,\eta]}(\cdot)$ for $\eta \in \mathbb{R}$.

Let the family of functions be reduced to those with $\eta \in I$, $I = [a, b]$, where $a \in \mathbb{R} \cup \{-\infty\}$ and $b \in \mathbb{R} \cup \{+\infty\}$. The conic extension of the generator $\{u(x) = -[\max(0, \eta - x)]^{p-1} : \eta \in I\}$ is obtained by using non-negative measures on I.

Let $\mathcal{M}_p^+(I)$ stand for the set of finite non-negative countably additive measures on I for which the integral $\int_I (\eta - x)_+^{p-1} \; \mu(\mathrm{d}\eta)$ is finite for every $x \in \mathbb{R}$. Then the set of functions

$$\mathcal{U}_p(I) = \Big\{u(x) = -\int_I (\eta - x)_+^{p-1} \; \mu(\mathrm{d}\eta), \; x \in \mathbb{R}; \; \mu \in \mathcal{M}_{p-1}^+(I)\Big\} \tag{5.26}$$

constitutes a generator for the relation $\succeq_{(p)}^{I}$.

Notice that the functions $u(\cdot)$ are well defined for all finite non-negative countably additive measures on I if $I = [a, b]$ is a compact interval. In that case, we may omit the subscript and write $\mathcal{M}^+([a, b])$.

We emphasize the one-to-one correspondence between the utility functions $u \in \mathcal{U}_p(I)$ for $p \geq 1$ and regular countably additive measures $\mu \in \mathcal{M}_{p-1}^+(I)$ in (5.26) such that the following key equation holds for all $X \in \mathcal{L}_{p-1}(\Omega, \mathcal{F}, P)$:

$$\mathbb{E}[u(X)] = -\int_a^b \mathbb{E}[\max(0, \eta - X)^{p-1}]\, \mu(\mathrm{d}\eta). \tag{5.27}$$

This observation will be very useful in our analysis of optimization problems involving dominance relations.

We establish an equivalent description of the generators, which reflects some of the classical results in the theory of stochastic dominance.

Theorem 5.12.

(i) *The set* $\mathcal{U}_1(I)$ *consists of all bounded non-decreasing functions, which satisfy* $u(t) = u(a)$ *for* $t \leq a$ *and* $u(t) = 0$ *for* $t \geq b$*;*

(ii) *The set* $\mathcal{U}_2(I)$ *consists of all non-decreasing concave functions, which for* $a > -\infty$ *satisfy* $u(t) = u(a) + c(t-a)$ *for* $t \leq a$*, with some constant* $c > 0$*, and* $u(t) = 0$ *for* $t \geq b$*; for* $a = -\infty$*,*

$$\lim_{t \to -\infty} \frac{|u(t)|}{|t|} < \infty;$$

(iii) *For a natural number* $k \geq 2$*, the set of functions* $\mathcal{U}_k(I)$ *consists of functions* $u(\cdot)$ *which are almost everywhere* $(k-1)$*-times continuously differentiable*[†] *and satisfy the following condition: a nonnegative, nonincreasing, left-continuous bounded function* $\varphi : \mathbb{R} \to \mathbb{R}$ *exists such that*

$$\begin{aligned}
&u^{(k-1)}(t) = (-1)^k \varphi(t), \quad \text{for almost all } t \in [a, b],\\
&u^{(k-1)}(t) = (-1)^k \varphi(a), \quad \text{for } t < a,\\
&u(t) = 0, \quad \text{for } t \geq b,\\
&u^{(i)}(b) = 0, \quad i = 1, \dots, k-2.
\end{aligned}$$

Proof. It follows from (5.26) for $p = 1$ that it is sufficient to include into the generator the function $u(t) = -\mu\big([a, t)\big)$, $t \in \mathbb{R}$, which proves (i). The functions are bounded because each μ is finite.

Consider now the case of $p = 2$. It is evident that the formula (5.26) defines nondecreasing concave functions $u(\cdot)$ with $u(b) = 0$. If $a > -\infty$, the conditions in (ii) are evident. For $a = -\infty$, we notice that $\mu \in \mathcal{M}_1^+\big((-\infty, b]\big)$, and thus

$$\lim_{t \to -\infty} \frac{1}{|t|} \int_{-\infty}^{t} |\eta|\, \mu(\mathrm{d}\eta) < \infty,$$

[†] Here, "almost everywhere" is in the sense of the Lebesgue measure.

which completes the verification of the conditions in (ii). To prove the equivalence, let $u(\cdot)$ satisfy the conditions in (ii). This implies that it is a nondecreasing concave function, whose left derivative,

$$u'_{-}(t) = \lim_{\tau \uparrow t} \frac{u(t) - u(\tau)}{t - \tau}$$

is well-defined, nonincreasing, and continuous from the left. Hence, a unique regular nonnegative measure μ exists such that

$$\mu\big([t,b]\big) = u'_{-}(t), \quad t \in \mathbb{R}.$$

For $x > b$, we have $u'_{-}(x) = 0$, hence $\mu[t,x] = \mu[t,b]$ for those x. Further, $u'_{-}(x)$ is constant for all $x \le a$, hence $\mu[x,b] = \mu[a,b]$ for those x. This means that the measure μ has its support on I. Therefore, $\mu \in \mathcal{M}_1^+(I)$. We have

$$u(x) = u(a) + \int_a^x u'_{-}(t)\, \mathrm{d}t = u(a) + \int_a^x \mu([t,b])\, \mathrm{d}t.$$

Since $u(b) = 0$, we obtain $u(a) = -\int_a^b \mu([t,b])\, \mathrm{d}t$. Hence, for $x \in I$, we have

$$u(x) = -\int_x^b \mu([t,b])\, \mathrm{d}t = -\int_x^b \int_t^b \mu(\mathrm{d}\eta)\, \mathrm{d}t = -\int_a^b \max(0, \eta{-}x)\, \mu(\mathrm{d}\eta).$$

We conclude that $u \in \mathcal{U}_2(I)$.

More generally, we consider an integer $k \ge 2$, the case (iii). According to (5.26), any utility function in $\mathcal{U}_{k-1}$ is associated with a non-negative measure $\mu \in \mathcal{M}_{k-1}^+(I)$ and has the form:

$$u(x) = -\int_a^b \max(0, \eta - x)^{k-1}\, \mu(\mathrm{d}\eta) = -\int_{\infty}^b \max(0, \eta - x)^{k-1}\, \mu(\mathrm{d}\eta).$$

Setting $\varphi(t) \triangleq (k-1)!\mu\big([t,b]\big)$, we obtain

$$u(x) = \frac{1}{(k-1)!} \int_{\infty}^b \max(0, \eta - x)^{k-1}\, \mathrm{d}\varphi(\eta).$$

Differentiation $k-1$ times yields

$$u^{(k-1)}(x) = (-1)^{k-1} \int_x^b \mathrm{d}\varphi(\eta) = (-1)^k \varphi(x),$$

as claimed. The other conditions of (iii) are evident. □

The description provided in Theorem 5.12 provides additional insight into the nature of the second-order stochastic dominance relation. According to Theorem 1.28, each utility function from the generator of the order, being non-decreasing and concave, corresponds to a risk-averse preference relation from Definition 1.27 (note that we are dealing with preferences to larger outcomes now, and risk aversion is associated with concavity). The stochastic dominance relation is not a preference relation in the sense of § 1.1, because it is not complete. However, the analysis above shows that it can be represented by a family of risk-averse preference relations. This explains its popularity in economics, finance, and insurance.

Turning to the case when small outcomes of the random quantity are preferred, we establish the counterpart characterizations.

First, we analyze the generator of the order $\preceq^I_{(p)}$; we denote it by $\mathcal{U}^\flat_p(I)$. In this case, we aim to identify the set of functions such that

$$X \preceq^I_{(p)} Y \iff \mathbb{E}[u(X)] \le \mathbb{E}[u(Y)] \quad \text{for all } u \in \mathcal{U}^\flat_p(I).$$

Analogously to the description (5.26) for the stochastic dominance relations, we have

$$\mathcal{U}^\flat_p(I) = \Big\{u : \mathbb{R} \to \mathbb{R} : u(x) = \int_I \max(0, x-\eta)^{p-1} \, d\mu(\eta) \quad x \in \mathbb{R}; \\ \mu \in \mathcal{M}^+_{p-1}(I)\Big\} \quad (5.28)$$

Let $I = [a, b] \subseteq \mathbb{R}$ be given.

Corollary 5.13. (i) *The set $\mathcal{U}^\flat_1([a,b])$ consists of all non-decreasing right-continuous functions, which satisfy $u(t) = 0$ for $t \le a$ and $u(t) = u(b)$ for $t \ge b$;*

(ii) *The set $\mathcal{U}^\flat_2([a,b])$ consists of all non-decreasing convex functions, which satisfy $u(t) = u(b) + c(t-b)$ with some constant $c > 0$ for $t \ge b$ and $u(t) = 0$ for $t \le a$;*

(iii) *For a natural number $k \ge 2$, the set of functions $\mathcal{U}^\flat_k([a,b])$ consists of functions satisfying the following condition: a nonnegative, non-decreasing, right-continuous bounded function $\varphi : \mathbb{R} \to \mathbb{R}$ exists such that u is almost everywhere $k-1$-time continuously differentiable and*

$$\begin{aligned} u^{(k-1)}(t) &= \varphi(t), \quad \text{for a.a. } t \in [a,b], \\ u^{(k-1)}(t) &= \varphi(a), \quad \text{for } t < a, \\ u(t) &= 0, \quad \text{for } t \le a, \\ u^{(i)}(a) &= 0, \quad i = 1, \dots, k-2. \end{aligned}$$

Proof. For (i), we identify $u(x) = \mu[a, x)$ for $\mu \in \mathcal{M}_{p-1}^{+}(I)$. For (ii), formula $u(x) = \int_I \max(0, x-\eta)^{p-1}\, d\mu(\eta)$ in (5.28) provides a non-decreasing convex function. On the other hand, if $u(\cdot)$ satisfies the conditions in (ii), then a unique regular nonnegative measure μ exists such that

$$\mu([a, t]) = u'_{+}(t).$$

For $x \le a$, we have $u'_{-}(t) = 0$. Further $u'_{+}(t)$ is constant for all $x \ge b$, hence $\mu[x, b] = \mu[a, b]$ for those x. This means that the measure μ has its support on I. The proof of statement (iii) follows the same arguments. □

5.1.5 Relations to Distortions

We return to the *dual utility theory* or *rank dependent utility theory*, whose mathematical foundations were discussed in Section 1.3.

Every decision maker using that framework has a certain *distortion function* $w : [0, 1] \to \mathbb{R}$. The decision maker prefers (weakly) a random outcome X over another random outcome Y, if

$$\int_0^1 F_X^{(-1)}(p)\, \mathrm{d}w(p) \ge \int_0^1 F_Y^{(-1)}(p)\, \mathrm{d}w(p).$$

The theoretical justification of this approach are provided by Theorem 1.17. Note that we are dealing with the preference to larger outcomes now, and that is why the direction of the inequality in the comparison is reversed. Accordingly, the risk-averse preference relations are represented by concave and nondecreasing functions $w(\cdot)$.

Here, we provide an equivalent characterization of the stochastic dominance relations of the first and second orders by distortion functions, in a similar spirit to the characterization by expected utility functionals.

We introduce the set $\mathcal{W}_1$ of all continuous nondecreasing functions $w : [0, 1] \to \mathbb{R}$, satisfying the conditions

$$\limsup_{\varepsilon \downarrow 0} \frac{w(\varepsilon)}{\varepsilon} < \infty \quad \text{and} \quad \limsup_{\varepsilon \downarrow 0} \frac{w(1) - w(1-\varepsilon)}{\varepsilon} < \infty, \tag{5.29}$$

and the set $\mathcal{W}_2 \subset \mathcal{W}_1$ of functions, which are additionally concave.

Theorem 5.14. *For any two random variables $X, Y \in \mathcal{L}_1(\Omega, \mathcal{F}, P)$, the following statements are true:*

(i) *$X \succeq_{(1)} Y$ if and only if for all $w \in \mathcal{W}_1$*

$$\int_0^1 F_X^{(-1)}(p)\,\mathrm{d}w(p) \ge \int_0^1 F_Y^{(-1)}(p)\,\mathrm{d}w(p); \tag{5.30}$$

(ii) $X \succeq_{(2)} Y$ *if and only if* (5.30) *is satisfied for all* $w \in \mathcal{W}_2$.

Proof. Statement (i) follows from the Stjelties integration of both sides of the inequality (5.18) with respect to $\mathrm{d}w(p)$. The integrals are well-defined, due to the monotonicity and continuity of $w(\cdot)$. They take finite values for all integrable random variables X, in view of the identity

$$\int_0^1 F_X^{(-1)}(p)\,\mathrm{d}p = \mathbb{E}[X] \tag{5.31}$$

and the bounds on the slope of the function $w(\cdot)$ at the ends of the interval [0, 1]. We turn to the statement (ii). It follows from (5.22) that the relation $X \succeq_{(2)} Y$ is equivalent to the requirement that the inequality

$$\int_0^1 F_X^{(-2)}(t)\,\nu(\mathrm{d}t) \ge \int_0^1 F_Y^{(-2)}(t)\,\nu(\mathrm{d}t) \tag{5.32}$$

holds for every finite non-negative regular measure ν on [0, 1].

Given such a measure ν, we define the function $\xi_\nu(\tau) \triangleq \nu([\tau, 1])$ (the ν-measure of the set $[\tau, 1]$) and the function $w : [0, 1] \to \mathbb{R}$ as follows:

$$w(p) \triangleq -\int_p^1 \xi_\nu(\tau)\,\mathrm{d}\tau.$$

Since $\nu \ge 0$, the function $\xi_\nu(\cdot)$ is nonnegative and nonincreasing, which implies that $w(\cdot)$ is nondecreasing and concave. Moreover, the right derivative $w'_+(0) = \nu([0, 1])$ is finite, and thus $w \in \mathcal{W}_2$.

Conversely, for any $w \in \mathcal{W}_2$, the left derivative of w,

$$w'_-(p) = \lim_{\tau \uparrow p} \frac{w(p) - w(\tau)}{p - \tau}, \quad p \in (0, 1],$$

is well-defined, nondecreasing, and continuous from the left. Hence, we can associate with w a unique regular nonnegative measure ν on [0, 1] by defining

$$\nu([p, 1]) = \begin{cases} w'_-(p) & \text{for } p \in (0, 1], \\ w'_+(0) & \text{for } p = 0. \end{cases}$$

This correspondence between nonnegative regular measures on [0, 1] and functions in $\mathcal{W}_2$ is one-to-one. We use it to prove the equivalence of (5.30) and (5.32). Changing the order of integration in the integrals in (5.32), we obtain

$$\begin{aligned}\int_0^1 F_X^{(-2)}(t)\,\nu(\mathrm{d}t) &= \int_0^1 \int_0^t F_X^{(-1)}(p)\,\mathrm{d}p\,\nu(\mathrm{d}t)\\ &= \int_0^1 \int_p^1 \nu(\mathrm{d}t)\,F_X^{(-1)}(p)\,\mathrm{d}p = \int_0^1 \nu([p,1])\,F_X^{(-1)}(p)\,\mathrm{d}p\\ &= \int_0^1 F_X^{(-1)}(p)\,\mathrm{d}w(p). \end{aligned}\tag{5.33}$$

Consequently, inequality (5.32) for all regular measures ν is equivalent to inequality (5.30) for all functions in $\mathcal{W}_2$. □

The functions $w(\cdot)$ are rank-dependent (dual) utility functions.

Consider now the relaxation of the stochastic order relations, which result from omitting some probability levels in the relations between the inverse functions and the Lorenz functions. More specifically, let $J = [\alpha, \beta] \subseteq [0, 1]$. We consider the relaxation

$$F_X^{(-1)}(p) \geq F_Y^{(-1)}(p) \quad \text{for all } p \in J \quad (\succeq_{(1)} \text{ relaxed to } J); \tag{5.34}$$

$$F_X^{(-2)}(p) \geq F_Y^{(-2)}(p) \quad \text{for all } p \in J \quad (\succeq_{(2)} \text{ relaxed to } J); \tag{5.35}$$

Definition 5.15. The sets $\mathcal{W}_1(J)$ and $\mathcal{W}_2(J)$ are defined as follows:

(a) The set $\mathcal{W}_1(J)$ contains all continuous non-decreasing functions $w : [0, 1] \to \mathbb{R}$, which satisfy $w(t) = w(\alpha)$ for $t \in [0, \alpha]$ and $w(t) = 0$ for $t \in (\beta, 1]$. Additionally, the corresponding condition in (5.29) is satisfied, if $\alpha = 0$ or $\beta = 1$;
(b) The set $\mathcal{W}_2(J) \subset \mathcal{W}_1(J)$ contains all non-decreasing concave functions $w : [0, 1] \to \mathbb{R}$, which satisfy

$$\begin{aligned} &w(p) = 0 \text{ for all } p \in [\beta, 1];\\ &w(p) = w(\alpha) + c(p - \alpha) \text{ with some } c > 0 \text{ for all } p \in [0, \alpha]. \end{aligned}$$

It is evident that both $\mathcal{W}_1(J)$ and $\mathcal{W}_2(J)$ are convex cones in $\mathcal{C}([0, 1])$ and $\mathcal{W}_1(J) \subset \mathcal{W}_1$, $\mathcal{W}_2(J) \subset \mathcal{W}_2$.

Lemma 5.16. *The sets $\mathcal{W}_1(J)$ and $\mathcal{W}_2(J)$ characterize the relaxed relations* (5.34) *and* (5.35), *respectively.*

Proof. The statements are established by the same arguments as in Theorem 5.14 and their detailed exposition is omitted. □

We also summarize our observations about the correspondence of regular countably additive measures on $J = [\alpha, \beta] \subset J = [\alpha, \beta] \subset [0, 1]$, distortion functions, and coherent measures of risk.

Lemma 5.17. *There is one-to-one correspondence between distortion functions $w \in \mathcal{W}_2(J)$ and measures $\nu \in \mathcal{M}^+(J)$ such that*

$$\int_J F_X^{(-2)}(t)\, \nu(\mathrm{d}t) = \int_J F_X^{(-1)}(p)\, \mathrm{d}w(p). \tag{5.36}$$

Furthermore, for every function $w \in \mathcal{W}_2(J)$, a spectral risk measure $\varrho(\cdot)$ and a constant $\kappa \geq 0$ exist such that for all random variables $X \in \mathcal{L}_1(\Omega, \mathcal{F}, P)$

$$\int_J F_X^{(-1)}(p)\, \mathrm{d}w(p) = -\kappa\varrho(X), \tag{5.37}$$

and the spectrum of ϱ has its support on J. Conversely, for every spectral measure of risk ϱ with spectral support on J, a function $w \in \mathcal{W}_2(J)$ exists such that equation (5.37) *holds with $\kappa = 1$.*

Proof. For any fixed $w \in \mathcal{W}_2(J)$, consider the functional

$$\mathcal{E}[X] = \int_\alpha^\beta F_X^{(-1)}(p)\, \mathrm{d}w(p). \tag{5.38}$$

We consider the left derivative of w, which is well-defined, non-increasing, and continuous from the left; it is given by the formula

$$w'_-(p) = \lim_{\tau \uparrow p} \frac{w(p) - w(\tau)}{p - \tau}, \quad p \in J.$$

Therefore, a unique regular nonnegative measure ν on J exists defined as follows:

$$\nu([p, 1]) = \begin{cases} w'_-(p) & \text{for } p \in [\alpha, \beta],\ p > 0, \\ w'_+(0) & \text{for } p = 0 \text{ if } \alpha = 0. \end{cases}$$

Notice that the measure ν is nonnegative because $w(\cdot)$ is concave. If $[\alpha, \beta]$ is a strict subset of $[0, 1]$, then the support of ν can be extended to the interval $[0, 1]$ by assigning zero measure to all measurable sets not intersecting $[\alpha, \beta]$. Abusing notation, we keep the same notation ν for the extended measure. Concave functions are almost everywhere continuously differentiable and we can use the measure ν in (5.38) to obtain the following:

$$
\begin{aligned}
\mathcal{E}[X] &= \int_\alpha^\beta F_X^{(-1)}(p)\,\nu([p,1])\,\mathrm{d}p = \int_0^1 \int_p^1 \nu(\mathrm{d}t)\,F_X^{(-1)}(p)\,\mathrm{d}p \\
&= \int_0^1 \int_0^t F_X^{(-1)}(p)\,\mathrm{d}p\,\nu(\mathrm{d}t) = \int_0^1 F_X^{(-2)}(t)\,\nu(\mathrm{d}t) \\
&= \int_0^1 \frac{1}{t} F_X^{(-2)}(t)\,t\,\nu(\mathrm{d}t).
\end{aligned}
\tag{5.39}
$$

The third equation is obtained via integration by parts.

The penultimate equation shows the correspondence between distortions and measures in $\mathcal{M}^+([\alpha,\beta])$ such that

$$
\int_\alpha^\beta F_X^{(-1)}(p)\,\mathrm{d}w(p) = \int_0^1 F_X^{(-2)}(t)\,\nu(\mathrm{d}t). \tag{5.40}
$$

To proceed to measures of risk, we define the ν-associated constant,

$$
\kappa_\nu = \int_0^1 t\,\nu(\mathrm{d}t), \tag{5.41}
$$

and observe that $\kappa_\nu = 0$ means $w(p) \equiv 0$ for all $p \in [0,1]$. If $\kappa_\nu > 0$, then we define the measure $\mu(\mathrm{d}t) = t\nu(\mathrm{d}t)/\kappa_\nu$, and observe that μ is a probability measure on $[0,1]$ with support on $[\alpha,\beta]$. In view of Remark 2.20,

$$
\mathcal{E}[X] = -\kappa_\nu \int_0^1 \mathrm{AVaR}_t^-(X)\,\mu(\mathrm{d}t). \tag{5.42}
$$

Having in mind the Kusuoka representation of measures of risk, we conclude that $\varrho(\cdot) = -\frac{1}{\kappa_\nu}\mathcal{E}(\cdot)$ is a spectral measure of risk.

Now we fix an arbitrary spectral measure of risk, represented as

$$
\varrho(X) = \int_\alpha^\beta \mathrm{AVaR}_t^-(X)\,\mu(\mathrm{d}t), \tag{5.43}
$$

with some probability measure μ on $[\alpha,\beta] \subset (0,1]$. We define the measure ν on $[\alpha,\beta]$ by the relation $\nu(\mathrm{d}t) = (1/t)\mu(\mathrm{d}t)$. Consider now the function $\xi(\tau) = \nu([\tau,\beta])$, which assigns to $\tau \in [\alpha,\beta]$ the ν-measure of the set $[\tau,\beta]$. The function $w : [\alpha,\beta] \to \mathbb{R}$ is defined as follows:

$$
w(p) = -\int_p^\beta \xi(\tau)\,\mathrm{d}\tau.
$$

Clearly, $\xi(\cdot)$ is a nonnegative, non-increasing function on the interval $[\alpha,\beta]$, which entails that the function $w(\cdot)$ is concave. Additionally, $w(\cdot)$ is nondecreasing by construction. Using the function w, we proceed as in (5.39), but in the reverse order, to derive the representation (5.37) of the risk measure $\varrho(\cdot)$. □

Similar results are valid for the orders $\preceq_{(p)}$ reflecting preference to small outcomes. Given $[\alpha, \beta] \subseteq [0, 1], \alpha < \beta$, we define the set $\mathcal{W}_2^\flat([\alpha, \beta])$ as the set of convex and non-decreasing functions $w : [0, 1] \to \mathbb{R}$ such that $w(p) = 0$ for all $p \in [0, \alpha]$ and $w(p) = w(\beta) + c(p - \beta)$, with some $c > 0$, for all $p \in [\beta, 1]$. Additionally, $w(\cdot)$ satisfies the condition $\limsup_{\varepsilon \downarrow 0} \frac{w(\varepsilon)}{\varepsilon} < \infty$ if $\alpha = 0$ and $\limsup_{\varepsilon \downarrow 0} \frac{w(1)-w(1-\varepsilon)}{\varepsilon} < \infty$, if $\beta = 1$.

Theorem 5.18. *Let $J = [\alpha, \beta] \subset [0, 1]$ be given. One-to-one correspondence between measures $\nu \in \mathcal{M}^+(J)$ and distortion functions $w \in \mathcal{W}_2^\flat(J)$ exists such that*

$$\int_\alpha^\beta \bar{F}_X^{(-2)}(p)\, d\nu(p) = \int_0^1 F_X^{(-1)}(p)\, dw(p) \tag{5.44}$$

holds for all integrable random variables X.

Proof. Given a measure $\nu \in \mathcal{M}^+(J)$, we can extend it to the interval [0, 1] by assigning zero mass to the measurable sets not interesting $[\alpha, \beta]$ if any such exist. Abusing notation, we denote the extension by ν as well. The function $w : [0, 1] \to \mathbb{R}$ is defined as follows:

$$w(p) = \int_0^p \nu([0, t])\, dt.$$

The derivative $w'(p) = \nu([0, p])$, which entails that the function $w'(\cdot)$ is non-negative and nondecreasing because $\nu \geq 0$. Hence, $w(\cdot)$ is nondecreasing and convex. The left derivative $w'_-(1) = \nu([0, 1])$ is finite, implying that w is sub-differentiable at $p = 1$. If $\beta < 1$, then the derivative $w'(p) = \nu(J)$ for all $p \in [\beta, 1)$. If $\alpha > 0$, then the derivative $w'(p) = 0$ for all $p \in (0, \alpha)$ and $w(\alpha) = \int_0^\alpha \nu([0, t])\, d\tau = 0$. Hence, $w \in \mathcal{W}_2^\flat(J)$.

For the opposite direction, let $w \in \mathcal{W}_2^\flat(J)$. Hence, the right derivative of w,

$$w'_+(p) = \lim_{\tau \downarrow p} [w(\tau) - w(p)]/(\tau - p), \quad p \in [0, 1),$$

is well-defined, nondecreasing, and right-continuous. We infer the existence of a unique regular nonnegative measure ν on [0, 1] satisfying

$$\nu([0, p]) = \begin{cases} w'_+(p) & \text{for } p \in [0, 1), \\ w'_-(1) & \text{for } p = 1. \end{cases}$$

If $J = [\alpha, \beta]$ with $\alpha > 0$, then $w'_+(p) = 0$ for $p \in [0, \alpha)$. If $\beta < 1$, then $\nu([0, p]) = c$ remains constant for $p \in [\beta, 1]$, which implies that the measure ν has support on $J = [\alpha, \beta]$. The following chain of equations holds:

$$\int_\alpha^\beta \bar{F}_X^{(-2)}(p)\, \mathrm{d}\nu(p) = \int_0^1 \int_t^1 F_X^{(-1)}(p)\, \mathrm{d}p\, \mathrm{d}\nu(t) = \int_0^1 \int_0^p \mathrm{d}\nu(t)\, F_X^{(-1)}(p)\, \mathrm{d}p$$
$$= \int_0^1 \nu([0, p])\, F_X^{(-1)}(p)\, \mathrm{d}p = \int_0^1 F_X^{(-1)}(p)\, \mathrm{d}w(p).$$

This completes the proof. □

Theorem 5.18 implies the following characterization of the (relaxed) increasing convex order via distortions.

Corollary 5.19. *For two integrable random variables X and Y, the relation $X \preceq_{(2)}^J Y$ holds if and only if for all distortion functions $w \in \mathcal{W}_2^\flat(J)$*

$$\int_0^1 F_X^{(-1)}(p)\, \mathrm{d}w(p) \le \int_0^1 F_Y^{(-1)}(p)\, \mathrm{d}w(p). \tag{5.45}$$

5.1.6 Relations to Measures of Risk

In this section, we assume that higher random outcomes are preferred, that is, the random variables under consideration express gains.

We start with the following definition.

Definition 5.20. We say that a risk measure $\varrho : \mathcal{L}_1(\Omega, \mathcal{F}, P) \to \overline{\mathbb{R}}$ is consistent with the stochastic order "$\succeq$", if $X \succeq Y$ implies that $\varrho[X] \le \varrho[Y]$.

We start from the first-order stochastic dominance relation.

Theorem 5.21. *Assume that the probability space $(\Omega, \mathcal{F}, P)$ is either standard and atomless or $\Omega = \{\omega_1, \dots, \omega_n\}$ with $P(\omega_i) = \frac{1}{n}$, $i = 1, \dots, n$. If a risk measure $\varrho : \mathcal{L}_1(\Omega, \mathcal{F}, P) \to \overline{\mathbb{R}}$ is law-invariant, then it is consistent with the first-order stochastic dominance if and only if it satisfies the monotonicity axiom.*

Proof. On a standard atomless probability space, a uniform random variable exists: a measurable function $U : \Omega \to \mathbb{R}$ such that $P(U \le t) = t$ for all $t \in [0, 1]$. Consider the random variables $\hat{X} = F_X^{-1}(U)$ and $\hat{Y} = F_Y^{-1}(U)$. If $X \succeq_{(1)} Y$, then $\hat{X}(\omega) \ge \hat{Y}(\omega)$ for all $\omega \in \Omega$, and thus the monotonicity axiom implies $\varrho(\hat{X}) \le \varrho(\hat{Y})$. The variables $\hat{X}$ and X (respectively, $\hat{Y}$ and Y)

have the same distribution by construction, and the risk measure $\varrho(\cdot)$ is law-invariant. Hence, we conclude that $\varrho(X) \le \varrho(Y)$, which shows the consistency of $\varrho(\cdot)$ with the first-order stochastic dominance. In the discrete space with equally likely elementary events, we proceed analogously, but with the discrete uniform random variable with values i/n, $i = 1, \dots, n$. □

Interestingly, we are also able to show the consistency with the second-order dominance for a large class of risk measures. First, we observe that the second-order stochastic dominance relation in the inverse form can be viewed as a continuum of relations between the Average Value at Risk for both variables.

In § 5.1.5, we observed the relation of stochastic dominance to Average Value-at-Risk. Recall that an integrable random variable X dominates in the second-order the random variable Y on $J \subseteq [0, 1]$ if (5.35) holds, i.e.,

$$F_X^{(-2)}(p) \ge F_X^{(-2)}(p) \quad \text{for all } p \in J.$$

The definition of the Average Value at Risk implies that the last relation is equivalent to

$$\text{AVaR}_p^-(X) \le \text{AVaR}_p^-(Y) \quad \text{for all } p \in J. \tag{5.46}$$

This helps address the consistency of law-invariant measures of risk with the second-order stochastic dominance.

Theorem 5.22. *Assume that $(\Omega, \mathcal{F}, P)$ is either standard and atomless or $\Omega = \{\omega_1, \dots, \omega_n\}$ with $P[\omega_i] = \frac{1}{n}$, $i = 1, \dots, n$. Any proper lower semicontinuous law-invariant coherent risk measure $\varrho : \mathcal{L}_1(\Omega, \mathcal{F}, P) \to \overline{\mathbb{R}}$ is consistent with the second order stochastic dominance.*

Proof. Under the assumptions of the theorem, any risk measure ϱ has a Kusuoka representation by virtue of Theorem 2.39, i.e.,

$$\varrho(Z) = \sup_{\lambda \in \Lambda_\varrho} \int_0^1 \text{AVaR}_\alpha^-(Z)\, \lambda(d\alpha),$$

where Λ_ϱ is a set of probability measures on the interval $(0, 1]$. If $X \succeq_{(2)} Y$ holds for some $X, Y \in \mathcal{L}_1(\Omega, \mathcal{F}, P)$, then inequality (5.46) holds with $J = (0, 1]$. Integrating (5.46) with a non-negative measure λ and taking the supremum preserves the inequality. Hence $\varrho(X) \le \varrho(Y)$ as required. □

The question of consistency of risk measures with stochastic orders, when the probability space is not atomless, is rather difficult and requires specialized analysis in each particular case.

Here, we show that the mean-semideviation risk measure of order $p \ge 1$ is consistent with the stochastic dominance relation of order $p + 1$. This result implies, in particular, that all mean-semideviation risk measures are consistent with second-order stochastic dominance.

Recall that we consider preference to larger outcomes, and the risk measure has the form:

$$\varrho(X) = -\mathbb{E}[X] + \varkappa \big\| (\mathbb{E}[X] - X)_+ \big\|_p, \quad \varkappa \in [0,1].$$

We denote the expected value and the lower semideviation of a random variable $X \in \mathcal{L}_p(\Omega, \mathcal{F}, P)$ with $p \geq 1$, by

$$\mu_X \triangleq \mathbb{E}[X] \quad \text{and} \quad \sigma_X^{(p)} \triangleq \big\| (\mathbb{E}[X] - X)_+ \big\|_p,$$

respectively.

Theorem 5.23. *For any random variables $X, Y \in \mathcal{L}_p(\Omega, \mathcal{F}, P)$, and for any $p \geq 1$, the relation $X \succeq_{(p+1)} Y$ implies*

$$\mu_X - \varkappa \sigma_X^{(p)} \geq \mu_Y - \varkappa \sigma_Y^{(p)} \quad \textit{for all } \varkappa \in [0,1].$$

Proof. Due to Theorem 5.5, we have $\mu_X \geq \mu_Y$. It is sufficient to show the inequality for $\varkappa = 1$. Suppose $p > 1$. Using the Newton-Leibniz formula, we obtain

$$\Big(\mathbb{E}\big[(\mu_X - X)_+^p\big]\Big)^{1/p} - \Big(\mathbb{E}\big[(\mu_Y - X)_+^p\big]\Big)^{1/p} = \int_{\mu_Y}^{\mu_X} \frac{\mathbb{E}\big[(\eta - X)_+^{p-1}\big]}{\Big(\mathbb{E}\big[(\eta - X)_+^p\big]\Big)^{(p-1)/p}}\, d\eta.$$

Consider the fraction under the integral on the right-hand side and denote $Z = (\eta - X)_+$. The numerator of this fraction can be bounded by the Hölder inequality as follows:

$$\mathbb{E}[Z^{p-1}] = \mathbb{E}[Z^{p-1}\mathbb{1}] \leq \|Z^{p-1}\|_{p/(p-1)} \|\mathbb{1}\|_p = \big(\|Z\|_p\big)^{p-1}.$$

The right-hand side is equal to the denominator. Therefore, the fraction under the integral is bounded above by 1. Hence,

$$\Big(\mathbb{E}\big[(\mu_X - X)_+^p\big]\Big)^{1/p} - \Big(\mathbb{E}\big[(\mu_Y - X)_+^p\big]\Big)^{1/p} \leq \mu_X - \mu_Y.$$

Since $X \succeq_{(p+1)} Y$, we have $\mathbb{E}\big[(\mu_Y - X)_+^p\big] \leq \mathbb{E}\big[(\mu_Y - Y)_+^p\big]$, which combined with the last inequality yields:

$$\sigma_X^{(p)} - \sigma_Y^{(p)} \leq \mu_X - \mu_Y,$$

as claimed.

If $p = 1$, then $X \succeq_{(2)} Y$ implies

$$F_X^{(2)}(\mu_X) - F_Y^{(2)}(\mu_Y) \leq \int_{\mu_Y}^{\mu_X} F_X(\eta)\, \mathrm{d}\eta \leq \mu_X - \mu_Y,$$

which is equivalent to $\sigma_X^{(1)} - \sigma_Y^{(1)} \leq \mu_X - \mu_Y$. □

Now we turn again to the stochastic order relations expressing preference to small outcomes. Let $J = [\alpha, \beta] \subseteq [0, 1]$ be given. Analogously to the arguments above, we obtain the following statement using measures of risk for losses.

Corollary 5.24. *For every function $w \in \mathcal{W}_2^{\flat}(J)$, a spectral risk measure ϱ and a constant $\kappa \geq 0$ exist such that for all random variables $X \in \mathcal{L}_1(\Omega, \mathcal{F}, P)$*

$$\int_J F_X^{(-1)}(p)\, \mathrm{d}w(p) = \kappa\varrho(X), \tag{5.47}$$

and the spectrum of ϱ has a support on J. Conversely, for every spectral measure of risk ϱ with a spectral support on J, a function $w \in \mathcal{W}_2^{\flat}(J)$ exists such that equation (5.47) holds with $\kappa = 1$.

Proof. From the proof of Theorem 5.18, we use the correspondence between a function $w \in \mathcal{W}_2^{\flat}(J)$ and a measure $\nu \in \mathcal{M}^+(J)$ in the following equations

$$\begin{aligned}\int_\alpha^\beta \bar{F}_X^{(-2)}(p)\, \nu(\mathrm{d}p) &= \int_0^1 \frac{1}{1-t} \int_t^1 F_X^{-1}(p)\, \mathrm{d}p\, (1-t)\, \nu(\mathrm{d}t)\\ &= \int_0^1 \mathrm{AVaR}_t^+(X)\, (1-t)\, \nu(\mathrm{d}t) = \kappa_\nu \int_0^1 \mathrm{AVaR}_t^+(X)\, \mu(\mathrm{d}t).\end{aligned}$$

with $\kappa_\nu = \int_0^1 (1-t)\, \nu(\mathrm{d}t)$ and $\mu(\mathrm{d}t) = (1-t)\nu(\mathrm{d}t)/\kappa_\nu$. □

This observation helps us to obtain a characterization of the relation $X \preceq_{(2)}^J Y$ on $J \subset [0, 1]$ by law-invariant coherent measures of risk ϱ having a Kusuoka representation with a set $\mathcal{M}_\varrho$ such that all probability measures in $\mathcal{M}_\varrho$ are supported on J. Denoting the set of those risk measures by $\mathcal{R}(J)$, we obtain the following statement.

Theorem 5.25. *For two integrable random variables X and Y,*

$$X \preceq_{(2)}^J Y \iff \varrho(X) \leq \varrho(Y) \quad \textit{for all } \varrho \in \mathcal{R}(J). \tag{5.48}$$

Proof. Theorem 5.18 implies that $X \preceq_{(2)}^{J} Y$ if and only if for all $\nu \in \mathcal{M}^{+}(J)$

$$\int_{\alpha}^{\beta} \bar{F}_X^{(-2)}(p)\ \nu(\mathrm{d}p) \leq \int_{\alpha}^{\beta} \bar{F}_X^{(-2)}(p)\ \nu(\mathrm{d}p),$$

hold. This inequality is equivalent to

$$\kappa_\nu \int_0^1 \mathrm{AVaR}_t^{+}(X)\ \mu(\mathrm{d}t) \leq \kappa_\nu \int_0^1 \mathrm{AVaR}_t^{+}(Y)\ \mu(\mathrm{d}t),$$

with $\kappa_\nu = \int_0^1 (1-t)\ \nu(\mathrm{d}t)$ and $\mu(\mathrm{d}t) = (1-t)\nu(\mathrm{d}t)/\kappa_\nu$.

For any measure $\varrho \in \mathcal{R}(J)$, we assume without loss of generality, that all $\nu \in \mathcal{M}_\varrho$ are not identically zero and hence $\kappa_\nu > 0$ for all $\nu \in \mathcal{M}_\varrho$. We obtain that $X \preceq_{(2)}^{J} Y$ implies

$$\int_0^1 \mathrm{AVaR}_t^{+}(X)\ \mu(\mathrm{d}t) \leq \int_0^1 \mathrm{AVaR}_t^{+}(Y)\ \mu(\mathrm{d}t) \quad \text{for all } \mu \in \mathcal{M}_\varrho.$$

Taking the supremum over μ entails that $\varrho(X) \leq \varrho(Y)$. For the reverse statement, it is sufficient to consider spectral risk measures. Then the relation $X \preceq_{(2)}^{J} Y$ is obtained by using Theorem 5.18 and Corollary 5.24. □

5.1.7 Acceptance Sets

We take a closer look at the sets defined by the stochastic dominance relations. Let a benchmark random variable $Y \in \mathcal{L}_{p-1}(\Omega, \mathcal{F}, P)$, $p \geq 1$, and an interval $I \subseteq \mathbb{R}$ be given. We shall consider the sets

$$\mathcal{A}_p(Y, I) = \left\{X \in \mathcal{L}_{p-1}(\Omega, \mathcal{F}, P) : X \succeq_{(p)}^{I} Y\right\}.$$

Recall that the notation $\succeq_{(p)}^{I}$ stands for the relation $F_X^{(p)}(\eta) \leq F_Y^{(p)}(\eta)$ for all $\eta \in I$. When $I = \mathbb{R}$, then we write $\mathcal{A}_p(Y)$. Theorem 5.3 implies the following inclusion for any $p > 1$ and $r \in [1, p]$:

$$\mathcal{A}_r(Y) \cap \mathcal{L}_{p-1}(\Omega, \mathcal{F}, P) \subset \mathcal{A}_p(Y).$$

As in the proof of Theorem 5.3, we obtain the same inclusion for the restricted stochastic dominance relations for any $p > 1$ and $r \in [1, p]$ when the interval $I = (-\infty, b]$ with $b \in \mathbb{R}$, i.e.,

$$\mathcal{A}_r(Y, I) \cap \mathcal{L}_{p-1}(\Omega, \mathcal{F}, P) \subset \mathcal{A}_p(Y, I). \tag{5.49}$$

The inclusion is not guaranteed when $I = [a, b]$ with $a > -\infty$.

We also observe the following useful topological property.

Lemma 5.26. *The multifunctions $Y \mapsto \mathcal{A}_1(Y)$ and $Y \mapsto \mathcal{A}_1(Y, I)$ for any $I = [a, b) \subseteq \mathbb{R}$ have a closed graph with respect to the convergence in distribution.*

Proof. Let $\lim_{n\to\infty} Y_n = Y$, $\lim_{n\to\infty} X_n = X$ and $X_n \succeq_{(p)} Y_n$. Consider first the case of $p = 1$. Due to the convergence in distribution $P(X_n \le \eta) \le P(Y_n \le \eta)$ implies $P(X \le \eta) \le P(Y \le \eta)$ for all $\eta \in I$, which are continuity points of the cumulative distribution functions of X and Y. Assume that $\bar{\eta} \in \mathbb{R}$ is a point of discontinuity of X and/or Y. The cumulative distribution functions are right-continuous and, thus, we can choose a sequence $\eta_k > \bar{\eta}$ converging to $\bar{\eta}$, such that for all $k \in \mathbb{N}$, η_k is a continuity point of both $F_X(\cdot)$ and $F_Y(\cdot)$. We have $F_X(\eta_k) \le F_Y(\eta_k)$ for all $k \in \mathbb{N}$. Passing to the limit with $k \to \infty$, we obtain $F_X(\bar{\eta}) \le F_Y(\bar{\eta})$, which proves the claim when $p = 1$. □

Notice that the multifunction $Y \mapsto \mathcal{A}_1(Y, [a, b])$ has a closed graph with respect to the convergence in distribution for any $[a, b] \subseteq \mathbb{R}$ such that b is a point of continuity of both $F_X(\cdot)$ and $F_Y(\cdot)$.

Lemma 5.27. *For any $I \subseteq \mathbb{R}$ and $p \ge 2$, the multifunctions $Y \mapsto \mathcal{A}_p(Y, I)$ are convex-valued and have a closed graph in $\mathcal{L}_{p-1}(\Omega, \mathcal{F}, P)$.*

Proof. Let $X, Z \in \mathcal{A}_p(Y, I)$ and $t \in (0, 1)$. For $p \ge 2$, we have established in Lemma (5.4) that for all $\eta \in \mathbb{R}$

$$F^{(p)}_{tX+(1-t)Z}(\eta) \le tF^{(p)}_X(\eta) + (1-t)F^{(p)}_Z(\eta).$$

Since $F^{(p)}_X(\eta) \le F^{(p)}_Y(\eta)$ and $F^{(p)}_Z(\eta) \le F^{(p)}_Y(\eta)$ for all $\eta \in I$, we obtain

$$F^{(p)}_{tX+(1-t)Z}(\eta) \le F^{(p)}_Y(\eta) \quad \text{for all } \eta \in I.$$

Hence, $tX + (1-t)Z \in \mathcal{A}_p(Y, I)$, as claimed.

Observe that for any $\eta \in I$, the operator $Z \mapsto F^{(p)}_Z(\eta)$ is continuous in $\mathcal{L}_{p-1}(\Omega, \mathcal{F}, P)$ due to the continuity of the mappings $z \to (\eta - z)_+^{p-1}$ and $Z \to \mathbb{E}[Z]$. Hence, the closed-graph property follows. □

Most interesting, from the point of view of applications, are the sets $\mathcal{A}_1(Y)$ and $\mathcal{A}_2(Y)$, defined by the full first- and second-order stochastic dominance relations. To study their relations in the space $\mathcal{L}_1(\Omega, \mathcal{F}, P)$, we introduce the set

$$\mathcal{A}_1^\sharp(Y) = \mathcal{A}_1(Y) \cap \mathcal{L}_1(\Omega, \mathcal{F}, P).$$

In view of (5.49), we have $\mathcal{A}_1^\sharp(Y) \subset \mathcal{A}_2(Y)$, and since the set $\mathcal{A}_2(Y)$ is convex and closed, we also have

$$\overline{\operatorname{conv}}\big(\mathcal{A}_1^\sharp(Y)\big) \subseteq \mathcal{A}_2(Y). \tag{5.50}$$

Here the symbol $\overline{\text{conv}}(A)$ denotes the closure of the convex hull of the set A. We are interested in the inverse inclusion.

First, we consider the special case of a finite probability space. We have the following result.

Theorem 5.28. *Assume that* $\Omega = \{\omega_1, \ldots, \omega_n\}$ *and* $P(\omega_k) = 1/n$, $k = 1, \ldots, n$. *Then*

$$\text{conv}\left(\mathbb{A}_1(Y)\right) = \mathbb{A}_2(Y).$$

Proof. We can represent every random variable Z on Ω as a vector $z \in \mathbb{R}^n$ by setting $z_i = Z(\omega_i)$, $i = 1, \ldots, n$. Suppose $X \in \mathbb{A}_2(Y)$. Since the probabilities of all elementary events are equal, the second-order stochastic dominance relation coincides with the concept of *weak majorization*:

$$\left[X \succeq_{(2)} Y\right] \Longleftrightarrow \left[\sum_{k=1}^{l} x_{[k]} \geq \sum_{k=1}^{l} y_{[k]}, \quad l = 1, \ldots, n\right],$$

where $x_{[k]}$ denotes the kth smallest component of x.

It follows from the theorem by Hardy, Littlewood, and Polya that weak majorization is equivalent to the existence of a doubly stochastic matrix Π such that $x \geq \Pi y$. Applying Birkhoff's Theorem, we infer that permutation matrices $Q^1, \ldots, Q^M$ exist such that $\Pi \in \text{conv}\{Q^1, \ldots, Q^M\}$. Defining vectors $z^j = Q^j y$, we conclude that nonnegative reals $\alpha_1, \ldots, \alpha_M$ totaling 1 exist, such that $x \geq \sum_{j=1}^{M} \alpha_j z^j$. Identifying with the vectors z^j random variables Z^j on $(\Omega, \mathcal{F}, P)$ we also see that

$$X(\omega) \geq \sum_{j=1}^{M} \alpha_j Z^j(\omega),$$

for all $\omega \in \Omega$. The distribution of Z^j is identical to the distribution of Y because each vector z^j is a permutation of the vector y and all probabilities are equal. Thus $Z^j \succeq_{(1)} Y, \quad j = 1, \ldots, M$. Let us define

$$\hat{Z}^j(\omega) = Z^j(\omega) + \left(X(\omega) - \sum_{k=1}^{M} \alpha_k Z^k(\omega)\right), \quad \omega \in \Omega, \quad j = 1, \ldots, M.$$

Then the last two inequalities render $\hat{Z}^j \in \mathbb{A}_1(Y)$, $j = 1, \ldots, M$, and $X(\omega) = \sum_{j=1}^{M} \alpha_j \hat{Z}^j(\omega)$, as required. □

The result of the theorem is not true for general probability spaces, as the following example illustrates.

Example 5.29. Suppose $\Omega = \{\omega_1, \omega_2\}$, $P(\omega_1) = 1/3$, $P(\omega_2) = 2/3$ and $Y(\omega_1) = -1$, $Y(\omega_2) = 1$. Then it is easy to see that $X \succeq_{(1)} Y$ if and only if $X(\omega_1) \geq -1$ and $X(\omega_2) \geq 1$. Thus $\mathcal{A}_1(Y)$ is convex.

Consider the random variable $Z = \mathbb{E}[Y] = 1/3$. It dominates Y in the second order, but it does not belong to $\operatorname{conv}\big(\mathcal{A}_1(Y)\big) = \mathcal{A}_1(Y)$. □

This example shows that the probability space must be sufficiently rich for the convexification phenomenon to occur. If we could define a new probability space $\Omega' = \{\omega_1, \omega_{21}, \omega_{22}\}$, in which the event ω_2 is split in two equally likely events ω_{21}, ω_{22}, then we could use Theorem 5.28 to obtain the equality $\operatorname{conv}\big(\mathcal{A}_1(Y)\big) = \mathcal{A}_2(Y)$. However, in the context of stochastic optimization, the probability space is fixed at the outset and we have to work with sets of random variables as elements of the functional space $\mathcal{L}_p(\Omega, \mathcal{F}, P; \mathbb{R}^n)$, rather than with sets of distributions. Therefore, the change of the probability space is excluded.

Theorem 5.30. *If $(\Omega, \mathcal{F}, P)$ is a standard atomless probability space, then*

$$\mathcal{A}_2(Y) = \overline{\operatorname{conv}}\big(\mathcal{A}_1^{\#}(Y)\big).$$

Proof. Suppose $X \succeq_{(2)} Y$. As the probability space is atomless, a uniform random variable U exists, such that $X = F_X^{(-1)}(U)$. Define $\Omega_i = \big\{i/N < U \leq (i+1)/N\big\}$, $i = 0, \dots, N-1$. By construction, $P(\Omega_i) = 1/N$. Let $\mathcal{B}$ be the finite σ-algebra generated by this partition. Consider the discrete random variable $\hat{X}_\varepsilon = \mathbb{E}\big[X \mid \mathcal{B}\big]$. Jensen's inequality implies that

$$\mathbb{E}\big[\big(\eta - \mathbb{E}\big[X \mid \mathcal{B}\big]\big)_+\big] \leq \mathbb{E}[(\eta - X)_+].$$

Hence $\hat{X}_\varepsilon$ dominates X in the second order, and thus $\hat{X}_\varepsilon \succeq_{(2)} Y$. Due to the integrability of X, for every $\varepsilon > 0$, we can choose N such that $\|\hat{X}_\varepsilon - X\| \leq \varepsilon$.

The function $F^{(2)}_{\hat{X}_\varepsilon}(\cdot)$ is convex and piecewise linear. Let $\delta > 0$ be its first nonzero slope.

Now, we define the random variable $\tilde{Y} = F_Y^{-1}(U)$. It has the same distribution as Y, which implies that $\hat{X} \succeq_{(2)} X \succeq_{(2)} \tilde{Y}$. We partition Ω into MN subsets such that

$$\Omega'_{ij} = \Big\{\frac{i}{N} + \frac{j}{MN} < U \leq \frac{i}{N} + \frac{j+1}{MN}\Big\},\qquad i = 0, \dots, N-1,\ j = 0, \dots, M-1, \tag{5.51}$$

and define the σ-subalgebra $\mathcal{B}'$ generated by this partition. Let $\hat{Y}_\varepsilon = \mathbb{E}\big[\tilde{Y} \mid \mathcal{B}'\big]$. Again, we can choose M large enough, so that $\|\hat{Y}_\varepsilon - \tilde{Y}\| \leq \varepsilon\delta$. Using the subadditivity of the shortfall function, we obtain

$$\mathbb{E}[(\eta-\tilde{Y})_+] \le \mathbb{E}[(\eta-\hat{Y}_\varepsilon)_+] + \mathbb{E}[(\hat{Y}_\varepsilon-\tilde{Y})_+] \le \mathbb{E}[(\eta-\hat{Y}_\varepsilon)_+] + \mathbb{E}[|\hat{Y}_\varepsilon-\tilde{Y}|].$$

It follows that

$$F^{(2)}_{\hat{Y}_\varepsilon}(\eta) = \mathbb{E}[(\eta-\hat{Y}_\varepsilon)_+] \ge \mathbb{E}[(\eta-\tilde{Y})_+] - \mathbb{E}[|\hat{Y}_\varepsilon-\tilde{Y}|] \ge F^{(2)}_Y(\eta) - \varepsilon\delta.$$

Having in mind that $\hat{X}_\varepsilon \succeq_{(2)} Y$, we can continue the last chain of inequalities as follows

$$F^{(2)}_{\hat{Y}_\varepsilon}(\eta) \ge F^{(2)}_{\hat{X}_\varepsilon}(\eta) - \varepsilon\delta, \quad \text{for all } \eta \in \mathbb{R}.$$

As $F^{(2)}_{\hat{Y}_\varepsilon}(\cdot)$ is nonnegative and the smallest nonzero slope of $F^{(2)}_{\hat{X}_\varepsilon}(\cdot)$ is δ, shifting $F^{(2)}_{\hat{X}_\varepsilon}(\cdot)$ to the right by ε yields a lower bound for $F^{(2)}_{\hat{Y}_\varepsilon}(\cdot)$. Thus the last inequality implies that

$$F^{(2)}_{\hat{Y}_\varepsilon}(\eta) \ge F^{(2)}_{\hat{X}_\varepsilon}(\eta-\varepsilon) \quad \text{for all} \quad \eta \in \mathbb{R}.$$

This is quivalent to,

$$\hat{X}_\varepsilon + \varepsilon \succeq_{(2)} \hat{Y}_\varepsilon.$$

We interpret the random variables $\hat{X}_\varepsilon$ and $\hat{Y}_\varepsilon$ as defined on a finite probability space with MN equally probable elementary events (5.51). By virtue of Theorem 5.28, we obtain

$$\hat{X}_\varepsilon + \varepsilon \in \operatorname{conv}\big(A_1^\sharp(\hat{Y}_\varepsilon)\big).$$

When $\varepsilon \to 0$, then $\hat{X}_\varepsilon \to X$, and $\hat{Y}_\varepsilon \to \tilde{Y}$, in the space $\mathcal{L}_1(\Omega, \mathcal{F}.P)$. Using Lemma 5.27, we obtain the result. □

We know that the set $\mathbb{A}_2(Y)$ can also be described as follows:

$$\mathbb{A}_2(Y) = \big\{X \in \mathcal{L}_1(\Omega, \mathcal{F}, P) : F^{(-2)}_X(p) \ge F^{(-2)}_Y(p)\ \forall\ p \in (0,1]\big\}.$$

However, no direct correspondence can be established between the set

$$\mathbb{A}_2(Y,[a,b]) = \big\{X \in \mathcal{L}_1(\Omega, \mathcal{F}, P) : F^{(2)}_X(\eta) \le F^{(2)}_Y(\eta)\ \forall\ \eta \in [a,b]\big\}$$

and the set

$$\mathbb{B}(Y,[\alpha,\beta]) = \big\{X \in \mathcal{L}_1(\Omega, \mathcal{F}, P) : F^{(-2)}_X(p) \ge F^{(-2)}_Y(p)\ \forall\ p \in [\alpha,\beta]\big\}.$$

We can demonstrate the following one-sided relation.

Lemma 5.31. *If $[a,b]$ contains all p-quantiles of Y for $p \in [\alpha,\beta]$, then*

$$\mathbb{A}_2(Y,[a,b]) \subset \mathbb{B}(Y,[\alpha,\beta]).$$

Proof. Suppose $p \in [\alpha, \beta]$ and $X \in \mathcal{A}_2(Y, [a, b])$. Using Theorem 5.9,

$$\begin{aligned} F_X^{(-2)}(p) &= \sup_{\eta \in \mathbb{R}} \left\{ \eta p - F_X^{(2)}(\eta) \right\} \\ &\geq \sup_{a \leq \eta \leq b} \left\{ \eta p - F_X^{(2)}(\eta) \right\} \geq \sup_{a \leq \eta \leq b} \left\{ \eta p - F_Y^{(2)}(\eta) \right\}. \end{aligned} \tag{5.52}$$

Since $\sup_{\eta \in \mathbb{R}} \left\{ \eta p - F_Y^{(2)}(\eta) \right\}$ is attained at a p-quantile of Y (see (5.11)) and we have assumed $p \in [\alpha, \beta]$, we can continue inequality (5.52) as follows:

$$\sup_{a \leq \eta \leq b} \left\{ \eta p - F_Y^{(2)}(\eta) \right\} = \sup_{\eta \in \mathbb{R}} \left\{ \eta p - F_Y^{(2)}(\eta) \right\} = F_Y^{(-2)}(p).$$

Thus, $F_X^{(-2)}(p) \geq F_Y^{(-2)}(p)$ for all $p \in [\alpha, \beta]$, and $X \in B(Y, [\alpha, \beta])$. □

The following lemma establishes the properties of the set $B(Y, [\alpha, \beta])$.

Lemma 5.32. (a) *For every $p \in (0, 1)$, the mapping $X \mapsto F_X^{(-2)}(p)$ is continuous, concave and positively homogeneous on $\mathcal{L}_1(\Omega, \mathcal{F}, P)$.*

(b) *For any interval $J \subseteq (0, 1]$, the multifunction $Y \to \mathbb{B}(Y, J)$ is convex-valued and has a closed graph in $\mathcal{L}_1(\Omega, \mathcal{F}, P)$.*

Proof. Since the mapping $X \mapsto F_X^{(-2)}(p)$ is identical to $X \mapsto -p\,\mathrm{AVaR}_p^-(X)$ defined in Remark 2.20, the concavity and the positive homogeneity follow from the coherence of $\mathrm{AVaR}_p^-(\cdot)$. If $X, Z \in \mathbb{B}\big(Y, J\big)$, then the concavity of $X \mapsto F_X^{(-2)}(p)$ implies that $tX + (1-t)Z \in \mathbb{B}\big(Y[\alpha, \beta]\big)$ entailing the convexity of the set $\mathbb{B}\big(Y, J\big)$. The continuity of $X \mapsto \mathrm{AVaR}_p^-(X)$ implies the continuity of the mapping $X \mapsto F_X^{(-2)}(p)$ in $\mathcal{L}_1(\Omega, \mathcal{F}, P)$, which in turn implies that the graph of $Y \to \mathbb{B}(Y, J)$ is closed. □

5.2 Optimization Problems and Optimality Conditions

One way to control risk is to introduce constraints on the probability distributions of the relevant outcomes. Frequently, a decision maker can point to a benchmark random outcome Y with an acceptable distribution. Hence, it makes sense to control risk by requiring that the random outcome of interest is ''better'' than the benchmark in some sense. Let a benchmark random outcome $Y \in \mathcal{L}_p(\Omega, \mathcal{F}, P)$ with $p \geq 1$ be given. The decision space $\mathcal{X}$ is assumed to be a separable Banach space. We shall analyze the following optimization problem:

$$\begin{aligned} \min\ & f(x) \\ \text{s.t. } & G(x) \succeq Y, \\ & x \in \mathcal{X}. \end{aligned} \tag{5.53}$$

Here $\mathcal{X}$ is a closed convex subset of the space $\mathfrak{X}$. We also assume that $f : \mathfrak{X} \to \mathbb{R}$ is a continuous functional, and the relation $\succeq$ is a suitably chosen stochastic order. The mapping G assigns to each decision $x \in \mathcal{X}$ an integrable random variable with a finite p moment, i.e., for each $x \in \mathcal{X}$, $G(x) \in \mathcal{L}_p(\Omega, \mathcal{F}, P)$ holds.

We call the constraint $G(x) \succeq Y$ a stochastic-order (or ordering) constraint or stochastic dominance constraint. Our analysis will provide insight into the ways this constraint reflects the risk-aversion of the decision-maker.

If we use the relation $\succeq_{(p)}$ in the stochastic-order constraint, then the relation $G(x) \succeq_{(p)} Y$ is equivalent to $G(x) \in \mathcal{A}_p(Y)$. As we have shown, the sets $\mathcal{A}_p(Y)$ are convex and closed when $p \geq 2$. We shall analyze the problem under convexity assumptions for the functionals involved, as well as for non-convex but Fréchet-differentiable functionals. We shall provide separate analysis for the cases when $G(x) \in \mathcal{A}_1(Y)$.

Given a bounded interval $[a, b] \subset \mathbb{R}$, we consider the formulation of (5.53) with stochastic dominance relation of order two or higher in the interval $[a, b]$. The optimization problem reads as follows.

$$\min f(x) \tag{5.54}$$

$$\text{subject to } \mathbb{E}\big[\big(\max(0, \eta - G(x))\big)^p\big] \leq \mathbb{E}\big[\big(\max(0, \eta - Y)\big)^p\big] \quad \forall \eta \in [a, b], \tag{5.55}$$

$$x \in \mathcal{X}. \tag{5.56}$$

In this formulation, $[a, b]$ is always a bounded interval.

Example 5.33 (Portfolio Optimization). We consider investing in securities $j = 1, \ldots, n$ with random return rates R_j. We have a benchmark return rate Y, which may be the return rate of an index, an existing portfolio, or other. Our decision variables x_j, $j = 1, \ldots, n$, indicate the fraction of capital invested in the jth security and let $\mathcal{X}$ be a set with some investment constraints. The return rate of our portfolio is $R(x) = \sum_{j=1}^{n} x_j R_j = x^\top R$. A risk-averse portfolio optimization problem can be formulated as follows.

$$\begin{aligned} \max\ & \mathbb{E}\big[x^\top R\big] \\ \text{s.t. } & x^\top R \succeq_{(2)} Y \\ & x \in \mathcal{X}. \end{aligned} \tag{5.57}$$

We have shown that $x^\top R \succeq_{(2)} Y$ is equivalent to any of the statements below:

(a) For all $\eta \in \mathbb{R}$, $\mathbb{E}\big[(\eta - x^\top R)_+\big] \le \mathbb{E}\big[(\eta - Y)_+\big]$, which can be interpreted as a continuum of shortfall constraints for every target level $\eta \in \mathbb{R}$.

(b) For all $p \in (0,1)$, $F^{(-2)}_{x^\top R}(p) \ge F^{(-2)}_Y(p)$ must hold, which is equivalent to a continuum of Average Value-at-Risk constraints for every risk level $p \in [0,1]$.

(c) It holds $\mathbb{E}\big[u(x^\top R)\big] \ge \mathbb{E}[u(Y)]$ for all concave nondecreasing functions $u : \mathbb{R} \to \mathbb{R}$ (von Neuman-Morgenstern utility functions) such that $\mathbb{E}[u(x^\top R)]$ is finite.

(d) It holds $\int_0^1 F^{(-1)}_{x^\top R}(p)\,\mathrm{d}w(p) \ge \int_0^1 F^{(-1)}_Y(p)\,\mathrm{d}w(p)$ for all concave nondecreasing functions $w : [0,1] \to \mathbb{R}$ (rank dependent utility functions) for which the integrals are well-defined. □

The last two statements show that this approach facilitates group decisions for risk-averse decision makers. All decision makers who use utility functions from the set $\mathcal{U}_2$ or dual utility functions in $\mathcal{W}_2$ would prefer the optimal portfolio of problem (5.57) over the benchmark. In many practical situations, it is easier to point to a benchmark outcome with an acceptable distribution rather than to choose a suitable utility function or a distortion function. Furthermore, the dominance constraint may be restricted to portions of the distributions, as we shall see next. It allows for risk-shaping by a fine and more precise control of the distributions of the random quantity of interest; in the example, this quantity is the return rate of the optimized portfolio.

5.2.1 The Convex Case

Constraint (5.55) is equivalent to $G(x) \in \mathcal{A}_{p+1}\big(Y, [a,b]\big)$, that is, $G(x)$ is stochastically larger than Y on the interval $[a,b]$ with respect to stochastic dominance of order $p+1 \ge 2$. We adopt the following general assumptions.

Assumption 5.34.

(i) *The function* $f(\cdot)$ *is convex and continuous.*

(ii) *The operator* $G : \mathcal{X} \to \mathcal{L}_p(\Omega, \mathcal{F}, P)$ *is norm-to-norm continuous and concave in the sense of Definition* 3.1.

(iii) *The set* X *is a closed a convex subset of* $\mathcal{X}$.

The following constraint qualification condition will be used.

Definition 5.35. Problem (5.54)–(5.56) satisfies the *uniform dominance condition* if a point $\tilde{x} \in X$ exists such that

$$\sup_{\eta \in [a,b]} \mathbb{E}\big[\max(0, \eta - G(\tilde{x}))^p - \max(0, \eta - Y)^p\big] < 0.$$

We introduce the utility-modified functional $L_{p+1} : \mathbb{R}^n \times \mathcal{U}_{p+1}([a,b]) \to \mathbb{R}$ associated with problem (5.54)–(5.56), defined as follows:

$$L_{p+1}(x,u) = f(x) + \mathbb{E}\big[u(Y) - u(G(x))\big].$$

Theorem 5.36. *Suppose Assumption* 5.34 *and the uniform dominance condition are satisfied for problem* (5.54)–(5.56). *If $\hat{x}$ is an optimal solution to that problem, then a utility function $\hat{u} \in \mathcal{U}_{p+1}([a,b])$ exists such that*

$$L_{p+1}(\hat{x},\hat{u}) = \min_{x\in\mathcal{X}} L_{p+1}(x,\hat{u}), \tag{5.58}$$

$$\mathbb{E}[\hat{u}(G(\hat{x}))] = \mathbb{E}[\hat{u}(Y)]. \tag{5.59}$$

Conversely, if for some function $\hat{u} \in \mathcal{U}_{p+1}([a,b])$ an optimal solution $\hat{x}$ of (5.58) *satisfies* (5.55) *and* (5.59), *then $\hat{x}$ is an optimal solution of problem* (5.54)–(5.56).

Proof. We consider the operator $\mathcal{G} : \mathcal{L}_p(\Omega,\mathcal{F},P) \to \mathcal{C}([a,b])$ defined as

$$[\mathcal{G}(Z)](\eta) = \mathbb{E}[(\max(0,\eta - Z))^p - \max(0,\eta - Y)^p], \quad \eta \in [a,b]. \tag{5.60}$$

We shall show that the composition $\mathcal{G}(G(\cdot))$ is a continuous operator with respect to the supremum norm in $\mathcal{C}([a,b])$. For all $\eta \in [a,b]$, the mapping

$$(\eta, Z) \mapsto \mathbb{E}\big[\big(\max(0,\eta - Z)\big)^p - \big(\max(0,\eta - Y)\big)^p\big]$$

is continuous on $[a,b] \times \mathcal{L}_p(\Omega,\mathcal{F},P)$ when $p \geq 1$. Hence, the mapping

$$Z \mapsto \sup_{\eta\in[a,b]} \mathbb{E}\big[\big(\max(0,\eta - Z)\big)^p - \big(\max(0,\eta - Y)\big)^p\big]$$

is continuous on $\mathcal{L}_p(\Omega,\mathcal{F},P)$. Having in mind the continuity assumption on $G(\cdot)$, we infer that the composition $\mathcal{G}(G(\cdot))$ is norm-to-norm continuous. Let K be the cone of non-positive functions in $\mathcal{C}([a,b])$. Problem (5.54)–(5.56) can be reformulated as follows:

$$\min f(x) \ \text{ subject to } \ \mathcal{G}(G(x)) \in K, \ x \in \mathcal{X}. \tag{5.61}$$

For every $\eta \in [a,b]$ the function $x \mapsto \eta - G(x)$ is convex, for almost all $\omega \in \Omega$, and the function $t \mapsto \max(0,t)$ is convex and nondecreasing. Therefore, the composition $\mathbb{E}\big[\max(0,\eta - G(x))^p\big]$ is a convex function of x for $p \geq 1$. It follows that the operator $\mathcal{G} \circ G$ is convex with respect to the cone K, that is, for any $x^1, x^2 \in \mathcal{X}$ and all $\lambda \in [0,1]$,

$$(\mathcal{G}\circ G)(\lambda x^1 + (1-\lambda)x^2) - \big[\lambda(\mathcal{G}\circ G)(x^1) + (1-\lambda)(\mathcal{G}\circ G)(x^2)\big] \in K.$$

The dual space to $\mathcal{C}([a,b])$ is the space $\mathcal{M}^+([a,b])$ of regular finite countably additive measures. The Lagrangian associated with problem (5.61) is $\mathfrak{L} : \mathfrak{X} \times \mathcal{M}^+([a,b]) \to \mathbb{R}$,

$$\mathfrak{L}(x,\nu) = f(x) + \int_{[a,b]} [\mathcal{G} \circ G(x)](\eta)\ \nu(\mathrm{d}\eta). \tag{5.62}$$

The uniform dominance condition is equivalent to the following generalized Slater condition: a point $\tilde{x} \in \mathcal{X}$ exists such that $(\mathcal{G} \circ G)(\tilde{x}) \in \operatorname{int} K$. The necessary conditions of optimality in Banach spaces state the existence of a nonnegative measure $\hat{\nu} \in \mathcal{M}^+([a,b])$ such that

$$\mathfrak{L}(\hat{x},\hat{\nu}) = \min_{x \in \mathcal{X}} \mathfrak{L}(x,\hat{\nu}), \tag{5.63}$$

$$\int_{[a,b]} \mathbb{E}\big[(\eta - G(\hat{x}))_+^p - (\eta - Y)_+^p\big]\ \hat{\nu}(\mathrm{d}\eta) = 0. \tag{5.64}$$

Using the description of the generator and the key relation (5.27), we associate a function $u_\nu : \mathbb{R} \to \mathbb{R}$ in $\mathcal{U}_{p+1}([a,b])$ with the measure ν, and we obtain that for any random variable Z

$$\int_{[a,b]} \mathbb{E}\big[\max(0,\eta - Z)^p\big]\,\nu(\mathrm{d}\eta) = -\mathbb{E}\big[u_\nu(Z)\big].$$

Therefore, the Lagrangian $\mathfrak{L}(x,\nu)$ can be represented as

$$\mathfrak{L}(x,\nu) = f(x) - \mathbb{E}\big[u_\nu(G(x))\big] + \mathbb{E}\big[u_\nu(Y)\big] = L_{p+1}(x,u_\nu).$$

Setting $\hat{u} = u_{\hat{\nu}}$, we obtain (5.58). The complementarity condition (5.64) takes on the form (5.59) by the same argument.

The converse statement follows from the convexity of problem (5.54)–(5.56). If the point $\hat{x}$ satisfies conditions (5.58), (5.59) and the dominance constraint, then $\hat{x}$ is an optimal solution of that optimization problem. □

Theorem 5.36 shows that the optimization problem with stochastic dominance constraint is equivalent to an optimization problem with an expected utility function. However, the utility function is unknown and is implicitly defined by the benchmark.

For every function $u \in \mathcal{U}_{p+1}([a,b])$ with $p \geq 1$, the problem

$$\min_{x \in \mathcal{X}} \Big\{ f(x) - \mathbb{E}\big[u(G(x))\big] + \mathbb{E}\big[u(Y)\big] \Big\} \tag{5.65}$$

is a relaxation of problem (5.54)–(5.56), and its optimal value

$$L_D(u) = \inf_{x \in \mathcal{X}} L_{p+1}(x, u)$$

is always smaller than or equal to the optimal value of (5.54)–(5.56). Indeed, any feasible solution x of (5.54)–(5.56) is also feasible for the relation (5.65), and the dominance relation (5.55) implies that $\mathbb{E}\big[u(G(x))\big] \geq \mathbb{E}\big[u(Y)\big]$.

We define the dual problem to problem (5.54)–(5.56) as follows

$$\max_{u \in \mathcal{U}_{p+1}([a,b])} L_D(u). \tag{5.66}$$

The set $\mathcal{U}_{p+1}([a,b])$ is a closed convex cone in $\mathcal{C}([a,b])$ and $L_D(\cdot)$ is a concave functional, so (5.66) is a convex optimization problem. Employing general duality relations in convex optimization, we obtain the following duality result.

Theorem 5.37. *Assume that the uniform dominance condition is satisfied and problem* (5.54)–(5.56) *has an optimal solution. Then problem* (5.66) *has an optimal solution and the optimal values of both problems coincide. Furthermore, the set of optimal solutions of* (5.66) *is the set of functions* $\hat{u} \in \mathcal{U}_{p+1}([a,b])$ *satisfying* (5.58)–(5.59) *at an optimal solution* $\hat{x}$ *of* (5.54)–(5.56).

Finally, we obtain the following properties of the set of optimal solutions.

Corollary 5.38. *Under the assumptions of Theorem* 5.36, *the set* $\hat{U}$ *of functions in* $\mathcal{U}_{p+1}([a,b])$ *satisfying* (5.58)–(5.59) *is convex, bounded, and weakly* closed in the following sense: if a sequence of functions* $u^k \in \hat{U}$ *and* $u^k \in \mathcal{U}_{p+1}([a,b])$ *are such that*

$$\lim_{k\to\infty} \mathbb{E}[u^k(Z)] = \mathbb{E}[u(Z)] \quad \textit{for all } Z \in \mathcal{L}_p(\Omega, \mathcal{F}, P),$$

then $u \in \hat{U}$. *The set* $\hat{U}$ *is the same for all optimal solutions of* (5.54)–(5.56).

Proof. Indeed, under the assumptions of Theorem 5.36, the set of Lagrange multipliers $\hat{\mu}$ corresponding to the Lagrangian $\mathfrak{L}$ is convex, bounded, and weakly* closed in the dual space $\mathcal{M}^+([a,b])$. Moreover, the same set of Lagrange multipliers corresponds to every optimal solution of problem (5.61). Using the key relation (5.27), we obtain the claim. □

We now focus on deriving optimality conditions in a differential form. Consider the multifunction $DF : \mathcal{L}_1(\Omega, \mathcal{F}, P) \times [a,b] \times \Omega \rightrightarrows \mathbb{R}$ defined as follows:

$$DF(Z, \eta, \omega) = \begin{cases} \{-1\} & \text{if } Z(\omega) < \eta, \\ [-1, 0] & \text{if } Z(\omega) = \eta, \\ \{0\} & \text{if } Z(\omega) > \eta. \end{cases} \tag{5.67}$$

The differentiability properties of the functional $Z \mapsto F_Z^{(p+1)}(\eta)$ are summarized in the following statement.

Lemma 5.39. *The mapping $Z \to F_Z^{(p+1)}(\eta)$ for $p \geq 1$ is subdifferentiable at every $Z \in \mathcal{L}_p(\Omega, \mathcal{F}, P)$ and its subdifferential is described as follows:*

$$\partial_Z F_Z^{(p+1)}(\eta) = \begin{cases} \{\xi \in \mathcal{L}_\infty(\Omega, \mathcal{F}, P) : \xi(\cdot) \in DF(Z, \eta, \cdot)\} & \text{if } p = 1, \\ \{-\frac{1}{\Gamma(p)} \max(0, \eta - Z)^{p-1}\} & \text{if } p > 1. \end{cases}$$

Proof. For a fixed η and $p \geq 1$, the mapping $Z \to F_Z^{(p+1)}(\eta)$ is a convex functional on $\mathcal{L}_p(\Omega, \mathcal{F}, P)$, as shown in Lemma 5.4. Furthermore, it is an integral functional with the normal convex integrand $t \mapsto \max(0, \eta - t)^p$. For $p = 1$, Strassen's Theorem 3.13 implies that the subdifferential of $F_Z^{(2)}(\cdot)$ at a point $Z \in \mathcal{L}_1(\Omega, \mathcal{F}, P)$ is the collection of $\xi(\omega)$ of all measurable selections $\xi(\cdot)$ of $DF(Z, \eta, \cdot)$. Note that each $\xi \in \mathcal{L}_\infty(\Omega, \mathcal{F}, P)$. For $p > 1$, the integrand is differentiable with derivative $-p \max(0, \eta - t)^{p-1}$. Hence, for $p > 1$,

$$\partial_Z F_Z^{(p+1)}(\eta) = -\frac{p}{\Gamma(p+1)} \max(0, \eta - Z)^{p-1} = -\frac{1}{\Gamma(p)} \max(0, \eta - Z)^{p-1}.$$

Since $Z \in \mathcal{L}_p(\Omega, \mathcal{F}, P)$, we infer that $\partial_Z F_Z^{(p+1)}(\eta) \subset \mathcal{L}_{\frac{p}{p-1}}(\Omega, \mathcal{F}, P)$. □

Recall that the normal cone to the convex set $\mathcal{X} \subset \mathfrak{X}$ at a point $x \in \mathcal{X}$ is given by the formula:

$$\mathcal{N}_{\mathcal{X}}(x) = \{d \in \mathfrak{X}^* : \langle d, y - x \rangle \leq 0 \quad \forall y \in \mathcal{X}\}.$$

Here, $\mathfrak{X}^*$ is the topological dual to $\mathfrak{X}$.

Assumption 5.40. *We assume that $G : \mathcal{X} \to \mathcal{L}_{p-1}(\Omega, \mathcal{F}, P)$ is defined as*

$$[G(x)](\omega) = g(x, \omega),$$

where $g : \mathfrak{X} \times \Omega \to \mathbb{R}$ is a Carathéodory function and $g(\cdot, \omega)$ is concave for all $\omega \in \Omega$.

For a fixed ω, the subdifferential $\partial g(\hat{x}, \omega)$ at $\hat{x}$ is as follows

$$\partial g(\hat{x}, \omega) = \{s \in \mathfrak{X}^* : g(x, \omega) \geq g(\hat{x}, \omega) + \langle s(\omega), x - \hat{x} \rangle \ \forall x \in \mathfrak{X}\}.$$

We use Definition 3.2 of a subgradient of a lattice-valued convex operator.

Theorem 5.41. *Under the assumptions of Theorem* 5.36 *and Assumption* 5.40, *relation* (5.58) *is equivalent to the existence of a subgradient $S \in \partial[G(\hat{x})]$ such that*

$$\mathbb{E}\big[S^*\hat{u}'(G(\hat{x}))\big] \in \partial_x f(\hat{x}) + \mathcal{N}_{\mathcal{X}}(\hat{x}), \tag{5.68}$$

where $\hat{u} \in \mathcal{U}_{p+1}([a,b])$ *is the function appearing in* (5.58), *and* $\hat{u}'(\cdot)$ *denotes a subgradient of it.*

Suppose additionally that either $p > 1$ *or* $G(\hat{x})$ *has a continuous distribution, then* (5.58) *is equivalent to the existence of a subgradient* $S \in \partial[G(\hat{x})]$ *and a utility function* $\tilde{u} \in \mathcal{U}_p([a,b])$ *such that*

$$\mathbb{E}\big[S^*\tilde{u}(G(\hat{x}))\big] \in \partial_x f(\hat{x}) + \mathcal{N}_{\mathcal{X}}(\hat{x}).$$

Proof. Theorem 3.11 implies that the operator $G : \mathcal{X} \to \mathcal{L}_p(\Omega, \mathcal{F}, P)$ is subdifferentiable at $\hat{x}$ and its subdifferential is given by

$$\partial[G(\hat{x})] = \big\{S : \mathcal{X} \to \mathcal{L}_p(\Omega, \mathcal{F}, P) : [Sd](\omega) = \langle s(\omega), d\rangle \\ \text{for some } s \in \mathcal{L}_p(\Omega, \mathcal{F}, P; \mathcal{X}^*);\ s(\omega) \in \partial g(\hat{x}, \omega)\big\}.$$

The Lagrangian $\mathfrak{L}$ defined in (5.62) is a convex functional of the first argument and we can calculate its subdifferential. For $p = 1$, we have

$$\begin{aligned}\partial_x\big[\mathcal{G}(G(\hat{x}))\big](\eta) &= \big\{-\mathbb{E}[S\xi] : S \in \partial G(\hat{x}),\ \xi \in \partial_Z F^{(2)}_{G(x)}(\eta)\big\}\\ &= \big\{-\mathbb{E}[s(\omega)\xi(\hat{x}, \eta, \omega)] : s(\omega) \in \partial g(\hat{x}, \omega),\\ &\qquad \xi(\hat{x}, \eta, \omega) \in DF(G(\hat{x}), \eta, \omega)\big\}.\end{aligned}$$

The following chain of equations holds:

$$\begin{aligned}\partial_x \int_a^b [\mathcal{G}(G(\hat{x}))](\eta)\ \hat{\nu}(\mathrm{d}\eta) &= \int_a^b \partial_x[\mathcal{G}(G(\hat{x}))](\eta)\ \hat{\nu}(\mathrm{d}\eta)\\ &= -\int_a^b \mathbb{E}[s(\omega)\xi(\hat{x}, \eta, \omega)]\ \hat{\nu}(\mathrm{d}\eta).\end{aligned}$$

The term $\mathbb{E}\big[s(\omega)\int_a^b \xi(\hat{x}, \eta, \omega)\ \hat{\nu}(\mathrm{d}\eta)\big]$ can be interpreted as $\mathbb{E}\big[S^*\hat{u}'_{\hat{\nu}}(G(\hat{x}))\big]$, which means that the first claim is satisfied for $p = 1$. Further, we observe that

$$\begin{aligned}\partial_x \int_a^b [\mathcal{G}(G(\hat{x}))](\eta)\ \hat{\nu}(\mathrm{d}\eta) &= -\int_a^b \mathbb{E}[s(\omega)\xi(\hat{x}, \eta, \omega)]\ \hat{\nu}(\mathrm{d}\eta)\\ &= -\mathbb{E}\Big[s(\omega)\int_a^b \mathbb{1}_{g(\hat{x},\omega)\le\eta}\ \hat{\nu}(\mathrm{d}\eta)\Big] = \mathbb{E}\big[s(\omega)\tilde{u}_{\hat{\nu}}\big(G(\hat{x})(\omega)\big)\big].\end{aligned}$$

The second equality holds by the assumption that $G(\hat{x})$ has continuous distribution and the last equality follows from the form of the generator given in (5.26) with $\tilde{u}_{\hat{\nu}}(\cdot) \in \mathcal{U}_1([a,b])$. Thus, condition (5.63) and the following two conditions are all equivalent:

$$0 \in \partial_x f(\hat{x}) - \mathbb{E}\big[s(\omega)\tilde{u}_{\hat{\nu}}\big(G(\hat{x})(\omega)\big)\big] + \mathcal{N}_{\mathcal{X}}(\hat{x});$$
$$0 \in \partial_x f(\hat{x}) - \mathbb{E}\big[S^*\tilde{u}_{\hat{\nu}}\big(G(\hat{x})\big)\big] + \mathcal{N}_{\mathcal{X}}(\hat{x}). \tag{5.69}$$

This proves the statement for $p = 1$.

For $p > 1$, following the same line of argument, we obtain the equation:

$$\partial_x \int_a^b [\mathcal{G}(\hat{x})](\eta)\, \hat{\nu}(\mathrm{d}\eta) = -p \int_a^b \mathbb{E}\,[s(\omega)(\eta - G(\hat{x})(\omega))_+^{p-1}]\, \hat{\nu}(\mathrm{d}\eta). \tag{5.70}$$

Again, we observe that also the following is true:

$$-\, p \int_a^b \mathbb{E}\,[s(\omega) \max(0, \eta - G(\hat{x})(\omega))^{p-1}]\, \hat{\nu}(\mathrm{d}\eta) = \mathbb{E}\big[S^*\hat{u}'(G(\hat{x})\big],$$

where $\hat{u}(\cdot)$ is chosen in the proof of Theorem 5.36. Now, we rescale the measure $\hat{\nu}$ as follows: $\tilde{\nu} = p\hat{\nu}$. Then we associate a utility function $\tilde{u}(\cdot) \in \mathcal{U}_p([a, b])$ with the measure $\tilde{\nu}$ according to (5.26). Using this function at the right-hand side of (5.70), we obtain

$$\begin{aligned}\partial_x \int_a^b [\mathcal{G}(\hat{x})](\eta)\, \hat{\nu}(\mathrm{d}\eta) &= -\mathbb{E}\big[s(\omega) \int_a^b (\eta - G(\hat{x})(\omega))_+^{p-1}\, \tilde{\nu}(\mathrm{d}\eta)\big] \\ &= \mathbb{E}\big[s(\omega)\tilde{u}\big(G(\hat{x})\big)\big],\end{aligned}$$

which completes the proof for $p > 1$. □

If the benchmark Y has a discrete distribution with finitely many realizations, then we can represent the stochastic dominance constraint by finitely many inequalities.

Lemma 5.42. *Assume that Y has a discrete distribution with realizations y_i, $i = 1, \dots, m$, where $a \le y_i \le b$ for all i. Then constraints* (5.55) *for $p = 1$ are equivalent to*

$$\mathbb{E}\,[(y_i - G(x))_+] \le \mathbb{E}\,[(y_i - Y)_+], \quad i = 1, \dots, m. \tag{5.71}$$

Moreover, in that case $\mathbb{E}\,[(\eta - G(x))_+] \le \mathbb{E}\,[(\eta - Y)_+]$ holds for all $\eta \in \mathbb{R}$.

Proof. Without loss of generality, we assume that $y_1 < y_2 < \cdots < y_m$. We shall prove that (5.71) imply that

$$F_{G(x)}^{(2)}(\eta) \le F_Y^{(2)}(\eta) \quad \text{for all } \eta \in \mathbb{R}.$$

We observe that the function $F_Y^{(2)}(\cdot)$ is convex, piecewise linear, and has break points at y_i, $i = 1, \dots, m$. Three cases, depending on the value of η are possible.

Case 1: If $\eta \le y_1$, then we have

$$0 \le F^{(2)}_{G(x)}(\eta) \le F^{(2)}_{G(x)}(y_1) \le F^{(2)}_Y(y_1) = 0.$$

Therefore the required inequality holds as an equation for such η.

Case 2: Let $\eta \in [y_i, y_{i+1}]$ for some i. Since, for any x, the function $F^{(2)}_{G(x)}(\cdot)$ is convex, inequalities (5.71) for i and $i+1$ imply that for all $\eta \in [y_i, y_{i+1}]$ the following holds

$$\begin{aligned} F^{(2)}_{G(x)}(\eta) &\le \lambda F^{(2)}_{G(x)}(y_i) + (1-\lambda) F^{(2)}_{G(x)}(y_{i+1}) \\ &\le \lambda F^{(2)}_Y(y_i) + (1-\lambda) F^{(2)}_Y(y_{i+1}) = F^{(2)}_Y(\eta), \end{aligned}$$

where $\lambda = (y_{i+1} - \eta)/(y_{i+1} - y_i)$.

Case 3: For $\eta > y_m$ we have

$$\begin{aligned} F^{(2)}_Y(\eta) &= F^{(2)}_Y(y_m) + \eta - y_m \\ &\ge F^{(2)}_{G(x)}(y_m) + \int_{y_m}^{\eta} F_{G(x)}(\alpha)\, d\alpha = F^{(2)}_{G(x)}(\eta), \end{aligned}$$

as required. □

If the probability space Ω is finite, i.e., $\Omega = \{\omega_1, \ldots, \omega_m\}$, then we can formulate (5.55) by using finitely many linear constraints at the expense of introducing additional decision variables. Let $p_k = P(\omega_k)$, $v_i = \mathbb{E}[(y_i - Y)_+]$, and $g_k(x) = g(x, \omega_k)$. We introduce the variables z_{ik} for i and k ranging from 1 to m. Then the following finite system of linear inequalities is equivalent to the stochastic dominance relation:

$$g_k(x) + z_{ik} \ge y_i, \quad i = 1, \ldots, m, \quad k = 1, \ldots, m, \tag{5.72}$$

$$\sum_{k=1}^{m} p_k z_{ik} \le v_i, \quad i = 1, \ldots, m, \tag{5.73}$$

$$z_{ik} \ge 0, \quad i = 1, \ldots, m, \quad k = 1, \ldots, m. \tag{5.74}$$

This formulation can be used for the numerical solution of dominance-constrained problems. However, it leads to a very large, possibly prohibitively large optimization problem, when m is large. In Chapter 7, we discuss more efficient numerical methods for solving optimization problems with constraints of this type.

Notice that the uniform dominance condition cannot be satisfied when $a \le y_1$ because $F^{(2)}_Y(y_1) = 0$. However, when *all* distributions are finite, then the functions $F^{(2)}_{G(x)}(y_i)$, $i = 1, \ldots, m$, which appear at the left-hand side of constraints (5.71), are convex polyhedral functions of the realizations $g^k(x)$,

$i = 1, \ldots, m$. Consequently, in the discrete case, Theorem 5.37 is true under the following constraint qualification condition: a point $\tilde{x} \in \operatorname{relint} \mathcal{X}$ exists such that

$$F^{(2)}_{G(\tilde{x})}(y_i) \le F^{(2)}_Y(y_i), \quad \text{for all} \quad y_i \in [a, b].$$

In this case, we may have an interval $[a, b]$ covering all possible realizations y_i of the reference outcome.

5.2.2 The Non-Convex Differentiable Case

In this section, we make the following general assumptions. The space $\mathfrak{X}$ is a separable Banach space, the probability space is $(\Omega, \mathcal{F}, P)$, and in the formulation of problem (5.54)–(5.56) the interval $[a, b]$ is bounded and $p \ge 1$.

Assumption 5.43.

(i) *The functional $f(\cdot)$ in the objective is Fréchet differentiable with derivative $f'(\cdot)$.*

(ii) *The operator $G(\cdot)$ has the form $\big[G(x)\big](\omega) = g(x, \omega)$, $x \in \mathfrak{X}$, $\omega \in \Omega$.*

(iii) *At every $x \in \mathfrak{X}$ the function $g(\cdot, \omega)$ is Fréchet differentiable for almost all ω, with the derivative $g'(x, \omega)$, and that $\|g'(x, \omega)\| \le \ell(\omega)$ with some Lipschitz constant $\ell \in \mathcal{L}_p(\Omega, \mathcal{F}, P)$.*

(iv) *The set $\mathcal{X}$ is closed and convex.*

The last assumption corresponds to Assumption 3.18. Under the conditions 5.43, it is established in (3.27) that the Fréchet derivative $G'(x)$ of $G(\cdot)$ has the form

$$\big[G'(x)\big](\omega) = g'(x, \omega) \quad a.s.$$

The set of feasible solutions for problem (5.54)–(5.56) is defined as:

$$\mathcal{A} = \big\{x \in \mathcal{X} : G(x) \in \mathcal{A}_{p+1}(Y, [a, b])\big\}.$$

We use again the operator $\mathcal{G}$ as defined in (5.60):

$$[\mathcal{G}(Z)](\eta) = \mathbb{E}[\max(0, \eta - Z)^p - \max(0, \eta - Y)^p], \quad \eta \in [a, b].$$

Given $x \in \mathcal{X}$, we define the set of active constraints:

$$I_0(x) = \big\{\eta \in [a, b] : \mathcal{G}\big(G(x)\big)(\eta) = 0\big\}.$$

We have observed that $\mathcal{G}(\cdot)(\eta)$ is convex and continuous with respect to the $\mathcal{L}_p$-norm. It is also subdifferentiable at any Z and the form of its subdifferential $\partial\mathcal{G}[Z] \subset \mathcal{L}_q(\Omega, \mathcal{F}, P)$ for $\frac{1}{p} + \frac{1}{q} = 1$ is given in Lemma 5.39.

The following constraint qualification condition will be used in the setting of this section, when Assumption 5.43 are adopted.

Definition 5.44. The set $\mathcal{A}$ satisfies the *differential constraint qualification condition* at the point $\tilde{x} \in \mathcal{A}$ if a point $x^{\mathrm{S}} \in \mathcal{X}$ and a constant $\delta > 0$ exist such that for all $\eta \in I_0(\tilde{x})$

$$\sup_{\zeta \in \partial \mathcal{G}[G(\tilde{x})](\eta)} \mathbb{E}\big[\zeta(\omega) g'(\tilde{x}, \omega)(x^{\mathrm{S}} - \tilde{x})\big] \leq -\delta. \tag{5.75}$$

Condition (5.44) may be interpreted as a generalization of Robinson's constraint qualification condition to the non-smooth composite constraint in our optimization problem. We shall describe the tangent cone $\mathcal{T}_{\mathcal{A}}(x)$ of the feasible set $\mathcal{A}$ at a given point $\bar{x}$.

Theorem 5.45. *Suppose Assumption* 5.43 *(ii)–(iv) and the differential constraint qualification condition are satisfied at a point* $\bar{x} \in \mathcal{A}$. *Then* $\mathcal{T}_{\mathcal{A}}(\bar{x})$ *is the set of vectors* $d \in \mathcal{T}_{\mathcal{X}}(\bar{x})$ *satisfying the inequalities*

$$\mathbb{E}\big[\zeta(\omega) g'(\bar{x}, \omega) d\big] \leq 0, \tag{5.76}$$

for all $\zeta \in \partial\big[\mathcal{G}(G(\bar{x})\big](\eta)$ *and all* $\eta \in I_0(\bar{x})$.

Proof. Suppose $d \in \mathcal{T}_{\mathcal{A}}(\bar{x})$. This means that sequences of points $\{x^k\} \subset \mathcal{A}$ and scalars $\{\tau_k\} \downarrow 0$ exist such that:

$$\lim_{k \to \infty} \frac{1}{\tau_k}(x^k - \bar{x}) = d. \tag{5.77}$$

Consider $\zeta_\eta \in \partial\big[\mathcal{G}(G(\bar{x})\big](\eta)$, for $\eta \in [a, b]$. Using the feasibility of x^k and the convexity of $[\mathcal{G}(\cdot)](\eta)$, we obtain

$$\begin{aligned} 0 \geq \mathcal{G}\big(G(x^k)\big)(\eta) &\geq \mathcal{G}\big(G(\bar{x})\big)(\eta) + \mathbb{E}\big[\zeta_\eta\big(G(x^k) - G(\bar{x})\big)\big] \\ &= \mathcal{G}\big(G(\bar{x})\big)(\eta) + \mathbb{E}\big[\zeta_\eta\big(G'(\bar{x})(x^k - \bar{x}) + o(x^k, \bar{x})\big]. \end{aligned} \tag{5.78}$$

Here, $\|o(x^k, \bar{x})\| / \|x^k - \bar{x}\| \to 0$ as $k \to \infty$.

If $\eta \in I_0(\bar{x})$, inequality (5.78) implies that

$$0 \geq \mathbb{E}\big[\zeta_\eta\big(G'(\bar{x})(x^k - \bar{x}) + o(x^k, \bar{x})\big)\big].$$

Dividing by τ_k, we obtain

$$0 \geq \lim_{k \to \infty} \mathbb{E}\Big[\zeta_\eta\Big(G'(\bar{x})\frac{x^k - \bar{x}}{\tau_k} + \frac{o(x^k, \bar{x})}{\tau_k}\Big)\Big]. \tag{5.79}$$

The boundedness of the derivative of $g(\cdot,\omega)$ implies that

$$\|g'(\bar{x},\omega)x-\bar{x}+o(x^k,\bar{x}))\| = \|g(x^k,\omega)-g(\bar{x},\omega)\| \le \ell(\omega)\|x^k-\bar{x}\|.$$

Hence, we may use the Lebesgue dominated convergence theorem to exchange the expectation and the limit in (5.79) and obtain

$$0 \ge \mathbb{E}\Big[\zeta_\eta \lim_{k\to\infty}\Big(G'(\bar{x})\frac{x^k-\bar{x}}{\tau_k}+\frac{o(x^k,\bar{x})}{\tau_k}\Big)\Big] = \mathbb{E}\big[\zeta_\eta\big(G'(\bar{x})d\big)\big],$$

which is (5.76).

We shall prove the converse implication. Define $d^0 = x^{\mathrm{S}}-\bar{x}$. Consider an arbitrary direction $d \in \mathcal{T}_{\mathcal{X}}(\bar{x})$ satisfying (5.76). We shall prove that $d \in \mathcal{T}_{\mathcal{A}}(\bar{x})$. Since d is tangent to the convex set $\mathcal{X}$, for every $t>0$, a point $\bar{x}(t)\in\mathcal{X}$ exists such that

$$\bar{x}(t)=\bar{x}+td+z(t)\in\mathcal{X} \quad\text{and}\quad \lim_{t\downarrow 0}\frac{1}{t}\|z(t)\|=0.$$

Choose $\alpha\in(0,1)$. We consider the convex combinations of points $\bar{x}(t)$ and $\bar{x}+td^0$ with coefficients $1-\alpha$ and α, i.e., the points

$$x(t)=\bar{x}+(1-\alpha)td+(1-\alpha)z(t)+\alpha td^0. \tag{5.80}$$

The points $x(t)$ belong to $\mathcal{X}$ due to the convexity of $\mathcal{X}$. We consider a sequence $\{\tau_k\}\downarrow 0$ and define $t_k=\tau_k/(1-\alpha)$, $\varepsilon=\alpha/(1-\alpha)$. Using these sequences in formula (5.80), we obtain the sequence of points

$$x^k=\bar{x}+\tau_k d+z^k+\varepsilon\tau_k d^0,$$

with $z^k=(1-\alpha)z(t_k)$. Observe that $\frac{1}{\tau_k}\|z^k\|\to 0$, as $k\to\infty$. We shall show that for all sufficiently large k the points x^k are feasible for sufficiently small ε, which would imply that $d\in T_{\mathcal{A}}(\bar{x})$.

The Fréchet differentiability of G and the fact that $\|z^k\|$ is infinitely smaller than τ_k, imply that

$$\big\|G(x^k)-G(\bar{x})-\tau_k G'(\bar{x})(d+\varepsilon d^0)\big\| \le o(\tau_k).$$

We now derive for any $x\in\mathcal{X}$ an estimate:

$$\begin{aligned}\max_{\eta\in[a,b]}\big|\mathcal{G}(G(x^k))(\eta)-\mathcal{G}(G(\bar{x}))(\eta)\big| &= \\ \max_{\eta\in[a,b]}\Big|\mathbb{E}\big[(\eta-G(x^k))_+^p\big]-\mathbb{E}\big[(\eta-G(\bar{x}))_+^p\big]\Big| & \\ \le \max_{\eta\in[a,b]}\mathbb{E}\Big[\big|(\eta-G(x^k))_+^p-(\eta-G(\bar{x}))_+^p\big|\Big]&.\end{aligned}$$

For $p = 1$, we have

$$\begin{aligned}\max_{\eta\in[a,b]} &\big|\mathcal{G}(G(x^k))(\eta) - \mathcal{G}(G(\bar{x}))(\eta)\big| \\ &\le \max_{\eta\in[a,b]} \mathbb{E}\Big[\big|(\eta - G(x^k))_+ - (\eta - G(\bar{x}))_+\big|\Big] \\ &\le \mathbb{E}\Big[\big|G(x^k) - G(\bar{x})\big|\Big] = \big\|G(x^k) - G(\bar{x})\big\|_1.\end{aligned}$$

For $p > 1$, the function $\mathbb{E}\big[\max(0, \eta - \cdot)^p\big]$ is differentiable, and hence

$$\begin{aligned}\mathbb{E}\big[\max(0, \eta - G(x^k))^p\big] &= \mathbb{E}\big[\max(0, \eta - G(\bar{x}))^p\big] + \\ &\mathbb{E}\Big[\max(0, \eta - G(\bar{x}))^{p-1}\big(G(x^k) - G(\bar{x})\big)\Big] + o_1(G(x^k), G(\bar{x})).\end{aligned}$$

Using this inequality together with the Hölder inequality, we obtain

$$\begin{aligned}\max_{\eta\in[a,b]} &\big|\mathcal{G}(G(x^k))(\eta) - \mathcal{G}(G(\bar{x}))(\eta)\big| \\ &\le \max_{\eta\in[a,b]} \mathbb{E}\Big[(\eta - G(\bar{x}))_+^{p-1}\big(G(x^k) - G(\bar{x})\big)\Big] + o_1(G(x^k), G(\bar{x})) \\ &\le \max_{\eta\in[a,b]} \big\|(\eta - G(\bar{x}))_+\big\|_p^{p-1}\big\|G(x^k) - G(\bar{x})\big\|_p + o_1(G(x^k), G(\bar{x})).\end{aligned}$$

Denote $\varkappa_p = \max_{\eta\in[a,b]} \|\max(0, \eta - G(\bar{x}))\|_p^{p-1} + 1$. Putting all estimates together, we arrive at the following inequality:

$$\max_{\eta\in[a,b]} \big|\mathcal{G}(G(x^k))(\eta) - \mathcal{G}(G(\bar{x}))(\eta)\big| \le \varkappa_p \big\|G(x^k) - G(\bar{x})\big\|_p.$$

Hence, the following holds:

$$\max_{\eta\in[a,b]} \mathcal{G}\big(G(x^k)\big)(\eta) \le \max_{\eta\in[a,b]} \mathcal{G}\big(G(\bar{x}) + \tau_k G'(\bar{x})(d + \varepsilon d^0)\big)(\eta) + \varkappa_p\, o(\tau_k). \tag{5.81}$$

Observe that

$$\max_{\eta\in[a,b]} \mathcal{G}(G(\bar{x}))(\eta) \le 0, \tag{5.82}$$

because the point $\bar{x}$ is feasible. If inequality (5.82) is strict, then the points x^k for large k are feasible, which implies that $d + \varepsilon d^0$ is a tangent direction. Passing to the limit with $\varepsilon \downarrow 0$, we infer that $d \in \mathcal{T}_{\mathcal{A}}(\bar{x})$, which completes the proof in that case.

Consider now the case of $\max_{\eta\in[a,b]} \mathcal{G}\big(G(\bar{x})\big)(\eta) = 0$.

Define the function $\psi : \mathcal{L}_p(\Omega, \mathcal{F}, P) \to \mathbb{R}$ as follows:

$$\psi(Z) = \max_{\eta\in[a,b]} \mathcal{G}\big(G(\bar{x}) + Z\big)(\eta).$$

The function $\psi(\cdot)$ is continuous and convex, which entails that it is directionally differentiable. We rewrite inequality (5.81) and continue the estimate as follows:

$$\max_{\eta\in[a,b]} \mathcal{G}\big(G(x^k)\big)(\eta) \le \psi\big(\tau_k G'(\bar{x})(d+\varepsilon d^0)\big) + \varkappa_p o(\tau_k)$$
$$= \psi(0) + \tau_k \psi'\big(0; G'(\bar{x})(d+\varepsilon d^0)\big) + o_2(\tau_k). \quad (5.83)$$

Here $\psi'(0;h)$ stands for the directional derivative of ψ at 0 in direction h, and $o_2(\tau_k)/\tau_k \to 0$, as $\tau_k \downarrow 0$.

Since we consider the case $\psi(0) = 0$, the maximum in (5.82) is attained on the set $I_0(\bar{x})$. The directional derivative has the form:

$$\psi'(0;Z) = \sup_{\eta\in I_0(\bar{x})} \sup_{\lambda\in\partial\mathcal{G}(G(\bar{x}))(\eta)} \mathbb{E}[\lambda Z].$$

Our estimate (5.83) can thus be continued as follows:

$$\max_{\eta\in[a,b]} \mathcal{G}\big(G(x^k)\big)(\eta) \le \tau_k \sup_{\eta\in I_0(\bar{x})} \sup_{\lambda\in\partial\mathcal{G}(G(\bar{x}))(\eta)} \mathbb{E}\big[\lambda\, G'(\bar{x})(d+\varepsilon d^0)\big] + o_2(\tau_k).$$

Using (5.76) and (5.75) in the last inequality, we obtain

$$\max_{\eta\in[a,b]} \mathcal{G}\big(G(x^k)\big)(\eta) \le -\tau_k\varepsilon\delta + o_2(\tau_k).$$

For k large enough, $\varepsilon \ge o_2(\tau_k)/(\tau_k\delta)$. Therefore, we have

$$\mathcal{G}\big(G(x_k)\big)(\eta) \le 0 \quad \text{for all} \quad \eta \in [a,b].$$

We conclude that for all $\varepsilon > 0$ a number $\bar{k}(\varepsilon)$ exists such that for all $k \ge \bar{k}(\varepsilon)$ the points x^k belong to $\mathcal{X}$. We fix a sequence $\varepsilon_i \to 0$, define $k_i = \bar{k}(\varepsilon_i)$ and consider points

$$x^{k_i} = \bar{x} + \tau_{k_i} d + \delta_{k_i} + \varepsilon_i \tau_{k_i} d^0.$$

They are feasible by construction. Moreover,

$$\frac{x^{k_i} - \bar{x}}{\tau_{k_i}} = d + \varepsilon_i d^0 + \frac{1}{\tau_{k_i}}\delta_{k_i} \to d \quad \text{as} \quad i \to \infty.$$

Consequently, d is indeed a tangent direction to $\mathcal{X}$. □

We notice that our description of the tangent cone involves only the derivative of the mapping G. This allows us to work with a local convex approximation of the feasible set $\mathcal{A}$, which is obtained by using a linearization of the constraint mapping around the point $\bar{x}$ instead of the mapping itself. We denote that convex approximation by $\mathcal{A}^{\text{ca}}(\bar{x})$; it is defined as follows:

$$\mathcal{A}^{\mathrm{ca}}(\bar{x}) = \Big\{x \in \mathcal{X} : \mathcal{G}\big(G(\bar{x}) + G'(\bar{x})(x - \bar{x})\big)(\eta) \le 0, \ \forall\ \eta \in [a, b]\Big\}.$$

The tangent cones to $\mathcal{A}$ and to $\mathcal{A}^{\mathrm{ca}}(\bar{x})$ at the point $\bar{x}$ are identical. We capture this fact in the next corollary.

Corollary 5.46. *Assume the conditions of Theorem* 5.45. *Then*

$$\mathcal{T}_{\mathcal{A}}(\bar{x}) = \mathcal{T}_{\mathcal{A}^{\mathrm{ca}}(\bar{x})}(\bar{x}).$$

Now, we are ready to formulate the optimality conditions for the non-convex problem. Let $\hat{x}$ be a local minimum of the non-convex problem (5.54)–(5.56). We shall use the optimality conditions for convex problems applied to the following local convex approximation of our original problem:

$$\min\ f(\hat{x}) + f'(\hat{x})(x - \hat{x}) \tag{5.84}$$
$$\text{s.t. } \mathcal{G}\big(G(\hat{x}) + G'(\hat{x})(x - \hat{x})\big)(\eta)) \le 0 \quad \text{for all} \quad \eta \in [a, b], \tag{5.85}$$
$$x \in \mathcal{X}. \tag{5.86}$$

Theorem 5.47. *Let Assumptions* 5.43 *be satisfied. If the point $\hat{x}$ is a local minimum of* (5.54)–(5.56) *and the differential constraint qualification condition is satisfied at $\hat{x}$, then a measure $\hat{\mu} \in \mathcal{M}^+([a, b])$ exists such that*

$$0 \in f'(\hat{x}) + \mathbb{E}\Big[g'(\hat{x}, \omega) \int_a^b \partial\mathcal{G}\big(G(\hat{x})\big)(\eta)\ \hat{\mu}(\mathrm{d}\eta)\Big] + \mathcal{N}_{\mathcal{X}}(\hat{x}), \tag{5.87}$$

$$\int_a^b \mathcal{G}\big(G(\hat{x})\big)(\eta)\ \hat{\mu}(\mathrm{d}t) = 0. \tag{5.88}$$

Equivalently, a utility function $\hat{u} \in \mathcal{U}_{p+1}(Y, [a, b])$ exists such that

$$0 \in f'(\hat{x}) - \mathbb{E}\big[g'(\hat{x}, \omega)\, \hat{u}'\big(G(\hat{x})(\omega)\big)\big] + \mathcal{N}_{\mathcal{X}}(\hat{x}); \tag{5.89}$$
$$\mathbb{E}\big[\hat{u}(G(\hat{x}))\big] = \mathbb{E}\big[\hat{u}(Y)\big]. \tag{5.90}$$

Proof. If $\hat{x}$ is a local minimum of problem (5.54)–(5.56), then the directional derivative of the objective $f'(\hat{x}; d) \ge 0$ for all $d \in \mathcal{T}_{\mathcal{A}}(\hat{x})$. Thus $\hat{x}$ is also a global minimum of problem (5.84)–(5.86).

In order to apply the optimality conditions for the convex problem, the constraint qualification condition for problem (5.84)–(5.86) formulated in Definition 5.35 has to hold. Definition 5.44 implies that a point $x^{\mathrm{S}} \in \mathcal{X}$ exists such that

$$\sup_{\lambda \in \partial\mathcal{G}(G(\hat{x}))(\eta)} \mathbb{E}\big[\lambda G'(\hat{x})(x^{\mathrm{S}} - \hat{x})\big] \le -\delta,$$

for all $t \in I_0(\hat{x})$. We consider again the convex continuous function

$$\psi(Z) = \max_{\eta \in [a,b]} \mathscr{G}(G(\hat{x}) + Z)(\eta).$$

We have $\psi(0) \leq 0$ because $\hat{x}$ is feasible for problem (5.84)–(5.86). If $\psi(0) < 0$, then the uniform dominance condition of Definition 5.35 is satisfied at $\hat{x}$.

If $\psi(0) = 0$, the active set of parameters η in the calculation of $\psi(0)$ is exactly $I_0(\hat{x})$. Therefore the directional derivative at 0 in the direction $G'(\hat{x})(x^{\mathrm{S}} - \hat{x})$ has the form

$$\psi'\big(0; G'(\hat{x})(x^{\mathrm{S}} - \hat{x})\big) = \sup_{\eta \in I_0(\hat{x})} \sup_{\lambda \in \partial \mathscr{G}(G(\hat{x}))(\eta)} \mathbb{E}\big[\lambda(\omega) g'(\hat{x}, \omega)(x^{\mathrm{S}} - \hat{x})\big].$$

The differential constraint qualification condition implies that

$$\psi'\big(0; G'(\hat{x})(x^{\mathrm{S}} - \hat{x})\big) \leq -\delta < 0.$$

For a small $\tau > 0$, we define the point $x(\tau) = (1 - \tau)\hat{x} + \tau x^{\mathrm{S}}$ and obtain

$$\begin{aligned} \max_{\eta \in [a,b]} \mathscr{G}\big(G(\hat{x}) + G'(\hat{x})(x(\tau) - \hat{x})\big)(\eta) &= \psi\big(G'(\hat{x})(x(\tau) - \hat{x})\big) \\ &= \psi\big(\tau G'(\hat{x})(x^{\mathrm{S}} - \hat{x})\big) < 0, \end{aligned}$$

provided $\tau > 0$ is sufficiently small. Hence, the point $x(\tau)$ satisfies the uniform dominance condition of Definition 5.35 for problem (5.84)–(5.86). Consequently, we may apply Theorem 5.41 to problem (5.84)–(5.86) to complete the proof of the statement. □

5.3 Inverse Formulations

Now we focus on the second-order stochastic dominance relation formulated with the use of the absolute Lorenz function. Throughout this section, we adopt again Assumption 5.34, i.e., the space $\mathcal{X}$ is a separable Banach space, the functional f is convex and continuous, the operator $G(\cdot)$ is norm-to-norm continuous, and the functions $[G(\cdot)](\omega)$ are concave for P- almost all $\omega \in \Omega$.

We analyze the following optimization problem:

$$\min f(x) \tag{5.91}$$

$$\text{subject to } F_{G(x)}^{(-2)}(p) \geq F_Y^{(-2)}(p), \quad \forall p \in [\alpha, \beta], \tag{5.92}$$

$$x \in \mathcal{X}. \tag{5.93}$$

We assume that $\alpha > 0$ and $\beta < 1$ in this formulation.

In Lemma 5.32, we have shown that for every $p \in (0,1)$ the mapping $Z \mapsto F_Z^{(-2)}(p)$ is concave and positively homogeneous on $\mathcal{L}_1(\Omega, \mathcal{F}, P)$, and the set $\mathcal{B}(Y, [\alpha, \beta])$ is convex. Furthermore, the mapping $Z \mapsto F_Z^{(-2)}(p)$ is monotonic with respect to the a.s. order and the mapping $G(\cdot)$ is concave. Hence, the composition $x \to F_{G(x)}^{(-2)}(p)$ is concave. This implies that problem (5.91)–(5.93) with the adopted assumptions is convex.

The counterpart of the uniform dominance condition of Definition 5.35 is the following constraint qualification condition.

Definition 5.48. Problem (5.91)–(5.93) satisfies the *uniform inverse dominance condition* if a point $\tilde{x} \in \mathcal{X}$ exists such that

$$\inf_{p\in[\alpha,\beta]} \left\{ F_{G(\tilde{x})}^{(-2)}(p) - F_Y^{(-2)}(p) \right\} > 0.$$

We define the distortion-modified function, $\Phi : \mathcal{X} \times \mathcal{W}_2([\alpha, \beta]) \to \mathbb{R}$, associated with problem (5.91)–(5.93) by setting

$$\Phi(x, w) = f(x) - \int_\alpha^\beta F_{G(x)}^{(-1)}(p)\, \mathrm{d}w(p) + \int_\alpha^\beta F_Y^{(-1)}(p)\, \mathrm{d}w(p). \tag{5.94}$$

It is well-defined due to the identity (5.36) in Lemma 5.16 and the fact that $w(\cdot)$ has bounded slopes.

We shall formulate optimality conditions for problem (5.91)–(5.93) in two forms. The results expose the relations of the problem to distortion functions and measures of risk.

Theorem 5.49. *Let Assumption* 5.34 *and the uniform inverse dominance condition be satisfied. If $\hat{x}$ is an optimal solution of* (5.91)–(5.93) *then a function $\hat{w} \in \mathcal{W}_2([\alpha, \beta])$ exists such that*

$$\Phi(\hat{x}, \hat{w}) = \min_{x\in\mathcal{X}} \Phi(x, \hat{w}), \tag{5.95}$$

$$\int_\alpha^\beta F_{G(\hat{x})}^{(-1)}(p)\, \mathrm{d}\hat{w}(p) = \int_\alpha^\beta F_Y^{(-1)}(p)\, \mathrm{d}\hat{w}(p). \tag{5.96}$$

Conversely, if for some function $\hat{w} \in \mathcal{W}_2([\alpha, \beta])$ an optimal solution $\hat{x}$ of (5.95) *satisfies the constraint* (5.92) *and the condition* (5.96), *then $\hat{x}$ is an optimal solution of problem* (5.91)–(5.93).

Proof. We consider the operator $\tilde{\mathcal{G}} : \mathcal{L}_p(\Omega, \mathcal{F}, P) \to \mathcal{C}([\alpha, \beta])$ defined as

$$[\tilde{\mathcal{G}}(Z)](p) = F_Y^{(-2)}(p) - F_Z^{(-2)}(p), \quad p \in [\alpha, \beta]. \tag{5.97}$$

The space $\mathcal{C}([\alpha, \beta])$ is equipped with the supremum norm. We shall show that the composition $\tilde{\mathcal{G}}(G(\cdot))$ is a norm-to-norm continuous operator. For all $p \in$

$[\alpha, \beta]$, the mapping

$$(p, Z) \mapsto F_Z^{(-2)}(p)$$

is continuous on $[\alpha, \beta] \times \mathcal{L}_1(\Omega, \mathcal{F}, P)$ because

$$|F_Z^{(-2)}(p) - F_{Z'}^{(-2)}(p')| \leq |F_Z^{(-2)}(p) - F_Z^{(-2)}(p')| + |F_Z^{(-2)}(p') - F_{Z'}^{(-2)}(p')|.$$

When (p', Z') converges to (p, Z), the first term on the right-hand side of the last displayed inequality converges to zero due to the continuity of the Lorenz function of Z, while the send term converges to zero due to Lemma 5.32. Hence, the mapping

$$Z \mapsto \sup_{p \in [\alpha, \beta]} F_Z^{(-2)}(p)$$

is continuous on $\mathcal{L}_1(\Omega, \mathcal{F}, P)$. This together with the continuity of $G(\cdot)$, implies that the composition $\tilde{\mathcal{G}}(G(\cdot))$ is norm-to-norm continuous. Let K be the cone of non-positive functions in $\mathcal{C}([\alpha, \beta])$. Problem (5.91)–(5.93) can be reformulated as follows:

$$\min f(x) \text{ subject to } \tilde{\mathcal{G}}(G(x)) \in K, \ x \in \mathcal{X}. \tag{5.98}$$

Since the mapping $Z \mapsto F_Z^{(-2)}(p)$ is concave by virtue of Lemma 5.32 and monotonic (with respect to the a.s. order) by definition, we obtain that $x \mapsto F_{G(x)}^{(-2)}(p)$ is concave for every $p \in [\alpha, \beta]$. Hence, the operator $\tilde{\mathcal{G}} \circ G$ is convex with respect to the cone K. We use again the pairing of $\mathcal{C}([\alpha, \beta])$ and the space $\mathcal{M}^+([\alpha, \beta])$ of regular finite countably additive measures to obtain the Lagrangian associated with problem (5.98). It is $\bar{\mathfrak{L}} : \mathcal{X} \times \mathcal{M}^+([\alpha, \beta]) \to \mathbb{R}$, defined as follows

$$\bar{\mathfrak{L}}(x, \nu) = f(x) + \int_{[\alpha,\beta]} [\tilde{\mathcal{G}} \circ G(x)](p) \ \nu(\mathrm{d}p). \tag{5.99}$$

In this setting, the inverse uniform dominance condition is equivalent to the generalized Slater condition: a point $\tilde{x} \in \mathcal{X}$ exists such that $(\tilde{\mathcal{G}} \circ G)(\tilde{x}) \in \operatorname{int} K$. The necessary conditions of optimality in Banach spaces state the existence of a nonnegative measure $\hat{\nu} \in \mathcal{M}^+([\alpha, \beta])$ such that

$$\bar{\mathfrak{L}}(\hat{x}, \hat{\nu}) = \min_{x \in \mathcal{X}} \bar{\mathfrak{L}}(x, \hat{\nu}), \tag{5.100}$$

and

$$\int_{[\alpha,\beta]} F_Y^{(-2)}(p) - F_{G(\hat{x})}^{(-2)}(p) \ \hat{\nu}(\mathrm{d}\eta) = 0. \tag{5.101}$$

Using the correspondence established in Lemma 5.17, we associate a function $w_\nu \in \mathcal{W}_2([\alpha, \beta])$ with the measure ν and transform the conditions (5.100)-(5.101) to (5.95)-(5.96).

The converse statement follows from the convexity of problem (5.91)–(5.93). If the point $\hat{x}$ satisfies conditions (5.100), (5.101) and the inverse dominance constraint (5.92), then $\hat{x}$ is an optimal solution of that optimization problem. □

Theorem 5.49 shows the relation of problem (5.91)–(5.93) to decision problems using distortions.

Notice that the set $\widehat{W}$ of functions in $\mathcal{W}_2([\alpha, \beta])$ satisfying (5.95)–(5.96) is convex, bounded and weakly* closed in the following sense: if a sequence of functions $w^k \in \widehat{W}$ and $w \in \mathcal{W}_2([\alpha, \beta])$ are such that for all $Z \in \mathcal{L}^1(\Omega, \mathcal{F}, P)$,

$$\lim_{k\to\infty} \int_0^1 F_Z^{(-1)}(p)\, \mathrm{d}w^k(p) = \int_0^1 F_Z^{(-1)}(p)\, \mathrm{d}w(p),$$

then $w \in \widehat{W}$. The set $\widehat{W}$ is the same for all optimal solutions of (5.91)–(5.93). These statements follow from the key relation (5.36) in Lemma 5.17.

We can also develop a duality theory based on the distortion-modified function (5.94). For every function $w \in \mathcal{W}_2([\alpha, \beta])$, the optimization problem

$$\min_{x\in\mathcal{X}} \left\{ f(x) - \int_0^1 F_{G(x)}^{(-1)}(p)\, \mathrm{d}w(p) + \int_0^1 F_Y^{(-1)}(p)\, \mathrm{d}w(p) \right\} \tag{5.102}$$

is a Lagrangian relaxation of problem (5.91)–(5.93). Its optimal value,

$$\Psi(w) = \inf_{x\in\mathcal{X}} \Phi(x, w),$$

is always lower than or equal to the optimal value of (5.91)–(5.93). Indeed, any feasible solution x of (5.91)–(5.93) is feasible for (5.102), and the dominance constraint (5.92) implies that

$$\int_0^1 F_{G(x)}^{(-1)}(p)\, \mathrm{d}w(p) \geq \int_0^1 F_Y^{(-1)}(p)\, \mathrm{d}w(p),$$

due to Theorem 5.14.

We define the dual problem as

$$\max_{w\in\mathcal{W}_2([\alpha,\beta])} \Psi(w). \tag{5.103}$$

The set $\mathcal{W}_2([\alpha, \beta])$ is a closed convex cone in $\mathcal{C}([0, 1])$ and $\Psi(\cdot)$ is a convex functional, so problem (5.103) is a convex optimization problem. Employing the duality relations in convex programming, we obtain the following result.

Theorem 5.50. *Suppose Assumption* 5.34 *and the uniform inverse dominance condition are satisfied. If problem* (5.91)–(5.93) *has an optimal solution, then problem* (5.103) *has an optimal solution and the optimal values of both problems coincide. Furthermore, the set of optimal solutions of* (5.103) *is the set of functions* $\hat{w} \in \mathcal{W}_2([\alpha, \beta])$ *satisfying* (5.95)–(5.96) *for an optimal solution* $\hat{x}$ *of* (5.91)–(5.93).

Now, we formulate and prove new optimality conditions for problem (5.91)–(5.93) which elicits new relations of the stochastic dominance constraints to spectral risk measures. In the setting of this section, due to the preference for larger values of $Z = G(x)$, a spectral risk measure has the form (5.43).

Theorem 5.51. *Suppose Assumption* 5.34 *and the uniform inverse dominance condition are satisfied. If* $\hat{x}$ *is an optimal solution of* (5.91)–(5.93) *then a spectral risk measure* $\hat{\varrho}(\cdot)$ *and a constant* $\kappa \geq 0$ *exist such that* $\hat{x}$ *is also an optimal solution of the problem*

$$\min_{x \in \mathcal{X}} \left\{ f(x) + \kappa \hat{\varrho}\big(G(x)\big) \right\}, \tag{5.104}$$

$$\kappa \hat{\varrho}\big(G(\hat{x})\big) = \kappa \hat{\varrho}(Y). \tag{5.105}$$

Conversely, if for some spectral risk measure $\hat{\varrho}(\cdot)$ *and some* $\kappa \geq 0$ *an optimal solution* $\hat{x}$ *of* (5.104) *satisfies* (5.92) *with* $[\alpha, \beta] = [0, 1]$ *and* (5.105), *then* $\hat{x}$ *is an optimal solution of* (5.91)–(5.93) *with* $[\alpha, \beta] = [0, 1]$.

Proof. The optimal solution of problem (5.91)–(5.93) satisfies (5.95) and (5.96) with some distortion function $\hat{w}$. The last term in $\Phi(x, \hat{w})$ does not depend on x and may be skipped. We apply Lemma 5.17 to associate with $w(\cdot)$ a spectral measure of risk $\hat{\varrho}(\cdot)$ and a constant $\kappa \geq 0$ such that the second term of $\Phi(x, \hat{w})$ can be written as follows:

$$\int_\alpha^\beta F^{(-1)}_{G(\hat{x})}(p)\, dw(p) = -\kappa \hat{\varrho}\big(G(\hat{x})\big).$$

A similar relation can be written for Y. Therefore, we can transform the complementarity condition (5.96) and re-write it in the form (5.105).

This proves that (5.104)–(5.105) are necessary conditions of optimality. We shall establish now the sufficient conditions.

If $\kappa = 0$, then the solution of problem $\min_{x \in \mathcal{X}} f(x)$ satisfies (5.92), which means that it is optimal for problem (5.91)–(5.93). Assume now that a positive constant κ and a spectral risk measure $\hat{\varrho}(\cdot)$ exist such that conditions (5.104) and (5.105) are satisfied for some feasible point $\hat{x}$. Then, for every $x \in \mathcal{X}$,

$$f(x) + \kappa \hat{\varrho}\big(G(x)\big) \geq f(\hat{x}) + \kappa \hat{\varrho}\big(G(\hat{x})\big) = f(\hat{x}) + \kappa \hat{\varrho}(Y).$$

Using Lemma 5.17 again, we associate a function $w \in \mathcal{W}_2([0, 1])$ with the risk measure $\hat{\varrho}(\cdot)$. We define $\hat{w}(\cdot) = \kappa w(\cdot)$ and substitute in equality (5.37) to obtain

$$f(x) - \int_0^1 F_{G(x)}^{(-1)}(p)\, \mathrm{d}\hat{w}(p) \geq f(\hat{x}) - \int_0^1 F_Y^{(-1)}(p)\, \mathrm{d}\hat{w}(p).$$

or equivalently

$$f(x) - \int_0^1 F_{G(x)}^{(-1)}(p)\, \mathrm{d}\hat{w}(p) + \int_0^1 F_Y^{(-1)}(p)\, \mathrm{d}\hat{w}(p) \geq f(\hat{x}).$$

If $G(x) \succeq_{(2)} Y$, then Theorem 5.14 implies that

$$\int_0^1 F_Y^{(-1)}(p)\, \mathrm{d}\hat{w}(p) - \int_0^1 F_{G(x)}^{(-1)}(p)\, \mathrm{d}\hat{w}(p) \leq 0.$$

Putting the last two inequalities together, we obtain $f(x) \geq f(\hat{x})$. This means that $\hat{x}$ is an optimal solution of problem (5.91)–(5.93) with $[\alpha, \beta] = [0, 1]$. □

Observe that if the dominance constraint (5.92) is active, then the associated measure $\nu \in \mathcal{M}^+[\alpha, \beta]$ will not be identical to zero. Then the constant κ defined in (5.41) is positive, and we can eliminate it from the complementarity condition (5.105). In that case, the complementarity condition takes on the form

$$\hat{\varrho}\big(G(\hat{x})\big) = \hat{\varrho}(Y).$$

Furthermore, it is straightforward to observe that the support of the measure μ in the spectral representation of $\hat{\varrho}(\cdot)$ is included in $[\alpha, \beta]$.

Consider now the special case when $f(x) = -\mathbb{E}[G(x)]$.

Corollary 5.52. *Under the assumptions of Theorem* 5.51, *if $\hat{x}$ is an optimal solution of* (5.91)–(5.93) *then a spectral risk measure $\bar{\varrho}$ exists such that $\hat{x}$ is also an optimal solution to the problem*

$$\min_{x \in \mathcal{X}} \bar{\varrho}\big(G(x)\big), \tag{5.106}$$

and

$$-\mathbb{E}\big[G(\hat{x})\big] + \bar{\kappa}\bar{\varrho}\big(G(\hat{x})\big) = -\mathbb{E}[Y] + \bar{\kappa}\bar{\varrho}(Y), \tag{5.107}$$

with some $\bar{\kappa} \geq 1$.

Proof. Using formula (5.42) from the proof of Lemma 5.17 we obtain

$$-\mathbb{E}[G(\hat{x})] + \mathcal{E}[G(\hat{x})] = -\operatorname{AVaR}_1^-\big(G(\hat{x})\big) - \kappa \int_\alpha^\beta \operatorname{AVaR}_t^-\big(G(\hat{x})\big)\, \mu(\mathrm{d}t).$$

Now we extend the measure μ to the entire interval $[0, 1]$ by assigning zero measure to all measurable subsets of $[0, 1]$ which do not intersect $[\alpha, \beta]$. Further, we define $\bar{\mu} = (\kappa\mu + \delta_1)/(\kappa + 1)$, where δ_1 is an atom of mass 1 at 1. Clearly, $\bar{\mu}$ is a probability measure as well. We conclude that $-\mathbb{E}[G(\hat{x})] + \mathcal{E}[G(\hat{x})] = -(1+\kappa)\bar{\varrho}\big(G(\hat{x})\big)$, with $\bar{\varrho}\big(G(\hat{x})\big) = \int_0^1 \mathrm{AVaR}_t^-\big(G(\hat{x})\big)\,\bar{\mu}(\mathrm{d}t)$. Therefore (5.104) takes on the form (5.106). Substituting the new measure of risk $\bar{\varrho}$ in (5.105), we get

$$-\mathbb{E}[G(\hat{x})] + (1+\kappa)\bar{\varrho}\big(G(\hat{x})\big) = -\mathbb{E}[Y] + (1+\kappa)\bar{\varrho}(Y).$$

□

In fact, the spectral measure $\bar{\varrho}(\cdot)$ has an atom $\lambda \in (0, 1)$ at 1, and $\bar{\kappa}$ satisfies $\lambda(1 + \bar{\kappa}) = 1$.

One can also ask if it is possible to identify a suitable benchmark so that the dominance-constrained problem would be equivalent to the following problem:

$$\min_{x\in\mathcal{X}} \big\{f(x) + \kappa\varrho\big(G(x)\big)\big\}, \tag{5.108}$$

with $\kappa > 0$ and some coherent law-invariant measure of risk $\varrho(\cdot)$. In other words, we are interested in constructing a problem of form (5.91)–(5.93) with identical solutions as problem (5.108). We denote the solution of the latter problem by $\hat{x}$.

We assume that $\varrho(\cdot)$ is a law-invariant risk measure with the Kusuoka representation (for Z representing "profits"):

$$\varrho(Z) = \sup_{\lambda\in\Lambda} \int_0^1 \mathrm{AVaR}_t^-(Z)\,\lambda(\mathrm{d}t), \tag{5.109}$$

where $\Lambda \subset \mathcal{P}((0, 1])$. The set of probability measures $\mathcal{P}((0, 1])$ is a subset of the space $\mathcal{M}^+((0, 1])$ of regular countably additive measures on the interval $(0, 1]$.

Theorem 5.53. *Assume that $\hat{x}$ is a solution of problem* (5.108)*, where ϱ has a representation* (5.109) *such that the set Λ is compact in the weak topology in $\mathcal{M}^+([0, 1])$. Then a measure $\hat{\lambda} \in \Lambda$ exists such that*

$$\varrho\big(G(\hat{x})\big) = \int_0^1 \mathrm{AVaR}_t^-\big(G(\hat{x})\big)\,\hat{\lambda}(\mathrm{d}t),$$

and for every $Y \in \mathcal{L}_1(\Omega, \mathcal{F}, P)$ satisfying the conditions

$$\begin{aligned} F_Y^{(-2)}(t) &\le F_{G(\hat{x})}^{(-2)}(t), \quad \textit{for all } t \in [0, 1], \\ F_Y^{(-2)}(t) &= F_{G(\hat{x})}^{(-2)}(t), \quad \textit{for all } t \in \mathrm{supp}(\hat{\lambda}), \end{aligned} \tag{5.110}$$

the point $\hat{x}$ *is also an optimal solution of problem* (5.91)–(5.93) *with* $[\alpha, \beta] = [0, 1]$.

Proof. Problem (5.108) is equivalent to the following problem:

$$\min_{x\in\mathcal{X}} \big\{f(x) + \kappa\varrho\big(G(x)\big)\big\} = \inf_{x\in\mathcal{X}} \sup_{\lambda\in\Lambda} \left\{ f(x) + \kappa \int_0^1 \mathrm{AVaR}_t^-\big(G(x)\big)\,\lambda(\mathrm{d}t)\right\}. \tag{5.111}$$

The minimum on the left-hand side is achieved at the point $\hat{x}$ by assumption. Since $\mathrm{AVaR}_t^-(\cdot)$ is coherent, for every $\lambda \in \mathcal{M}^+([0,1])$ the functional

$$x \mapsto f(x) + \kappa \int_0^1 \mathrm{AVaR}_t^-\big(G(x)\big)\,\lambda(\mathrm{d}t)$$

is convex and continuous in the space $\mathcal{L}_1(\Omega, \mathcal{F}, P)$. Furthermore, for every $Z \in \mathcal{L}_1(\Omega, \mathcal{F}, P)$, the function $t \to \mathrm{AVaR}_t^-(Z)$ is continuous on $(0, 1]$. Therefore, the functional $\lambda \mapsto \int_0^1 \mathrm{AVaR}_t^-\big(G(x)\big)\,\lambda(\mathrm{d}t)$ is linear and continuous in the weak topology on $\mathcal{M}^+([0,1])$. The set Λ is assumed compact in the weak topology. Hence, the assumptions of the asymmetric min-max theorem are satisfied and we can exchange the supremum and the infimum operations in (5.111). Under the assumptions of the theorem, the supremum with respect to $\lambda \in \Lambda$ is achieved. We obtain

$$\begin{aligned}\inf_{x\in\mathcal{X}} \sup_{\lambda\in\Lambda} &\left\{ f(x) + \kappa \int_0^1 \mathrm{AVaR}_t^-\big(G(x)\big)\,\lambda(\mathrm{d}t)\right\} \\ &= \max_{\lambda\in\Lambda} \inf_{x\in\mathcal{X}} \left\{ f(x) + \kappa \int_0^1 \mathrm{AVaR}_t^-\big(G(x)\big)\,\lambda(\mathrm{d}t)\right\}.\end{aligned} \tag{5.112}$$

Let $\hat{\lambda} \in \Lambda$ be the measure at which the supremum on the right-hand side is achieved. It follows that the pair $(\hat{x}, \hat{\lambda})$ is a saddle point of the function $f(x) + \kappa \int_0^1 \mathrm{AVaR}_t^-\big(G(x)\big)\,\lambda(\mathrm{d}t)$. Consequently, $\hat{x}$ is a solution of the problem

$$\min_{x\in\mathcal{X}} \left\{ f(x) + \kappa \int_0^1 \mathrm{AVaR}_t^-\big(G(x)\big)\,\hat{\lambda}(\mathrm{d}t)\right\}.$$

Now, we define the spectral measure

$$\hat{\varrho}(X) = \int_0^1 \mathrm{AVaR}_t^-(X)\,\hat{\lambda}(\mathrm{d}t),$$

and we conclude that $\hat{x}$ is also a solution of the problem

$$\min_{x \in \mathcal{X}} \left\{ f(x) + \kappa \hat{\varrho}\big(G(x)\big) \right\}.$$

We apply the sufficient conditions of Theorem 5.51 and see that (5.110) are equivalent to constraint (5.92) with $[\alpha, \beta] = [0, 1]$ and (5.105). □

We also note that due to Lemma 5.17, there is a one-to-one correspondence between spectral measures of risk and rank-dependent utility functions. Each spectrum $\lambda \in \Lambda$ gives rise to a rank-dependent utility function $w(\cdot)$ such that

$$\int_0^1 \mathrm{AVaR}_t^-(Z)\, \lambda(\mathrm{d}t) = -\int_0^1 F_Z^{(-1)}(p)\, \mathrm{d}w(p).$$

We denote the collection of rank-dependent utility functions obtained via the measures from Λ by $\mathcal{W}_\Lambda$, and the function corresponding to the measure $\hat{\lambda}$ by $\hat{w}$.

Due to (5.112), the problem (5.108) corresponds to the following min-max problem:

$$\begin{aligned} \min_{x \in \mathcal{X}} \left\{ f(x) + \kappa \varrho\big(G(x)\big) \right\} &= \max_{\lambda \in \Lambda} \min_{x \in \mathcal{X}} \left\{ f(x) + \kappa \int_0^1 \mathrm{AVaR}_t^-\big(G(x)\big)\, \mu(\mathrm{d}t) \right\} \\ &= \max_{w \in \mathcal{W}_\Lambda} \inf_{x \in \mathcal{X}} \left\{ f(x) - \kappa \int_0^1 F_{G(x)}^{(-1)}(p)\, \mathrm{d}w(p) \right\}. \end{aligned}$$

Therefore, the function $\hat{w}$ is a solution of a certain dual problem, similar to (5.103). There are two differences: the set $\mathcal{W}_2$ is replaced by its subset $\kappa \mathcal{W}_\Lambda$, and the linear terms in λ involving the benchmark Y are absent.

5.4 First-Order Stochastic Dominance Constraints

In this section, we discuss the following general problem:

$$\min\ f(x) \tag{5.113}$$

$$\text{subject to } P[G(x) \le \eta] \le P[Y \le \eta], \quad \eta \in [a, b], \tag{5.114}$$

$$x \in \mathcal{X}. \tag{5.115}$$

Here, $\mathcal{X}$ is a convex closed subset of a separable Banach space $\mathfrak{X}$. The random variable $Y \in \mathcal{L}_0(\Omega, \mathcal{F}, P; \mathbb{R})$ is a fixed reference outcome. For example, one may consider $Y = G(\bar{x})$.

Constraint (5.114) is a relaxed form of the first order stochastic dominance relation $G(x) \succeq_{(1)}^{I} Y$ with $I = [a, b]$, where we assume that both a and b are finite real numbers. Observe that this constraint is a continuum of chance constraints. We adopt the following assumptions about problem (5.113)–(5.115) in this section.

Assumption 5.54.

(i) *The function* $f(\cdot)$ *is continuously differentiable on* $\mathfrak{X}$ *in the Fréchet sense.*

(ii) *For every* $x \in \mathcal{X}$, *the distribution of* $G(x)$ *has a probability density function* $\varphi(x, \cdot)$, *where* $\varphi : \Upsilon \times \mathbb{R} \to \mathbb{R}_+$ *with* Υ *being an open subset of* $\mathfrak{X}$.

(iii) *The function* $\varphi(\cdot, \cdot)$ *is continuously differentiable with respect to the first argument for every value of the second argument. Furthermore, for every* $x^0 \in \mathcal{X}$ *a neighborhood of* x^0 *exists, such that for all* x *of that neighborhood, the derivative* $D_x\varphi(x, \cdot)$ *is bounded by an integrable function* $\ell(x^0; \cdot)$.

(iv) *The cumulative distribution function of* Y *is continuous on* $[a, b]$.

In what follows, we identify each derivative $D_x\varphi(x, z)$ with an element of the dual space $\mathfrak{X}^*$.

Recall that the generator $\mathcal{U}_1(I)$ of the relation $\succeq^I_{(1)}$ consists of nondecreasing and left-continuous functions $u(\cdot)$ satisfying the following conditions:

$$u(t) = 0 \text{ for all } t \geq b;$$
$$u(t) = u(a), \text{ for all } t \leq a.$$

For any real random variable Z and any measure $\mu \in \mathcal{M}^+([a, b])$, we have

$$\int_a^b P\big[Z \leq \eta\big]\, \mu(\mathrm{d}\eta) = -\mathbb{E}\big[u_\mu(Z)\big], \tag{5.116}$$

where $u_\mu(z) = -\mu([z, b])$.

We use the utility-modified functional $L_1 : \mathcal{X} \times \mathcal{U}_1([a, b]) \to \mathbb{R}$, associated with problem (5.113)–(5.115), which is defined as follows:

$$L_1(x, u) = f(x) - \mathbb{E}\big[u(G(x)) - u(Y)\big].$$

Assumption 5.54 (ii) and (iii) about the density $\varphi(\cdot)$ of $G(x)$ allows us to differentiate the probability function at the left-hand side of constraint (5.114) as follows:

$$S(x; \eta) \triangleq D_x\big[P\big(G(x) \leq \eta\big)\big] = D_x \int_{-\infty}^{\eta} \varphi(x, z)\, \mathrm{d}z = \int_{-\infty}^{\eta} D_x\varphi(x, z)\, \mathrm{d}z.$$

Furthermore, using (5.116), we conclude that the functional $L_1(\cdot, u)$ is differentiable for all $u \in \mathcal{U}_1([a, b])$ with the derivative given by the formula

$$\begin{aligned} D_x L_1(x, u) &= f'(x) - D_x\mathbb{E}\big[u(G(x))\big] \\ &= f'(x) + D_x \int_a^b P\big(G(x) \leq \eta\big)\, \mu(\mathrm{d}\eta) \\ &= f'(x) + \int_a^b \int_{-\infty}^{\eta} D_x\varphi(x, z)\, \mathrm{d}z\, \mu(\mathrm{d}\eta), \end{aligned}$$

where μ is the measure in $\mathcal{M}^+([a, b])$ corresponding to the function u.

Definition 5.55. Problem (5.113)–(5.115) satisfies the *differential uniform dominance condition* at the point $\hat{x} \in \mathcal{X}$ if a point $x^0 \in \mathcal{X}$ exists such that

$$\sup_{a \le \eta \le b} \Big\{ P\big(G(\hat{x}) \le \eta\big) + S(\hat{x};\eta)(x^0 - \hat{x}) - P\big(Y \le \eta\big) \Big\} < 0.$$

Theorem 5.56. *Suppose conditions* 5.54 *are satisfied. If $\hat{x}$ is an optimal solution of problem* (5.113)–(5.115) *and the differential uniform dominance condition is satisfied at the point $\hat{x}$, then a non-zero function $\hat{u} \in \mathcal{U}_1([a,b])$ exists such that*

$$0 \in D_x L(\hat{x},\hat{u}) + \mathcal{N}_{\mathcal{X}}(\hat{x}), \tag{5.117}$$

$$\mathbb{E}\big[\hat{u}(G(x))\big] = \mathbb{E}\big[\hat{u}(Y)\big]. \tag{5.118}$$

Proof. We consider the operator $\mathcal{G} : \mathcal{X} \to \mathcal{C}([a,b])$ defined as follows:

$$[\mathcal{G}(x)](\eta) = P\big(G(x) \le \eta\big) - v(\eta), \quad \eta \in [a,b], \tag{5.119}$$

where $v(\eta) = P\big(Y \le \eta\big)$. Let K be the cone of nonpositive functions in $\mathcal{C}([a,b])$. Problem (5.113)–(5.115) can be formulated as follows:

$$\min f(x) \tag{5.120}$$

$$\text{subject to } \mathcal{G}(x) \in K, \tag{5.121}$$

$$x \in \mathcal{X}. \tag{5.122}$$

Applying the Riesz representation theorem as in Theorem 5.36, we introduce the Lagrangian $\Lambda : \mathcal{X} \times \mathcal{M}^+([a,b]) \to \mathbb{R}$ by setting

$$\mathfrak{L}_1(x,\mu) = f(x) + \int_a^b \mathcal{G}(x)(\eta)\,\mu(\mathrm{d}\eta). \tag{5.123}$$

We shall show that the Lagrangian is differentiable with respect to the first argument. For each fixed η we can differentiate $\mathcal{G}(x)(\eta)$ as follows:

$$D_x\big[\mathcal{G}(x)(\eta)\big] = D_x\big[P[G(x) \le \eta]\big] = \int_{-\infty}^{\eta} D_x\varphi(x,z)\,\mathrm{d}z. \tag{5.124}$$

To prove that $D_x\big[\mathcal{G}(x)(\cdot)\big]$ is the derivative of $\mathcal{G}$ let us consider the error for $\eta \in [a,b]$:

$$\begin{aligned} E(x,x^1,\eta) &= \mathcal{G}(x^1)(\eta) - \mathcal{G}(x)(\eta) - \int_{-\infty}^{\eta} \langle D_x\varphi(x,z), x^1 - x\rangle\,\mathrm{d}z \\ &= P[G(x^1) \le \eta] - P[G(x) \le \eta] - \int_{-\infty}^{\eta} \langle D_x\varphi(x,z), x^1 - x\rangle\,\mathrm{d}z \\ &= \int_{-\infty}^{\eta} \big(\varphi(x^1,z) - \varphi(x,z) - \langle D_x\varphi(x,z), x^1 - x\rangle\big)\,\mathrm{d}z \\ &\le 2\|x^1 - x\| \int_{-\infty}^{\eta} \ell(z,x)\,\mathrm{d}z. \end{aligned}$$

Evidently, the error $E(x^1, x, \cdot)$ is a continuous function of its third argument. Furthermore, the integrand is bounded by an integrable function, due to Assumption 5.54(iii). Hence, we can apply the Lebesgue dominated convergence theorem to show that for each $\eta \in [a, b]$

$$\lim_{\|x^1-x\|\to 0} \frac{E(x, x^1, \eta)}{\|x^1 - x\|} = \int_{-\infty}^{\eta} \lim_{\|x^1-x\|\to 0} \frac{o(\|x^1 - x\|)}{\|x^1 - x\|} \, dz = 0.$$

Therefore, the error E is uniformly small with respect to $\|x^1 - x\|$ on the interval $[a, b]$. Consequently $\mathcal{G}(\cdot)$ is Fréchet differentiable and its derivative, as a continuous linear operator from $\mathfrak{X}$ to $\mathcal{C}([a, b])$, has the form (5.124). We infer that $\mathfrak{L}_1(\cdot, \mu)$ is continuously differentiable with the derivative

$$D_x \mathfrak{L}_1(x, \mu) = f'(x) + \int_a^b \int_{-\infty}^{\eta} D_x \varphi(x, z) \, dz \, \mu(d\eta). \tag{5.125}$$

Notice that the differential uniform dominance condition is equivalent to Robinson's constraint qualification condition:

$$0 \in \operatorname{int}\Big\{\mathcal{G}(\hat{x}) + D_x\mathcal{G}(\hat{x})(\mathcal{X} - \hat{x}) + K\Big\}. \tag{5.126}$$

Indeed, by definition, the uniform dominance condition implies Robinson's condition. On the other hand, if Robinson's condition holds, then a $\varepsilon > 0$ exists such that the function identically equal to $-\varepsilon$ is an element of the set at the right-hand side of (5.126). Then a point x^0 exists such that

$$[\mathcal{G}(\hat{x})](\eta) + \big[D_x\mathcal{G}(\hat{x})(\eta)\big](x^0 - \hat{x}) \le -\varepsilon \quad \text{for all } \eta \in [a, b].$$

Consequently, the uniform dominance condition is satisfied.

Hence, we can apply the necessary optimality conditions for problem (5.120)–(5.122). We infer that a nonnegative measure $\hat{\mu} \in \mathcal{M}^+([a, b])$ exists such that

$$-\, D_x \mathfrak{L}_1(\hat{x}, \hat{\mu}) \in \mathcal{N}_{\mathcal{X}}(\hat{x}), \tag{5.127}$$

$$\int_a^b \mathcal{G}(\hat{x})(\eta) \, \hat{\mu}(d\eta) = 0. \tag{5.128}$$

Using the correspondence between measures and utility functions, we obtain the equation

$$\begin{aligned}\int_a^b \mathcal{G}(x)(\eta) \, \hat{\mu}(d\eta) &= \int_a^b \big(F_{G(x)}(\eta) - F_Y(\eta)\big) \, \hat{\mu}(d\eta) \\ &= -\mathbb{E}\big[\hat{u}(G(x))\big] + \mathbb{E}\big[\hat{u}(Y)\big],\end{aligned}$$

where $\hat{u}(x)$ is an element of $\mathcal{U}_1([a,b])$. The correspondence between nonnegative measures $\mu \in \mathcal{M}^+([a,b])$ and utility functions $u \in \mathcal{U}_1([a,b])$ and the last equation imply that (5.128) is equivalent to (5.118). Moreover,

$$\mathfrak{L}_1(x,\mu) = L_1(x,u)$$

and, therefore, (5.127) is equivalent to (5.117). □

Remark 5.57. The set $\hat{U}$ of functions in $\mathcal{U}_1([a,b])$ satisfying (5.117)–(5.118) for the local minimum $\hat{x}$ is convex, bounded and weakly* closed in the following sense: if a sequence of functions $\{u^k\} \subset \hat{U}$ and $u \in \mathcal{U}_1([a,b])$ are such that

$$\lim_{k\to\infty} \mathbb{E}\big[u^k(Z)\big] = \mathbb{E}\big[u(Z)\big] \quad \text{for all bounded real random variables } Z,$$

then $u \in \hat{U}$. This statement follows from the fact that the set of Lagrange multipliers $\hat{\mu}$ corresponding to the Lagrangian $\mathfrak{L}_1$ is convex, bounded and weakly* closed in the dual space $\mathcal{M}^+([a,b])$ and the correspondence given by (5.27). □

We can show that problem (5.113)–(5.115) has an implicit finite representation when the decisions are finite-dimensional. In that case, it is sufficient to consider piecewise-constant utility functions only.

Theorem 5.58. *Assume that $\mathcal{X} = \mathbb{R}^n$ and let the assumptions of Theorem* 5.56 *be satisfied. Then a piecewise constant utility function $u(\cdot) \in \mathcal{U}_1([a,b])$ exists that satisfies the necessary optimality conditions* (5.117)–(5.118). *Moreover, the function $u(\cdot)$ has at least one and at most $n+2$ jump points: there exist numbers $\eta_k \in [a,b]$, $k = 1,\dots,m \le n+2$, such that $u(\cdot)$ is constant on the intervals $(-\infty,\eta_1]$, $(\eta_1,\eta_2]$, …, (η_m,∞).*

Proof. Consider the mapping $\mathcal{G}$ defined by (5.119) and recall that the derivative of the Lagrangian $\mathfrak{L}_1$ has the form (5.125). We observe that the necessary condition of optimality (5.127) can be re-written in the following equivalent form

$$\int_a^b D\mathcal{G}(\hat{x})(\eta)\,\mu(\mathrm{d}\eta) = g \in -f'(\hat{x}) - \mathcal{N}_{\mathcal{X}}(\hat{x}), \tag{5.129}$$

$$\int_a^b \mathcal{G}(\hat{x})(\eta)\,\mu(\mathrm{d}\eta) = 0. \tag{5.130}$$

Every nonnegative solution μ of equations (5.129)–(5.130) can be used as the Lagrange multiplier satisfying conditions (5.127)–(5.128) at $\hat{x}$. Let us denote $c = \hat{\mu}([a,b])$ and observe that $c > 0$ because $\hat{\mu} \neq 0$. Normalizing by c, we can express the left-hand sides of (5.129)–(5.130) in an equivalent way

$$\begin{bmatrix} g/c \\ 0 \end{bmatrix} \in \operatorname{conv} \left\{ \begin{bmatrix} D\mathcal{G}(\hat{x})(\eta) \\ \mathcal{G}(\hat{c})(\eta) \end{bmatrix} : \eta \in [a,b] \right\} \subset \mathbb{R}^{n+1}.$$

Using Carathéodory's Theorem we infer that there exist numbers $\eta_k \in [a,b]$, and $\alpha_k \geq 0$, $k = 1, \dots, m$, with $m \leq n+2$, such that

$$\begin{bmatrix} g/c \\ 0 \end{bmatrix} = \sum_{k=1}^{m} \alpha_k \begin{bmatrix} D\mathcal{G}(\hat{x})(\eta_k) \\ \mathcal{G}(\hat{x})(\eta_k) \end{bmatrix},$$

$$\sum_{k=1}^{m} \alpha_k = 1.$$

Now, we define the atomic measure ν having atoms of mass $c\alpha_k$ at points η_k, $k = 1, \dots, m$. The measure ν satisfies (5.129)–(5.130), which are equivalent to (5.127)–(5.128).

The form of the generator (5.26) entails that the utility function

$$u(t) = -\nu[t,b] \quad t \in \mathbb{R},$$

satisfies the assertion of the theorem. □

It follows from Theorem 5.58 that for a finite-dimensional decision space, at most $n+2$ target values η_k and target probabilities $v_k = P\big(Y \leq \eta_k\big)$, $k = 1, \dots, n+2$, exist which are critical for problem (5.113)–(5.115). They define a relaxation of that problem with finitely many chance constraints:

$$\begin{aligned} &\min\ f(x) \\ &\text{subject to } P\big[G(x) \leq \eta_k\big] \leq v_k, \quad k = 1, \dots, n+2, \\ &\qquad x \in \mathcal{X}. \end{aligned}$$

The necessary conditions of optimality for this relaxation yield a solution of the optimality conditions of the original problem (5.113)–(5.115). However, this representation is implicit and the critical target values and target probabilities are not known in advance.

Now, we focus on sufficient conditions of optimality for problem (5.113)–(5.115). In our analysis of such conditions, we assume that $f(\cdot)$ is convex and $G(\cdot)$ has the following form:

$$G(x) = g(x, Z), \tag{5.131}$$

where $g : \mathcal{X} \times \mathbb{R}^s \to \mathbb{R}$, and $Z : \Omega \to \mathbb{R}^s$ is a random vector.

Let us recall the notion of an r-concave function.

Definition 5.59. For a number $r \in [-\infty, \infty]$, a function $\varphi : \mathbb{R}^s \to \mathbb{R}_+$ is called *r-concave* if for all $x, y \in \mathbb{R}^s$ and all $\lambda \in [0,1]$,

$$\varphi(\lambda x + (1-\lambda)y) \geq \mathfrak{m}_r(\varphi(x), \varphi(y), \lambda),$$

where $\mathfrak{m}_r : \mathbb{R}_+ \times \mathbb{R}_+ \times [0,1] \to \mathbb{R}$ is defined as follows:

$$\mathfrak{m}_r(a, b, \lambda) = 0 \quad \text{for} \quad ab = 0 \text{ and } r \leq 0;$$

otherwise,

$$\mathfrak{m}_r(a, b, \lambda) = \begin{cases} a^\lambda b^{1-\lambda} & \text{if } r = 0, \\ \max\{a, b\} & \text{if } r = \infty, \\ \min\{a, b\} & \text{if } r = -\infty, \\ (\lambda a^r + (1-\lambda)b^r)^{1/r} & \text{otherwise.} \end{cases}$$

If $r = -\infty$, we call φ *quasi-concave*, for $r = 0$ it is known as *log-concave*, and for $r = 1$ the function φ is concave in the usual sense.

Lemma 5.60. *Assume that G has the structure* (5.131)*, the function g is quasi-concave, and the random vector Z has an r-concave probability density function with $r \geq -1/s$. Then the set defined by* (5.114) *is convex and closed.*

Proof. Let us fix $\eta \in [a, b]$ and consider the function

$$P\big[G(x) \leq \eta\big] = P\big[g(x, Z) - \eta \leq 0\big] = 1 - P\big[g(z, V) - \eta > 0\big].$$

We can formulate constraint (5.114) as follows:

$$P\big[g(x, Z) - \eta > 0\big] \geq P\big[Y > \eta\big].$$

Denote the set of x satisfying this inequality by $X(\eta)$. The probability distribution of V is α-concave, where $\alpha = r/(1 + sr)$. We shall show that the function

$$h(x) = P\big[g(x, Z) - \eta > 0\big]$$

at the left-hand side of the last inequality is quasi-concave. This would imply that the set $X(\eta)$ is convex and closed. Consider two points $x^1, x^2 \in X(\eta)$ and $\lambda \in (0, 1)$. We define the sets

$$A_i = \{v \in \mathbb{R}^s : g(x^i, v) - \eta > 0\}, \quad i = 1, 2,$$

and let the set B be the Minkowski sum $B = \lambda A_1 + (1-\lambda)A_2$. We evaluate

$$h(\lambda x^1 + (1-\lambda)x^2) = P\big[g(\lambda x^1 + (1-\lambda)x^2, V) - \eta > 0\big].$$

If $w \in B$, then $v = \lambda v^1 + (1-\lambda)v^2$ for some $v^1 \in A_1$ and $v^2 \in A_2$. Due to the quasi-concavity of g, we have

$$g(\lambda x^1 + (1-\lambda)x^2, \lambda v^1 + (1-\lambda)v^2) \geq \min\big(g(x^1, v^1), g(x^2, v^2)\big) > \eta.$$

Hence $w \in \{v \in \mathbb{R}^s : g(\lambda x^1 + (1-\lambda)x^2, v) - \eta > 0\}$, which entails that

$$h(\lambda x^1 + (1-\lambda)x^2) \geq P(B).$$

Due to the α-concavity of the measure, the following relations hold:

$$h(\lambda x^1 + (1-\lambda)x^2) \geq P(B) \geq \mathfrak{m}_\alpha(P(A_1), P(A_2), \lambda) = \\ \mathfrak{m}_\alpha(h(x^1), h(x^2), \lambda) \geq P\big[Y > \eta\big].$$

The last inequality follows from the fact that $x^1, x^2 \in X(\eta)$. This shows that the set $X(\eta)$ is convex. It is closed due to the continuity of the α-concave functions. The set defined by (5.114) is the intersection of the sets $X(\eta)$ for $\eta \in [a, b]$. Therefore, it is convex and closed. □

Now we are ready to establish the sufficient conditions of optimality for problem

Theorem 5.61. *Assume that a point $\hat{x}$ satisfies* (5.113)–(5.115). *Suppose a function $\hat{u} \in \mathcal{U}_1([a, b])$ exists, $\hat{u} \neq 0$, such that conditions* (5.117)–(5.118) *are satisfied. If the function $f(\cdot)$ is convex and the function $G(\cdot)$ satisfies the assumptions of Lemma* 5.60, *then $\hat{x}$ is an optimal solution of problem* (5.113)–(5.115).

Proof. Using the correspondence between measures and functions in $\mathcal{U}_1([a, b])$, we observe that optimality conditions (5.127)–(5.128) for problem (5.120)–(5.122) are satisfied. Lemma 5.60 entails that the feasible set defined by (5.120)–(5.122) is convex and closed because it is the intersection of the closed convex defined by (5.114) and $\mathcal{X}$, which is assumed closed and convex.

Let the operator $\mathcal{G}$ and the cone K be defined as in the proof of Theorem 5.56. Consider a feasible direction d at the point $\hat{x}$. By the convexity of the feasible set

$$\mathcal{G}(\hat{x} + \tau d) \in K,$$

for all sufficiently small $\tau > 0$. Since $\mathcal{G}$ is differentiable,

$$\big[D\mathcal{G}(\hat{x})\big] d \in \mathcal{T}_K(\mathcal{G}(\hat{x})),$$

where $\mathcal{T}_K(z)$ denotes the tangent cone to K at z. We observe that $t \in \mathbb{R}$ exists such that

$$\big[D\mathcal{G}(\hat{x})\big] d + t\mathcal{G}(\hat{x}) \in K \tag{5.132}$$

because of the form of the tangent cone at a point $w \in K$:

$$\mathcal{T}_K(w) = K + \{tw : t \in \mathbb{R}\}.$$

Condition (5.127) implies that a point $q \in -\mathcal{N}_{\mathcal{X}}(\hat{x})$ exists such that

$$f'(\hat{x}) + \int_a^b D\mathcal{G}(\hat{x})(\eta)\ \mu(\mathrm{d}\eta) = q.$$

Applying both sides of this equation to the direction d and using the fact that $q \in -\mathcal{N}_{\mathcal{X}}(\hat{x})$ and $d \in T_{\mathcal{X}}(\hat{x})$ we obtain:

$$f'(\hat{x})(d) + \int_a^b \big[D\mathcal{G}(\hat{x})(\eta)\big](d)\,\mu(\mathrm{d}\eta) \geq 0. \tag{5.133}$$

Condition (5.128), relation (5.132), and the nonnegativity of μ imply that

$$\int_a^b \big(D\mathcal{G}(\hat{x})(\eta)\big)(d)\,\mu(\mathrm{d}\eta) = \int_a^b \Big(\big[D\mathcal{G}(\hat{x})(\eta)\big](d) + t\mathcal{G}(\hat{x})(\eta)\Big)\,\mu(\mathrm{d}\eta) \leq 0.$$

Substituting into (5.133) we conclude that

$$f'(\hat{x})(d) \geq 0,$$

for every feasible direction d at $\hat{z}$. Hence for feasible point x, using the convexity of f, we obtain the inequality

$$f(x) \geq f(\hat{x}) + f'(\hat{x})(d) \geq f(\hat{x}).$$

This proves that $\hat{x}$ is optimal. □

In view of Theorem 5.30, we can consider the following relaxation of problem (5.113)–(5.115):

$$\min f(x) \tag{5.134}$$
$$\text{subject to } G(x) \succeq_{(2)} Y \tag{5.135}$$
$$x \in \mathcal{X}. \tag{5.136}$$

If the function $f(\cdot)$ is convex, $G(\cdot)$ is concave, and the set $\mathcal{X}$ is convex, then problem (5.134)–(5.136) is a convex optimization problem. However, it may not be the tightest convexification of (5.113)–(5.115). We can only state the following inclusion for atomless probability spaces:

$$\begin{aligned}
&\overline{\text{conv}}\big\{x \in \mathcal{X} : G(x) \succeq_{(1)} Y\big\}\\
&\quad = \overline{\text{conv}}\big\{x \in \mathcal{X} : G(x) \in \mathcal{A}_1^{\sharp}(Y)\big\}\\
&\quad \subseteq \big\{x \in \mathcal{X} : G(x) \in \overline{\text{conv}}\big[\mathcal{A}_1^{\sharp}(Y)\big]\big\}\\
&\quad = \big\{x \in \mathcal{X} : G(x) \in \mathcal{A}_2(Y)\big\} = \big\{x \in \mathcal{X} : G(x) \succeq_{(2)} Y\big\}.
\end{aligned}$$

For example, if the assumptions of Theorem 5.61 are satisfied, then the feasible set of problem (5.113)–(5.115) is already convex. However, it may be smaller than the set $\big\{x \in \mathcal{X} : G(x) \succeq_{(2)} Y\big\}$.

Chapter 6
Multivariate and Sequential Stochastic Orders

6.1 Definition and Representations of Multivariate Orders

We shall discuss problems with stochastic orders that compare random vectors instead or scalar-valued outcomes. Let $(\Omega, \mathcal{F}, P)$ be a probability space. For any measurable mapping $X : \Omega \to \mathbb{R}^m$, we use the norm $\|X\|_p$ defined by

$$\|X\|_p = \left(\mathbb{E}\left[|X(\omega)|^p\right]\right)^{1/p}, \quad p \in [1, \infty),$$

where $|\cdot|$ denotes a norm in $\mathbb{R}^m$. The standard symbol $\mathcal{L}_p(\Omega, \mathcal{F}, P; \mathbb{R}^m)$, denotes the space of random elements X from Ω to $\mathbb{R}^m$ with $\|X\|_p < \infty$.

Given two elements $X, Y \in \mathcal{L}_p(\Omega, \mathcal{F}, P; \mathbb{R}^m)$, the simplest way to define a stochastic order relation between these vectors is to require a univariate stochastic order relation for each coordinate

$$X_i \succeq Y_i, \quad i = 1, \ldots, m. \tag{6.1}$$

The analysis in Chapter 5 extends to this case in a straightforward manner. This approach, however, ignores the dependence between the coordinates of the two vectors. In the case of the first-order dominance, we could directly extend Definition 5.2 to the distribution functions of vectors. This is, however, associated with conceptual difficulties and no natural extension to the second and higher orders stochastic dominance is available.

Here, we adopt the following definition using linear scalarization with positive coefficients.

Definition 6.1. A random vector $X \in \mathcal{L}_p(\Omega, \mathcal{F}, P; \mathbb{R}^m)$ dominates (linearly) in order p a random vector $Y \in \mathcal{L}_p(\Omega, \mathcal{F}, P; \mathbb{R}^m)$, if for all $c \in \mathbb{R}^m_+$

$$c^\top X \succeq_{(p)} c^\top Y. \tag{6.2}$$

D. Dentcheva, A. Ruszczyński, *Risk-Averse Optimization and Control*, Springer Series in Operations Research and Financial Engineering,
https://doi.org/10.1007/978-3-031-57988-2_6

We denote this relation by $X \succeq^{\text{lin}}_{(p)} Y$.

Remark 6.2. It is clear that the set of the vectors c in (6.2) may be reduced by normalization to the scalarizations $c \in \mathbb{S}^m$, where $\mathbb{S}^m$ is the simplex:

$$\mathbb{S}^m = \{c \in \mathbb{R}^m_+ : c_1 + c_2 + \cdots + c_m = 1\}.$$

This restriction does not change the relation $\succeq^{\text{lin}}_{(p)}$.

Indeed, if (6.2) is true for all $c \in \mathbb{S}^m$, then it is equivalent to the relation between the shortfall functions for all scalarizations, i.e., for all $\eta \in \mathbb{R}$ and for all $c \in \mathbb{S}^m$;

$$\mathbb{E}\big[\max(0, \eta - c^\top X)^{p-1}\big] \le \mathbb{E}\big[\max(0, \eta - c^\top Y)^{p-1}\big]. \tag{6.3}$$

If $\tilde{c} \in \mathbb{R}^m_+$, then we consider $c' = \frac{\tilde{c}}{\sum_{i=1}^m \tilde{c}_i}$, which is an element of $\mathbb{S}^m$, and observe that (6.3) for $c = c'$ implies that

$$\mathbb{E}\big[\max(0, \eta' - \tilde{c}^\top X)^{p-1}\big] \le \mathbb{E}\big[\max(0, \eta' - \tilde{c}^\top Y)^{p-1}\big],$$

with $\eta' = \eta \sum_{i=1}^m \tilde{c}_i$. As η runs over all of $\mathbb{R}$, so does η'. □

We notice that the linear order in Definition 6.1 implies the coordinate order but is not equivalent to it. The multivariate stochastic order relation are sometimes associated with scalarizations taken from the entire space $\mathbb{R}^m$ or from a given compact set $\tilde{S} \subset \mathbb{R}^m$. We use only scalarizations c with nonnegative components because we want to preserve the preference to large vs. small outcomes for every component of the random vector.

Additionally, we allow for relaxing the univariate comparison (6.2) by restricting the required inequality (6.3) to some compact interval $[a_c, b_c] \subset \mathbb{R}$ instead of the entire real line. Assume that $\mathfrak{J} : \mathbb{S}^m \rightrightarrows \mathbb{R}$ is a Lebesgue-measurable multifunction whose images are closed bounded intervals and let graph($\mathfrak{J}$) denote its graph:

$$\text{graph}(\mathfrak{J}) = \{(c, \eta) : \eta \in \mathfrak{J}(c),\ c \in \mathbb{S}^m\}.$$

Definition 6.3. A random vector $X \in \mathcal{L}_p(\Omega, \mathcal{F}, P; \mathbb{R}^m)$ is stochastically larger in order p than a random vector $Y \in \mathcal{L}_p(\Omega, \mathcal{F}, P; \mathbb{R}^m)$ with respect to $\mathfrak{J}$ (denoted $X \succeq^{\mathfrak{J}}_{(p)} Y$), if

$$F^{(p)}_{c^\top X}(\eta) \le F^{(p)}_{c^\top Y}(\eta) \quad \text{for all } (c, \eta) \in \text{graph}(\mathfrak{J}).$$

Remark 6.4. A few remarks are in order here.

1. Although we consider only scalarizations in the simplex $\mathbb{S}^m$, our analysis could be carried out similarly when $\mathbb{S}^m$ is substituted by any compact polyhedral set of scalarizations with non-negative components.

2. We note that the intervals $[a_c, b_c]$ could be chosen very large so that $P\big(c^\top X \notin [a_c, b_c]\big) \le \varepsilon$ for a very small $\varepsilon > 0$. Additionally, if the probability space Ω is finite, then the intervals $[a_c, b_c]$ may be chosen to contain the entire supports of $c^\top X$ and $c^\top Y$.
3. The mapping $\mathfrak{I}$ helps us to add flexibility in the choice of the part of the distribution that is restricted by the order relation. For example, if the random vector X consisted of two components, one of which is of bounded support and another one of unbounded support with high variance, fixing the interval $[a, b]$ to be the same for both scalarizing vectors $c^1 = (1, 0)$ and $c^2 = (0, 1)$, might substantially impair our ability to impose meaningful restrictions on the distribution of the second component. One way to define the mapping $\mathfrak{I}$ would be to select intervals $[a_i, b_i]$ associated with the unit vectors $e^i \in \mathbb{R}^m$, which would provide a constraint on the marginal distributions of the random vector. Then for every other vector $c \in \mathbb{S}^m$, one could choose $a_c = \sum_{i=1}^m c_i a_i$ and $b_c = \sum_{i=1}^m c_i b_i$. Other possibly non-linear combinations of the marginal boundaries might be chosen as well. □

Using (5.26), we obtain the following statement.

Corollary 6.5. *For two random vectors $X, Y \in \mathcal{L}_{p-1}(\Omega, \mathcal{F}, P; \mathbb{R}^m)$, the relation $X \succeq_{(p)}^{\mathfrak{I}} Y$ is equivalent to*

$$\mathbb{E}\big[u(c^\top X)\big] \ge \mathbb{E}\big[u(c^\top Y)\big] \quad \textit{for all } u \in \mathcal{U}_p(\mathfrak{I}(c)) \textit{ and all } c \in \mathbb{S}^m. \tag{6.4}$$

Consider a mapping $\psi_{\mathfrak{I}} : \mathbb{S}^m \to \mathcal{U}_p(\mathfrak{I}(c))$ such that the functions $(c, x) \to [\psi_{\mathfrak{I}}(c)](c^\top x)$ are Lebesgue measurable on $\mathbb{S}^m \times \mathbb{R}^m$. The collection of all such mappings $\psi_{\mathfrak{I}}$ is denoted by $\Psi_p(\mathfrak{I})$.

Every pair consisting of a mapping $\psi \in \Psi_p(\mathfrak{I})$ and a measure μ from $\mathcal{M}^+(\mathbb{S}^m)$ gives rise to a function $\varphi_{\psi,\mu} : \mathbb{R}^m \to \mathbb{R}$ defined as follows:

$$\varphi_{\psi,\mu}(x) = \int_S [\psi(c)](c^\top x)\, \mu(\mathrm{d}c).$$

Recall that $\mathcal{M}^+(\mathbb{S}^m)$ stands for the regular nonnegative and finite countably additive measures on $\mathbb{S}^m$. We consider the collection of all functions of that form:

$$\boldsymbol{\mathcal{U}}_p(\mathfrak{I}) = \{\varphi_{\psi,\mu} : \psi \in \Psi_p(\mathfrak{I}),\ \mu \in \mathcal{M}^+(\mathbb{S}^m)\}$$

and we shall demonstrate that $\boldsymbol{\mathcal{U}}_p(\mathfrak{I})$ is a generator of the order $\succeq_{(p)}^{\mathfrak{I}}$.

Proposition 6.6. *For each pair of random vectors $X, Y \in \mathcal{L}_{p-1}(\Omega, \mathcal{F}, P; \mathbb{R}^m)$ the relation $X \succeq_{(p)}^{\mathfrak{I}} Y$ is equivalent to*

$$\mathbb{E}[\varphi(X)] \ge \mathbb{E}[\varphi(Y)] \quad \textit{for all } \varphi \in \boldsymbol{\mathcal{U}}_p(\mathfrak{I}). \tag{6.5}$$

Proof. Assume $X \succeq^{\mathfrak{I}}_{(p)} Y$ and take any $\psi \in \Psi_p(\mathfrak{I})$. It follows from Corollary 6.5 that

$$\mathbb{E}\big([\psi(c)](c^\top X)\big) \geq \mathbb{E}\big([\psi(c)](c^\top Y)\big) \quad \text{for all } c \in \mathbb{S}^m.$$

Integrating these inequalities with respect to $\mu \in \mathcal{M}^+(\mathbb{S}^m)$ and exchanging the integral with respect to μ with the expectation, we obtain (6.5). When we apply constant mappings $\psi_{\mathfrak{I}}(c) = u \in \mathcal{U}_p(\mathfrak{I}(c))$ and atomic measures μ in the definition of the function $\varphi(\cdot)$, which is used in (6.5), we infer the converse implication. □

6.2 Optimality Conditions for Problems with Multivariate Stochastic Dominance Constraints

6.2.1 Optimality Conditions

We assume that the decision space $\mathcal{X}$ is a separable Banach space and a benchmark random vector $Y \in \mathcal{L}_{p-1}(\Omega, \mathcal{F}, P; \mathbb{R}^m)$ with an acceptable distribution is given. Furthermore, let a Lebesgue measurable multivariate mapping $\mathfrak{I} : \mathbb{S}^m \rightrightarrows \mathbb{R}$ with convex and compact images and a compact graph be fixed.

We consider the following optimization problem for $p \geq 2$ with a direct formulation of the multivariate stochastic order constraint.

$$\min f(x) \tag{6.6}$$

$$\text{subject to } F^{(p)}_{c^\top G(x)}(\eta) \leq F^{(p)}_{c^\top Y}(\eta) \text{ for all } (c, \eta) \in \operatorname{graph}(\mathfrak{I}), \tag{6.7}$$

$$x \in X. \tag{6.8}$$

In this section, we adopt the counterpart to Assumption 5.34.

Assumption 6.7.

(i) *The functional* $f : \mathcal{X} \to \mathbb{R}$ *is continuous and convex.*

(ii) *The operator* $G : \mathcal{X} \to \mathcal{L}_{p-1}(\Omega, \mathcal{F}, P; \mathbb{R}^m)$ *is a norm-continuous and concave in the following sense: for all* $x, y \in \mathcal{X}$ *and* $\alpha \in (0,1)$, *the inclusion* $G(\alpha x + (1-\alpha)y) - \alpha G(x) - (1-\alpha)G(y) \in \mathbb{R}^m_+$ *holds* P*-a.s..*

(iii) *The set* $X \subset \mathcal{X}$ *is closed and convex.*

We define the utility-modified functional $L_p : \mathcal{X} \times \boldsymbol{\mathcal{U}}_p(\mathfrak{I}) \to \mathbb{R}$ associated with problem (6.6)–(6.8):

$$L_p(x, \varphi) = \mathbb{E}\left[f(x) + \big(\varphi(Y) - \varphi(G(x))\big)\right].$$

The uniform dominance condition (5.35) is generalized to the multivariate case as follows.

Definition 6.8. Problem (6.6)–(6.8) satisfies the *multivariate uniform dominance condition* if a point $\tilde{x} \in \mathcal{X}$ exists such that

$$\inf_{(c,\eta)\in\text{graph}(\mathfrak{I})} \left\{ F^{(p)}_{c^\top Y}(\eta) - F^{(p)}_{c^\top G(\tilde{x})}(\eta) \right\} > 0.$$

Now, we shall establish necessary and sufficient optimality conditions for problem (6.6)–(6.8).

Theorem 6.9. *Assume that condition* (6.8) *is satisfied. If* $\hat{x}$ *is an optimal solution of* (6.6)–(6.8), *then a function* $\hat{\varphi} \in \mathcal{U}_p(\mathfrak{I})$ *exists such that*

$$L_p(\hat{x},\hat{\varphi}) = \min_{x\in\mathcal{X}} L(x,\hat{\varphi}), \tag{6.9}$$

$$\mathbb{E}[\hat{\varphi}(G(\hat{x}))] = \mathbb{E}[\hat{\varphi}(Y)]. \tag{6.10}$$

Conversely, if for some function $\hat{\varphi} \in \mathcal{U}_p(\mathfrak{I})$ *an optimal solution* $\hat{x}$ *of* (6.9) *satisfies* (6.7) *and* (6.10), *then* $\hat{x}$ *is an optimal solution of* (6.6)–(6.8).

Proof. We proceed as in the proof of Theorem 5.36 for the univariate case. We define the operator $\mathfrak{G}_p : \mathfrak{X} \to \mathcal{C}(\text{graph}(\mathfrak{I}))$ by setting

$$[\mathfrak{G}_p(x)](c,\eta) = \mathbb{E}\big[\max(0,\eta - c^\top G(x))^{p-1}\big] - \mathbb{E}\big[\max(0,\eta - c^\top Y)^{p-1}\big] \quad \text{for all } (c,\eta) \in \text{graph}(\mathfrak{I}). \tag{6.11}$$

Now problem (6.6)–(6.8) can be re-written in the following general form:

$$\begin{aligned} &\min f(x)\\ &\text{subject to } \mathfrak{G}_p(x) \in K,\\ &x \in \mathcal{X}, \end{aligned}$$

where K is the cone of non-positive continuous functions defined on graph($\mathfrak{I}$). Recall that graph($\mathfrak{I}$) is compact and each image $\mathfrak{I}(c)$ is a closed and bounded interval by assumption. With the continuity assumption on G, we establish that $\mathfrak{G}_p(x)$ is norm-to-norm continuous in the same way as in the proof of Theorem 5.36. Analogously, the operator $\mathfrak{G}_p$ is convex with respect to the cone K. We use again the Riesz representation theorem on duality of the space $\mathcal{C}(\text{graph}(\mathfrak{I}))$ of continuous functions on graph($\mathfrak{I}$) and the set of regular countably additive non-negative measures in $\mathcal{M}^+(\text{graph}(\mathfrak{I}))$. The Lagrangian $\mathfrak{L}_p : \mathfrak{X} \times \mathcal{M}^+(\text{graph}(\mathfrak{I})) \to \mathbb{R}$ of the general problem has the form

$$\mathfrak{L}_p(x,\nu) = f(x) + \int_{\text{graph}(\mathfrak{I})} [\mathfrak{G}_p(x)](c,\eta)\ \nu(\mathrm{d}(c,\eta)). \tag{6.12}$$

The multivariate uniform dominance condition (6.8) allows the application of the necessary conditions of optimality associated with the Lagrangian (6.12). We infer the existence of a nonnegative measure $\hat{\nu} \in \mathcal{M}^+(\text{graph}(\mathfrak{F}))$ such that

$$\mathfrak{L}_p(\hat{x}, \hat{\nu}) = \min_{x \in \mathcal{X}} \mathfrak{L}_p(x, \hat{\nu}) \tag{6.13}$$

and

$$\int_{\text{graph}(\mathfrak{F})} [\mathfrak{G}_p(\hat{x})](c, \eta)\ \hat{\nu}(\mathrm{d}(c, \eta)) = 0. \tag{6.14}$$

We shall transform these conditions to (6.9)–(6.10).

Since the graph of $\mathfrak{F}(\cdot)$ is a compact set, we can consider $\mathfrak{F}(\mathbb{S}^m) \subset [a, b]$ for sufficiently large interval $[a, b]$. Any nonnegative measure $\nu \in \mathcal{M}^+(\text{graph}(\mathfrak{F}))$ can be extended to $\tilde{\nu} \in \mathcal{M}^+(\mathbb{S}^m \times [a, b])$. Let ν_c be the conditional measure of $\tilde{\nu}$ on $\mathbb{R}$ given $c \in \mathbb{S}^m$ (the kernel), and μ the marginal measure of $\tilde{\nu}$ on $\mathbb{S}^m$. The integral over graph$(\mathfrak{F})$ can be written as an iterated integral

$$\int_{\text{graph}(\mathfrak{F})} \mathbb{E}\big[\max(0, \eta - c^\top Y)^{p-1}\big]\ \mathrm{d}\tilde{\nu} =$$

$$\int_{\mathbb{S}^m} \Big(\int_a^b \mathbb{E}\big[\max(0, \eta - c^\top Y)^{p-1}\big]\ \nu_c(\mathrm{d}\eta)\Big)\ \mu(\mathrm{d}c). \tag{6.15}$$

Since for every Lebesgue measurable set $B \subset [a, b]$ the mapping $c \to \nu_c(B)$ is measurable with respect to the Borel σ-algebra on S, then the mapping

$$(c, x) \mapsto -\int_{c^\top x}^b \nu_c([\tau, b])\ \mathrm{d}\tau$$

is measurable. We associate a function $u_c : \mathbb{R} \to \mathbb{R}$ with the conditional measure ν_c according to relation (5.27). We obtain that the mapping $c \mapsto u_c$ is an element of the family $\Psi_p(\mathfrak{F})$. Now, we can re-write (6.15) as follows:

$$\int_{\text{graph}(\mathfrak{F})} \mathbb{E}\big[\max(0, \eta - c^\top Y)^{p-1}\big]\ \mathrm{d}\nu = -\int_{\mathbb{S}^m} \mathbb{E}\,[u_c(c^\top Y)]\ \mu(\mathrm{d}c)$$

$$= -\mathbb{E}\int_{\mathbb{S}^m} u_c(c^\top Y)\ \mu(\mathrm{d}c) = -\mathbb{E}\big[\varphi(Y)\big]. \tag{6.16}$$

We observe that for all $c \in \mathbb{S}^m$, $\nu_c([a, b] \setminus [a_c, b_c]) = 0$ by the definition of the measure ν and its extension $\tilde{\nu}$. Hence, the function

$$\varphi(y) = \int_{\mathbb{S}^m} u_c(c^\top y)\, \mu(\mathrm{d}c) \in \boldsymbol{\mathcal{U}}_p(\mathfrak{F}).$$

In this way, we have established a correspondence between nonnegative measures in $\mathcal{M}^+(\text{graph}(\mathfrak{F}))$ and functions in $\boldsymbol{\mathcal{U}}_p(\mathfrak{F})$. Using that correspondence for the measure $\hat{\nu}$, we obtain a function $\hat{\varphi} \in \boldsymbol{\mathcal{U}}_p(\mathfrak{F})$. Then we transform condition (6.13) to (6.9), and condition (6.14) to (6.10).

Now, we turn to the converse statement. If $\varphi \in \boldsymbol{\mathcal{U}}_p(\mathfrak{F})$ then a mapping $\psi_{\mathfrak{F}} : c \mapsto \mathcal{U}_p(\mathfrak{F}(c))$ and a measure $\mu \in \mathcal{M}^+(\mathbb{S})$ exist such that

$$\varphi(y) \equiv \int_{\mathbb{S}^m} [\psi_{\mathfrak{F}}(c)](c^\top y)\, \mu(\mathrm{d}c).$$

Consider the left derivative of $[\psi_{\mathfrak{F}}(c)](\cdot)$, for a fixed c:

$$[\psi_{\mathfrak{F}}(c)]'_-(t) = \lim_{\tau \uparrow t} \big([\psi_{\mathfrak{F}}(c)](t) - [\psi_{\mathfrak{F}}(c)](\tau)\big)/(t-\tau).$$

It is well-defined, nonincreasing, and continuous from the left by the definition of the family $\mathcal{U}_p(\mathfrak{F}(c))$. We obtain a regular finite nonnegative measure ν_c satisfying

$$\nu_c([t, b_c]) = [\psi_{\mathfrak{F}}(c)]'_-(t).$$

We extend ν_c to the entire interval $[a,b]$ and denote the extension by $\tilde{\nu}_c$. Due to the measurability of $\psi_{\mathfrak{F}}$, the mapping $c \to \tilde{\nu}_c(B)$ is measurable, for every measurable set $B \subset [a,b]$.

We can now define a measure ν on graph$(\mathfrak{F})$ as the composition of the measure μ on S and the extended measures $\tilde{\nu}_c$ for $c \in S$, by viewing μ as the marginal measure, and $\tilde{\nu}_c$ as a kernel.

Consequently, the correspondence between measures in $\mathcal{M}^+(\text{graph}(\mathfrak{F}))$ and functions in $\varphi \in \boldsymbol{\mathcal{U}}_p(\mathfrak{F})$ is one-to-one and formula (6.16) is always valid. Therefore, the minimizer $\hat{x}$ of (6.9) is also the minimizer of $\mathfrak{L}_p(x, \hat{\nu})$, where $\hat{\nu}$ is derived from $\hat{\varphi}$ in the way described above. It follows from sufficient conditions of optimality that if $\hat{x}$ satisfies (6.7) and (6.10) then it is optimal for (6.6)–(6.8).

Let us define the dual functional $L_D : \boldsymbol{\mathcal{U}}_p(\mathfrak{F}) \to \overline{\mathbb{R}}$ associated with problem (6.6)–(6.8) as follows:

$$L_D(\varphi) = \inf_{x \in \mathcal{X}} L(x, \varphi) = \inf_{x \in \mathcal{X}} \big\{f(x) - \mathbb{E}\big[\varphi(G(x)) - \varphi(Y)\big]\big\}. \tag{6.17}$$

We also define the dual problem:

$$\max\,\{L_D(\varphi) : \varphi \in \boldsymbol{\mathcal{U}}_p(\mathfrak{F})\}. \tag{6.18}$$

As a direct consequence of Theorem 6.9, we obtain the following duality theorem.

Theorem 6.10. *Suppose Assumption* 6.7 *and the uniform dominance condition* (6.8) *are satisfied. If problem* (6.6)–(6.8) *has an optimal solution, then the dual problem* (6.18) *has an optimal solution as well, and the optimal values of both problems coincide. Furthermore, for every solution* $\hat{\varphi}$ *of the dual problem, any optimal solution* $\hat{x}$ *of* (6.9) *satisfying* (6.7) *and* (6.10), *is an optimal solution of the primal problem* (6.6)–(6.8).

This result provides a theoretical foundation for the possible development of specialized dual decomposition methods for optimization involving multivariate stochastic dominance constraints.

6.2.2 Optimality Conditions in a Subgradient Form

In this section, we assume for simplicity that $\mathfrak{I}(c) = [a, b]$ for all $c \in S$. Additionally, we adopt an assumption about the structure of the operator G.

Assumption 6.11. *Suppose* $G : \mathcal{X} \to \mathcal{L}_{p-1}(\Omega, \mathcal{F}, P; \mathbb{R}^m)$ *is defined as*

$$[G(x)](\omega) = g(x, \omega),$$

where $g : \mathcal{X} \times \Omega \to \mathbb{R}^m$ *is such that each component* $g_i(\cdot, \cdot)$, $i = 1, \ldots m$ *is a Carathéodory function and* $g_i(\cdot, \omega)$ *is concave for all* $\omega \in \Omega$. *Furthermore, assume that* $\mathbb{E}[G(\cdot)]$ *is a continuous function.*

If Assumption 6.11 is satisfied, then Theorem 3.11 implies that the operator $G : \mathcal{X} \to \mathcal{L}_{p-1}(\Omega, \mathcal{F}, P; \mathbb{R}^m)$ is subdifferentiable at $\hat{x}$ and its subdifferential is given by the formula

$$\partial[G(\hat{x})] = \big\{S : \mathcal{X} \to \mathcal{L}_{p-1}(\Omega, \mathcal{F}, P; \mathbb{R}^m) : [Sd](\omega) = s(\omega)d,\ \omega \in \Omega,$$
$$\text{for some } s \in \mathcal{L}_{p-1}\big(\Omega, \mathcal{F}, P; (\mathcal{X}^*)^m\big) \text{ such that } s(\omega) \in \partial g(\hat{x}, \omega) \text{ a.s.}\big\}.$$

Theorem 6.12. *If the assumptions of Theorem* 6.9 *and Assumption* 6.11 *are satisfied, then condition* (6.9) *is equivalent to the existence of an operator* $S \in \partial[G(\hat{x})]$ *such that*

$$0 \in \partial f(\hat{x}) - \mathbb{E}[S^* \partial\hat{\varphi}(G(\hat{x}))] + \mathcal{N}_{\mathcal{X}}(\hat{x}), \tag{6.19}$$

where $\hat{\varphi} \in \mathcal{U}_p([a, b])$ *is the function appearing in* (6.9) *and* $\partial\hat{\varphi}(G(\hat{x}))$ *is the concave subdifferential of* $\hat{\varphi}(\cdot)$ *calculated at* $G(\hat{x})$.

Proof. Using Lemma 5.39, we infer that the functional $Z \mapsto F^{(p)}_{c^\top Z}(\eta)$ for $Z \in \mathcal{L}_{p-1}(\Omega, \mathcal{F}, P; \mathbb{R}^m)$ is subdifferentiable and its subdifferential is described as follows:

$$\partial_Z F^{(p)}_{c^\top Z}(\eta) = \begin{cases} \{c\xi : \xi \in DF(c^\top Z, \eta, \omega)\} & \text{for } p = 2, \\ \{-\frac{1}{\Gamma(p)} \max(0, \eta - c^\top Z)^{p-1} c\} & \text{for } p > 2. \end{cases}$$

Recall that the mapping DF is defined in (5.67) and $DF(c^\top Z, \eta, \omega)$ is given by

$$DF(c^\top Z, \eta, \omega) = \begin{cases} \{-1\} & \text{if } c^\top Z(\omega) < \eta, \\ [-1, 0] & \text{if } c^\top Z(\omega) = \eta, \\ \{0\} & \text{if } c^\top Z(\omega) > \eta. \end{cases} \tag{6.20}$$

The Lagrangian $\mathfrak{L}_p$ defined in (6.12) is a convex functional of the first argument and condition (6.13) is equivalent to the condition

$$0 \in \partial_x \mathfrak{L}_p(\hat{x}, \hat{\nu}) + \mathcal{N}_{\mathcal{X}}(\hat{x}).$$

We shall calculate the subdifferential $\partial_x \mathfrak{L}_p(\hat{x}, \hat{\nu})$. For $p = 2$, we have

$$\partial_x [\mathfrak{G}_p(\hat{x})](c, \eta) = \big\{\mathbb{E}[S^* \xi c] : S \in \partial G(\hat{x}),\ \xi \in \partial_X F^{(2)}_X(\eta)\big\}.$$

The following chain of equations holds

$$\begin{aligned} \partial_x \int_{\mathbb{S}^m \times [a,b]} [\mathfrak{G}_p(\hat{x})](c, \eta)\ \hat{\nu}(\mathrm{d}(c, \eta)) &= \int_{\mathbb{S}^m \times [a,b]} \partial_x [\mathfrak{G}_p(\hat{x})](c, \eta)\ \hat{\nu}(\mathrm{d}(c, \eta)) \\ &= \int_{\mathbb{S}^m \times [a,b]} \mathbb{E}[s(\omega)\xi(\hat{x}, c, \eta, \omega)c]\ \hat{\nu}(\mathrm{d}(c, \eta)) \\ &= \mathbb{E}\Big[s(\omega) \int_{\mathbb{S}^m \times [a,b]} c \mathbb{1}_{\{c^\top g(\hat{x}, \omega) < \eta\}}\ \hat{\nu}(\mathrm{d}(c, \eta))\Big] \end{aligned} \tag{6.21}$$

We disintegrate again $\hat{\nu}$ into a marginal measure $\hat{\mu}$ on $\mathbb{S}^m$ and conditional measures $\hat{\nu}_c$ on $[a, b]$ given $c \in \mathbb{S}^m$. Define the function

$$\varphi_{\hat{\mu}}(y) = -\int_{\mathbb{S}^m} \max(0, \eta - c^\top y)\ \hat{\mu}(\mathrm{d}c).$$

We observe that $\varphi_{\hat{\mu}} \in \mathcal{U}_2([a, b])$ and the expression on the right-hand side of (6.21) is an element of $\partial_x \mathbb{E}\big[\varphi_{\hat{\mu}}(G(\hat{x}))\big]$.

Thus, condition (6.13) and the following condition are equivalent:

$$0 \in \partial f(\hat{x}) - \mathbb{E}\big[S^* \partial \varphi_{\hat{\nu}}(G(\hat{x}))\big] + \mathcal{N}_{\mathcal{X}}(\hat{x}). \tag{6.22}$$

As condition (6.13) is equivalent to (6.9), the statement of the theorem follows for $p = 2$. For $p > 2$, using the same line of argument, we obtain the relation:

$$
\begin{aligned}
&\partial_x \int_{\mathbb{S}^m\times[a,b]} [\mathfrak{G}_p(\hat{x})](c,\eta)\ \hat{\nu}(\mathrm{d}(c,\eta)) = \\
&-(p-1) \int_{\mathbb{S}^m\times[a,b]} \mathbb{E}\big[s(\omega)c\max(0,\eta - c^\top G(\hat{x})(\omega))^{p-2}\big]\ \hat{\nu}(\mathrm{d}(c,\eta)).
\end{aligned}
\tag{6.23}
$$

We identify the function

$$
\varphi_{\hat{\mu}}(y) = -\int_{S} \max(0,\eta - c^\top y)^{p-1}\ \hat{\mu}(\mathrm{d}c)\ \in \boldsymbol{\mathcal{U}}_p([a,b])
$$

and observe that the following is true:

$$
\begin{aligned}
-(p-1) &\int_{\mathbb{S}^m\times[a,b]} \mathbb{E}\big[s(\omega)c\max(0,\eta - c^\top G(\hat{x})(\omega))^{p-2}\big]\ \hat{\nu}(\mathrm{d}(c,\eta)) \\
&\in \mathbb{E}[S^*\ \partial\varphi_{\hat{\mu}}(G(\hat{x}))].
\end{aligned}
$$

This completes the proof for $p > 2$. □

6.2.3 Optimality Conditions for the Inverse Formulation

Now we fix a Lebesgue measurable multivariate mapping $\mathfrak{J} : \mathbb{S}^m \rightrightarrows [0,1]$ with a compact graph whose images are convex closed intervals included in $(0,1)$:

$$
\operatorname{graph}(\mathfrak{J}) = \big\{(c,p) :\ p \in \mathfrak{J}(c),\ c \in \mathbb{S}^m\big\}.
$$

We shall analyze the optimization problem for $p = 2$ with the inverse formulation of the multivariate stochastic order constraint based on the Lorenz functions of the random variables involved.

We introduce the following inverse relations of the multivariate order associated with the mapping $\mathfrak{J}$.

Definition 6.13. For two random vectors $Z, Y \in \mathcal{L}_1(\Omega, \mathcal{F}, P; \mathbb{R}^m)$, the relation $Z \succeq_{(2)}^{\mathfrak{J}} Y$ holds if and only if

$$
F_{c^\top Z}^{(-2)}(p) \geq F_{c^\top Y}^{(-2)}(p) \quad \text{for all } (c,p) \in \operatorname{graph}(\mathfrak{J}).
$$

Let us consider a mapping ϑ that assigns to each $c \in \mathbb{S}^m$ a function $\vartheta_c \in \mathcal{W}_2(\mathfrak{J}(c))$. The collection of all such mappings ϑ is denoted by $\Theta(\mathfrak{J})$.

Given a finite measure $\theta \in \mathcal{M}^+(\text{graph}(\mathfrak{J}))$, we disintegrate it into a conditional measure ν_c on $\mathfrak{J}(c)$ and a marginal measure $\mu \in \mathcal{M}^+(\mathbb{S}^m)$. There is one-to-one correspondence of the conditional measure ν_c to distortion functions $w \in \mathcal{W}_2(\mathfrak{J}(c))$, as it was shown in Lemma 5.17. For all $\vartheta \in \Theta(\mathfrak{J})$ and $\mu \in \mathcal{M}^+(\mathbb{S}^m)$, we define $\varphi_{\vartheta,\mu}$ as follows:

$$\varphi_{\vartheta,\mu}(Z) = \int_{\mathbb{S}^m} \int_{\mathfrak{J}(c)} F^{(-1)}_{c^\top Z}(p)\, \mathrm{d}\vartheta_c(p)\, \mu(\mathrm{d}c). \tag{6.24}$$

Hence, we obtain the following equivalent representation of the multivariate relation $\succeq^{\mathfrak{J}}_{(2)}$.

Corollary 6.14. *For $Z, Y \in \mathcal{L}_1(\Omega, \mathcal{F}, P; \mathbb{R}^m)$, the relation $Z \succeq^{\mathfrak{J}}_{(2)} Y$ holds if and only if for all $\vartheta \in \Theta(\mathfrak{J})$ and for all $\mu \in \mathcal{M}^+(S)$*

$$\varphi_{\vartheta,\mu}(Z) \geq \varphi_{\vartheta,\mu}(Y). \tag{6.25}$$

We also apply Lemma 5.17 to establish the one-to one correspondence between these distortion functions, measures $\theta \in \mathcal{M}^+(\text{graph}(\mathfrak{J}))$, and coherent measures of risk. On the one hand, we have

$$\int_{\text{graph}(\mathfrak{J})} F^{(-2)}_{c^\top Z}(p)\, \theta(\mathrm{d}(c,p)) = \int_{\mathbb{S}^m} \int_0^1 F^{(-2)}_{c^\top Z}(p)\, \nu_c(\mathrm{d}p)\, \mu(\mathrm{d}c), \tag{6.26}$$

where the measure θ is disintegrated to a conditional measure ν_c and a marginal measure μ. The conditional measure ν_c is associated with the distortion function $\vartheta_c \in \mathcal{W}_2(\mathfrak{J}(c))$ via the relation

$$\nu_c([p,1]) = \begin{cases} (\vartheta_c)'_-(p) & \text{for } p \in \mathfrak{J}(c),\ p > 0, \\ (\vartheta_c)'_+(0) & \text{for } p = 0 \text{ if } 0 \in \mathfrak{J}(c). \end{cases}$$

Additionally, each function $\vartheta_c \in \mathcal{W}_2(\mathfrak{J}(c))$ corresponds to a spectral risk measure ϱ_c and a constant $\kappa_c \geq 0$ such that for all random variables $Z \in \mathcal{L}_1(\Omega, \mathcal{F}, P; \mathbb{R}^m)$ the following relation holds:

$$\begin{aligned} \int_{\mathbb{S}^m} \int_0^1 F^{(-2)}_{c^\top Z}(p)\, \nu_c(\mathrm{d}p)\, \mu(\mathrm{d}c) &= -\int_{\mathbb{S}^m} \kappa_c \varrho_c[c^\top Z]\, \mu(\mathrm{d}c) \\ &= -\int_{\mathbb{S}^m} \varrho_c[c^\top Z]\, \bar{\mu}(\mathrm{d}c). \end{aligned} \tag{6.27}$$

Here, ϱ_c is a spectral coherent measure of risk whose spectrum has a support on $\mathfrak{J}(c)$, and κ_c is a normalizing constant for the measure ν_c. For the last equality, we define the measure $\bar{\mu} \in \mathcal{M}^+(\mathbb{S}^m)$ by setting $\bar{\mu}(\mathrm{d}c) = \kappa_c \mu(\mathrm{d}c)$.

Consider now the optimization problem with the inverse multivariate stochastic dominance relation:

$$\min f(x) \tag{6.28}$$

$$\text{subject to } F^{(-2)}_{c^\top G(x)}(p) \geq F^{(-2)}_{c^\top Y}(p), \quad \forall\,(c,p) \in \operatorname{graph}(\mathfrak{J}), \tag{6.29}$$

$$x \in \mathcal{X}. \tag{6.30}$$

Here $G : \mathfrak{X} \to \mathcal{L}_{p-1}(\Omega, \mathcal{F}, P; \mathbb{R}^m)$ is a norm-continuous operator, which is also concave in the sense of Definition 3.1. The objective functional $f(\cdot)$ is continuous and convex. The constraint qualification condition for this formulation takes on the following form.

Definition 6.15. Problem (6.28)–(6.30) satisfies the *multivariate inverse uniform dominance condition* if a point $\tilde{x} \in \mathcal{X}$ exists such that

$$\inf_{(c,p)\in\operatorname{graph}(\mathfrak{J})} \left\{ F^{(-2)}_{c^\top G(\tilde{x})}(p) - F^{(-2)}_{c^\top Y}(p) \right\} > 0.$$

We define the distortion-adjusted functional $\tilde{\Phi}$ on $\mathfrak{X} \times \mathcal{W}_2 \times \mathcal{M}^+(\mathbb{S}^m)$ as follows

$$\tilde{\Phi}(x, \vartheta, \mu) = f(x) + \varphi_{\vartheta,\mu}(Y) - \varphi_{\vartheta,\mu}(G(x)). \tag{6.31}$$

The necessary and sufficient conditions of optimality for problem (6.28)–(6.30) take on the following form.

Theorem 6.16. *Let Assumption* 6.7 *and the dominance condition of Definition* 6.15 *be satisfied for problem* (6.28)–(6.30). *If $\hat{x}$ is a solution of problem* (6.28)–(6.30), *then a mapping $\hat{\vartheta} : \mathbb{S}^m \to \mathcal{W}_2(\mathfrak{J})$ and a measure $\hat{\mu} \in \mathcal{M}^+(\mathbb{S}^m)$ exist such that*

$$\tilde{\Phi}(\hat{x}, \hat{\vartheta}, \hat{\mu}) = \min_{x\in\mathcal{X}} \tilde{\Phi}(x, \hat{\vartheta}, \hat{\mu}), \tag{6.32}$$

$$\varphi_{\hat{\vartheta},\hat{\mu}}(Y) = \varphi_{\hat{\vartheta},\hat{\mu}}(G(\hat{x})). \tag{6.33}$$

If for some $\hat{\vartheta} : \mathbb{S}^m \to \mathcal{W}(\mathfrak{J})$, $\hat{\mu} \in \mathcal{M}_+(\mathbb{S}^m)$, and a solution $\hat{x}$ of the problem at the right-hand side of (6.32), *the order constraint* (6.29) *and* (6.33) *are satisfied, then $\hat{x}$ is a solution of problem* (6.28)–(6.30).

Proof. We consider the mapping $\mathfrak{G} : \mathfrak{X} \to C(\operatorname{graph}(\mathfrak{J}))$ defined as

$$[\mathfrak{G}(x)](c,p) = F^{(-2)}_{\langle c,Y\rangle}(p) - F^{(-2)}_{\langle c,G(x)\rangle}(p), \quad (c,p) \in \operatorname{graph}(\mathfrak{J}). \tag{6.34}$$

The operator $\mathfrak{G}$ is continuous with respect to the supremum norm on the space $\mathcal{C}(\text{graph}(\mathfrak{F}))$ due to the continuity assumption on G and the continuity of the Lorenz function with respect to the random variable. The arguments are analogous to those used in the proof of Theorem 5.49. Denoting the cone of nonpositive functions in $\mathcal{C}(\text{graph}(\mathfrak{F}))$ by K, we reformulate problem (6.28)–(6.30) as follows:

$$\min f(x) \text{ subject to } \mathfrak{G}(x) \in K, \; x \in \mathcal{X}. \tag{6.35}$$

Further, we need prove that $\mathfrak{G}$ is a convex mapping with respect to the cone K. For every $(c, p) \in \text{graph}(\mathfrak{F})$ the function $x \mapsto \langle c, G(x)\rangle$ is concave, for almost all $\omega \in \Omega$ by Assumption 6.11. For an integrable random variable Z, the functional $Z \mapsto F_Z^{(-2)}(p)$ is concave and positively homogeneous for every $p \in (0, 1)$ as shown in Lemma 5.32. Furthermore, for integrable Z, V, the relation $Z \geq V$ a.s. implies $F_Z^{(-2)}(p) \geq F_V^{(-2)}(p)$. Therefore, for $\alpha \in [0, 1]$, we have for every $p \in (0, 1)$

$$\begin{aligned} F^{(-2)}_{\langle c,G(\alpha x+(1-\alpha)y)\rangle}(p) &\geq F^{(-2)}_{\alpha\langle c,G(x)\rangle+(1-\alpha)\langle c,G(y)\rangle}(p), \\ &= \alpha F^{(-2)}_{\langle c,G(x)\rangle}(p) + (1-\alpha) F^{(-2)}_{\langle c,G(y)\rangle}(p). \end{aligned}$$

Thus, the mapping $[\mathfrak{G}(\cdot)](c, p)$ is convex for all $(c, p) \in \text{graph}(\mathfrak{F})$ implying that the operator $\mathfrak{G}$ is convex with respect to the cone K. Now, we invoke the Lagrangian $\bar{\mathfrak{L}} : \mathfrak{X} \times \mathcal{M}^+(\text{graph}(\mathfrak{F})) \to \mathbb{R}$ associated with problem (6.35), which is defined as follows:

$$\bar{\mathfrak{L}}(x, \theta) = f(x) + \int_{\text{graph}(\mathfrak{F})} [\mathfrak{G}(x)](c, p)\, \theta(\mathrm{d}(c, p)). \tag{6.36}$$

The multivariate uniform inverse dominance condition in Definition 6.15 is equivalent to the generalized Slater condition: a point $\tilde{x} \in \mathcal{X}$ exists such that $\mathfrak{G}(\tilde{x}) \in \text{int}K$. Hence, we may apply the necessary conditions of optimality in Banach spaces to infer the existence of a nonnegative measure $\hat{\theta} \in \mathcal{M}^+(\text{graph}(\mathfrak{F}))$ such that

$$\bar{\mathfrak{L}}(\hat{x}, \hat{\theta}) = \min_{x \in \mathcal{X}} \bar{\mathfrak{L}}(x, \hat{\theta}) \quad \text{and} \tag{6.37}$$

$$\int_{\text{graph}(\mathfrak{F})} [F^{(-2)}_{c^\top Y}(p) - F^{(-2)}_{c^\top G(\hat{x})}(p)]\, \hat{\theta}(\mathrm{d}(c, p)) = 0. \tag{6.38}$$

We disintegrate $\hat{\theta}$ into conditional measures $\hat{\nu}_c \in \mathcal{M}^+(\mathfrak{F}(c))$ and a marginal measure $\hat{\mu} \in \mathcal{M}^+(\mathbb{S}^m)$. Extending $\hat{\nu}_c$ to the entire interval $(0, 1]$, allows us to rewrite the relations (6.26) and (6.24) as follows:

$$\int\limits_{\text{graph}(\mathfrak{F})} [\mathfrak{G}(x)](c,p)\,\hat{\theta}(\mathrm{d}(c,p)) = \int\limits_{\mathbb{S}^m}\int_0^1 F^{(-1)}_{c^\top Y}(p)\,\mathrm{d}\hat{\vartheta}_c(p)\,\hat{\mu}(\mathrm{d}c)$$

$$-\int\limits_{\mathbb{S}^m}\int_0^1 F^{(-1)}_{c^\top G(\hat{x})}(p)\,\mathrm{d}\hat{\vartheta}_c(p)\,\hat{\mu}(\mathrm{d}c) = \varphi_{\hat{\vartheta},\hat{\mu}}(Y) - \varphi_{\hat{\vartheta},\hat{\mu}}(G(\hat{x})).$$

Hence, conditions (6.37) and (6.38) are equivalent to (6.32) and (6.33). The converse statement follows from the convexity of the problem (6.28)–(6.30) and the one-to-one correspondence in (6.26). □

Further application of the correspondence between measures $\theta \in \mathcal{M}^+(\text{graph}(\mathfrak{F}))$ and sets of spectral measures of risk ϱ_c and a measure $\bar{\mu} \in \mathcal{M}^+(\mathbb{S}^m)$, as shown in (6.27), provide the following equivalent representation of the optimality conditions for problem (6.28)–(6.30).

Corollary 6.17. *Assume that the dominance condition of Definition* 6.15 *holds for problem* (6.28)–(6.30). *If* $\hat{x}$ *is a solution of problem* (6.28)–(6.30), *then a set of coherent measures of risk* ϱ_c, $c \in \mathbb{S}^m$, *and a measure* $\hat{\mu} \in \mathcal{M}^+(\mathbb{S}^m)$ *exist such that* $\hat{x}$ *is also a solution of*

$$\min_{x\in\mathcal{X}} \Big\{ f(x) + \int\limits_{\mathbb{S}^m} \varrho_c[c^\top G(\hat{x})]\,\hat{\mu}(\mathrm{d}c) \Big\} \tag{6.39}$$

and, additionally,

$$\int\limits_{\mathbb{S}^m} \varrho_c[c^\top G(\hat{x})]\,\hat{\mu}(\mathrm{d}c) = \int\limits_{\mathbb{S}^m} \varrho_c[c^\top Y]\,\hat{\mu}(\mathrm{d}c). \tag{6.40}$$

If for some set of spectral measures of risk $\hat{\varrho}_c$, $c \in \mathbb{S}^m$, *and a measure* $\hat{\mu} \in \mathcal{M}^+(\mathbb{S}^m)$, *we have a solution* $\hat{x}$ *of problem* (6.39), *such that the order constraint* (6.29) *and the complementarity condition* (6.40) *are satisfied, then* $\hat{x}$ *is a solution of problem* (6.28)–(6.30).

6.3 Sequential Orders and Time Consistency

For the space $(\Omega, \mathcal{F}, P)$, assume that a filtration $\mathbb{F} = \{\mathcal{F}_t\}_{t=1,\dots,T}$, with $\mathcal{F}_1 \subset \mathcal{F}_2 \subset \cdots \subset \mathcal{F}_T$, and with $\mathcal{F}_1 = \{\emptyset, \Omega\}$ and $\mathcal{F}_T = \mathcal{F}$, is generated by a process $\{D_t\}_{t=1,\dots,T}$. We shall compare random sequences adapted to this filtration. They are elements of the space $\mathcal{Z} = \mathcal{L}_p(\Omega, \mathcal{F}_1, P) \times \cdots \times \mathcal{L}_p(\Omega, \mathcal{F}_T, P)$. We denote $\mathcal{Z}_t = \mathcal{L}_p(\Omega, \mathcal{F}_t, P)$ and $\mathcal{Z}_{s:t} = \mathcal{Z}_s \times \cdots \times \mathcal{Z}_t$ for $1 \le s \le t \le T$.

While our goal is to construct stochastic orders between sequences, the orders will, in fact, compare the distributions of the sequences, as was the case of the stochastic orders in the spaces of random variables. In the dynamic setting, we shall also compare conditional distributions.

We denote the history of the process $\{D_t\}_{t=1,\dots,T}$ until time t by $D_{1:t}$. To alleviate notation, for a sequence $\mathbb{X} = (X_1, X_2, \dots, X_T)$ adapted to the filtration $\mathbb{F}$ we shall write $\mathbb{X} \,|D_{1:t}$ to represent a random sequence whose distribution is the conditional distribution of $\mathbb{X}$, given $D_{1:t}$. In particular, its history up to time t is $(X_1(D_1), X_2(D_{1:2}), \dots, X_t(D_{1:t}))$, and for $t < j \le T$, $X_j \mid D_{1:t}$ is a random variable distributed according to the conditional distribution of X_j, given $D_{1:t}$.

We use the symbol $\trianglerighteq_t$ for an order between the conditional distributions of random sequences for each fixed $D_{1:t}$. A dynamic order $\trianglerighteq$ among random sequences in the space $\mathcal{Z}$ is a sequence of orders $\trianglerighteq_t$, $t = 1, \dots, T-1$.

Given two sequences $\mathbb{X} = (X_1, X_2, \dots, X_T)$ and $\mathbb{Y} = (Y_1, Y_2, \dots, Y_T)$ in $\mathcal{Z}$, we should be able to compare them at any time $t = 1, \dots, T$, given $\mathcal{F}_t$, in a consistent way. This issue is of crucial importance in a sequential decision-making process. We introduce the following notion of a time-consistent comparison of sequences.

Definition 6.18. A stochastic order between elements of $\mathcal{Z}$ is called *time consistent*, if for all sequences $\mathbb{X}, \mathbb{Y} \in \mathcal{Z}$, for all $1 \le t < T-1$, and for all $D_{1:t}$ the relation $\mathbb{X} \mid (D_{1:t}, D_{t+1}) \trianglerighteq_{t+1} \mathbb{Y} \mid (D_{1:t}, D_{t+1})$ for almost all[†] D_{t+1} implies that $\mathbb{X} \mid D_{1:t} \trianglerighteq_t \mathbb{Y} \mid D_{1:t}$.

Definition 6.18 is in harmony with the notion of time consistency of dynamic risk measures.

Two functions of the sequence play an important role in our theory. First, we consider the cumulative sum of $\mathbb{X}$ until time t, denoted $\mathbb{x}_t$, as a function of the path $D_{1:t}$:

$$\mathbb{x}_t(D_{1:t}) = (X_1 + X_2 + \cdots + X_t)(D_{1:t}), \quad t = 1, \dots, T. \tag{6.41}$$

In particular, the total reward at time T, $\mathbb{x}_T$, has realizations $(X_1 + X_2 + \cdots + X_T)(D_{1:T})$. We define the one-step projected future value for the sequence $\mathbb{X}$ at time t, when $D_{1:t}$ is fixed, in the following way:

$$\begin{aligned} \mathbb{X}_{t+1}|D_{1:t} =& X_{t+1}|D_{1:t} + \mathbb{E}_{t+1}\Big[X_{t+2}|D_{1:t} + \\ & \mathbb{E}_{t+2}\big[X_{t+3}|D_{1:t} + \cdots + \mathbb{E}_{T-1}[X_T|D_{1:t}]\big]\Big]. \end{aligned} \tag{6.42}$$

The shorthand notation $\mathbb{E}_\tau[Z] = \mathbb{E}[Z \,|\mathcal{F}_\tau]$ will help declutter the notation.

We shall identify a stochastic order relation $\succeq$ between scalar random variables with its generating family $\mathcal{U}$ of functions $u : \mathbb{R} \to \mathbb{R}$, and shall denote the relation between two scalar variables Z, Z' as $Z \succeq_{\mathcal{U}} Z'$.

[†] "Almost all" is understood with respect to the conditional probability, given $D_{1:t}$.

Consider a mapping $\mathbb{U}$ that provides for each time t, given $D_{1:t}$, a generator $\mathbb{U}_t(D_{1:t})$ of some univariate stochastic order. We can view $\mathbb{U}$ as a multifunction, adapted to the filtration $\{\mathcal{F}_t\}$, whose values are cones in a suitable space of functions.

We shall call $\mathbb{U}$ a *sequential generator* provided that the generators $\mathbb{U}_t(D_{1:t})$ reflect the same preference (either to small or to large outcomes), and that $\mathbb{E}[u(Z)] < \infty$ for all $u \in \mathbb{U}_t(D_{1:t})$ for a.a. $D_{1:t}$, for all $t = 1, \dots, T$, and for all $Z \in \mathcal{L}_p(\Omega, \mathcal{F}, P)$.

Definition 6.19. Given a sequential generator $\mathbb{U}$ expressing preference to large outcomes, we call a sequence $\mathbb{X} \in \mathcal{Z}$ stochastically larger than $\mathbb{Y} \in \mathcal{Z}$ with respect to $\mathbb{U}$ (denoted $\mathbb{X} \succeq_{\mathbb{U}} \mathbb{Y}$) if at any time $1 \le t < T$ the following holds

$$x_t(D_{1:t}) + X_{t+1} \,|D_{1:t} \succeq_{\mathbb{U}_t(D_{1:t})} y_t(D_{1:t}) + Y_{t+1} \,|D_{1:t}. \tag{6.43}$$

Analogously, for a sequential generator $\mathbb{U}^\flat$ expressing preference to small outcomes, we call a sequence $\mathbb{X} \in \mathcal{Z}$ stochastically smaller than $\mathbb{Y} \in \mathcal{Z}$ with respect to $\mathbb{U}^\flat$ (denoted $\mathbb{X} \preceq_{\mathbb{U}^\flat} \mathbb{Y}$) if at any time $1 \le t < T$

$$x_t(D_{1:t}) + X_{t+1} \,|D_{1:t} \preceq_{\mathbb{U}_t^\flat(D_{1:t})} y_t(D_{1:t}) + Y_{t+1} \,|D_{1:t}. \tag{6.44}$$

In other words, the accumulated reward at time t and the projected future reward of $\mathbb{X}$ is compared to that of $\mathbb{Y}$ with respect to the chosen univariate order. Notice that X_1 is deterministic for the trivial $\mathcal{F}_1$. Hence, we write simply x_1 and X_2 instead of $x(D_1)$ and $X_2|D_1$, respectively.

It is not difficult to see that the stochastic orders introduced in Definition 6.19 are reflexive and transitive relations on $\mathcal{Z} \times \mathcal{Z}$.

Remark 6.20. Some modifications of Definition 6.19 are possible, which do not change its essence. For example, we could include a discount for the future rewards or employ other time-consistent projections of the future rewards instead of conditional expectations. □

We show next that Definition 6.19 allows us to construct time-consistent orders.

Theorem 6.21. *Assume that all images $\mathbb{U}_t(D_{1:t})$ ($\mathbb{U}_t^\flat(D_{1:t})$) of the sequential generator $\mathbb{U}$ ($\mathbb{U}^\flat$) are contained in the space of monotonically non-decreasing functions and they also contain a non-constant linear utility function. Then the dynamic order $\succeq_{\mathbb{U}}$ ($\preceq_{\mathbb{U}^\flat}$) in Definition* 6.19 *is time-consistent.*

Proof. We shall demonstrate the statement when large values are preferred. When the stochastic order relations express preferences to small outcomes the argument is the same.

Assume that for some t, the conditions in Definition 6.18 are satisfied. The definition of the order entail for any history $D_{1:t}$:

$$x_{t-1}(D_{1:t-1}) + X_t(D_{1:t}) + \mathbb{X}_{t+1} \mid D_{1:t} \succeq_{U_t(D_{1:t})} \\ y_{t-1}(D_{1:t-1}) + Y_t(D_{1:t}) + \mathbb{Y}_{t+1} \mid D_{1:t}.$$

Due to the presence of a linear utility in $\mathbb{U}_t(D_{1:t-1})$, which is increasing, this relation implies the inequality

$$\mathbb{E}_t[x_{t-1}(D_{1:t-1}) + (X_t + \mathbb{X}_{t+1})|D_{1:t}] \geq \\ \mathbb{E}_t[y_{t-1}(D_{1:t-1}) + (Y_t + \mathbb{Y}_{t+1})|D_{1:t}].$$

Taking the conditional expectation of both sides with respect to $\mathcal{F}_{t-1}$ we obtain

$$x_{t-1}(D_{1:t-1}) + \mathbb{E}[X_t + \mathbb{X}_{t+1}|D_{1:t-1}] \geq \\ y_{t-1}(D_{1:t-1}) + \mathbb{E}[Y_t + \mathbb{Y}_{t+1}|D_{1:t-1}],$$

which is the same as

$$x_{t-1}(D_{1:t-1}) + \mathbb{X}_t|D_{1:t-1} \geq y_{t-1}(D_{1:t-1}) + \mathbb{Y}_t|D_{1:t-1}.$$

This implies that

$$x_{t-1}(D_{1:t-1}) + \mathbb{X}_t \mid D_{1:t-1} \succeq_{U_t(D_{1:t-1})} x_{t-1}(D_{1:t-1}) + \mathbb{Y}_t \mid D_{1:t-1},$$

because of the monotonicity of the utility functions in all generators. □

We obtain the following corollary when using the stochastic dominance orders discussed in detail in the previous chapter.

Corollary 6.22. *If for all $t = 1, \dots, T_1$, $\mathbb{U}_t(D_{1:t}) = \mathcal{U}_{q(t)}$ with $q(t) \in [1, p]$, then the dynamic order $\succeq_U$ in Definition* 6.19 *is time-consistent. If for all $t = 1, \dots, T_1$, $\mathbb{U}_t^\flat(D_{1:t}) = \mathcal{U}_{q(t)}^\flat$ with $q(t) \in [1, p]$, then the dynamic order $\preceq_{U^\flat}$ in Definition* 6.19 *is time-consistent.*

The corollary (ii) follows in the same way as Theorem 6.21 because the images of the sequential generator $\mathbb{U}$ are independent of the history $D_{1:t}$ and contain monotonic utility functions having in mind Theorem 5.5 and Corollary 5.8.

Alternatively, we may consider sequential orders that are based on sets of distortion functions. To this end, we consider a mapping $\mathbb{W}$ that provides for each time t given $D_{1:t}$ a set of distortion functions $\mathbb{W}_t(D_{1:t})$ with $w \in \mathbb{W}_t(D_{1:t})$, $w : (0, 1) \to \mathbb{R}$, being all non-decreasing for a.a. $D_{1:t}$ and for all $t = 1 \dots, T$, and such that the Stieltjes integral $\int_0^1 F_Z^{(-1)}(\alpha)\, \mathrm{d}w(\alpha)$ is well-defined and finite for all $Z \in \mathcal{L}_p(\Omega, \mathcal{F}, P)$.

We shall call $\mathbb{W}$ a *sequential distortion generator* and $\mathcal{W} = \mathbb{W}_t(D_{1:t})$ will be called simply a *distortion generator*.

Definition 6.23. Given a distortion generator $\mathcal{W}$ expressing preference to large values, for two random variables Z, Z', we say that $Z \succeq_{\mathcal{W}} Z'$ if

$$\int_0^1 F_Z^{(-1)}(p)\,dw(p) \geq \int_0^1 F_{Z'}^{(-1)}(p)\,dw(p) \quad \text{for all } w \in \mathbb{W}.$$

When $\mathcal{W}$ expresses preference to small values, we say that $Z \preceq_{\mathcal{W}} Z'$ if the inequality above is reversed.

Definition 6.24. Given a sequential distortion generator $\mathbb{W}$ expressing preference to large values, we call a sequence $\mathbb{X} \in \mathcal{Z}$ stochastically larger than $\mathbb{Y} \in \mathcal{Z}$ with respect to $\mathbb{W}$ (denoted $\mathbb{X} \succeq_{\mathbb{W}} \mathbb{Y}$) if at any time $1 \leq t < T$ the following holds

$$x_t(D_{1:t}) + \mathbb{X}_{t+1}\,|D_{1:t} \succeq_{\mathbb{W}_t(D_{1:t})} y_t(D_{1:t}) + \mathbb{Y}_{t+1}\,|D_{1:t}. \tag{6.45}$$

Given a sequential distortion generator $\mathbb{W}^\flat$ expressing preference to small values, a sequence $\mathbb{X} \in \mathcal{Z}$ stochastically smaller than $\mathbb{Y} \in \mathcal{Z}$ with respect to $\mathbb{W}^\flat$ (denoted $\mathbb{X} \preceq_{\mathbb{W}^\flat} \mathbb{Y}$) if at any time $1 \leq t < T$ the following holds

$$x_t(D_{1:t}) + \mathbb{X}_{t+1}\,|D_{1:t} \preceq_{\mathbb{W}_t^\flat(D_{1:t})} y_t(D_{1:t}) + \mathbb{Y}_{t+1}\,|D_{1:t}. \tag{6.46}$$

We provide conditions for the time-consistency of this orders as well.

Theorem 6.25. *Assume that all images $\mathbb{W}_t(D_{1:t})$ of the sequential distortion generator $\mathbb{W}$ are contained in the space of monotonically non-decreasing functions and they all contain a non-constant linear function. Then the dynamic orders constructed in Definition 6.24 are time-consistent.*

Proof. It is sufficient to show the statement when large values are preferred. Assume that for some t, the conditions in Definition 6.19 are satisfied, that is, for any history $D_{1:t}$:

$$x_{t-1}(D_{1:t-1}) + X_t(D_{1:t}) + \mathbb{X}_{t+1}|D_{1:t} \succeq_{\mathbb{W}_t(D_{1:t})}$$
$$y_{t-1}(D_{1:t-1}) + Y_t(D_{1:t}) + \mathbb{Y}_{t+1}|D_{1:t-1}.$$

The definition of the order entails that for every distortion $w \in \mathbb{W}_t(D_{1:t})$, the following relation holds

$$\int_0^1 F^{(-1)}_{x_{t-1}(D_{1:t-1})+X_t(D_{1:t})+\mathbb{X}_{t+1}|D_{1:t}}(p)\,dw(p) \geq$$
$$\int_0^1 F^{(-1)}_{y_{t-1}(D_{1:t-1})+Y_t(D_{1:t})+\mathbb{Y}_{t+1}|D_{1:t}}(p)\,dw(p).$$

Due to the presence of linear in $\mathbb{W}_t(D_{1:t})$, which is increasing, this relation implies the inequality

$$\begin{aligned}
&\mathbb{E}_t\big[\mathbb{x}_{t-1}(D_{1:t-1}) + (X_t + \mathbb{X}_{t+1})|D_{1:t}\big] \\
&\qquad = \int_0^1 F^{(-1)}_{\mathbb{x}_{t-1}(D_{1:t-1})+X_t(D_{1:t})+\mathbb{X}_{t+1}|D_{1:t}}(p)\,\mathrm{d}p \\
&\qquad \geq \int_0^1 F^{(-1)}_{\mathbb{y}_{t-1}(D_{1:t-1})+Y_t(D_{1:t})+\mathbb{Y}_{t+1}|D_{1:t}}(p)\,\mathrm{d}p \\
&\qquad\qquad\qquad = \mathbb{E}_t\big[\mathbb{y}_{t-1}(D_{1:t-1}) + (Y_t + \mathbb{Y}_{t+1})|D_{1:t}\big].
\end{aligned}$$

Taking the conditional expectation with respect to $\mathcal{F}_{t-1}$ and proceeding as in the proof of Theorem 6.21 we obtain the relation

$$\mathbb{x}_{t-1}(D_{1:t-1}) + \mathbb{X}_t \mid D_{1:t-1} \succeq_{W_t(D_{1:t-1})} \mathbb{x}_{t-1}(D_{1:t-1}) + \mathbb{Y}_t \mid D_{1:t-1},$$

as required. □

Similarly as before, we obtain the following special time-consistent order.

Corollary 6.26. *If for all $t = 1, \ldots, T-1$, $W_t(D_{1:t}) = \mathcal{W}_2$, then the sequential order $\succeq_W$ in Definition 6.24 is time-consistent. The same is true if $W_t(D_{1:t}) = \mathcal{W}_2^\flat$.*

Notice that in the case when the sequential order is defined as in Definition 6.24 and $W_t(D_{1:t}) = \mathcal{W}_2$ for all time periods or it is defined as in Definition 6.19 with $U_t(D_{1:t}) = \mathcal{U}_2$ for all time periods, we deal with the same stochastic comparison and we can use the notation

$$\mathbb{x}_t(D_{1:t}) + \mathbb{X}_{t+1}\,|D_{1:t} \succeq_{(2)} \mathbb{y}_t(D_{1:t}) + \mathbb{Y}_{t+1}\,|D_{1:t}. \tag{6.47}$$

This is due to the fact that second-order stochastic dominance can be characterized equivalently by the set of utility functions $\mathcal{U}_2$ as well as by the set of distortions $\mathcal{W}_2$.

Similarly, if the sequential order is defined by using $W_t(D_{1:t}) = \mathcal{W}_2^\flat$ for all time periods or by using $U_t(D_{1:t}) = \mathcal{U}_2^\flat$ for all time periods, we deal with the increasing convex order and we can use the notation

$$\mathbb{x}_t(D_{1:t}) + \mathbb{X}_{t+1}\,|D_{1:t} \preceq_{(2)} \mathbb{y}_t(D_{1:t}) + \mathbb{Y}_{t+1}\,|D_{1:t}. \tag{6.48}$$

This is due to the equivalence of the characterization by the set of utility functions $\mathcal{U}_2^\flat$ to that using distortions from the set $\mathcal{W}_2^\flat$.

We explore the consistency of the sequential order with law-invariant dynamic measures of risk. To this end, we recall Corollary 4.6.

Theorem 6.27. *Assume that $\succeq_U$ is given by $U_t(D_{1:t}) = \mathcal{U}_2$ for all $t = 1, \ldots, T-1$. Let $\varrho[\cdot]$ be any dynamic measure of risk expressing preferences to large values, where the last one-step conditional measure of risk $\rho_{T-1}(\cdot)$ is consistent with the second order stochastic dominance. Then the relation $\mathbb{X} \succeq_U \mathbb{Y}$*

implies $\varrho[\mathbb{X}] \le \varrho[\mathbb{Y}]$. *Analogously, if* $\succeq_W$ *is given by* $\mathbb{W}_t(D_{1:t}) = \mathcal{W}_2$, *then the relation* $\mathbb{X} \succeq_W \mathbb{Y}$ *implies* $\varrho[\mathbb{X}] \le \varrho[\mathbb{Y}]$ *for such measures.*

Proof. As already noted the second-order stochastic dominance can be characterized equivalently by the set of utility functions $\mathcal{U}_2$ as well as by the set of distortions $\mathcal{W}_2$. Hence, it is sufficient to consider the case of $\mathbb{X} \succeq_U \mathbb{Y}$. Due to the assumed consistency of the one-step conditional measure $\rho_{T-1}(\cdot)$ the second-order stochastic dominance, and using Definition 6.19, we obtain for almost all $D_{1:T-1}$:

$$\rho_{T-1}\big[x_{T-1}(D_{1:T-1}) + X_T | D_{1:T-1}\big] \le \rho_{T-1}\big[y_{T-1}(D_{1:T-1}) + Y_T | D_{1:T-1}\big],$$

which by translation is equivalent to

$$x_{T-1}(D_{1:T-1}) + \rho_{T-1}\big[X_T | D_{1:T-1}\big] \le y_{T-1}(D_{1:T-1}) + \rho_{T-1}\big[Y_T | D_{1:T-1}\big].$$

Fixing any history $D_{1:T-2}$, we obtain that

$$\begin{aligned} &x_{T-2}(D_{1:T-2}) + X_{T-1}|D_{1:T-1} + \rho_{T-1}[X_T|D_{1:T-1}] \le \\ &\qquad y_{T-2}(D_{1:T-1}) + Y_{T-1}|D_{1:T-1} + \rho_{T-1}[Y_T|D_{1:T-1}] \end{aligned}$$

holds. The monotonicity of the conditional measure of risk $\rho_{T-2}(\cdot)$ and the translation property yield:

$$\begin{aligned} &x_{T-1}(D_{1:T-2}) + \rho_{T-2}\big[X_{T-1}|D_{1:T-2} + \rho_{T-1}[X_T|D_{1:T-1}]\big] \le \\ &\qquad y_{T-1}(D_{1:T-2}) + \rho_{T-2}\big[Y_{T-1}|D_{1:T-2} + \rho_{T-1}[Y_T|D_{1:T-1}]\big]. \end{aligned}$$

Proceeding this way, we obtain the statement for the case of $\mathbb{X} \succeq_U \mathbb{Y}$. □

It is clear from the proof of Theorem 6.27 that analogous statement is true also in the situation when small values are preferred. We reflect this observation in the following corollary.

Corollary 6.28. *Suppose the dynamic order* $\preceq_U$ ($\preceq_W$) *is given by* $\mathbb{U}_t(D_{1:t}) = \mathcal{U}_2^\flat$ ($\mathbb{W}_t(D_{1:t}) = \mathcal{W}_2^\flat$) *for all* $t = 1, \dots, T-1$. *If* $\varrho[\cdot]$ *is a dynamic measure of risk reflecting preferences to small values with* $\rho_{T-1}(\cdot)$ *being consistent with the increasing convex order, then the relation* $\mathbb{X} \preceq_U \mathbb{Y}$ *(equivalently,* $\mathbb{X} \preceq_W \mathbb{Y}$*) implies* $\varrho[\mathbb{X}] \le \varrho[\mathbb{Y}]$.

In the last corollary, we have used the fact that both $\preceq_U$ and $\preceq_W$ defined in the statement are equivalent to the relation (6.48) for all $t = 1, \dots, T-1$.

In what follows, we shall focus on sequential orders when small values are preferred.

6.4 Multistage Stochastic Optimization Problems with Sequential Stochastic Order Constraints

We focus on multistage stochastic optimization problems with time horizon $T \geq 2$, in which we introduce distributional constraints in the form of stochastic order. We start from the following general risk-neutral formulation:

$$\begin{aligned} \min\ & \mathbb{E}\Big[\sum_{t=1}^{T} f_t(x_t, D_t)\Big] \\ \text{s.t. } & B_t(D_t)x_{t-1} + A_t(D_t)x_t = b_t(D_t), \quad t = 2, \dots, T, \\ & A_1(D_1)x_1 = b_1(D_1), \\ & x_t \in \mathcal{X}_t(D_t), \quad t = 1, \dots, T. \end{aligned} \tag{6.49}$$

Here, the random process $D_1, D_2, \dots, D_T$ comprises the random data of the problem with D_t being a random vector defined on a probability space $(\Omega, \mathcal{F}, P)$ with realizations in $\mathbb{R}^d$. We assume that the distribution of the data process does not depend on the decisions. The history of the data process up to time t is denoted by $D_{1:t}$. Let $\mathcal{F}_t$ be the σ-algebra generated by $D_{1:t}$; $\mathcal{F}_t \subset \mathcal{F}$.

For simplicity, we set $\mathcal{F}_1 = \{\emptyset, \Omega\}$ and $\mathcal{F}_T = \mathcal{F}$. We identify our decision space at time t as $\mathfrak{X}_t = \mathcal{L}_p(\Omega, \mathcal{F}_t, P; \mathbb{R}^n)$, $t = 1, \dots, T$, with $p \geq 1$ since the decision vector x_t in each instance depends on the observed data $D_{1:t}$. We denote $\mathfrak{X} = \mathfrak{X}_1 \times \cdots \times \mathfrak{X}_T$.

The objective functions $f_t : \mathbb{R}^n \times \mathbb{R}^d \to \mathbb{R}$ are Carathéodory functions.

The sets $\mathcal{X}_t$ are $\mathcal{F}_t$-measurable random sets with closed and convex images in $\mathbb{R}^n$. Since D_1 is observed before the decision process starts, the set $\mathcal{X}_1 \subseteq \mathbb{R}^n$ and the constraint $A_1 x_1 = b_1$ are deterministic. We shall omit in the notation of A_t, B_t, b_t, and $\mathcal{X}_t$ the explicit dependence on the random data process as it will not lead to ambiguity.

While optimizing the random cost on average, we also would like to restrict our risk of excessive cost by imposing constraints on the distribution of the cost sequence. We formulate a risk-averse version of problem (6.49) by using a distributional constraint in the form of a sequential order relation to a benchmark process. Let $\mathbb{Y} \in \mathsf{Z}$ be an adapted benchmark process and a sequential generator $\mathbb{U}$ with $\mathcal{F}_t$-measurable images $\mathbb{U}_t(D_{1:t})$, $t = 1, \dots, T-1$ be given. We shall use a stochastic order expressing preference for small outcomes such as (5.17). The constraint has the form

$$\big\{f_t(x_t, D_t)\big\}_{t=1}^{T} \preceq_{\mathbb{U}} \mathbb{Y}.$$

The benchmark process may be derived using an existing policy, which we desire to improve not only on average but taking risk into account.

We shall analyze the risk-averse multistage problem focusing on the following formulation:

$$\begin{aligned}
&\min\ \mathbb{E}\Big[\sum_{t=1}^{T} f_t(x_t, D_t)\Big] \\
&\text{s.t. } B_t x_{t-1} + A_t x_t = b_t, \quad t = 2, \dots, T, \\
&\quad A_1 x_1 = b_1, \\
&\quad x_t \in \mathcal{X}_t, \quad t = 1, \dots T, \\
&\quad \{f_t(x_t, D_t)\}_{t=1}^{T} \preceq_U \mathbb{Y}.
\end{aligned} \tag{6.50}$$

Let us derive a dynamic programming formulation for problem (6.50) introducing auxiliary variables σ_t, $t = 1, \dots, T$ and additional relations to describe their evolution. The variables σ_t track the difference between the cumulative costs and the cumulative benchmark.

At the last stage, we solve the problem:

$$\begin{aligned}
&\min_{x_T, \sigma_T}\ f_T(x_T, D_T) \\
&\text{s.t. } \sigma_T = \sigma_{T-1} + f_T(x_T, D_T) - y_T, \\
&\quad B_T x_{T-1} + A_T x_T = b_T, \\
&\quad x_T \in \mathcal{X}_T.
\end{aligned} \tag{6.51}$$

The optimal value of problem (6.51) is denoted by $Q_T(x_{T-1}, \sigma_{T-1}, D_{1:T})$; it depends on our decisions until time $T-1$ and on the data process $D_{1:T}$. At time T, it depends on D_T only, but at earlier stages the dependence on the history is essential. The conditional expectation of $Q_T(x_{T-1}, \sigma_{T-1}, D_{1:T})$ with respect to $\mathcal{F}_{T-1}$ is denoted

$$\mathcal{Q}_T(x_{T-1}, \sigma_{T-1}, D_{1:T-1}) = \mathbb{E}_{T-1}\big[Q_T(x_{T-1}, \sigma_{T-1}, D_{1:T})\big].$$

The variable σ_T does not play a role in this problem, but we introduce it for the sake of consistency of notation with the earlier stages.

For $t = T-1, \dots, 2$, we solve

$$\begin{aligned}
&\min_{x_t, \sigma_t, z_t}\ z_t + \mathcal{Q}_{t+1}(x_t, \sigma_t, D_{1:t}), \\
&\text{s.t. } f_t(x_t, D_t) \le z_t \\
&\quad \sigma_t = \sigma_{t-1} + z_t - y_t, \\
&\quad \sigma_t + Q_{t+1}(x_t, \sigma_t, D_{1:t+1})\,|D_{1:t} \preceq_{U_t(D_{1:t})} \mathbb{Y}_{t+1}\,|D_{1:t}, \\
&\quad B_t x_{t-1} + A_t x_t = b_t, \quad x_t \in \mathcal{X}_t.
\end{aligned} \tag{6.52}$$

The auxiliary decision variables σ_t are functions of the data process and become elements of $\mathcal{Z}_t$ by construction. At the optimal solution, we have $z_t = f_t(x_t, D_t)$. The optimal value of (6.52) is denoted by $Q_t(x_{t-1}, \sigma_{t-1}, D_{1:t})$, its $\mathcal{F}_t$-conditional expectation is

$$\mathcal{Q}_t(x_{t-1}, \sigma_{t-1}, D_{1:t-1}) = \mathbb{E}_t\big[Q_t(x_{t-1}, \sigma_{t-1}, D_{1:t})\big].$$

At the first stage, we solve the problem:

$$\begin{aligned}
\min_{x_1, \sigma_1}\ & z_1 + \mathbb{E}\big[Q_2(x_1, D_{1:2})\big] \\
\text{s.t.}\ & f_1(x_1, D_1) \le z_1, \\
& \sigma_1 = z_1 - y_1, \\
& \sigma_1 + Q_2(x_1, D_{1:2}) \preceq_{U_1(D_1)} \mathbb{Y}_2, \\
& A_1 x_1 = b_1, \quad x_1 \in \mathcal{X}_1.
\end{aligned} \tag{6.53}$$

The equivalence of problem (6.50) with the dynamic programming formulation (6.51)–(6.53) follows directly from the definitions. In an attempt to declutter notation, we shall indicate that $D_{1:t}$ is fixed by noting $D^t_{1:t+1}$, i.e.,

$$Q_{t+1}(x_t, \sigma_t, D^t_{1:t+1}) = Q_{t+1}(x_t, \sigma_t, D_{1:t+1})\,|D_{1:t}.$$

We also use $\mathbb{Y}_{t,t+1} = \mathbb{Y}_{t+1}\,|D_{1:t}$.

6.4.1 Optimality Conditions

In this section, we focus on multistage stochastic optimization problems with time horizon $T \ge 2$ and sequential generator given by $\mathbb{U}$ such that

$$\mathbb{U}_t(D_{1:t}) = \mathcal{U}^{\flat}_{p_t}(I(D_t)),$$

where $I(D_t(\omega)) \subseteq \mathbb{R}$ is a nonempty compact interval and $I(D_t) \rightrightarrows \mathbb{R}$ is a Lebesgue measurable multifunction; $p_t \ge 2$ is a real number, and $\mathcal{U}^{\flat}_{p_t}(I(D_t))$ is defined in (5.28).

We shall assume that all functions $f_t(\cdot, D_t)$ are convex for all $t = 1, \dots, T$. If we deal with a long time horizon $T > 2$, we may want to choose p_t as an increasing sequence of time satisfying $p_t \le p + 1$. Another option in long time horizon might be to choose an intermediate time $\bar{t} \ge 2$ such that the generators for stages later than $\bar{t}$ contain only the identity function, i.e., $\mathbb{U}_t(D_{1:t}) = \{u : \mathbb{R} \to \mathbb{R} : u(r) = r\}$ for $t = \bar{t}+1, \dots, T$. This implies that for $t = \bar{t}+1, \dots, T$, the constraint

$$\sigma_t + Q_{t+1}(x_t, \sigma_t, D^t_{1:t+1}) \preceq_{U_t(D_{1:t})} \mathbb{Y}_{t,t+1},$$

reduces to the inequality

$$\sigma_t + \mathbb{E}_t\big[Q_{t+1}(x_t, \sigma_t, D^t_{1:t+1})\big] \le \mathbb{E}_t[\mathbb{Y}_{t,t+1}].$$

We represent the order constraint using the operators $\mathcal{G}_t : \mathcal{X}_t \times \mathcal{L}_p(\Omega, \mathcal{F}_{t+1}, P) \to \mathcal{C}[I(D_t)]$, $t = 1, \dots, T-1$, which are defined as follows

$$[\mathcal{G}_t(x,\sigma)](\eta) = \mathbb{E}\big[\max\big(0, -\eta + \sigma + Q_{t+1}(x, \sigma, D^t_{1:t+1})\big)^{p_t - 1} - \max\big(0, -\eta + \mathbb{Y}_{t,t+1}\big)^{p_t - 1}\big].$$

The notation $K(D_t)$ stands for the cone of non-positive functions in $\mathcal{C}[I(D_t)]$. Problem (6.50) now reads:

$$\begin{aligned}
\min\ & \mathbb{E}\Big[\sum_{t=1}^{T} z_t\Big] \\
\text{s.t. } & B_t x_{t-1} + A_t x_t = b_t \quad 2 = 1, \dots T, \\
& A_1 x_1 = b_1, \quad \sigma_1 = g_1(D_1) - y_1, \\
& x_t \in \mathcal{X}_t, \quad t = 1, \dots T, \\
& f_t(x_t, D_t) \le z_t, \quad t = 1, \dots T, \\
& \sigma_t = \sigma_{t-1} + z_t - y_t \quad t = 2, \dots T-1, \\
& \mathcal{G}_t(x_t, \sigma_t) \in K(D_t) \quad t = 1, \dots T-1.
\end{aligned} \tag{6.54}$$

Consider the dynamic programming formulation of problem (6.54). The optimization problem at the last stage is (6.51). For $t = T-1, \dots, 2$, we have

$$\begin{aligned}
\min_{x_t, \sigma_t, z_t}\ & z_t + \mathcal{Q}_{t+1}(x_t, \sigma_t, D_{1:t}) \\
\text{s.t. } & f_t(x_t, D_t) \le z_t, \\
& B_t x_{t-1} + A_t x_t = b_t, \\
& \sigma_t = \sigma_{t-1} + z_t - y_t, \\
& \mathcal{G}_t(x_t, \sigma_t) \in K(D_t), \\
& x_t \in \mathcal{X}_t.
\end{aligned} \tag{6.55}$$

At the first stage, we solve the problem:

$$\begin{aligned}
\min_{x_1, z_1, \sigma_1}\ & z_1 + \mathbb{E}\big[Q_2(x_1, D_{1:2})\big] \\
\text{s.t. } & f_1(x_1, D_1) \le z_1, \\
& \sigma_1 = z_1 - y_1, \\
& \mathcal{G}_1(x_1, \sigma_1) \in K_1, \\
& x_1 \in \mathcal{X}_1.
\end{aligned} \tag{6.56}$$

Observe that at the optimal solution, the first constraint is always active, that is, $z_t = f_t(x_t, D_t)$. For $t = 1, \dots, T-1$, we denote the set

$$\mathcal{Y}_t = \{(x_t, \sigma_t, z_t) : x_t \in \mathcal{X}_t,\ \mathcal{G}_t(x_t, \sigma_t) \in K(D_t),\ f_t(x_t, D_t) \le z_t\}.$$

Due to our assumptions, the sets $\mathcal{Y}_t$ are convex. We define the function ψ_t by setting

$$\psi_t(x_t, \sigma_t, z_t, D_{1:t}) = z_t + \mathcal{Q}_{t+1}(x_t, \sigma_t, D_{1:t}) + \mathbb{I}_{\mathcal{Y}_t}(x_t, \sigma_t, z_t),$$

where $\mathbb{I}_{\mathcal{Y}_t}(\cdot)$ is the indicator function of convex analysis. Assigning Lagrange multipliers π_{t1} and π_{t2} to the constraints describing the dynamics of the problem, the Lagrangian of problem (6.55) with this partial relaxation takes on the form:

$$\begin{aligned} L_t(x_t, \sigma_t, z_t, \pi_t, D_{1:t}) &= \psi_t(x_t, \sigma_t, z_t, D_{1:t}) \\ &\quad + \pi_{t1}^\top \big(B_t x_{t-1} + A_t x_t - b_t\big) + \pi_{t2}(\sigma_t - \sigma_{t-1} - z_t + y_t). \end{aligned}$$

The dual functional is

$$\begin{aligned} \Delta_t(\pi_t, D_{1:t}) &= \inf_{x_t, \sigma_t, z_t} L_t(x_t, \sigma_t, z_t, \pi_t, D_{1:t}) \\ &= - \sup_{x_t, \sigma_t, z_t} \Big\{\pi_{t1}^\top \big(A_t x_t\big) + \pi_{t2}(\sigma_t - z_t) - \psi_t(x_t, \sigma_t, z_t, D_{1:t})\Big\} \\ &\qquad - \pi_{t1}^\top \big(B_t x_{t-1} - b_t\big) - \pi_{t2}(\sigma_{t-1} - y_t) \\ &= -\psi_t^*(A_t^\top \pi_{t1}, \pi_{t2}, D_{1:t}) + \pi_{t1}^\top \big(b_t - B_t x_{t-1}\big) + \pi_{t2}(y_t - \sigma_{t-1}). \end{aligned}$$

The dual problem based on this Lagrangian becomes

$$\max_{\pi_t} -\psi_t^*(A_t^\top \pi_{t1}, \pi_{t2}, D_{1:t}) + \pi_{t1}^\top \big(b_t - B_t x_{t-1}\big) + \pi_{t2}(y_t - \sigma_{t-1}). \tag{6.57}$$

Problem (6.55) and its dual problem (6.57) are both convex problems. We assume conditions under which no duality gap between those problems exists.

Assumption 6.29. *One of the following conditions is satisfied:*

(A1) *For all $t = 1, \dots, T$, the functions $f_t(x_t, D_t)$ are random polyhedral, $\mathcal{X}_t$ are random polyhedral sets, and the number of scenarios is finite;*

(A2) *For all sufficiently small perturbations of vectors b_t and the benchmark $\mathbb{Y}$, the corresponding optimal value $Q_t(x_{t-1}, \sigma_{t-1}, D_{1:t})$ is finite.*

Let $\mathfrak{D}_t\ (x_{t-1}, \sigma_{t-1}, D_{1:t})$ stand for the set of optimal solutions of the dual problem (6.57). The following statement is true.

Theorem 6.30. *Assume that either condition* (A1) *holds and* $Q_t(x_{t-1}, \sigma_{t-1}, D_{1:t})$ *is finite or condition* (A2) *holds. Then:*

(i) *no duality gap exists between problems* (6.55) *and* (6.57), *meaning that*

$$Q_t(x_{t-1}, \sigma_{t-1}, D_{1:t}) = \sup_{\pi_t} \big\{\pi_{t1}^\top(b_t - B_t x_{t-1}) \\ + \pi_{t2}(y_t - \sigma_{t-1}) - \psi_t^*(A_t^\top \pi_{t1}, \pi_{t2}, D_{1:t})\big\}; \quad (6.58)$$

(ii) *the function* $Q_t(\cdot, \cdot, D_{1:t})$ *is subdifferentiable at* (x_{t-1}, σ_{t-1}) *and*

$$\partial Q_t\big(x_{t-1}, \sigma_{t-1}, D_{1:t}\big) = \\ \Big\{ - \big[\pi_{t1}^\top B_t, \pi_{t2}\big]^\top : \pi_t \in \mathcal{D}_t\big(x_{t-1}, \sigma_{t-1}, D_{1:t}\big)\Big\}. \quad (6.59)$$

Proof. First, we observe that the functions $Q_t(\cdot, \cdot, D_{1:t})$ are convex. Indeed, looking at problem (6.51), we notice that $Q_T(x_{T-1}, \sigma_{T-1}, D_{1:T})$ does not depend on the second argument and

$$Q_T(x_{T-1}, \sigma_{T-1}, D_{1:T}) = \inf_{x_T} \phi_T(x_T, x_{T-1}, D_{1:T}),$$

with

$$\phi_T(x_T, x_{T-1}, D_{1:T}) = \begin{cases} f_T(x_T, D_T) + \mathbb{I}_{\mathcal{X}_T}(x_T) & \text{if } A_T x_T = b_T - B_T x_{T-1}, \\ +\infty & \text{otherwise.} \end{cases}$$

The convexity of $f_T(\cdot, D_T)$ and $\mathcal{X}_T$ imply that $\phi_T(x_T, x_{T-1}, D_{1:T})$ is jointly convex with respect to the first two arguments. Hence $Q_T(\cdot, \cdot, D_{1:T})$ is convex. Similar, in the abstract formulation of (6.52), we have

$$Q_{T-1}(x_{T-2}, \sigma_{T-2}, D_{1:T-1}) = \\ \inf_{x_{T-1}, \sigma_{T-1}, z_{T-1}} \phi_{T-1}(x_T, x_{T-1}, \sigma_{T-1}, \sigma_{T-2}, z_{T-1}, D_{1:T-1}),$$

where

$$\phi_{T-1}(x_{T-1}, x_{T-2}, \sigma_{T-1}, \sigma_{T-2}, z_{T-1}, D_{1:T-1}) \\ = \begin{cases} z_{T-1} + \mathbb{I}_{\mathcal{Y}_{T-1}}(x_{T-1}, \sigma_{T-1}, z_{T-1}) & \text{if } A_{T-1}x_{T-1} = b_{T-1} - B_{T-1}x_{T-2}, \\ & \sigma_{T-1} = \sigma_{T-2} + z_{T-1} - y_{T-1}, \\ +\infty & \text{otherwise.} \end{cases}$$

The function ϕ_{T-1} is jointly convex in the first five arguments, provided the joint convexity of the indicator function $\mathbb{I}_{\mathcal{Y}_{T-1}}(x_{T-1}, \sigma_{T-1}, z_{T-1})$. It is a routine check to see that the indicator function of the set $\{(x, z) : f_{T-1}(x, D_{T-1}) \le z\}$

is jointly convex in (x, z). We only need show that the operator $\mathscr{G}_{T-1}(\cdot, \cdot)$ is jointly convex. To this end, we observe that the mapping

$$(x, \sigma) \to \max\big(0, \sigma - \eta + Q_T(x, \sigma, D_{1:T})\big)$$

is convex due to the convexity of $Q_T(\cdot, \cdot, D_{1:T})$. The convexity gets preserved when $\max\big(0, \sigma - \eta + Q_T(x, \sigma, D_{1:T})\big)$ is raised to power $p_t - 1 \geq 1$. Hence, the expectation mapping

$$(x, \sigma) \to \mathbb{E}\big[\big(\sigma - \eta + Q_T(x, \sigma, D_{1:T})\big)_+^{p_t - 1}\big]$$

is convex entailing that $\mathscr{G}_{T-1}(\cdot, \cdot)$ is convex. We conclude that ϕ_{T-1} is jointly convex in the first five arguments, which implies that $Q_{T-1}(\cdot, \cdot, D_{1:T-1})$ is convex.

Proceeding the same way, we establish that $Q_t(\cdot, \cdot, D_{1:t})$ are convex for all $t = 2, \ldots, T$. Now, we consider the optimal value function

$$\vartheta_t(\theta, \zeta) = \inf_{x_t, \sigma_t, z_t} \big\{\psi_t(x_t, \sigma_t, z_t, D_{1:t}) : A_t x_t = \theta, \ \sigma_t = \zeta\big\}.$$

Since the function $\psi_t(\cdot, D_{1:t})$ is convex, the function $\vartheta_t(\cdot)$ is also convex. Condition (A2) means that $\vartheta_t(\theta, \zeta)$ is finite-valued for all (θ, ζ) in a neighborhood of $\bar{\theta} = b_t - B_t x_{t-1}, \bar{\zeta} = \sigma_{t-1} - y_t$. It follows that $\vartheta_t(\cdot, \cdot)$ is continuous and subdifferentiable at $(\bar{\theta}, \bar{\zeta})$. Using the relations between the conjugate functions and their subgradients we infer that (i) holds and the set of optimal solutions of the dual problem (6.57) coincides with the subdifferential of $\vartheta_t(\cdot, \cdot)$ at $(\bar{\theta}, \bar{\zeta})$. Then the formula for the subdifferential in (ii) follows by the chain rule.

If condition (A1) holds, then the functions $Q_t(\cdot, D_{1:t})$ are polyhedral, and, hence, $\vartheta_t(\cdot, \cdot)$ are polyhedral. It follows that $\vartheta_t(\cdot, \cdot)$ are lower semicontinuous and subdifferentiable at any point where it is finite valued. Then statements (i)-(ii) follow due to the conjugate duality relation. □

Now, we can derive optimality conditions for problem (6.55) which feature the utility functions from the generator of the order. Observe that condition (A2) implies that the uniform dominance condition of Definition 5.35 is satisfied for problem (6.55). Additionally, we assume the following conditions:

(A3) For $t = 2, \ldots, T$ and for a.a. $D_{1:t}$ the conditional expectation function $\mathcal{Q}_t(\cdot, \cdot, D_{1:t-1})$ is finite-valued.

Theorem 6.31. *Assume that either condition* (A1) *holds and* $Q_t(x_{t-1}, \sigma_{t-1}, D_{1:t})$ *is finite, or conditions* (A2) *and* (A3) *hold. Then a feasible policy* $(\bar{x}_t, \bar{\sigma}_t, \bar{z}_t)$ *is optimal if and only if for all* $t = 1, \ldots, T$*, measurable mappings*

$\bar{\pi}_t(D_{1:t})$ along with functions $u_{(D_t)} \in \mathcal{U}^{\flat}_{p_t}(I(D_t))$ exist such that the following conditions are satisfied a.s.[†]*:*

$$
\begin{aligned}
0 \in & \mathbb{E}_t\Big[B^\top \pi^1_{t+1}(D^t_{1:t+1})\partial u_{(D_t)}\big(Q_{t+1}(x_t,\sigma_t,D_{1:t+1})+\sigma_t\big)\Big] \\
& - \mathbb{E}_t\big[B_t^\top \pi_{t+1,1}(D_{1:t+1})\big] - A_t^\top \pi^1_t + (1-\pi^2_t)\partial f_t(x_t,D_t) \\
& \qquad + \mathcal{N}_{\mathcal{X}_t}(\bar{x}_t(D_{1:t})), \qquad\qquad (6.60)\\
0 = & \mathbb{E}_t\big[\pi^2_{t+1}(D_{1:t+1})\big] - \pi^2_t(D_{1:t}) - \\
& \mathbb{E}_t\Big[\big(\pi^2_{t+1}(D_{1:t+1})+1\big)\partial u_{(D_t)}\big(Q_{t+1}(x_t,\sigma_t,D_{1:t+1})+\sigma_t\big)\Big];
\end{aligned}
$$

Additionally, $\bar{z}_t = f(\bar{x}_t(D_{1:t}))$, $\pi^2_t \le 1$, and the following conditions hold:

$$
\mathbb{E}_t\Big[u_{(D_t)}\big(Q_{t+1}(x_t,\sigma_t,D_{1:t+1})+\sigma_t\big)\Big] = \mathbb{E}_t\big[u_{(D_t)}\big(\mathbb{Y}_{t,t+1}\big)\big]. \qquad (6.61)
$$

Proof. The Lagrangian associated with problem (6.55) has the form

$$
\begin{aligned}
& \mathfrak{L}_t(z_t,x_t,\sigma_t,\pi_t) \\
& = z_t + \mathcal{Q}_{t+1}(x_t,\sigma_t,D_{1:t}) + (\pi^1_t)^\top\big(b_t - B_t x_{t-1} - A_t x_t\big) \\
& + \pi^2_t(\sigma_t - \sigma_{t-1} - z_t + y_t) + \pi^3_t(f_t(x_t,D_t) - z_t) \\
& + \int_{I(D_t)} \mathcal{G}_t(x_t,\sigma_t)(\eta)\,\mu(\mathrm{d}\eta; D_t).
\end{aligned}
$$

Taking minimum with respect to z_t, we obtain $\pi^3_t = 1 - \pi^1_t$ and $\pi^2_t \le 1$. Therefore, we obtain:

$$
\begin{aligned}
\min_{z_t,x_t,\sigma_t} \mathfrak{L}_t(z_t,x_t,\sigma_t,\pi_t) = & \min_{x_t,\sigma_t}\Big\{\mathcal{Q}_{t+1}(x_t,\sigma_t,D_{1:t}) \\
& + (\pi^1_t)^\top\big(b_t - B_t x_{t-1} - A_t x_t\big) + \pi^2_t(\sigma_t - \sigma_{t-1} + y_t) \\
& + (1-\pi^2_t) f_t(x_t,D_t) + \int_{I(D_t)} \mathcal{G}_t(x_t,\sigma_t)(\eta)\,\mu(\mathrm{d}\eta, D_t)\Big\}.
\end{aligned}
$$

Due to the convexity of the optimization problem at the right-hand side, we obtain the following optimality conditions:

[†] For $t = T$, the respective term with $T+1$ is omitted.

$$0 \in \partial_{x_t} \mathcal{Q}_{t+1}(x_t, \sigma_t, D_{1:t}) - A_t^\top \pi_t^1 + \partial_{x_t} \int_{I(D_t)} \mathcal{G}_t(x_t, \sigma_t)(\eta)\, \mu(\mathrm{d}\eta, D_t)$$
$$+(1 - \pi_t^2)\partial f_t(x_t, D_t) + \mathcal{N}_{\mathcal{X}_t}(\bar{x}_t(D_{1:t})), \tag{6.62}$$

$$0 \in \partial_{\sigma_t} \mathcal{Q}_{t+1}(x_t, \sigma_t, D_{1:t}) + \pi_t^2 + \partial_{\sigma_t} \int_{I(D_t)} \mathcal{G}_t(x_t, \sigma_t)(\eta)\, \mu(\mathrm{d}\eta, D_t), \tag{6.63}$$

$$\int_{I(D_t)} \mathcal{G}_t(x_t, \sigma_t)(\eta)\, \mu(\mathrm{d}\eta, D_t) = 0, \tag{6.64}$$

$$(1 - \pi_t^2)(f_t(x_t, D_t) - z_t) = 0. \tag{6.65}$$

Inclusion (6.63) means that

$$-\pi_{t2} \in \partial_{\sigma_t} \mathcal{Q}_{t+1}(x_t, \sigma_t, D_{1:t}) + \partial_{\sigma_t} \int_{I(D_t)} \mathcal{G}_t(x_t, \sigma_t)(\eta)\, \mu(\mathrm{d}\eta, D_t).$$

Now, we use Theorem 6.30 to derive the conditions

$$0 \in -\mathbb{E}_t\big[B_t^\top \pi_{t+1}^1(D_{1:t+1})\big] - A_t^\top \pi_t^1 + \partial_{x_t} \int_{I(D_t)} \mathcal{G}_t(x_t, \sigma_t)(\eta)\, \mu(\mathrm{d}\eta, D_t)$$
$$+(1 - \pi_t^2)\partial f_t(x_t, D_t) + \mathcal{N}_{\mathcal{X}_t}(\bar{x}_t(D_{1:t})), \tag{6.66}$$

$$0 \in \pi_t^2 - \mathbb{E}_t\big[\pi_{t+1}^2(D_{1:t+1})\big] + \partial_{\sigma_t} \int_{I(D_t)} \mathcal{G}_t(x_t, \sigma_t)(\eta)\, \mu(\mathrm{d}\eta, D_t).$$

It remains to calculate the subdifferential of the integral term. To this end, we proceed as in Lemma 5.39.

We consider the case of $p_t = 2$ first. In that case, we apply Strassen's theorem to calculate

$$\begin{aligned} \partial_{x_t}[\mathcal{G}_t(x, \sigma)](\eta) &= \partial_{x_t} \mathbb{E}_t\big[\max\big(0, -\eta + \sigma_t + Q_{t+1}(x_t, \sigma_t, D_{1:t+1})\big)\big] \\ &= \mathbb{E}_t \partial_{x_t}\big[\max\big(0, -\eta + \sigma_t + Q_{t+1}(x_t, \sigma_t, D_{1:t+1})\big)\big] \\ &= \mathbb{E}_t\big[\partial_{x_t} Q_{t+1}(x_t, \sigma_t, D_{1:t+1}^t)\, \xi(x_t, \sigma_t, \eta, D_{1:t+1})\big], \end{aligned} \tag{6.67}$$

where $\xi(x_t, \sigma_t, \eta, D_{1:t+1}^t)$ is a measurable selection of the mapping

$$DF_t(x_t, \sigma_t, \eta, D_{t+1}^t) = \begin{cases} 1 & \text{if } Q_{t+1}(x_t, \sigma_t, D_{1:t+1}^t) + \sigma_t > \eta; \\ [0, 1] & \text{if } Q_{t+1}(x_t, \sigma_t, D_{1:t+1}^t) + \sigma_t = \eta; \\ 0 & \text{if } Q_{t+1}(x_t, \sigma_t, D_{1:t+1}^t) + \sigma_t < \eta. \end{cases}$$

Substituting (6.67) under the integral and using Theorem 6.30 again, we obtain

$$
\begin{aligned}
\partial_{x_t} \int\limits_{I(D_t)} \mathcal{G}_t(x_t,\sigma_t)(\eta)\,\mu(\mathrm{d}\eta, D_t) &= \int\limits_{I(D_t)} \partial_{x_t}\mathcal{G}_t(x_t,\sigma_t)(\eta)\,\mu(\mathrm{d}\eta, D_t) \\
&= \int\limits_{I(D_t)} \mathbb{E}_t\big[-B^\top \pi^1_{t+1}(D_{1:t+1})\,\xi(x_t,\sigma_t,\eta,D_{t+1})\big]\,\mu(\mathrm{d}\eta, D_t) \\
&= -\mathbb{E}_t\Big[B^\top \pi^1_{t+1}(D_{1:t+1}) \int\limits_{I(D_t)} \xi(x_t,\sigma_t,\eta,D_{t+1})\,\mu(\mathrm{d}\eta, D_t)\Big].
\end{aligned}
$$

Notice that the integral term on the right-hand side can be interpreted as the integral of the indicator function of the event $(-\infty, \sigma_t + Q_{t+1}(x_t,\sigma_t,D_{1:t+1}))$ depending on η. Hence that integral term can be transformed with the application of the utility functions from $\mathcal{U}^{\flat}_{p_t}(I(D_t))$ as follows:

$$
\begin{aligned}
&\mathbb{E}_t\Big[B^\top \pi^1_{t+1}(D_{1:t+1}) \int\limits_{I(D_t)} \xi(x_t,\sigma_t,\eta,D_{t+1})\,\mu(\mathrm{d}\eta, D_t)\Big] \\
&\qquad = \mathbb{E}_t\Big[B^\top \pi^1_{t+1}(D_{1:t+1})\partial u_{(D_t)}\big(\sigma_t + Q_{t+1}(x_t,\sigma_t,D_{1:t+1})\big)\Big]. \qquad (6.68)
\end{aligned}
$$

Turning to the subgradient with respect to σ_t, we get

$$
\begin{aligned}
\partial_{\sigma_t}[\mathcal{G}_t(x,\sigma)](\eta) &= \partial_{\sigma_t}\mathbb{E}_t\big[\max\big(0, -\eta + \sigma_t + Q_{t+1}(x_t,\sigma_t,D_{1:t+1})\big) \\
&= \mathbb{E}_t\big[(1 + \partial_{\sigma_t}\big[Q_{t+1}(x_t,\sigma_t,D_{1:t+1})\,\xi(x_t,\sigma_t,\eta,D_{1:t+1})\big],
\end{aligned}
$$

Hence, we can transform the partial subgradient with respect to σ_t of the integral over $\mathcal{G}_t$ as follows:

$$
\begin{aligned}
&\partial_{\sigma_t} \int\limits_{I(D_t)} \mathcal{G}_t(x_t,\sigma_t)(\eta)\,\mu(\mathrm{d}\eta, D_t) = \\
&\quad \int\limits_{I(D_t)} \mathbb{E}_t\big[\big(1 - \pi^2_{t+1}(D_{1:t+1})\big)\,\xi(x_t,\sigma_t,\eta,D_{1:t+1})\big]\,\mu(\mathrm{d}\eta, D_t) \\
&\qquad = \mathbb{E}_t\Big[\big(1 - \pi^2_{t+1}(D_{1:t+1})\big)\partial u_{(D_t)}\big(Q_{t+1}(x_t,\sigma_t,D_{1:t+1}) + \sigma_t\big)\Big].
\end{aligned}
$$

Substituting the obtained expression and (6.68) into (6.66), we obtain the final formula for the optimality conditions in the case $p_t = 2$.

It is evident that at the optimal solution $z_t = f_t(x_t, D_t)$ because the stochastic order constraint also prefers smaller values for σ_t and, hence for z_t. Thus the associated complementarity condition (6.65) can be dropped. The complementarity associated with the stochastic order constraint (6.64) is transformed to the form (6.61) by using the correspondence of μ to the utility function $u_{(D_t)}$ again.

Let us consider now the case of $p_t > 2$. In that case, we have the following:

$$\begin{aligned}\partial_{x_t}[\mathscr{G}_t(x,\sigma)](\eta) &= \partial_{x_t}\mathbb{E}_t\big[\max\big(0, -\eta + \sigma_t + Q_{t+1}(x_t, \sigma_t, D_{1:t+1})\big)^{p_t-1}\big] \\ &= (p_t - 1)\mathbb{E}_t\big[\partial_{x_t} Q_{t+1}(x_t, \sigma_t, D_{1:t+1}) \\ &\qquad \max\big(0, -\eta + \sigma_t + Q_{t+1}(x_t, \sigma_t, D_{1:t+1})\big)^{p_t-2}\big].\end{aligned}$$

Now proceeding analogously to the case of $p_t = 2$, using Theorem 6.30, and the correspondence of the measure $\tilde{\mu}(\mathrm{d}\eta, D_t) = (p_t - 1)\mu(\mathrm{d}\eta, D_t)$ to a utility function $u_{(D_t)}$, we obtain

$$\begin{aligned}\partial_{x_t}\int\limits_{I(D_t)} \mathscr{G}_t(x_t, \sigma_t)(\eta)\, \mu(\mathrm{d}\eta, D_t) &= \int\limits_{I(D_t)} \mathbb{E}_t\big[-B^\top \pi^1_{t+1}(D_{1:t+1}) \\ &\qquad \max\big(0, -\eta + \sigma_t + Q_{t+1}(x_t, \sigma_t, D_{1:t+1})\big)^{p_t-2}\big]\, \tilde{\mu}(\mathrm{d}\eta, D_t) \\ &= \mathbb{E}_t\Big[B^\top \pi^1_{t+1}(D_{1:t+1})\, \partial u_{(D_t)}\big(Q_{t+1}(x_t, \sigma_t, D_{1:t+1}) + \sigma_t\big)\Big].\end{aligned}$$

Similarly, for the subgradient component corresponding to σ_t, we get

$$\begin{aligned}\partial_{\sigma_t}[\mathscr{G}_t(x,\sigma)](\eta) &= \partial_{\sigma_t}\mathbb{E}_t\big[\max\big(0, -\eta + \sigma_t + Q_{t+1}(x_t, \sigma_t, D^t_{1:t+1})\big)^{p_t-2}\big] \\ &= (p_t - 1)\mathbb{E}_t\big[\big(1 + \partial_{\sigma_t}\big[Q_{t+1}(x_t, \sigma_t, D_{1:t+1})\big) \\ &\qquad \max\big(0, -\eta + \sigma_t + Q_{t+1}(x_t, \sigma_t, D_{1:t+1})\big)^{p_t-2}\big]\, \tilde{\mu}(\mathrm{d}\eta, D_t)\big].\end{aligned}$$

The same line of argument as in the case of $p_t = 2$ leads to the following:

$$\begin{aligned}\partial_{\sigma_t}\int\limits_{I(D_t)} \mathscr{G}_t(x_t, \sigma_t)(\eta)\, \mu(\mathrm{d}\eta, D_t) &= \int\limits_{I(D_t)} \mathbb{E}_t\big[\big(1 - \pi^2_{t+1}(D_{1:t+1})\big) \\ &\qquad \max\big(0, -\eta + \sigma_t + Q_{t+1}(x_t, \sigma_t, D_{1:t+1})\big)^{p_t-2}\, \tilde{\mu}(\mathrm{d}\eta, D_t) \\ &= \mathbb{E}_t\Big[\big(\pi^2_{t+1}(D_{1:t+1}) - 1\big)\, \partial u_{(D_t)}\big(Q_{t+1}(x_t, \sigma_t, D_{1:t+1}) + \sigma_t\big)\Big].\end{aligned}$$

Substituting the obtained expressions into (6.66), we obtain the formula for the optimality conditions also in the case $p_t > 2$.

The associated complementarity conditions are dealt with as in the case of $p_t = 2$. □

The optimality conditions of Theorem 6.31 show that the following problem with utility functions is equivalent to the sequential optimization problem in the sense that its solution satisfies the same optimality conditions:

$$
\begin{aligned}
\min_{x_t,\sigma_t,z_t}\ & z_t + \mathcal{Q}_{t+1}(x_t,\sigma_t,D_{1:t}) + \mathbb{E}_t\big[u_{(D_t)}\big(Q_{t+1}(x_t,\sigma_t,D_{1:t+1})+\sigma_t\big)\big]\\
\text{s.t. } & f_t(x_t,D_t) \le z_t,\\
& B_t x_{t-1} + A_t x_t = b_t,\\
& \sigma_t = \sigma_{t-1} + z_t - y_t,\\
& x_t \in \mathcal{X}_t.
\end{aligned}
\tag{6.69}
$$

However, the optimal values of problems (6.55) and (6.69) are different.

6.4.2 Relations to Dynamic Risk Measures

Consider now the case in which the sequential order is formulated by relations between Lorenz functions. We represent the order constraint by redefining the operator $\mathcal{G}_t : \mathcal{X}_t \times \mathcal{L}_p(\Omega,\mathcal{F}_{t+1},P) \to \mathcal{C}[I(D_t)]$, $t = 1,\dots,T-1$, as follows

$$
[\mathcal{G}_t(x,\sigma)](\alpha) = \overline{F}^{(-2)}_{\mathbb{Y}_{t,t+1}}(\alpha) - \overline{F}^{(-2)}_{\sigma+Q_{t+1}(x,\sigma,D^t_{1:t+1})}(\alpha),
$$

for all $\alpha \in J(D_t) \subset [0,1]$. The notation $K(D_t)$ stands for the cone of non-negative functions in $\mathcal{C}[J(D_t)]$ and the abstract formulation of the multistage problem looks exactly the same as before.

The constraint $\mathcal{G}_t(x,\sigma) \in K(D_t)$ is a convex constraint as well. Hence, we can proceed with the analysis of the multistage problem in exactly the same way as in the previous section. To infer that the conclusions of Theorem 6.30 are valid for the new recourse function, we only need show that the new operator $\mathcal{G}_t(x,\sigma)$ is jointly convex in (x,σ). We have observed that $Q_T(\cdot,\cdot,D^{T-1}_{1:T})$ is jointly convex, as the optimization problem at the last stage has not changed. Using backward inductions again, we assume that $Q_{t+1}(\cdot,\cdot,D^t_{1:t+1})$ is convex. This implies that the function $(x,\sigma) \mapsto \sigma + Q_{t+1}(x,\sigma,D^t_{1:t+1})$ is jointly convex. The mapping $X \mapsto -\overline{F}^{(-2)}_X(\alpha)$ is convex and monotonically decreasing with respect to the almost sure order; consequently, the composition $(x,\sigma) \mapsto \overline{F}^{(-2)}_{\sigma+Q_{t+1}(x,\sigma,D^t_{1:t+1})}(\alpha)$ is jointly convex, implying the desired property of $\mathcal{G}_t(\cdot,\cdot)$. We infer that Theorem 6.30 holds for the new problem as well.

Theorem 6.32. *Assume that either condition* (A1) *holds and* $Q_t(x_{t-1},\sigma_{t-1},D_{1:t})$ *is finite or conditions* (A2) *and* (A3) *hold. Then a feasible policy* $(\bar{x}_t,\bar{\sigma}_t,\bar{z}_t)$ *is optimal if and only if for all* $t = 1,\dots,T$, *measurable mappings* $\bar{\pi}_t(D_{1:t})$ *and* $\kappa(D_t)$ *along with a dynamic measure of risk* ϱ *exist such that the following conditions are satisfied*†*:*

† For $t = T$, the respective term with $T+1$ is omitted.

$$\begin{aligned}
0 \in & \mathbb{E}_t\Big[B^\top \pi^1_{t+1}(D_{1:t+1})\Big]\kappa(D_t)\, \partial \varrho_{(D_t)}\big(Q_{t+1}(x_t,\sigma_t,D^t_{1:t+1})+\sigma_t\big)\\
& - \mathbb{E}_t\big[B_t^\top \pi^1_{t+1}(D_{1:t+1})\big] - A_t^\top \pi^1_t + (1-\pi^2_t)\,\partial f_t(x_t,D_t)\\
& \qquad + \mathcal{N}_{\mathcal{X}_t}(\bar{x}_t(D_{1:t})),\\
0 = & \mathbb{E}_t\big[\pi^2_{t+1}(D_{1:t+1})|D_{1:t}\big] - \pi^2_t -\\
& \mathbb{E}_t\Big[\big(\pi^2_{t+1}(D_{1:t+1})+1\big)\Big]\kappa(D_t)\, \partial \varrho_{(D_t)}\big(Q_{t+1}(x_t,\sigma_t,D_{1:t+1})+\sigma_t\big).
\end{aligned} \tag{6.70}$$

Additionally, $\bar{z}_t = f(\bar{x}_t(D_{1:t}))$, $\pi^2_t \le 1$, *and the following complementarity condition holds:*

$$\kappa(D_t)\varrho_{(D_t)}\big(Q_{t+1}(x_t,\sigma_t,D_{1:t+1})+\sigma_t\big) = \kappa(D_t)\varrho_{(D_t)}\big(\mathbb{Y}_{t,t+1}\big). \tag{6.71}$$

Proof. In the proof of Theorem 6.31, we have established the following condition (the relations (6.66)):

$$\begin{aligned}
0 \in -\mathbb{E}_t\big[B_t^\top \pi^1_{t+1}(D_{1:t+1})\big] - A_t^\top \pi^1_t + \partial_{x_t} \int_{I(D_t)} \mathcal{G}_t(x_t,\sigma_t)(\eta)\, \mu(\mathrm{d}\eta, D_t)\\
+(1-\pi^2_t)\partial f_t(x_t,D_t) + \mathcal{N}_{\mathcal{X}_t}(\bar{x}_t(D_{1:t})),\\
0 \in \pi^2_t - \mathbb{E}_t\big[\pi^2_{t+1}(D_{1:t+1})\big] + \partial_{\sigma_t} \int_{I(D_t)} \mathcal{G}_t(x_t,\sigma_t)(\eta)\, \mu(\mathrm{d}\eta, D_t).
\end{aligned}$$

The statement follows by using the correspondence of measures in $\mathcal{M}_1^+(J(D_t))$ to spectral coherent measures of risk, leading to the following representation of the integral terms

$$\begin{aligned}
&\int_{I(D_t)} \mathcal{G}_t(x_t,\sigma_t)(\eta)\, \mu(\mathrm{d}\eta, D_t)\\
&\quad = \kappa(D_t)\, \varrho_{(D_t)}\big(Q_{t+1}(x_t,\sigma_t,D^t_{1:t+1})+\sigma_t\big) - \kappa(D_t)\, \varrho_{(D_t)}\big(\mathbb{Y}_{t,t+1}\big).
\end{aligned}$$

The remaining arguments are straightforward. □

This statement shows additional consistency property between sequential stochastic orders and dynamic measures of risk.

Chapter 7
Numerical Methods for Problems with Stochastic Dominance Constraints

In this chapter, we discuss numerical solution of problems with second-order stochastic dominance constraints. Because of the involved nature of these constraints, specialized methods need to be designed, which are not simple adaptations of general optimization techniques.

We focus on the second-order stochastic dominance constraints, as the most useful in applications to risk-averse decision making.

7.1 Reformulations of the Second Order Dominance

In this section, we present reformulations of the following optimization problem

$$\min f(x) \tag{7.1}$$
$$\text{subject to } \mathbb{E}[\max(0, \eta - G(x))] \le \mathbb{E}[\max(0, \eta - Y)] \quad \forall \eta \in I, \tag{7.2}$$
$$x \in \mathcal{X}. \tag{7.3}$$

Here $I \subset \mathbb{R}$, and $Y \in \mathcal{L}_1(\Omega, \mathcal{F}, P)$. The space $\mathfrak{X} = \mathbb{R}^n$, the functional $f : \mathfrak{X} \to \mathbb{R}$ is continuous, and the set $\mathcal{X} \subset \mathfrak{X}$ is closed. Additionally, $G(\cdot)$ is a continuous operator with respect to the $\mathcal{L}_1$-norm.

We have already indicated that a linear representation of the stochastic order exists when Ω is a finite set $\{\omega_1, \dots, \omega_m\}$; it is given by (5.72)–(5.74). However, this representation introduces m^2 new variables to the optimization problem, replaces (7.2) with $m(2m + 1)$ linear constraints, and is impractical.

Recall that if the benchmark random variable Y has a discrete distribution with m realization y_j, $j = 1, \dots, m$, then a finite number of constraints are sufficient to represent (7.2) even with $I = \mathbb{R}$, as shown in Lemma 5.42. The following set of inequalities:

$$\mathbb{E}[\max(0, y_j - G(x)] \le \mathbb{E}[\max(0, y_j - Y)] \quad \forall j = 1, \dots, m, \tag{7.4}$$

D. Dentcheva, A. Ruszczyński, *Risk-Averse Optimization and Control*, Springer Series in Operations Research and Financial Engineering,
https://doi.org/10.1007/978-3-031-57988-2_7

implies that $\mathbb{E}[\max(0, \eta - G(x)] \le \mathbb{E}[\max(0, \eta - Y)]$ for all $\eta \in \mathbb{R}$. It is straightforward to modify the result for a compact interval I.

Another reformulation of the constraint (7.53) using coupling is possible. For this purpose, we assume that we deal with random variables Z (obtained via $G(x)$) and Y that have finitely many realizations. Let us denote the realizations of Z by z_i and their respective probabilities by p_i, $i = 1, \dots, N$, while the realizations of Y are denoted y_j taken with probabilities q_j, $j = 1, \dots, S$. Without loss of generality, we may assume that $y_1 < y_2 < \cdots < y_S$. Let P_Z and P_Y denote the induced distributions of Z and Y on $\mathbb{R}$, that is, $P_Z = P \circ Z^{-1}$, $P_Y = P \circ Y^{-1}$.

Recall Strassen's theorem about the couplings: for two integrable random variables Z and Y with distributions P_Z and P_Y, the relation $Z \succeq_{(2)} Y$ is equivalent to the existence of random variables Z' and Y', defined on a common probability space, such that $Z' \sim P_Z$, $Y' \sim P_Y$, and $Z' \ge \mathbb{E}[Y' \mid Z']$ almost surely. This result implies the following statement for the special case of discrete distributions.

Lemma 7.1. *For the random variables Z and Y with discrete distributions as above, the relation $Z \succeq_{(2)} Y$ holds if and only if a matrix $\pi \in \mathbb{R}^{NS}$ with non-negative components exists such that*

$$\sum_{j=1}^{S} y_j \pi_{ij} \le z_i, \quad i = 1, \dots, N; \tag{7.5}$$

$$\sum_{j=1}^{S} \pi_{ij} = 1, \quad i = 1, \dots, N; \tag{7.6}$$

$$\sum_{i=1}^{N} p_i \pi_{ij} = q_j, \quad j = 1, \dots, S. \tag{7.7}$$

Proof. First, assume that $Z \succeq_{(2)} Y$ holds and let $\pi_{ij} = P'[Y' = y_j \mid Z' = z_i]$, for all $i = 1, \dots, N$, $j = 1, \dots, S$, where P' is the probability measure of the probability space on which Z' and Y' are defined. By construction, π is a Markov kernel: $\pi_{ij} \ge 0$ and (7.6) holds. Furthermore, we easily verify (7.7), as follows:

$$\begin{aligned} q_j = P(Y' = y_j) &= \sum_{i=1}^{N} P[Y' = y_j \mid Z' = z_i] P(Z' = z_i) \\ &= \sum_{i=1}^{N} p_i \pi_{ij} \quad \text{for all } j = 1, \dots, S. \end{aligned}$$

Finally, for each $i = 1, \dots, N$, we have

$$z_i \geq \mathbb{E}[Y'|Z' = z_i] = \sum_{j=1}^{S} y_j \pi_{ij}.$$

To prove the opposite inclusion, suppose π is a matrix that satisfies (7.5)–(7.7). We define the random variables

$$Z'(i,j) = z_i, \quad Y'(i,j) = y_j, \quad \text{for all } i = 1, \dots, N, \ j = 1, \dots, S.$$

due to (7.6) and (7.7), $Z' \sim P_Z$ and $Y' \sim P_Y$. Furthermore, (7.5) implies that $Z \succeq_{(2)} Y$. □

We can strengthen this formulation.

Theorem 7.2. *For two random variables Z and Y with discrete distributions as described, the relation $Z \succeq_{(2)} Y$ holds if and only if a matrix $\pi \in \mathbb{R}^{NS}$ with non-negative components exists such that* (7.5) *and* (7.6) *are satisfied in addition to the following system of equations:*

$$\sum_{i=1}^{N} p_i \sum_{j=1}^{k-1} (y_k - y_j)\pi_{ij} = \sum_{j=1}^{k-1} (y_k - y_j) q_j, \quad k = 2, \dots, S. \tag{7.8}$$

Proof. Suppose $Z \succeq_{(2)} Y$. Using Lemma 7.1, we obtain the existence of $\pi \in \mathbb{R}^{NS}$ such that

$$\sum_{i=1}^{N} p_i \sum_{j=1}^{k-1} (y_k - y_j)\pi_{ij} = \sum_{j=1}^{k-1} (y_k - y_j) \sum_{i=1}^{N} p_i \pi_{ij} = \sum_{j=1}^{k-1} (y_k - y_j) q_j.$$

Hence, (7.8) is satisfied.

Now, suppose a matrix $\pi \in \mathbb{R}^{NS}$ with non-negative components exists such that (7.5), (7.6), and (7.8) hold. Using these relations, we obtain the following chain of relations:

$$\begin{aligned}
\max(0, y_k - z_i) &\leq \max\Big(0, y_k - \sum_{j=1}^{S} y_j \pi_{ij}\Big) = \max\Big(0, \sum_{j=1}^{S} (y_k - y_j)\pi_{ij}\Big) \\
&\leq \sum_{j=1}^{S} \max(0, y_k - y_j)\pi_{ij} = \sum_{j=1}^{k-1} (y_k - y_j)\pi_{ij}.
\end{aligned}$$

Using the inequalities for all i multiplied by p_i, $i = 1, \dots, N$, we obtain

$$\begin{aligned}\mathbb{E}[\max(0, y_k - Z)] &\le \sum_{i=1}^{N} p_i \sum_{j=1}^{k-1} (y_k - y_j)\pi_{ij} \\ &= \sum_{j=1}^{k-1} (y_k - y_j) q_j = \mathbb{E}[\max(0, y_k - Y)].\end{aligned}$$

In the second relation, we used (7.8). Hence, $Z \succeq_{(2)} Y$. □

Let us further assume that the operator $G(\cdot)$ is represented by concave functions $g^i(x)$, $i = 1, \dots, N$. The observations in Theorem 7.2 lead us to the conclusion that problem (7.1)–(7.3) can be formulated in the following equivalent way under the assumptions in this section.

$$\begin{aligned}&\min_{x,\pi,v} f(x)\\ &\text{subject to} \sum_{j=1}^{S} y_j \pi_{ij} \le g^i(x), \quad i = 1, \dots, N;\\ &\sum_{j=1}^{S} \pi_{ij} = 1, \quad i = 1, \dots, N;\\ &\sum_{i=1}^{N} p_i \pi_{ij} = v_j, \quad j = 1, \dots, S;\\ &\sum_{j=1}^{k-1} (y_k - y_j) v_j = \sum_{j=1}^{k-1} (y_k - y_j) q_j, \quad k = 2, \dots, S;\\ &x \in \mathcal{X}.\end{aligned} \tag{7.9}$$

The advantage of this formulation is that it replaces the stochastic ordering relation by convex constraints, if $g^i(\cdot)$ are concave, or just linear constraints, if $g^i(\cdot)$ are linear. However, if the numbers of realizations N and S are large, this formulation may also be computationally expensive.

7.2 Event Cut Methods

In this section, we propose methods that replace the second-order stochastic dominance constraint $G(x) \succeq_{(2)} Y$ with inequalities of the form

$$\mathbb{E}\left[\mathbb{1}_{\mathcal{A}^j} G(x)\right] \ge b(\mathcal{A}^j, Y),$$

for a finite collection of events $\mathcal{A}^j \in \mathcal{F}$, $j = 1, \ldots, k$. We call these inequalities the *event cuts*. The methods differ in the way the sets $\mathcal{A}^j$ are constructed, and in the form of the right-hand side function $b(\mathcal{A}_j, Y)$, $j = 1, 2, \ldots$.

7.2.1 Shortfall Event Cut Method

The idea of this method is to construct the event cuts using the shortfall formulation (7.2). We do not assume that the random variables involved have finite distributions. Denote $v(\eta) = \mathbb{E}[\max(0, \eta - Y)]$ and let $I = [a, b]$. We start with the following basic observation.

Lemma 7.3. *Assume that the random variables Z and Y are elements of the space $\mathcal{L}_1(\Omega, \mathcal{F}, P)$. Then $Z \succeq_{(2)}^{I} Y$ if and only if for all events $\mathcal{A} \in \mathcal{F}$ and all $\eta \in [a, b]$,*

$$\mathbb{E}[(\eta - Z)\mathbb{1}_{\mathcal{A}}] \le v(\eta).$$

Proof. The statement follows from inequality (5.6) and the observation that $\mathbb{E}[(\eta - Z)\mathbb{1}_{\mathcal{A}}] \le \mathbb{E}[\max(0, \eta - Z)]$ holds for all events $\mathcal{A} \in \mathcal{F}$. The inequality becomes an equation when $\mathcal{A} = \{Z < \eta\}$. □

This observation suggests the following equivalent formulation of problem (7.1)–(7.3):

$$\begin{aligned} &\min f(x) \\ &\text{subject to} \int_{\mathcal{A}} \big(\eta - G(x)\big)\, \mathrm{d}P \le v(\eta), \quad \forall \eta \in [a, b],\ \forall \mathcal{A} \in \mathcal{F}, \\ &\qquad x \in \mathcal{X}. \end{aligned} \tag{7.10}$$

Observe that all inequalities in problem (7.10) are linear with respect to the random variable $G(x)$. In particular, if $G(\cdot)$ is linear, the constraints are linear with respect to x as well. Although the formulation (7.10) appears to be very difficult, because it involves all events $\mathcal{A}$, in the method below we use only finitely many specially constructed events.

Shortfall Event Cut Method

Step 0. Set $k = 1$, $\eta_0 \in [a, b]$, and $\mathcal{A}^0 = \Omega$.

Step 1. Solve the problem:

$$\begin{aligned} &\min f(x) \\ &\text{s.t. } \mathbb{E}\big[\mathbb{1}_{\mathcal{A}^j}(\eta_j - G(x))\big] \le v(\eta_j), \quad j = 0, \ldots, k-1, \\ &\quad x \in \mathcal{X}. \end{aligned} \tag{7.11}$$

Let x^k denote the solution of problem (7.11), and let $Z^k = G(x^k)$.

Step 2. Calculate the quantity

$$\delta_k = \sup_{\eta} \left\{ \mathbb{E}\left[(\eta - Z^k)_+\right] - v(\eta) : \ \eta \in [a, b] \right\}. \tag{7.12}$$

Step 3. If $\delta_k \leq 0$, stop; otherwise, continue.

Step 4. Determine $\eta_k \in [a, b]$ such that

$$\mathbb{E}\left[(\eta_k - Z^k)_+\right] - v(\eta_k) \geq \frac{\delta_k}{2}, \tag{7.13}$$

and save the event

$$\mathcal{A}^k = \left\{ Z^k < \eta_k \right\}.$$

Step 5. Increase k by one, and go to Step 1.

We call the inequalities in problem (7.11) the *shortfall event cuts*, which explains the name of the method.

The following convergence result holds.

Theorem 7.4. *Assume that problem* (7.1)–(7.3) *has a feasible solution, that the set $\mathcal{X}$ is compact, and the function $f : \mathcal{X} \to \mathbb{R}$ and the operator $G : \mathcal{X} \to \mathcal{L}_1(\Omega, \mathcal{F}, P)$ are continuous. Then the Shortfall Event Cut Method either stops at an optimal solution of problem* (7.1)–(7.3) *or it generates a sequence of points $\{x^k\}$ such that every accumulation point $\hat{x}$ of $\{x^k\}$ is an optimal solution of problem* (7.1)–(7.3).

Proof. Problem (7.11) is a relaxation of problem (7.10). Thus, it always has a feasible solution by assumption. Furthermore, the feasible set of problem (7.11) at iteration $k > 1$ is contained in the feasible set of the problem at iteration $k - 1$ and these sets are compact. Therefore, problem (7.11) is always solvable and its optimal value satisfies $f(x^k) \leq f^*$, with f^* being the optimal value of problem (7.1)–(7.3). If the method stops at Step 3 of iteration k, then the stochastic dominance constraint is satisfied at the point x^k. Thus, x^k is feasible also for problem (7.1)–(7.3) and, therefore, $f(x^k) \geq f^*$. The two relations together imply $f(x^k) = f^*$, showing the optimality of x^k.

It remains to consider the case in which the method generates an infinite sequence of points $\{x^k\}$. We shall show that $\lim_{k\to\infty} \delta_k = 0$. Suppose it is not true, and that an infinite set of indices $\mathcal{K}$ exists such that $\delta_k \geq \delta > 0$ for all $k \in \mathcal{K}$. Consider $k \in \mathcal{K}$ and $l \in \mathcal{K} : l > k$. Since the event cut generated at iteration k is satisfied at iteration l, then we have:

$$\mathbb{E}\left[(\eta_k - Y)_+ - \mathbb{1}_{\mathcal{A}^k}(\eta_k - Z^l)\right] \geq 0. \tag{7.14}$$

We estimate the second term using the definition of the event $\mathcal{A}^k$:

$$\begin{aligned}\mathbb{1}_{\mathcal{A}^k}(\eta_k - Z^l) &= \mathbb{1}_{\mathcal{A}^k}(\eta_k - Z^k) + \mathbb{1}_{\mathcal{A}^k}(Z^k - Z^l)\\ &= (\eta_k - Z^k)_+ + \mathbb{1}_{\mathcal{A}^k}(Z^k - Z^l) \le (\eta_k - Z^l)_+ + 2|Z^k - Z^l|.\end{aligned}$$

As the sequence $\{x^k\}$ is convergent, and the operator $G : \mathcal{X} \to \mathcal{L}_1(\Omega, \mathcal{F}, P)$ is continuous, for all sufficiently large k and all $l > k$, we have $\|Z^k - Z^l\| \le \delta/8$. Therefore,

$$\mathbb{E}\big[\mathbb{1}_{\mathcal{A}^k}(\eta_k - Z^l)\big] \le \mathbb{E}\big[(\eta_k - Z^l)_+\big] + \frac{\delta}{4}.$$

This estimate together with (7.14) yield the inequality

$$\mathbb{E}\big[(\eta_k - Y)_+ - (\eta_k - Z^l)_+\big] \ge -\frac{\delta}{4}.$$

On the other hand, (7.13) at iteration l requires that

$$\mathbb{E}\big[(\eta_l - Z^l)_+ - (\eta_l - Y)_+\big] \ge \frac{\delta_l}{2} \ge \frac{\delta}{2}.$$

Adding the last two inequalities and using the Lipschitz continuity of the mapping $\max(0, \cdot)$ (with the Lipschitz constant 1), we obtain for all sufficiently large $k, l \in \mathcal{K}$, $l > k$, the inequality

$$\frac{\delta}{4} \le \mathbb{E}\big[(\eta_l - Z^l)_+ - (\eta_k - Z^l)_+ + (\eta_k - Y)_+ - (\eta_l - Y)_+\big] \le 2|\eta_l - \eta_k|.$$

This is impossible because of the compactness of the set $[a, b]$. Thus, $\delta_k \to 0$.

As the set $\mathcal{X}$ is compact, the sequence $\{x^k\}$ has a convergent subsequence. Let x^* be an accumulation point of the sequence $\{x^k\}$ corresponding to some subsequence $\mathcal{K}_1 \subset \mathbb{N}$. The point $x^* \in \mathcal{X}$ because $\mathcal{X}$ is closed. Due to the choice of δ_k in Step 2, the following inequality holds:

$$\sup_{\eta}\Big\{\mathbb{E}\big[(\eta - G(x^k))_+ - (\eta - Y)_+\big] : \ \eta \in [a, b]\Big\} \le \frac{\delta_k}{2}.$$

Passing to the limit over the subsequence $\mathcal{K}_1$ and using the continuity of G, we obtain

$$\sup_{\eta}\Big\{\mathbb{E}\big[(\eta - G(x^*))_+ - (\eta - Y)_+\big] : \ \eta \in [a, b]\Big\} \le 0.$$

We conclude that the point x^* is feasible for problem (7.1)–(7.3), which entails $f(x^*) \ge f^*$. On the other hand $f(x^k) \le f^*$ for all iterations k. Consequently, $f(x^*) = f^*$ as stated. □

Consider the discrete probability space $\Omega = \{\omega_1, \dots, \omega_M\}$. Assume that the realizations of Y are $y_1, \dots, y_M$ with respective shortfalls $v_j = \mathbb{E}[(y_j - Y)_+]$, $j = 1, \dots, M$. Using Lemma 5.42, we can restrict the range of η in the method to the set $\{y_1, \dots, y_M\}$. Furthermore, only finitely many events exist and the method converges in finitely many iterations. Practical experience indicates that if M is large, then the number of event cuts generated is much smaller than M.

7.2.2 Inverse Event Cut Method

We start from the following basic property of stochastic dominance, which is of independent interest.

Theorem 7.5. *Suppose $Y \in \mathcal{L}_1(\Omega, \mathcal{F}, P)$. Then $X \succeq_{(2)} Y$ if and only if*

$$\mathbb{E}[X|\mathcal{A}] \geq \frac{1}{P(\mathcal{A})} F_Y^{(-2)}(P(\mathcal{A})) \quad \forall\, \mathcal{A} \in \mathcal{F} : P(\mathcal{A}) > 0. \tag{7.15}$$

Proof. Given a set $\mathcal{A} \in \mathcal{F}$ of positive probability, we define

$$\hat{\eta} = \inf\bigl\{t : P(X \leq t) \geq P(\mathcal{A})\bigr\} \text{ and } \mathcal{B} = \{X < \hat{\eta}\}.$$

The following relations are evident:

$$\mathbb{E}[\mathbb{1}_{\mathcal{A}} X] + \mathbb{E}[\mathbb{1}_{\mathcal{B}}(\hat{\eta} - X)] = \mathbb{E}[\mathbb{1}_{\mathcal{A}\setminus\mathcal{B}} X] + \mathbb{E}[\mathbb{1}_{\mathcal{A}\cap\mathcal{B}}\hat{\eta}] + \mathbb{E}[\mathbb{1}_{\mathcal{B}\setminus\mathcal{A}}(\hat{\eta} - X)].$$

Observing that the first term is bounded below by $\hat{\eta} P(\mathcal{A} \setminus \mathcal{B})$ and the third term is nonnegative, we infer

$$\mathbb{E}[\mathbb{1}_{\mathcal{A}} X] + \mathbb{E}[(\hat{\eta} - X)_+] \geq \hat{\eta} P(\mathcal{A}). \tag{7.16}$$

Corollary 5.11 (ii) implies that $F_X^{(-2)}(P(\mathcal{A})) = \hat{\eta} P(\mathcal{A}) - \mathbb{E}[(\hat{\eta} - X)_+]$. Substituting into (7.16) we conclude that

$$\mathbb{E}[\mathbb{1}_{\mathcal{A}} X] \geq F_X^{(-2)}(P(\mathcal{A})).$$

The relation (7.15) follows now from the characterization of the stochastic dominance by the Lorenz function (Corollary 5.10).

To prove the converse implication, consider the sets $\mathcal{A}_t = \{X \leq t\}$ and $\mathcal{B}_t = \{X < t\}$, for $t \in \mathbb{R}$. The left hand side of (7.15) at $\mathcal{A} = \mathcal{A}_t$ and $\mathcal{A} = \mathcal{B}_t$ takes on the forms

$$\mathbb{E}[X|\mathcal{A}_t] = \frac{1}{P(\mathcal{A}_t)} F_X^{(-2)}\bigl(P(\mathcal{A}_t)\bigr) \quad \text{and} \quad \mathbb{E}[X|\mathcal{B}_t] = \frac{1}{P(\mathcal{B}_t)} F_X^{(-2)}\bigl(P(\mathcal{B}_t)\bigr).$$

It follows from (7.15) that

$$F_X^{(-2)}(p) \geq F_Y^{(-2)}(p) \quad \text{for all} \quad p \in J_1 \cup J_2, \tag{7.17}$$

where

$$J_1 = \{P(\mathcal{A}_t) : t \in \mathbb{R}\} \cap (0,1) \quad \text{and} \quad J_2 = \{P(\mathcal{B}_t) : t \in \mathbb{R}\} \cap (0,1).$$

If $J_1 \cup J_2 = (0,1)$, the proof is complete. Suppose $p \notin J_1 \cup J_2$, $p \in (0,1)$. Define

$$t_* = \inf\{t : P(\mathcal{A}_t) \geq p\}.$$

Then $P(\mathcal{B}_{t*}) < p < P(\mathcal{A}_{t*})$, where the strict inequalities follow from $p \notin J_1 \cup J_2$. This implies that $F_X(\cdot)$ is discontinuous at t_*. Thus, the function $F_X^{(-1)}(\cdot)$ is constant on the interval $\big(P(\mathcal{B}_{t*}), P(\mathcal{A}_{t*})\big)$. Consequently, the function $F_X^{(-2)}(\cdot)$ is affine on this interval. Due to the convexity of the function $F_Y^{(-2)}(\cdot)$ and because inequality (7.17) holds at the ends of this interval, we infer that (7.17) extends to all p in this interval. Consequently, it holds for all $p \in [0,1]$. □

Corollary 7.6. *Suppose $Y \in \mathcal{L}_1(\Omega, \mathcal{F}, P)$. Then $X \succeq_{(2)} Y$ if and only if*

$$\mathbb{E}[X|X \leq t] \geq \frac{1}{F_X(t)} F_Y^{(-2)}\big(F_X(t)\big), \quad \forall\, t \in \mathbb{R},\ F_X(t) > 0. \tag{7.18}$$

Proof. The relation $X \succeq_{(2)} Y$ implies (7.18) by virtue of Theorem 7.5. To prove the converse implication, we observe that the proof of Theorem 7.5 shows that it is sufficient to verify the inequalities

$$\mathbb{E}[X|\mathcal{B}_t] \geq \frac{1}{P(\mathcal{B}_t)} F_Y^{(-2)}\big(P(\mathcal{B}_t)\big), \quad \forall\, t \in \mathbb{R},\ P(\mathcal{B}_t) > 0, \tag{7.19}$$

for the events $\mathcal{B}_t = \{X < t\}$. For any fixed t, we take a strictly increasing sequence $\{\tau_i\}$ convergent to t, and we note that

$$P(\mathcal{B}_t) = \lim_{i\to\infty} P(\mathcal{A}_{\tau_i}) \quad \text{and} \quad \mathbb{E}[X|\mathcal{B}_t] = \lim_{i\to\infty} \mathbb{E}[X|\mathcal{A}_{\tau_i}].$$

Passing to the limit in (7.18), we obtain (7.19). □

For a discrete distribution of X we consider its ordered realizations $x_{[j]}$ and their probabilities π_j. In the corollary below, we do not assume that Y is discrete.

Corollary 7.7. *Suppose X has a discrete distribution with N realizations, and $Y \in \mathcal{L}_1(\Omega, \mathcal{F}, P)$. Then $X \succeq_{(2)} Y$ if and only if*

$$\sum_{k=1}^{j} \pi_k x_{[k]} \geq F_Y^{(-2)}\Big(\sum_{k=1}^{j} \pi_k\Big), \quad j = 1, \ldots, N.$$

The statement follows from Corollary 7.6 taking into account that $\mathbb{E}[X|X \le t]$ is piecewise linear as a function of t, and $F_Y^{(-2)}\big(F_X(\cdot)\big)$ is convex.

We consider now the problem formulation

$$\min\ f(x) \tag{7.20}$$

$$\text{subject to } F^{(-2)}_{G(x)}(p) \ge F_Y^{(-2)}(p), \quad \forall p \in [\alpha, \beta], \tag{7.21}$$

$$x \in \mathcal{X}. \tag{7.22}$$

Using Theorem 7.5, we construct the following numerical method for solving problem (7.20)–(7.22).

Inverse event cut method

Step 0. Set $k = 1$, $\mathcal{A}^0 = \Omega$.

Step 1. Solve the problem:

$$\begin{aligned} &\min\ f(x)\\ &\text{subject to } \mathbb{E}\big[G(x)\,\big|\,\mathcal{A}^j\big] \ge \frac{1}{P(\mathcal{A}^j)} F^{(-2)}\big(Y; P(\mathcal{A}^j)\big), \quad j = 1, \dots, k-1,\\ &x \in \mathcal{X}. \end{aligned} \tag{7.23}$$

Let x^k denote the solution of problem (7.23), and let $Z^k = G(x^k)$.

Step 2. Consider the sets $\mathcal{B}_t^k = \{Z^k \le t\}$ and let

$$\delta_k = \sup_t \Big\{ \frac{1}{P(\mathcal{B}_t^k)} F_Y^{(-2)}\big(P(\mathcal{B}_t^k)\big) - \mathbb{E}[Z^k|\mathcal{B}_t^k] : P(\mathcal{B}_t^k) > 0 \Big\}. \tag{7.24}$$

If $\delta_k \le 0$, then stop; otherwise, continue.

Step 3. Find t_k such that $P(Z^k \le t_k) > 0$ as well as

$$\mathbb{E}\big[Z^k \,\big|\, \mathcal{B}_{t_k}^k\big] - \frac{1}{P(\mathcal{B}_{t_k}^k)} F_Y^{(-2)}\big(P(\mathcal{B}_{t_k}^k)\big) \le -\frac{\delta_k}{2}. \tag{7.25}$$

Step 4. Set $\mathcal{A}^{k+1} = \mathcal{B}_{t_k}^k$, increase k by one, and go to Step 1.

In Step 1, we are using only selected sets $\mathcal{A}^j$, and thus problem (7.23) is a relaxation of problem (7.20)–(7.22). If the functions $f(\cdot)$ is convex, the operator $G(\cdot)$ is concave, and the set $\mathcal{X}$ is convex, then problem (7.23) is a convex optimization problem.

We call the constraints in problem (7.23) the *inverse event cuts.*

Remark 7.8. The inverse event cuts are related to the event cuts used in § 7.2.1 via the duality relation of Theorem 2.15 and the stochastic dominance characterization of Corollary 5.10. The event cut added to problem (7.11) at iteration k of the Shortfall Event Cut Method has the form:

$$\mathbb{E}\big[\mathbb{1}_{\mathcal{A}^k}(\eta_k - G(x))\big] \le v(\eta_k).$$

Equivalently,

$$\mathbb{E}\big[\mathbb{1}_{\mathcal{A}^k} G(x)\big] \ge P(\mathcal{A}^k)\,\eta_k - v(\eta_k).$$

On the other hand, using the duality relation of Theorem 2.15, we can write the inverse event cut in problem (7.23) as

$$\mathbb{E}\big[\mathbb{1}_{\mathcal{A}^k} G(x)\big] \ge F^{(-2)}\big(Y; P(\mathcal{A}^k)\big) = \sup_{\eta}\big\{P(\mathcal{A}^k)\,\eta - v(\eta)\big\}.$$

It follows that the inverse event cut is the deepest among all event cuts of the form

$$\mathbb{E}\big[\mathbb{1}_{\mathcal{A}^k} G(x)\big] \ge P(\mathcal{A}^k)\,\eta - v(\eta), \quad \eta \in \mathbb{R}.$$

□

Due to Corollary 7.6, if $\delta_k \le 0$, then $Z^k \succeq_{(2)} Y$, and, hence, x^k is a solution of problem (7.20)–(7.22). Therefore, the stopping test of Step 2 yields an optimal solution. We now analyze convergence of the method, when an infinite number of iterations are made.

Theorem 7.9. *Assume that the set $\mathcal{X} \subset \mathbb{R}^n$ is compact, the function $f(\cdot)$ is continuous, and the operator $G(\cdot)$ is Lipschitz continuous on $\mathcal{X}$ as a mapping from $\mathbb{R}^n$ to $\mathcal{L}_\infty(\Omega, \mathcal{F}, P)$. If problem* (7.20)–(7.22) *has a nonempty feasible set, then either the inverse event cut method stops at a solution of* (7.20)–(7.22), *or every accumulation point of the sequence $\{x^k\}$ generated by that method is a solution of* (7.20)–(7.22).

Proof. Observe that the feasible set of (7.23) is nonempty and compact. Due to the continuity of $f(\cdot)$ and $G(\cdot)$, an optimal solution x^k at Step 1 of each iteration k exists.

First, we prove that $\delta_k \to 0$, as $k \to \infty$. Suppose it is not true and an infinite set of indices $\mathcal{K}$ exists such that $\delta_k \ge \delta > 0$ for all $k \in \mathcal{K}$. Consider $k \in \mathcal{K}$ and $l \in \mathcal{K} : l > k$. As the cut generated at iteration k is satisfied at iteration l we have:

$$\mathbb{E}[Z^l | Z^k \le t_k] - \frac{1}{P(Z^k \le t_k)} F_Y^{(-2)}\big(P(Z^k \le t_k)\big) \ge 0.$$

Combining this with (7.25), we obtain

$$\mathbb{E}[Z^l | Z^k \le t_k] - \mathbb{E}[Z^k | Z^k \le t_k] \ge \frac{\delta}{2}.$$

We can further transform the left-hand side of this inequality by using the Hölder inequality as follows:

$$
\begin{aligned}
\frac{\delta}{2} &\le \frac{1}{P(Z^k \le t_k)} \left| \int (Z^l(\omega) - Z^k(\omega)) \mathbb{1}_{\{Z^k \le t_k\}}(\omega) \, P(d\omega) \right| \\
&\le \frac{1}{P(Z^k \le t_k)} \|Z^l - Z^k\|_\infty \| \mathbb{1}_{\{Z^k \le t_k\}} \|_1 \\
&= \|Z^l - Z^k\|_\infty \le C \|x^l - x^k\|,
\end{aligned}
$$

where C is the Lipschitz constant of $G(\cdot)$ on the set $\mathcal{X}$. We conclude that

$$
\|x^l - x^k\| \ge \frac{\delta}{2C}, \quad \forall k, l \in \mathcal{K}, \ l > k.
$$

This is impossible, because of the compactness of the set $\mathcal{X}$. Thus, $\delta_k \to 0$. Relation (7.24) implies that

$$
\mathbb{E}[Z^k \mathbb{1}_{\{X_k \le t\}}] \ge F_Y^{(-2)}\big(P(Z^k \le t)\big) - \delta_k P(Z^k \le t) \quad \forall t \in \mathbb{R}.
$$

This is equivalent to

$$
F_{Z^k}^{(-2)}\big(P(Z^k \le t)\big) \ge F_Y^{(-2)}\big(P(Z^k \le t)\big) - \delta_k P(Z^k \le t) \quad \forall t \in \mathbb{R}. \tag{7.26}
$$

Let us fix k and apply the convexity argument as in the second part of the proof of Theorem 7.5, to prove that $P(Z^k \le t)$ in inequality (7.26) can be replaced by any $p \in [0, 1]$, i.e.,

$$
F_{Z^k}^{(-2)}(p) \ge F_Y^{(-2)}(p) - \delta_k p, \quad \forall p \in [0, 1]. \tag{7.27}
$$

Consider $p \in [0, 1]$ and choose $t = \inf\{\eta : P(Z^k \le \eta) \ge p\}$. If $P(Z^k \le t) = p$, then the inequality in (7.27) is identical with the one in (7.26) for this p. If $P(Z^k \le t) > p$, then $P(Z^k < t) < p < P(Z^k \le t)$. We consider an increasing sequence $\{\tau_i\}$ convergent to t and obtain $\lim_{i\to\infty} P(Z^k \le \tau_i) = P(Z^k < t)$. Using the continuity of $F_{Z^k}^{(-2)}(\cdot)$ and $F_Y^{(-2)}(\cdot)$, we conclude that

$$
F_{Z^k}^{(-2)}\big(P(Z^k < t)\big) \ge F_Y^{(-2)}\big(P(Z^k < t)\big) - \delta_k P(Z^k < t) \quad \forall t \in \mathbb{R}. \tag{7.28}
$$

The function $F_{Z^k}^{(-2)}(p)$ is affine for $p \in [P(Z^k < t), P(Z^k \le t)]$, while the function $F_Y^{(-2)}(p) - \delta_k p$ is convex with respect to p. Due to (7.26) and (7.28), the inequality in (7.27) holds at the ends of the interval, and thus remains valid for all p in this interval. This concludes the proof of (7.27).

Consider a convergent subsequence $\{x^k\}$, $k \in \mathcal{K}$, and the corresponding random variables $Z^k = G(x^k)$. Using the Lipschitz continuity of $G(\cdot)$ we conclude that $\{Z^k\}$, $k \in \mathcal{K}$, is convergent in $\mathcal{L}_\infty(\Omega, \mathcal{F}, P)$ to some random variable $Z^* = G(x^*)$, where x^* is the limit of $\{x^k\}$ on this subsequence. Since

$Z \mapsto F_Z^{(-2)}(p)$ is continuous in $\mathcal{L}_1(\Omega, \mathcal{F}, P)$, we can pass to the limit in (7.27) to conclude that $F_{Z^*}^{(-2)}(p) \geq F_Y^{(-2)}(p)$, for all $p \in [0, 1]$.

Therefore, x^* is feasible for problem (7.20)–(7.22). As $f(x^k) \leq f(x)$ for every feasible x in problem (7.20)–(7.22) and $f(\cdot)$ is continuous, we can pass to the limit in this inequality and conclude that x^* is an optimal solution. □

Now we shall analyze the case when $G(\cdot)$ is Lipschitz continuous as a mapping from $\mathbb{R}^n$ to $\mathcal{L}_1(\Omega, \mathcal{F}, P)$, rather than to $\mathcal{L}_\infty(\Omega, \mathcal{F}, P)$. We need to modify the method in order to ensure convergence.

Modified Inverse Event Cut Method

Step 0. Set $k = 0$, $\varepsilon_0 \in (0, 1)$.

Step 1. Solve problem (7.23) to obtain a point z^k, and let $Z^k = G(z^k)$.

Step 2a. Consider the sets $\mathcal{B}_t^k = \{Z^k \leq t\}$ and let

$$\delta_k = \sup_t \left\{ \frac{1}{P(\mathcal{B}_t^k)} F_Y^{(-2)}\big(P(\mathcal{B}_t^k)\big) - \mathbb{E}[Z^k | \mathcal{B}_t^k] : P(\mathcal{B}_t^k) \geq \varepsilon_k \right\}. \quad (7.29)$$

Step 2b. If $\delta_k \leq 0$, replace ε_k by $\varepsilon_k/2$ and go to Step 2a; otherwise, continue.

Step 3. Calculate t_k such that $P(Z^k \leq t_k) \geq \varepsilon_k$ as well as

$$\mathbb{E}[Z^k | \mathcal{B}_{t_k}^k] - \frac{1}{P(\mathcal{B}_{t_k}^k)} F_Y^{(-2)}\big(P(\mathcal{B}_{t_k}^k)\big) \leq -\frac{\delta_k}{2}. \quad (7.30)$$

Step 4. If $\delta_k < \varepsilon_k$ then set $\varepsilon_{k+1} = \min\{\delta_k, \varepsilon_k\}/2$; otherwise set $\varepsilon_{k+1} = \varepsilon_k$. Set $\mathcal{A}^{k+1} = \mathcal{B}_{t_k}^k$, increase k by one, and go to Step 1.

Observe that the method can cycle between Steps 2a and 2b without increasing the iteration index k. We shall show that this situation is equivalent to the stopping test of Step 2 of the Inverse Event Cut Method.

Theorem 7.10. *Assume that the set $\mathcal{X} \subset \mathbb{R}^n$ is compact, the function $f(\cdot)$ is continuous, and the operator $G(\cdot)$ is Lipschitz continuous on $\mathcal{X}$ as a mapping from $\mathcal{X}$ to $\mathcal{L}_1(\Omega, \mathcal{F}, P)$. If problem* (7.20)–(7.22) *has a nonempty feasible set, then one of the following two cases occurs:*

(i) *The Modified Inverse Event Cut Method generates a finite sequence $x^1, \ldots, x^k$, in which case x^k is an optimal solution of problem* (7.20)–(7.22)*;*

(ii) *The Modified Inverse Event Cut Method generates an infinite sequence $\{x^k\}$. In this case, the subsequence $\{x^k : \delta_k \in (0, \varepsilon_k)\}$ is infinite and each accumulation point of it is a solution of problem* (7.20)–(7.22).

Proof. First, we note that case (i) occurs when the algorithm repeats Step 2a and Step 2b infinitely many times without reaching Step 3. In this case ε_k decreases geometrically to zero, and in the limit, relation (7.29) becomes

$$0 \geq \delta_k = \sup_t \left\{ \frac{1}{P(\mathcal{B}_t^k)} F_Y^{(-2)}\big(P(\mathcal{B}_t^k)\big) - \mathbb{E}[Z^k | \mathcal{B}_t^k] : P(\mathcal{B}_t^k) > 0 \right\}.$$

This inequality coincides with the stopping rule of the LFAM. Using Corollary 7.6, we infer that the point x^k with $G(x^k) = Z^k$ is feasible. It is also optimal by the same argument as in Theorem 7.9.

In case (ii), we observe that the sequence $\{\varepsilon_k\}$ is non-increasing. Assume that $\varepsilon_k \geq \varepsilon > 0$ for all $k \geq 1$. This implies that $\delta_k \geq \varepsilon$ for all k. As the new inequality constraint generated at iteration k is satisfied at every iteration $l > k$, we have:

$$\mathbb{E}[Z^l | Z^k \leq t_k] - \frac{1}{P(Z^k \leq t_k)} F_Y^{(-2)}\big(P(Z^k \leq t_k)\big) \geq 0.$$

Combining this with (7.30), we obtain

$$\mathbb{E}[Z^l | Z^k \leq t_k] - \mathbb{E}[Z^k | Z^k \leq t_k] \geq \frac{\delta_k}{2}.$$

Using that $\delta_k \geq \varepsilon$ and $P(Z^k \leq t_k) \geq \varepsilon$, we transform the last displayed inequality as follows:

$$\begin{aligned}
\frac{\varepsilon}{2} &\leq \frac{1}{P(Z^k \leq t_k)} \left| \int (Z^l(\omega) - Z^k(\omega))\, \mathbb{1}_{\{Z^k \leq t_k\}}(\omega)\, P(\mathrm{d}\omega) \right| \\
&\leq \frac{1}{P(Z^k \leq t_k)} \| Z^l - Z^k \|_1 \| \mathbb{1}_{\{Z^k \leq t_k\}} \|_\infty \leq \frac{1}{\varepsilon} \| Z^l - Z^k \|_1 \\
&\leq \frac{C_1}{\varepsilon} \| z^l - z^k \|,
\end{aligned}$$

where C_1 is the Lipschitz constant of $G(\cdot)$ on $\mathcal{X}$ as a mapping to the space $\mathcal{L}_1(\Omega, \mathcal{F}, P)$. We conclude that

$$\| x^l - x^k \| \geq \frac{\varepsilon^2}{2C_1}, \qquad \forall k, l \in \mathcal{K},\ l > k.$$

This is impossible, because of the compactness of the set $\mathcal{X}$. Thus, $\varepsilon_k \to 0$ for $k \in \mathcal{K}$, where $\mathcal{K} = \big\{k : \delta_k \in (0, \varepsilon_k)\big\}$ is an infinite set. This implies that $\delta_k \to 0$ for $k \in \mathcal{K}$.

From (7.29), for all $t \in \mathbb{R}$ such that $P(Z^k \leq t) \geq \varepsilon_k$, we have

$$F_{Z^k}^{(-2)}\big(P(Z^k \leq t)\big) \geq F_Y^{(-2)}\big(P(Z^k \leq t)\big) - \delta_k P(Z^k \leq t). \tag{7.31}$$

The sequence $\{x^k\}_{k \in \mathcal{K}}$ has a convergent subsequence due to the compactness of $\mathcal{X}$. Consider any convergent subsequence $\{x^k\}_{k \in \mathcal{K}_1}$, $\mathcal{K}_1 \subset \mathcal{K}$, and the corresponding random variables $Z^k = G(x^k)$. By the Lipschitz continuity of $G(\cdot)$, we conclude that $\{Z^k\}_{k\in\mathcal{K}_1}$ is convergent in $\mathcal{L}_1(\Omega, \mathcal{F}, P)$ to some random variable $Z^* = G(x^*)$, where x^* is the limit of $\{x^k\}$ on this subsequence. As convergence in $\mathcal{L}_1(\Omega, \mathcal{F}, P)$ implies convergence in distribution, we have

$$\lim_{\substack{k\to\infty\\ k\in\mathcal{K}_1}} P(Z^k \le t) = P(Z^* \le t)$$

at all continuity points t of $F_{Z^*}(t)$. Consider any continuity point $\bar{t} \in \mathbb{R}$ of $F_{Z^*}(t)$ such that $F_{Z^*}(\bar{t}) > 0$. Then an index $k \in \mathcal{K}_1$ exists such that $F_{Z^*}(\bar{t}) = P(Z^* \le \bar{t}) > \varepsilon_k$. Thus, $P(Z^k \le \bar{t}) \ge \varepsilon_k$ for $k \in \mathcal{K}_1$ large enough. Furthermore, the function $(Z, p) \mapsto F_Z^{(-2)}(p)$ is continuous, which allows us to pass to the limit in (7.31) for $\bar{t}$ over the subsequence $\mathcal{K}_1$. We obtain for all $t \in \mathbb{R}$ such that $F_{Z^*}(\cdot)$ is continuous at t

$$F_{Z^*}^{(-2)}\big(P(Z^* \le t)\big) \ge F_Y^{(-2)}\big(P(Z^* \le t)\big). \tag{7.32}$$

Consider $p \in [0, 1]$ and choose $t = \inf\{\eta : P(Z^* \le \eta) \ge p\}$. If $P(Z^* \le t) = p$, then we take a decreasing sequence $\{\tau_i\}$ of continuity points of $F_{Z^*}(\cdot)$ convergent to t. From (7.32), we obtain

$$F_{Z^*}^{(-2)}(p) = \lim_{i\to\infty} F_{Z^*}^{(-2)}\big(P(Z^* \le \tau_i)\big) \ge \lim_{i\to\infty} F_Y^{(-2)}\big(P(Z^* \le \tau_i)\big) = F_Y^{(-2)}(p).$$

If $P(Z^k \le t) > p$, we can apply the same convexity argument as in the proof of Theorem 7.9 to infer from the last inequality that

$$F_{Z^*}^{(-2)}(p) \ge F_Y^{(-2)}(p), \quad \forall p \in [0, 1].$$

Therefore, x^* is feasible for problem (7.20)–(7.22). As in the proof of Theorem 7.9 it follows that x^* is an optimal solution. □

In a finite probability space Ω, both versions of the inverse event cut method are convergent in finitely many iterations because only finitely many different events S^k may occur.

We also note that we may modify Step 3 in the Inverse Event Cut Method as follows:

Step 3a: Calculate t_k such that $P(X^k \le t_k) > 0$ as well as

$$F_{X^k}^{(-2)}\big(P(\mathcal{B}_{t_k}^k)\big) - F_Y^{(-2)}\big(P(\mathcal{B}_{t_k}^k)\big) \le -\frac{\delta_k}{2} \tag{7.33}$$

are satisfied. The absence of the normalization by the probability $P(\mathcal{B}_t^k)$ may lead to selections of different events to add to the collection J_k. We call it the *un-*

scaled version of the Inverse Event Cut Method. Its convergence can be proved analogously to the proof of Theorem 7.9 under the assumptions of the compactness of the set $\mathcal{X}$, the continuity of the function $f(\cdot)$, the Lipschitz continuity of the operator $G(\cdot)$ on $\mathcal{X}$ as a mapping from $\mathbb{R}^n$ to $\mathcal{L}_1(\Omega, \mathcal{F}, P)$ and the feasibility of the problem.

7.3 Dual Methods

In this section, we present two methods exploiting duality relations associated with the stochastic dominance constraint. As it is infinitely dimensional and the dual variables are measures, the methods operate with iteratively refined finite-dimensional approximations of these measures.

7.3.1 *Progressive Bundle Method*

In this section, we additionally assume that $f(\cdot)$ is convex, the operator $G(\cdot)$ is concave, and that the set $\mathcal{X}$ is convex and compact. Recall that we have defined the mapping $\mathfrak{G} : \mathfrak{X} \to \mathcal{C}([a, b])$ for $p = 1$ as follows:

$$[\mathfrak{G}(x)](\eta) = \mathbb{E}[(\max(0, \eta - G(x))) - \max(0, \eta - Y)], \quad \eta \in [a, b].$$

Using the operator $\mathfrak{G}(\cdot)$, we have reformulated problem (7.1)–(7.3) in an equivalent way, namely

$$\min f(x) \text{ subject to } \mathfrak{G}(x) \in K, \ x \in \mathcal{X}, \tag{7.34}$$

with K being the cone of non-positive continuous functions on $[a, b]$. The dual space to $\mathcal{C}([a, b])$ is the space $\mathcal{M}^+([a, b])$ of regular countably additive measures and we have defined in (5.62) the following Lagrangian:

$$\Lambda(x, \nu) = f(x) + \int_a^b [\mathfrak{G}(x)](\eta) \, \nu(d\eta).$$

The dual function associated with this Lagrangian is

$$D(\nu) = \min_{x \in \mathcal{X}} \Lambda(x, \nu).$$

The dual problem becomes

$$\max D(\nu) \text{ subject to } \nu \in \mathcal{M}^+([a, b]). \tag{7.35}$$

We consider a reduced problem associated with a collection of points η_j, $j \in J_k$, for some index set $J_k \subset \mathbb{N}$, $k \in \mathbb{N}$:

$$\begin{aligned} &\min\ f(x)\\ &\text{s.t. } \mathbb{E}\big[(\eta_j - G(x))_+\big] \le \mathbb{E}\big[(\eta_j - Y)_+\big], \quad j \in J_k, \\ &\quad\ x \in \mathcal{X}. \end{aligned} \tag{7.36}$$

The Lagrangian of it is defined for any $x \in \mathcal{X}$ and $\mu^k \in \mathbb{R}_+^{|J_k|}$ as

$$\begin{aligned} \Lambda_k(x,\mu^k) &= f(x) + \sum_{j\in J_k} \mu_j^k \mathbb{E}\big[(\eta_j - G(x))_+ - (\eta_j - Y)_+\big] \\ &= f(x) + \mathbb{E}\Big[\sum_{j\in J_k} \mu_j^k (\eta_j - G(x))_+ - \sum_{j\in J_k} \mu_j^k (\eta_j - Y)_+\Big]. \end{aligned}$$

This Lagrange function $\Lambda_k(x,\mu^k)$ is equal to the Lagrangian $\Lambda(x,\tilde{\mu}^k)$, where $\tilde{\mu}^k$ is the extension of the atomic measure μ^k to the entire set $[a,b]$ by setting the measure of any event A in the Borel sigma-algebra on $[a,b]$ as follows:

$$\tilde{\mu}^k(A) = \sum_{j\in J_k} \mu_j^k \mathbb{1}_{A\cap\{\eta_j\}}. \tag{7.37}$$

The dual function $D_k : \mathbb{R}_+^{|J_k|} \to \mathbb{R}$ of the reduced problem calculated at the point μ^k is equal to the dual function of problem (7.34) at the point $\tilde{\mu}^k$:

$$D_k(\mu^k) = \min_{x\in\mathcal{X}} \Lambda_k(x,\mu^k) = D(\tilde{\mu}^k).$$

To simplify notation, we shall write μ^k also for the extended measure $\tilde{\mu}^k$ because it is not going to create any ambiguity.

Assuming that problem (7.1)–(7.3) has a feasible solution, the domain of the dual function is given by $\mu \in \mathcal{M}^+([a,b])$. We calculate the subgradients of the dual function $D_k(\mu)$ of the reduced problem as follows:

$$\Delta_j^k(\mu) = \mathbb{E}\big[\big(\eta_j - G(x_\mu^k)\big)_+ - \big(\eta_j - Y\big)_+\big], \quad j = 1,\dots,|J_k|,$$

where x_μ^k is such that $D_k(\mu) = \Lambda_k(x_\mu^k,\mu)$. Assume that μ^ℓ is an atomic measure on $[a,b]$ with non-negative atoms on the set $\{\eta_j,\ j \in J_\ell\}$. We can view it as a vector in $\mathbb{R}_+^{|J_\ell|}$. We define an extension $\mu^{\ell,k} \in \mathbb{R}_+^{|J_k|}$ of this measure to the set $\{\eta_j,\ j \in J_k \supset J_\ell\}$ by setting

$$\mu_j^{\ell,k} = \begin{cases} \mu_j^\ell & \text{if } j \in J_\ell \\ 0 & \text{if } j \notin J_\ell. \end{cases}$$

By construction, the value of the dual function at the measure μ^ℓ is the same as the value calculated at the extended measure $\mu^{\ell,k}$; however, the subgradients of the dual function at $\mu^{\ell,k}$ change substantially:

$$\Delta_j^{\ell,k}(\mu^{\ell,k}) = \mathbb{E}\big[\big(\eta_j - G(x^\ell_{\mu^\ell})\big)_+ - \big(\eta_j - Y\big)_+\big], \quad j = 1, \ldots, |J_k|.$$

We construct the following piecewise linear model $\vartheta^k : \mathbb{R}^{|J_k|} \to \mathbb{R}$ of the dual function:

$$\vartheta^k(\mu) = \min_{1\le\ell\le k} \big\{D(\tilde{\mu}^\ell) + (\Delta^{\ell,k})^\top(\mu - \mu^{\ell,k})\big\}.$$

The following method with parameters $r > 0$, $\gamma \in (0,1)$ and an accuracy parameter ε solves the dual problem (7.35).

Progressive Bundle Method

Step 0. Set $k = 1$, $J_1 = \{1\}$, $\eta_1 \in [a, b]$, $\mu^1 = \delta_{\eta_1}$.

Step 1. Calculate an optimal solution x^k of problem

$$\min_{x\in\mathcal{X}} \Lambda_k(x, \mu^k)$$

and define $Z^k = G(x^k)$.

Step 2. Calculate

$$\begin{aligned}
\Delta_j^{\ell,k} &= \Delta_j^{\ell,k-1}, \quad j = 1\ldots, k-1, \quad \ell = 1, \ldots, k-1;\\
\Delta_k^{\ell,k} &= \mathbb{E}\big[\big(\eta_k - Z^\ell\big)_+ - \big(\eta_k - Y\big)_+\big], \quad \ell = 1, \ldots, k-1;\\
\Delta_j^{k,k} &= \mathbb{E}\big[\big(\eta_j - Z^k\big)_+ - \big(\eta_j - Y\big)_+\big], \quad j = 1, \ldots, k.
\end{aligned}$$

Step 3. If $k = 1$ or if $D(\mu^k) \ge (1-\gamma)D(w^{k-1}) + \gamma\vartheta^{k-1}(\mu^k)$, then set $w^k = \mu^k$ (descent step); otherwise set $w^k = w^{k-1}$ (null step).

Step 4. Calculate a solution $(\theta_{k+1}, \mu^{k+1})$ of the master problem:

$$\begin{aligned}
&\max\ \theta - \frac{r}{2}\|\mu - w^k\|_2^2\\
&\text{subject to } D(\mu^\ell) + (\Delta^{\ell,k})^\top(\mu - \mu^{\ell,k}) \ge \theta, \quad \ell = 1, \ldots, k,\\
&\qquad \mu \in \mathbb{R}_+^{|J_k|}.
\end{aligned} \tag{7.38}$$

Set $\vartheta^k(\mu^{k+1}) = \theta_{k+1}$. Let π_ℓ be the optimal Lagrange multiplier associated with the ℓ-th constraint of problem (7.38). Define $\tilde{x}^k = \sum_{\ell=1}^k \pi_\ell x^\ell$ and $\tilde{Z}^k = G(\tilde{x}^k)$.

Step 5. Calculate the quantities

$$
\begin{aligned}
\delta_k &= \sup_{\eta} \Big\{ \mathbb{E}\big[(\eta - \tilde{Z}^k)_+ - (\eta - Y)_+ \big] : \eta \in [a, b] \Big\}, \\
\delta'_k &= \max_{j \in J_k} \Big(\mathbb{E}\Big[\big(\eta_j - \tilde{Z}^k\big)_+ - \big(\eta_j - Y\big)_+ \Big] \Big).
\end{aligned}
\tag{7.39}
$$

Step 6. If $D(w^k) \geq \theta^{k+1} - \varepsilon$ and $\delta_k \leq \varepsilon$, then stop; otherwise continue.

Step 7. If $\delta_k > \varepsilon$ and $\delta'_k \leq \frac{\delta_k}{4}$, then determine η_{k+1} such that

$$
\mathbb{E}\big[(\eta_{k+1} - \tilde{Z}^k)_+ - (\eta_{k+1} - Y)_+ \big] \geq \frac{\delta_k}{2} \tag{7.40}
$$

and set $J_{k+1} = J_k \cup \{k+1\}$; otherwise set $J_{k+1} = J_k$.
Increase $k \leftarrow k + 1$ and go to step 1.

The proposed method is a version of the bundle method applied to the dual function. In this version of the method, however, the master problem and the Lagrangian relaxation are solved on the intersection of $\mathcal{M}^+([a, b])$ with a finite-dimensional subspace. The dimension of the subspace increases from iteration to iteration. Furthermore, our model function ϑ is not necessarily an upper bound of the dual function and it is not clear how to obtain practicable estimates of the difference between the subgradients of the dual function and those calculated from the reduced function in the course of the procedure.

Theorem 7.11. *Assume that problem* (7.1)–(7.3) *has a feasible solution, the set* $\mathcal{X}$ *is convex and compact,* $f(\cdot)$ *is convex, and the operator* $G : \mathbb{R}^n \to \mathcal{L}_1(\Omega, \mathcal{F}, P)$ *is continuous and concave. Then the Progressive Bundle Method stops after finitely many iterations with approximate optimal solutions* $\tilde{x}^k$ *and* w^k *of problems* (7.1)–(7.3) *and* (7.35), *respectively, and a computable bound of the approximation accuracy.*

Proof. The assumptions of the theorem guarantee that the domain of the dual function is non-empty and the dual problem has a solution.

First, we consider the case, in which, the method does not generate new elements η after a finite number k_0 of iterations. In this case, the set J_k remains constant for all iterations $k \geq k_0$ and the Progressive Bundle Method becomes a standard bundle method applied to the dual function. The method converges due to the convergence theory of the bundle method. If we would require $D(w^k) = \theta^{k+1}$ in the stopping criterion, we would have

$$
\lim_{k \to \infty} (\mu^{k+1} - w^k) = 0, \quad \text{and} \quad \lim_{k \to \infty} \theta_k = D^*_k,
$$

where D^*_k is the optimal value of the dual problem to problem (7.36). Furthermore, any accumulation point of the sequence $\{w^k\}$ would be an optimal solution of that problem. This means that our method will stop after finitely many iterations at some iteration $k \geq k_0$ with $D(w^k) = D_k(w^k) \geq \theta^{k+1} - \varepsilon$.

We gather the active constraints at μ^{k+1} for the master problem (7.38) in the set

$$J_a^{k+1} = \{\ell \in \{1, \ldots, k\} : D(\mu^\ell) + (\Delta^{\ell,k})^\top(\mu^{k+1} - \mu^\ell) = \theta^{k+1}\}.$$

The optimality conditions for that problem at iteration k state that

$$0 \in \begin{bmatrix} r(\mu^{k+1} - w^k) \\ -1 \end{bmatrix} + \sum_{\ell \in J_a} \pi_\ell \begin{bmatrix} \Delta^{\ell,k} \\ 1 \end{bmatrix} + \begin{bmatrix} \mathcal{N}_{\mathbb{R}_+^{|J_k|}}(\mu^{k+1}) \\ 0 \end{bmatrix}, \tag{7.41}$$

where $\pi_\ell \geq 0$. The last row of the inclusion yields $\sum_{\ell \in J_a} \pi_\ell = 1$. Therefore, due to the convexity of $\mathcal{X}$, the point $\tilde{x}^k$ defined in Step 4 belongs to $\mathcal{X}$. This together with the stopping criterion of Step 6, implies that the point $\tilde{x}^k$ is ε-feasible.

Using the extensions of the measures μ^ℓ as defined in (7.37), we observe that

$$D(\mu^\ell) = f(x^\ell) + \int_a^b \mathfrak{G}(x^\ell)(\eta)\, \mu^\ell(\mathrm{d}\eta) = f(x^\ell) + \langle \Delta^{\ell,k}, \mu^\ell \rangle.$$

Substituting this for all active constraints in the set J_a^{k+1}, we obtain that

$$\vartheta^k(\mu^{k+1}) = f(x^\ell) + \langle \Delta^{\ell,k}), \mu^{k+1} \rangle, \quad \ell \in J_a^{k+1}.$$

Multiplying both sides by π_ℓ and summing up we infer that

$$\vartheta^k(\mu^{k+1}) = \sum_{\ell \in J_a} \pi_\ell f(x^\ell) + \Big\langle \sum_{\ell \in J_a} \pi_\ell \Delta^{\ell,k}, \mu^{k+1} \Big\rangle.$$

Using the convexity of $f(\cdot)$ and $\mathfrak{G}(\cdot)(\eta)$, we obtain

$$\vartheta^k(\mu^{k+1}) \geq f(\tilde{x}^k) + \int_a^b \mathfrak{G}(\tilde{x}^k)(\eta)\, \mu^{k+1}(\mathrm{d}\eta) \geq f(\tilde{x}^k) - \varepsilon\|\mu^{k+1}\|_2,$$

where $\|\mu^{k+1}\|_2$ stands for the Euclidean norm of the vector μ^{k+1}. The last inequality follows from the stopping criterion of Step 6 and the Cauchy-Schwarz inequality. Denoting the optimal value of problem (7.1)–(7.3) by f^*, and using the stopping criterion and the last inequality, we conclude that

$$\begin{aligned} f^* &\geq D(w^k) \geq \vartheta^k(\mu^{k+1}) - \varepsilon \\ &\geq f(\tilde{x}^k) - \varepsilon\|\mu^{k+1}\|_2 - \varepsilon \geq D^* - \varepsilon(1 + \|\mu^{k+1}\|_2). \end{aligned} \tag{7.42}$$

Consequently, the point $\tilde{x}^k$ is an $\tilde{\epsilon}$-optimal solution of the problem, where $\tilde{\epsilon} = \varepsilon(1 + \|\mu^{k+1}\|_2)$.

Let us denote the optimal value of the problem (7.38) at iteration k by ζ. We have

$$\begin{aligned}\frac{r}{2}\|\mu^{k+1} - w^k\|_2^2 &= \vartheta^k(\mu^{k+1}) - \zeta \\ &\le \vartheta^k(\mu^{k+1}) - \max_{\mu}\left\{D_k(\mu) - \frac{r}{2}\|\mu - w^k\|^2\right\} \\ &\le \vartheta^k(\mu^{k+1}) - D_k(w^k) \le \varepsilon. \end{aligned} \tag{7.43}$$

The last inequality follows from the stopping criterion in Step 6. Using this estimate and (7.42), we have

$$D(w^k) \ge D^* - \varepsilon(1 + \|\mu^{k+1}\|_2) \ge D^* - \varepsilon(1 + \|w^k\|_2 + \sqrt{2\varepsilon/r}).$$

Passing to the limit over a subsequence converging to the optimal point w^* of the dual approximate problem, we obtain the general estimate

$$D_k^* = D(w^*) \ge D^* - \varepsilon(1 + \|w^*\|_2 + \sqrt{2\varepsilon/r}),$$

where D_k^* is the optimal value of the approximate dual problem. This inequality provides an estimate of the approximation error of the dual method.

It remains to analyze the case in which the method generates an infinite sequence of points $\{\eta^k\}$ at iterations $k \in \mathcal{K} \subset \mathbb{N}$. Due to the assumptions of the theorem the set $G(\mathcal{X})$ is bounded in $\mathcal{L}_1(\Omega, \mathcal{F}, P)$. Therefore, the sequence $\{\delta_k\}$ is bounded. This together with the assumption that $|J_k| \to \infty$ implies $\limsup_{k\in\mathcal{K}} \delta_k = \bar{\delta} \ge \varepsilon > 0$. Due to the compactness of $\mathcal{X}$ and $[a, b]$, the sequences $\tilde{x}^k$ and $\{\eta_k\}$, $k \in \mathcal{K}$, have accumulations points, which we denote by $\bar{x}$ and $\bar{\eta}$, respectively. More precisely, a subsequence $\mathcal{K}_1 \subset \mathcal{K}$ exists such that

$$\lim_{k\in\mathcal{K}_1} G(\tilde{x}^k) = G(\bar{x}), \qquad \lim_{k\in\mathcal{K}_1} \eta_k = \bar{\eta}. \tag{7.44}$$

Passing to the limit over the subsequence $\mathcal{K}_1$ in (7.39), we obtain

$$\bar{\delta} \le \sup_{\eta}\left\{\mathbb{E}\left[\left(\eta - G(\bar{x})\right)_+ - \left(\eta - Y\right)_+\right] : \eta \in [a, b]\right\}.$$

Moreover, inequality (7.40) implies

$$\mathfrak{G}(\bar{x})(\bar{\eta}) = \mathbb{E}\left[(\bar{\eta} - G(\bar{x}))_+ - (\bar{\eta} - Y)_+\right] \ge \frac{\bar{\delta}}{2}.$$

Due to the test in Step 3, we also obtain the following estimate for every fixed $j \in J_k$, $j < k$, $k \in \mathcal{K}_1$:

$$\mathfrak{G}(\bar{x})(\eta_j) = \lim_{k\in\mathcal{K}_1} \mathbb{E}\left[(\eta_j - \tilde{Z}^k)_+ - (\eta_j - Y)_+\right] \le \limsup_{k\in\mathcal{K}_1} \frac{\delta_k}{4} = \frac{\bar{\delta}}{4}.$$

Putting these estimates together and using the Lipschitz continuity of the mapping $\max(0, \cdot)$, we infer the following chain of inequalities:

$$\begin{aligned}\frac{\bar{\delta}}{2} &\leq \mathfrak{G}(\bar{x})(\bar{\eta}) \\ &= \mathfrak{G}(\bar{x})(\eta_j) + \mathbb{E}\big[(\bar{\eta} - G(\bar{x}))_+ - (\eta_j - G(\bar{x}))_+ + (\bar{\eta} - Y)_+ - (\eta_j - Y)_+\big] \\ &\leq \mathfrak{G}(\bar{x})(\eta_j) + 2|\eta_j - \bar{\eta}| \leq \frac{\bar{\delta}}{4} + 2|\eta_j - \bar{\eta}|.\end{aligned}$$

This implies the following inequality for any fixed $j \in J_k$ for some $k \in \mathcal{K}_1$:

$$|\eta_j - \bar{\eta}| \geq \frac{\bar{\delta}}{4},$$

which contradicts (7.44). Therefore, the set J_k does not grow infinitely. Due to the accuracy enforcement $\delta_k \leq \varepsilon$, some iteration $\bar{k}$ exists, such that no new points will be added to the set J_k at $k \geq \bar{k}$. □

7.3.2 Progressive Augmented Lagrangian Method

Our motivation in devising numerical methods for problem (7.1)–(7.3) based on augmented Lagrangians is founded in the well-known disadvantages of methods based on piecewise linear approximations such as inefficiency of the initial iterations and numerical instability. In our problem, these are compounded by the continuum of constraints expressing the stochastic order relations. Various regularization approaches are available to help alleviate those numerical issues. We note that the stochastic dominance constraint can be naturally ill-conditioned in the portions addressing the lower tails of the distributions.

We consider the the reduced problem (7.36) and denote

$$\Delta(\eta, x) = \mathbb{E}\big[\big(\eta - G(x)\big)_+ - \big(\eta - Y\big)_+\big] \quad \eta \in J_k.$$

The augmented Lagrangian function of problem (7.36) is the following:

$$L_r^k(x, \mu) = f(x) + \frac{r}{2} \sum_{\eta \in J_k} \big[\max\big(0, \Delta(\eta, x) + \frac{\mu_j}{r}\big)\big]^2 - \frac{1}{2r} \|\mu_j\|^2,$$

where r is a positive parameter.

The following method solves the dual problem (7.35) while also providing a solution to problem (7.1)–(7.3). It uses a penalty parameter $r > 0$, a parameter $\tau \in (0, 1)$, and an accuracy parameter $\varepsilon > 0$.

Progressive Augmented Lagrangian method

Step 0. Set $k = 1$, $J_1 = \{1\}$, $\mu^1 = \delta_{\eta_1}$.

Step 1. Calculate the optimal solution x^k of the problem

$$\min_{x \in \mathcal{X}} L_r^k(x, \mu^k). \tag{7.45}$$

Define $Z^k = G(x^k)$.

Step 2. Calculate $\Delta_j^k = \mathbb{E}\big[(\eta_j - Z^k)_+ - (\eta_j - Y)_+\big]$, for all $j \in J_k$, and set

$$\mu_j^{k+1} = \max(0, \mu_j^k + r\Delta_j^k).$$

Step 3. Calculate the quantities

$$\begin{aligned} \delta_k &= \sup_{\eta} \Big\{ \mathbb{E}\big[(\eta - Z^k)_+ - (\eta - Y)_+\big] : \eta \in [a, b] \Big\}, \\ \delta_k' &= \max_{1 \le j \le |J_k|} \Delta_j^k. \end{aligned} \tag{7.46}$$

Step 4. If $\delta_k > \varepsilon$ and $\delta_k' \le \frac{\tau}{2}\delta_k$, then determine $\eta_{k+1} \in [a, b]$ such that

$$\hat{\Delta} = \mathbb{E}\big[(\eta_{k+1} - Z^k)_+ - (\eta_{k+1} - Y)_+\big] \ge \frac{\delta_k}{2}. \tag{7.47}$$

set $J_{k+1} = J_k \cup \{k+1\}$ and define $\mu_{|J_k|+1}^{k+1} = r\hat{\Delta}$, $\mu_{|J_k|+1}^k = 0$.
Otherwise, set $J_{k+1} = J_k$.

Step 5. If $\max_{1 \le j \le |J_{k+1}|} |\mu_j^{k+1} - \mu_j^k| \le \varepsilon$ and $\delta_k \le \varepsilon$, then stop. Otherwise, increase k by 1 and go to Step 1.

The quantities δ_k and δ_k' provide a measure of violation of the dominance constraint in the problem (7.1)–(7.3) and in the problem (7.36), respectively. The reduced problem is updated in Step 4 when the full dominance constraint is violated much more than the constraint in the reduced problem. Notice that in this method, we check the convergence of the multipliers as well as the satisfaction of the dominance constraint as a stopping criterion in Step 5. Hence, if the method stops at iteration k, then μ^k is an ε - solution of the dual problem and the point x^k is ε -feasible for problem (7.1)–(7.3).

Theorem 7.12. *Assume that problem* (7.1)–(7.3) *has a feasible solution, the set $\mathcal{X}$ is convex and compact, $f(\cdot)$ is convex and the operator $G : \mathbb{R}^n \to \mathcal{L}_1(\Omega, \mathcal{F}, P)$ is continuous and concave. Then the Progressive Augmented Lagrangian method generates a sequence of measures $\{\mu^k\}$, which converges to an optimal solution of problem* (7.35). *Furthermore, any accumulation point of the sequence $\{x^k\}$ is an optimal solution of problem* (7.1)–(7.3).

Proof. The assumptions of the theorem guarantee that the domain of the dual function is non-empty and the dual problem (7.35) has a solution. Observe that when the set J_k does not increase, then either $\delta_k \le \varepsilon$ or $\delta'_k > \frac{\tau}{2}\delta_k$. When the set J_k does not increase, then δ'_k eventually converges to zero because δ'_k is the violation of the constraints included in the augmented Lagrangian L_r^k. When the set J_k remains unchanged, then the method reduces to the standard augmented Lagrangian method applied to the reduced problem (7.36) where the constraints are determined by the points in J_k. Due to the convergence properties of the standard augmented Lagrangian method for convex problems, we infer that δ'_kconverges to zero. This means that if $\delta_k > \varepsilon$ eventualy $\delta'_k < \frac{\tau}{2}\delta_k$ and new points η will be added to J_k.

Suppose the method generates an infinite sequence of points $\{\eta_k\}$ at iterations $k \in \mathcal{K} \subset \mathcal{N}$. We obtain in the same way as in the proof of the Progressive Bundle Method that the sequence $\{\delta_k\}$ is bounded from above. This together with the assumption that $|J_k| \to \infty$ implies $\limsup_{k\in\mathcal{K}} \delta_k = \bar{\delta}$. Assume that $\bar{\delta} > \varepsilon$. Due to the compactness of $\mathcal{X}$ and $[a,b]$, the sequences $\{x^k\}$ and $\{\eta_k\}$, $k \in \mathcal{K}$, have accumulations points, which we denote by $\bar{x}$ and $\bar{\eta}$, respectively. We infer that a subsequence $\mathcal{K}_1 \subset \mathcal{K}$ exists such that

$$\lim_{k\in\mathcal{K}_1} G(x^k) = G(\bar{x}), \qquad \lim_{k\in\mathcal{K}_1} \eta_k = \bar{\eta}. \tag{7.48}$$

Passing to the limit over the subsequence $\mathcal{K}_1$ in (7.46), we obtain

$$\bar{\delta} \le \sup_{\eta} \Big\{ \mathbb{E}\big[\big(\eta - G(\bar{x})\big)_+ - \big(\eta - Y\big)_+ \big] : \eta \in [a,b] \Big\}.$$

Further, inequality (7.47) implies

$$\mathbb{E}\big[(\bar{\eta} - G(\bar{x}))_+ - (\bar{\eta} - Y)_+\big] \ge \frac{\bar{\delta}}{2}.$$

Due to the test at Step 3, we also obtain the following estimate for every fixed $j \in J_k$, $j < k$, $k \in \mathcal{K}_1$:

$$\lim_{k\in\mathcal{K}_1} \mathbb{E}\big[(\eta_j - X^k)_+ - (\eta_j - Y)_+\big] \le \limsup_{k\in\mathcal{K}_1} \frac{\tau}{2}\delta_k = \frac{\tau}{2}\bar{\delta}.$$

Putting these estimates together and using the Lipschitz-continuity of the mapping $\max(0,\cdot)$, we infer the inequality

$$\frac{\bar{\delta}}{2} \le \frac{\tau}{2}\bar{\delta} + 2|\eta_j - \bar{\eta}|,$$

in the same way as in the previous proof. This implies the following inequality for any fixed $j \in J_k$ for some $k \in \mathcal{K}_1$:

$$|\eta_j - \bar{\eta}| \geq \frac{(1-\tau)\bar{\delta}}{2}.$$

This inequality contradicts (7.48). Therefore, $\bar{\delta} = 0$, which entails that $\delta_k \leq \varepsilon$ for iterations k large enough. From that point on set J_k remains constant and the statement follows from the convergence properties of the standard augmented Lagrangian method for convex problems. □

7.4 Exact Penalty Method

In this section, we discuss the penalty function approach to problem (7.1)–(7.3). We assume that $\Omega = \{\omega_1, \ldots, \omega_N\}$ with $P(\omega_i) = p_i$ and $\mathcal{F}$ containing all subsets of Ω. Additionally, we assume that $G(x) = g(x, \omega)$ and denote $g_i(x) = g(x, \omega_i)$ and $y_i = Y(\omega_i)$, $i = 1, \ldots, N$. Let us define

$$v_j = \mathbb{E}[\max(0, y_j - Y)] = \sum_{i=1}^{N} p_i \max(0, y_j - y_i), \quad j = 1, \ldots, N.$$

We consider the following penalty functions:

$$P_1(x) = \sum_{j=1}^{N} \Big(\sum_{i=1}^{N} p_i \max(0, y_j - g_i(x)) - v_j \Big)_+,$$

$$P_\infty(x) = \max_{1 \leq j \leq N} \Big(\sum_{i=1}^{N} p_i \max(0, y_j - g_i(x)) - v_j \Big)_+.$$

The corresponding penalty problems are

$$\min_{x \in \mathcal{X}} f(x) + r P_1(x); \tag{7.49}$$

$$\min_{x \in \mathcal{X}} f(x) + r P_\infty(x), \tag{7.50}$$

with a penalty coefficient $r > 0$. We can apply the analysis of the exact penalty methods to solve problem (7.49) or problem (7.50). The following statement justifies this approach.

Lemma 7.13. *Assume that the objective function $f(\cdot)$ of problem (7.1)–(7.3) is Lipschitz continuous on the set $\mathcal{X}$ with the modulus κ, and $g_i(\cdot)$, $i = 1, \ldots, N$ are concave functions on $\mathcal{X}$. Let $\mathcal{X}$ be a compact subset of $\mathbb{R}^n$. Suppose a point x^0 exists, which is feasible for problem (7.1)–(7.3) and satisfies the uniform stochastic dominance condition (5.35) on $[a, b]$, where $y_j \in [a, b]$ for all $j = 1, \ldots, N$. Then a positive constant $\bar{r}$ exist such that for all $r > \bar{r}$ a point $\hat{x}$ is an* *optimal solution of problem* (7.1)–(7.3) *if and only if it solves problems* (7.49) *and* (7.50) *as well.*

Proof. The dominance condition is equivalent to Slater's condition in this setting. Furthermore, the functions $g_i(\cdot)$ are globally Lipschitz over $\mathcal{X}$ because $\mathcal{X}$ is compact. Without loss of generality, we may assume that they have a common Lipschitz constant $\kappa > 0$. Then the statement follows from Clarke's exact penalty function theorem. □

For a convex function $f(\cdot)$, we can approximate the objectives of the penalty problems by cutting planes, while maintaining separate approximations for the function f and for the penalty term. To simplify notation, we use the following:

$$A(x) = \Big\{j = 1, \ldots, N : \sum_{i=1}^{N} p_i \max(0, y_j - g_i(x)) - v_j > 0\Big\},$$

$$J(x) = \Big\{j = 1, \ldots, N : P_\infty(x) = \sum_{i=1}^{N} p_i \max(0, y_j - g_i(x)) - v_j > 0\Big\},$$

$$B_j(x) = \big\{i = 1, \ldots, N : y_j > g_i(x)\big\}.$$

We may choose subgradients of $P_1(x)$ from the following subset of the full subdifferential:

$$S_1(x) = \sum_{j \in A(x)} \sum_{i \in B_j(x)} p_i \partial(-g_i(x)).$$

The subdifferential of $P_\infty(x)$ contains a zero vector, if $J(x) = \emptyset$; otherwise we may choose subgradients from the following subset of the full subdifferential:

$$S_\infty(x) = \Big\{ \sum_{j \in J(x)} \alpha_j \sum_{i \in B_j(x)} p_i \partial(-g_i(x)) : \sum_{j \in J(x)} \alpha_j = 1,\ \alpha_j \geq 0,\ j \in J(x)\Big\}.$$

Assuming that f is a convex function, we can use the following method for solving problem (7.1)–(7.3). The method has a parameter $\ell > 1$.

Penalty Method

Step 0. Set $k = 1$, choose $r_1 > 0$ and $x^0 \in \mathcal{X}$. Calculate $s_f^0 \in \partial f(x^0)$, $s_p^0 \in S_\infty(x^0)$ and set $J_f^1 = J_p^1 = \{0\}$.

Step 1. Solve the problem

$$
\begin{aligned}
\min_{x, w_f, w_p} \quad & w_f + r_k w_p \\
\text{s.t.} \quad & f(x^j) + \langle s_f^j, x - x^j \rangle \le w_f, \quad j \in J_f^k, \\
& P_\infty(x^j) + \langle s_p^j, x - x^j \rangle \le w_p, \quad j \in J_p^k. \\
& x \in \mathcal{X}.
\end{aligned}
\tag{7.51}
$$

Let (x^k, w_f^k, w_p^k) denote the solution of problem (7.51).

Step 2. For x^k, calculate $P_\infty(x^k)$ and $f(x^k)$.
If $P_\infty(x^k) = 0$ and $f(x^k) = w_f^k$, then stop; otherwise continue.

Step 3. If $P_\infty(x^k) > 0$, then calculate $s_p^k \in S_\infty(x^k)$ and set $r_k = \ell r_{k-1}$ and $J_p^k = J_p^{k-1} \cup \{k\}$.
If $f(x^k) > w_f^k$, then calculate $s_f^k \in \partial f(x^k)$ and set $J_p^k = J_p^{k-1} \cup \{k\}$.

Step 4. Increase k by one, and go to Step 1.

Theorem 7.14. *Assume that problem* (7.1)–(7.3) *satisfies the uniform stochastic dominance condition* (5.35) *on* $[a, b]$, *where* $y_j \in [a, b]$ *for all* $j = 1, \ldots, N$. *Furthermore, suppose the set* $\mathcal{X}$ *is convex and compact,* $f(\cdot)$ *is convex and* $g_i(\cdot)$, $i = 1, \ldots, N$ *are concave function on* $\mathcal{X}$. *Then the penalty method stops after finitely many iterations with an optimal solution of problem* (7.1)–(7.3) *or it generates a sequence of points* $\{\mu^k\}$, *which converges to an optimal solution of that problem.*

Proof. If the method stops at Step 2 in iteration k, then x^k is a feasible solution of problem (7.1)–(7.3) because the penalty function is equal to zero. Hence $f(x^k) \ge f^*$, where f^* stands for the optimal value of problem (7.1)–(7.3). Furthermore, the approximation of the objective function $f(\cdot)$ given by w_f^k is exact at x^k. For all points x that are feasible for (7.1)–(7.3), we have

$$
f(x) \ge \max_{j \in J_f^k} \left(f(x^j) + \langle s_f^j, x - x^j \rangle \right)
$$

implying that $f^* \ge w_f^k$. Hence $f(x^k) \ge f^* \ge w_f^k = f(x^k)$. We conclude that if the method stops at Step 2, then it delivers an optimal solution of problem (7.1)–(7.3). Assume that the method generates an infinite sequence of points. Two cases are possible: a number k_0 exists such that the set J_p^k remains the same for all $k \ge k_0$, or such a number does not exist. In the case when J_p^k remains the same for all $k \ge k_0$, we observe that r_k and w_p^k remain constant and all points x^k for $k \ge k_0$ are feasible for (7.1)–(7.3). The statement then follows from the convergence properties of the cutting plane method. Consider now the case, when the set J_p^k grows infinitely. This means that in Step 3, the penalty parameter r_k is increased infinitely many times. According to Lemma 7.13, there is a finite value $\bar{r}$ such that for all $r > \bar{r}$ a point $\hat{x}$ is an optimal solution of problem (7.1)–(7.3) if and only if it solves problem (7.50) as well. Hence, the method will reach a value that is sufficiently large and from that point on, the convergence statement will follow from the convergence of the cutting plane method applied to problem (7.50). □

Remark 7.15. It is evident that we can replace $P_\infty(\cdot)$ by $P_1(\cdot)$ in the method and the convergence statement of the modified method will be the same.

7.5 Methods for Problems with Multivariate Dominance Constraints

In this section, we address the numerical solution of the problem

$$\min\ f(x) \tag{7.52}$$

$$\text{subject to } \mathbb{E}\big[(\eta - c^\top G(x))_+\big] \le \mathbb{E}\big[(\eta - c^\top Y)_+\big] \ \ \forall(\eta, c) \in \operatorname{graph}(\mathfrak{J}), \tag{7.53}$$

$$x \in \mathcal{X}. \tag{7.54}$$

Here $G : \mathcal{X} \to \mathcal{L}_1(\Omega, \mathcal{F}, P; \mathbb{R}^m)$ is an integrable random vector with m components, and Y is an m-dimensional integrable random vector serving as a benchmark. The set $\mathcal{X} \subseteq \mathbb{R}^n$ is a compact set and $\mathfrak{J} : \mathcal{S} \rightrightarrows \mathbb{R}$ is a multifunction with compact convex images and a compact graph.

We can adapt any of the methods in § 7.1 to solve this problem. All steps and the analysis are analogous except for the verification of the dominance constraint. Let us illustrate this point by discussing the adaptation of the Shortfall Event Cut Method for solving problem (7.52)–(7.54).

The method proceeds as follows.

Shortfall event cut method for multivariate dominance

Step 0. Set $k = 1$, $(\eta_0, c^0) \in \operatorname{graph}(\mathfrak{J})$, and $\mathcal{A}^0 = \Omega$.

Step 1. Solve the problem:

$$\begin{aligned} &\min\ f(x) \\ &\text{s.t. } \mathbb{E}\big[\mathbb{1}_{\mathcal{A}^j}(\eta_j - (c^j)^\top G(x))\big] \le \mathbb{E}\big[(\eta_j - (c^j)^\top Y)_+\big], \quad j = 0, \dots, k-1. \\ &x \in \mathcal{X}. \end{aligned} \tag{7.55}$$

Let x^k denote the solution of problem (7.55), and let $Z^k = G(x^k)$.

Step 2. Calculate the quantity

$$\delta_k = \sup_{\eta, c} \Big\{ \mathbb{E}\big[(\eta - \langle c, Z^k\rangle)_+ - (\eta - \langle c, Y\rangle)_+\big] : \ (\eta, c) \in \operatorname{graph}(\mathfrak{J}) \Big\}. \tag{7.56}$$

Step 3. If $\delta_k \le 0$, stop; otherwise, continue.
Step 4. Determine $(\eta_k, c^k) \in \operatorname{graph}(\mathfrak{I})$ such that

$$\mathbb{E}\big[(\eta - \langle c^k, Z^k\rangle)_+ - (\eta - \langle c^k, Y\rangle)_+\big] \ge \frac{\delta_k}{2}. \tag{7.57}$$

and save the event

$$\mathcal{A}^k = \big\{\langle c^k, Z^k\rangle < \eta_k\big\}.$$

Step 5. Increase k by one, and go to Step 1.

Problem (7.55) is a non-smooth convex optimization problem. More difficult is the constraint verification problem (7.56), which is non-convex. Notice that its objective function is a difference of two convex functions. Hence, we can adapt methods for "difference-of-convex" (DC) optimization to solve it. In what follows, we discuss this approach.

For any $Z \in \mathcal{L}_1(\Omega, \mathcal{F}, P; \mathbb{R}^m)$, we can write

$$F^{(2)}_{c^\top Z}(\eta) = \sup_{A \in \mathcal{F}} \mathbb{E}\big[\mathbb{1}_A(\eta - c^\top Z)\big].$$

By virtue of Strassen's theorem, we see that for any $(\eta, c) \in \operatorname{graph}(\mathfrak{I})$,

$$\begin{aligned}
&\partial_{(\eta,c)} F^{(2)}_{c^\top Z}(\eta) \\
&= \operatorname{conv}\left\{ \mathbb{E}\left[\mathbb{1}_A \begin{pmatrix} 1 \\ -Z \end{pmatrix}\right] : \mathbb{E}\big[\mathbb{1}_A(\eta - c^\top Z)\big] = F^{(2)}_{c^\top Z}(\eta),\ A \in \mathcal{F} \right\} \\
&= \operatorname{conv}\Bigg\{ \mathbb{E}\left[\mathbb{1}_A \begin{pmatrix} 1 \\ -Z \end{pmatrix}\right] : \\
&\qquad \big\{\omega \in \Omega : c^\top Z(\omega) < \eta\big\} \subseteq A \subseteq \big\{\omega \in \Omega : c^\top Z(\omega) \le \eta\big\} \Bigg\}.
\end{aligned}$$

Additionally, we note that for any $s \in \partial_{(\eta,c)} F^{(2)}_{c^\top Z}(\eta)$, it holds

$$s^\top \begin{pmatrix} \eta \\ c \end{pmatrix} = F^{(2)}_{c^\top Z}(\eta). \tag{7.58}$$

In order to calculate

$$\delta_k = \sup\big\{\mathbb{E}\big[(\eta - c^\top Z^k)_+ - (\eta - c^\top Y)_+\big] : (\eta, c) \in \operatorname{graph}(\mathfrak{I})\big\}, \tag{7.59}$$

we define the functions

$$g_k(\eta, c) = \mathbb{E}\big[(\eta - c^\top Z^k)_+\big] \quad \text{and} \quad h(\eta, c) = \mathbb{E}\big[(\eta - c^\top Y)_+\big].$$

Problem (7.59) is equivalent to the following problem

$$\max\left\{g_k(\eta,c)-h(\eta,c):(\eta,c)\in\operatorname{graph}(\mathfrak{I})\right\}.$$

Equivalently, we have to calculate

$$\max_{A\in\mathcal{F}}\sup_{\eta,c}\left\{P(A)\eta+c^\top\mathbb{E}\left[\mathbb{1}_A Z^k\right]-h(\eta,c):(\eta,c)\in\operatorname{graph}(\mathfrak{I})\right\}. \tag{7.60}$$

We approximate the outer maximization by considering only finite collections of events $\mathcal{F}^j\subset\mathcal{F}$, constructing the problem

$$\max_{B\in\mathcal{F}^j}\sup_{\eta,c}\left\{P(B)\eta+c^\top\mathbb{E}\left[\mathbb{1}_B Z^k\right]-h(\eta,c):(\eta,c)\in\operatorname{graph}(\mathfrak{I})\right\}. \tag{7.61}$$

The collection is enlarged until a close approximation of a solution of (7.60) is obtained. The following Multivariate Dominance Verification Method can be used in Step 2.

Multivariate Dominance Verification Method

Step 2a: Set $j=1$ and let $\mathcal{F}^1=\{\mathcal{A}^0,\dots,\mathcal{A}^{k-1}\}$.

Step 2b: Let $g_k^j(\eta,c)=\max_{B\in\mathcal{F}^j}\left(P(B)\eta-c^\top\mathbb{E}\left[\mathbb{1}_B X^k\right]\right)$. Calculate a solution (ξ_j,v^j) of problem (7.61).

Step 2c: If $g_k^j(\xi_j,v^j)=g_k(\xi_j,v^j)$, then stop. Otherwise, define the set

$$B^j=\{\omega:\langle v^j,X^k(\omega)\rangle\le\xi_j\}.$$

Set $\mathcal{F}^{j+1}=\mathcal{F}^j\cup\{B^j\}$, increase j by one and go to Step 2b.

We recall the notion of a critical point adopted in the DC-optimization theory.

Definition 7.16. Let g and h be proper, convex, lower semi-continuous functions on $\mathbb{R}^n$. A point $x^*\in\mathbb{R}^n$ is a *critical point of* $g-h$ if $\partial h(x^*)\cap\partial g(x^*)\neq\emptyset$. Given a convex closed set $\mathcal{X}\subseteq\mathbb{R}^n$, we say that $x^*\in\mathcal{X}$ is a *critical point of* $g-h$ *in* $\mathcal{X}$ if

$$\partial g(x^*)\cap\left(\partial h(x^*)+\mathcal{N}_{\mathcal{X}}(x^*)\right)\neq\emptyset.$$

Now we analyze the dominance verification method method.

Theorem 7.17. *The sequence* $\{(\xi_j,v^j)\}$ *generated by the Multivariate Dominance Verification Method has the following properties:*

(i) $\{g_k(\xi_j,v^j)-h(\xi_j,v^j)\}$ *is a nondecreasing sequence converging to a constant* $\beta\le\delta_k$.

(ii) *If* Ω *is finite, then the method terminates after finitely many iterations at a critical point of* g_k-h *in* $\operatorname{graph}(\mathfrak{I})$.

Proof. First, we show statement (i). Assume that the Multivariate Dominance Verification Method does not stop at iteration j. This means that $g_k^j(\xi_j, v^j) < g_k(\xi_j, v^j)$. In Step 2b of iteration $j+1$, we generate a point

$$(\xi_{j+1}, v^{j+1}) \in \arg\max \big\{g_k^{j+1}(\eta, c) - h(\eta, c) : (\eta, c) \in \operatorname{graph}(\mathfrak{I})\big\}.$$

Since $g_k^{j+1}(\eta, c) = \max\big\{g_k^j(\eta, c), P(B^j)\eta - c^\top \mathbb{E}\big[\mathbb{1}_{B^j} Z^k\big]\big\}$, one of the following cases will occur:

(a) $(\xi_{j+1}, v^{j+1}) \in \arg\max \big\{P(B^j)\eta - \big\langle c, \mathbb{E}\big[\mathbb{1}_{B^j} Z^k\big]\big\rangle - h(\eta, c) : (\eta, c) \in \operatorname{graph}(\mathfrak{I})\big\}$;

(b) $(\xi_{j+1}, v^{j+1}) \in \arg\max \big\{g_k^j(\eta, c) - h(\eta, c) : (\eta, c) \in \operatorname{graph}(\mathfrak{I})\big\}$.

We shall show that only case (a) can occur. Assuming that case (b) takes place, we infer that

$$\begin{aligned} g_k^j(\xi_{j+1}, v^{j+1}) - h(\xi_{j+1}, v^{j+1}) \geq \\ P(B^j)\xi_{j+1} - \big\langle v^{j+1}), \mathbb{E}\big[\mathbb{1}_{B^j} Z^k\big]\big\rangle - h(\xi_{j+1}, v^{j+1}). \quad (7.62) \end{aligned}$$

This relation implies the following

$$\begin{aligned} g_k^{j+1}(\xi_{j+1}, v^{j+1}) &= \max\Big\{g_k^j(\xi_{j+1}, v^{j+1}), P(B^j)\xi_{j+1} - \big\langle v^{j+1}), \mathbb{E}\big[\mathbb{1}_{B^j} Z^k\big]\big\rangle\Big\} \\ &= g_k^j(\xi_{j+1}, v^{j+1}). \end{aligned}$$

This together with the optimality of the point (ξ_{j+1}, v^{j+1}) implies

$$\begin{aligned} g_k^j(\xi_{j+1}, v^{j+1}) - h(\xi_{j+1}, v^{j+1}) &= g_k^{j+1}(\xi_{j+1}, v^{j+1}) - h(\xi_{j+1}, v^{j+1}) \\ &\geq g_k^{j+1}(\xi_j, v^j) - h(\xi_j, v^j). \quad (7.63) \end{aligned}$$

Since the method has not terminated at iteration j, we have

$$g_k^j(\xi_j, v^j) < g_k(\xi_j, v^j) = P(B^j)\xi_j - \langle v^j, \mathbb{E}\big[\mathbb{1}_{B^j} Z^k\rangle\big] = g_k^{j+1}(\xi_j, v^j).$$

Using this inequality and the optimality of (ξ_j, v^j) for the problem (7.61) in iteration j, we obtain

$$\begin{aligned} g_k^j(\xi_{j+1}, v^{j+1}) - h(\xi_{j+1}, v^{j+1}) &\leq g_k^j(\xi_j, v^j) - h(\xi_j, v^j) \\ &< g_k^{j+1}(\xi_j, v^j) - h(\xi_j, v^j). \end{aligned}$$

This relation contradicts (7.63). Thus, we infer that in fact case (a) occurs. Since $\big(P(B^j), -\mathbb{E}\big[\mathbb{1}_{B^j} Z^k\big]\big)^\top \in \partial g_k(\xi_j, v^j)$, we have

$$g_k(\xi_j, v^j) = P(B^j)\xi_j - \big\langle v^j, \mathbb{E}\big[\mathbb{1}_{B^j} Z^k\big]\big\rangle,$$

due to (7.58). The following chain of inequalities holds

$$\begin{aligned} g_k(\xi_{j+1}, v^{j+1}) - h(\xi_{j+1}, v^{j+1}) \\ &\geq P(B^j)\xi_{j+1} - (v^{j+1})^\top \mathbb{E}\big[\mathbb{1}_{B^j} Z^k\big] - h(\xi_{j+1}, v^{j+1}) \\ &\geq P(B^j)\xi_j - (v^j)^\top \mathbb{E}\big[\mathbb{1}_{B^j} Z^k\big] - h(\xi_j, v^j) \\ &= g_k(\xi_j, v^j) - h(\xi_j, v^j). \end{aligned}$$

Since both $g_k(\eta, c)$ and $h(\eta, c)$ are finite and continuous on the compact set $\operatorname{graph}(\mathfrak{F})$, the quantity δ_k is well-defined and finite. Furthermore,

$$g_k(\xi_j, v^j) - h(\xi_j, v^j) \leq \delta_k \quad \text{for all iterations } k \in \mathbb{N}.$$

The monotonic convergence theorem implies that the sequence $\{g_k(\xi_j, v^j) - h(\xi_j, v^j)\}$ converges to some $\beta \leq \delta_k$. This completes the proof of statement (i).

Let j be the first iteration at which $g_k^j(\xi_j, v^j) = g_k(\xi_j, v^j)$. Then, arguing as above, we can conclude that

$$(\xi_j, v^j) \in \arg\max \big\{P(B^{j-1})\eta - c^\top \mathbb{E}\big[\mathbb{1}_{B^{j-1}} Z^k\big] - h(\eta, c) : (\eta, c) \in \operatorname{graph}(\mathfrak{F})\big\}.$$

The first order necessary conditions of optimality for problem (7.60) imply

$$\big(P(B^{j-1}), -\mathbb{E}\big[\mathbb{1}_{B^{j-1}} Z^k\big]\big)^\top \in \partial h(\xi_j, v^j) + \mathcal{N}_{\operatorname{graph}(\mathfrak{F})}(\xi_j, v^j).$$

Additionally, since $g_k^{j-1}(\xi_j, v^j) < P(B^{j-1})\xi_j - (v^j)^\top \mathbb{E}\big[\mathbb{1}_{B^{j-1}} Z^k\big]$, we infer that

$$P(B^{j-1})\xi_j - (v^j)^\top \mathbb{E}\big[\mathbb{1}_{B^{j-1}} Z^k\big] = g_k^j(\xi_j, v^j) = g_k(\xi_j, v^j).$$

Therefore, we conclude that $\big(P(B^{j-1}), -\mathbb{E}\big[\mathbb{1}_{B^{j-1}} Z^k\big]\big)^\top \in \partial g_k(\xi_j, v^j)$. Thus,

$$\partial g_k(\xi_j, v^j) \cap \partial h(\xi_j, v^j) + \mathcal{N}_{\operatorname{graph}(\mathfrak{F})}(\xi_j, v^j) \neq \emptyset,$$

so (ξ_j, v^j) is a critical point of $g_k - h$ in $\operatorname{graph}(\mathfrak{F})$ as stated.

If Ω is finite, then the function $h_k(\eta, c)$ is polyhedral because only finitely many sets $A \in \mathcal{F}$ exist. Therefore, after finely many iterations $N < 2^{|\Omega|}$, we shall obtain $g_k^N(\eta, c) = g_k(\eta, c)$ for all $(\eta, c) \in \operatorname{graph}(\mathfrak{F})$. Thus, the point (ξ_N, v^N) is a critical point of $g_k - h$ in $\operatorname{graph}(\mathfrak{F})$. □

It is possible to extend the method and our arguments to more general probability spaces showing that any limit point (ξ^*, v^*) is a critical point of $g_k - h$ in $\operatorname{graph}(\mathfrak{F})$ for infinite Ω but we omit the details here.

In general, we can only guarantee that the limit point provided by Multivariate Dominance Verification Method is a critical point of $g_k - h$. If Ω is finite, then we can establish whether or not the critical point is in fact a local maximizer. When Ω is finite, the functions g_k and h are polyhedral. In that case, we can use the following result in DC-optimization.

Theorem 7.18. *Let $\mathcal{X} \subseteq \mathbb{R}^n$ be a closed convex set and let $g, h : \mathbb{R}^n \to \mathbb{R}$ be convex functions. If g is polyhedral, then $x^* \in \mathcal{X}$ is a local maximizer of $g - h$ in $\mathcal{X}$ if and only if $\partial g(x^*) \subseteq \partial h(x^*) + \mathcal{N}_{\mathcal{X}}(x^*)$.*

In our setting, this statement leads to the following result.

Corollary 7.19. *Let (ξ^*, v^*) be the critical point found by the Multivariate Dominance Verification Method, and let g_k^* be the polyhedral approximation when the method terminates. If $\partial g_k(\xi^*, v^*) \subseteq \partial g_k^*(\xi^*, v^*)$ then (ξ^*, v^*) is a local maximizer of $g_k - h$ in* $\operatorname{graph}(\mathfrak{F})$.

Proof. If (ξ^*, v^*) is the critical point found, then we know that (ξ^*, v^*) is in fact the global maximizer of the approximate problem

$$\max_{(\eta,c)\in\operatorname{graph}(\mathfrak{F})} g_k^*(\eta, c) - h(\eta, c).$$

By virtue of Theorem 7.18, $\partial g_k^*(\xi^*, v^*) \subseteq \partial h(\xi^*, v^*) + \mathcal{N}_{\operatorname{graph}(\mathfrak{F})}(\xi^*, v^*)$. By assumption, $\partial g_k(\xi^*, v^*) \subseteq \partial g_k^*(\xi^*, v^*)$, and, thus,

$$\partial g_k(\xi^*, v^*) \subseteq \partial h(\xi^*, v^*) + \mathcal{N}_{\operatorname{graph}(\mathfrak{F})}(\xi^*, v^*).$$

The point (ξ^*, v^*) is thus a local maximizer of $g_k - h$ in $\operatorname{graph}(\mathfrak{F})$, again by virtue of Theorem 7.18. □

We can use this criterion to modify our method as follows. Instead of exiting at Step 2c once (ξ^*, v^*) is found, we proceed with the following steps.

Step 2d: Let $E_0 = \{\omega \in \Omega : (v^j)^\top Z^k(\omega) = \xi_j\}$. Enumerate all $2^{|E_0|} - 1$ sets E such that

$$E_1 = \{\omega \in \Omega : \langle v^j, Z^k(\omega)\rangle < \xi_j\} \subseteq E \\ \subset B^j = \{\omega \in \Omega : \langle v^j, Z^k(\omega)\rangle \le \xi_j\}.$$

Step 2e: If there is some i such that $E_i \notin \mathcal{F}^j$, then let $\mathcal{F}^{j+1} = \mathcal{F}^j \cup \{E_i\}$, set $j \leftarrow j + 1$. Go to Step 2b. Otherwise, stop: (ξ_j, v^j) is a local maximizer of $g_k - h$ in $\operatorname{graph}(\mathfrak{F})$.

Theorem 7.20. *If Ω is finite, then the Multivariate Dominance Verification Method extended with Steps 2d and 2e will stop at a local maximizer of $g_k - h$ in* $\operatorname{graph}(\mathfrak{F})$ *in finitely many steps.*

Proof. From Theorem 7.17, we see that in finitely many steps a critical point, say (ξ_j, v^j), will be generated in Step 2c.

If it is discovered in Step 2e that there is some $E_i \notin \mathcal{F}^j$, then we conclude that $z = \mathbb{E}\big[\mathbb{1}_{E_i}(1, -Z^k)^\top\big] \in \partial g_k(\xi_j, v^j) \setminus \partial g_k^j(\xi_j, v^j)$. There are two possibilities:

(i) $z \in \partial h(\xi_j, v^j) + \mathcal{N}_{\text{graph}(\mathfrak{F})}(\xi_j, v^j)$;

(ii) $z \notin \partial h(\xi_j, v^j) + \mathcal{N}_{\text{graph}(\mathfrak{F})}(\xi_j, v^j)$.

In case (i), there is some $\gamma \in \partial h(\xi_j, v^j)$ and $q \in \mathcal{N}_{\text{graph}(\mathfrak{F})}(\xi_j, v^j)$, such that $z = \gamma + q$. Using the subgradient inequality, we infer for any $(\eta, c) \in \text{graph}(\mathfrak{F})$ the following

$$\begin{aligned}P(E_i)(\eta - \xi_j) - \big\langle c - v^j, \mathbb{E}\big[\mathbb{1}_{E_i} Z^k\big]\big\rangle &= z^\top \begin{pmatrix} \eta - \xi_j \\ c - v^j \end{pmatrix} \\ &= \gamma^\top \begin{pmatrix} \eta - \xi_j \\ c - v^j \end{pmatrix} + q^\top \begin{pmatrix} \eta - \xi_j \\ c - v^j \end{pmatrix} \le \gamma^\top \begin{pmatrix} \eta - \xi_j \\ c - v^j \end{pmatrix} \\ &\le h(\eta, c) - h(\xi_j, v^j).\end{aligned}$$

Using this relation and the fact $\big(P(E_i), -\mathbb{E}\big[\mathbb{1}_{E_i} Z^k\big]\big)^\top \in \partial g_k(\xi_j, v^j)$, we see that for any $(\eta, c) \in \text{graph}(\mathfrak{F})$,

$$\begin{aligned}P(E_i)\eta - c^\top \mathbb{E}\big[\mathbb{1}_{E_i} Z^k\big] - h(\eta, c) &\le P(E_i)\xi_j - (v^j)^\top \mathbb{E}\big[\mathbb{1}_{E_i} Z^k\big] - h(\xi_j, v^j) \\ &= g_k(\xi_j, v^j) - h(\xi_j, v^j).\end{aligned}$$

Therefore, Steps 2b and 2c would return $(\xi_{j+1}, v^{j+1}) = (\xi_j, v^j)$ as a critical point. This loop would continue until reaching some j where $\partial g_k^j(\xi_j, v^j) = \partial g_k(\xi_j, v^j)$ or until some E_i is found for which the case (ii) holds.

If case (ii) holds, then some $(\eta, c) \in \text{graph}(\mathfrak{F})$ exists such that

$$h(\eta, c) < h(\xi_j, v^j) + P(E_i)(\eta - \xi_j) - \big\langle c - v^j, \mathbb{E}\big[\mathbb{1}_{E_i} Z^k\big]\big\rangle.$$

This is equivalent to the following inequality:

$$P(E_i)\eta - \big\langle c, \mathbb{E}\big[\mathbb{1}_{E_i} Z^k\big]\big\rangle - h(\eta, c) > P(E_i)\xi_j - \big\langle v^j, \mathbb{E}\big[\mathbb{1}_{E_i} Z^k\big]\big\rangle - h(\xi_j, v^j).$$

In particular, we set $(\eta, c) = (\xi_{j+1}, v^{j+1})$, where

$$(\xi_{j+1}, v^{j+1}) \in \arg\max \big\{P(E_i)\eta - \big\langle c, \mathbb{E}\big[\mathbb{1}_{E_i} Z^k\big]\big\rangle - h(\eta, c) : (\eta, c) \in \text{graph}(\mathfrak{F})\big\}.$$

From this point, we continue along the steps of the main algorithm.

Again, we conclude that the Multivariate Dominance Verification Method will exit at step 2e with a local maximizer of $g_k - h$ in $\text{graph}(\mathfrak{F})$ after finitely many steps because only finitely many possible events are contained in $\mathcal{F}$. □

7.6 Two-Stage Problems with Stochastic Dominance Constraints

In this section, we focus on numerical methods for solving the two-stage stochastic optimization with an increasing convex order constraint on the recourse function. The problem has the following form:

$$\begin{aligned} \min_{x\in\mathbb{R}^n} \ & f(x) + \mathbb{E}[\mathcal{Q}(x, D)] \\ \text{s.t } & \mathcal{Q}(x, D) \preceq_{(2)} W, \\ & x \in \mathcal{X}. \end{aligned} \tag{7.64}$$

Here $f : \mathbb{R}^n \to \mathbb{R}$ is a convex function which may be interpreted as the expected cost of the first-stage decision $x \in \mathbb{R}^n$. The set $\mathcal{X} \subseteq \mathbb{R}^n$ is a closed convex set, which represents deterministic constraints imposed on x. The functional $\mathcal{Q}(x, D)$ stands for the optimal value of the following second-stage problem:

$$\begin{aligned} \min_{y\in\mathbb{R}^m} \ & \langle q, y\rangle \\ \text{s.t. } & Bx + Ay = b, \ y \in \mathcal{Y}. \end{aligned} \tag{7.65}$$

Here $D = (q, b, B, \mathcal{Y})$ comprises the data of the second-stage problem, and some or all elements of D are random. The matrix $A \in \mathbb{R}^m \times \mathbb{R}^d$ is considered deterministic. The vector q is m-dimensional, B is $m \times n$ matrix, b is d-dimensional vector, and $\mathcal{Y} \in \mathbb{R}^m$ is a random closed polyhedral subset of $\mathbb{R}^m$. The relation $\mathcal{Q}(x, D) \preceq_{(2)} Z$ is the increasing convex order.

For the numerical solution of the problem, we assume that the probability space Ω is finite with elements ω^s, $s = 1, \dots, S$. The probabilities of the elementary events are denoted by p_s, $s = 1, \dots, S$. We also use the notation $B^s = B(\omega^s)$, $b^s = b(\omega^s)$, $y^s = y(\omega^s)$, $q^s = q(\omega^s)$, and $\mathcal{Y}^s = \mathcal{Y}(\omega^s)$. Problem (7.64)–(7.65) takes on the form:

$$\begin{aligned} \min \ & f(x) + \sum_{s=1}^{S} p_s Q^s(x) \\ \text{s.t. } & Q(x, D) \preceq_{(2)} W, \\ & x \in \mathcal{X}, \end{aligned} \tag{7.66}$$

where $Q^s(x) = Q(x, D^s)$ are the realizations of the recourse function for the data realizations D^s:

$$Q^s(x) = \inf_y \{\langle q^s, y\rangle : Ay = b^s - B^s x, \ y \in \mathcal{Y}^s\}, \quad s = 1, \dots, S. \tag{7.67}$$

We base our methods on the multicut method for the risk-neutral two-stage problem. Recall that the idea of the method is to construct separate piecewise linear models of the functions $Q^s(\cdot)$ and polyhedral approximations of their domains. The construction of these approximations is carried out as follows.

Given a point x^k, suppose $Q^s(x^k) < \infty$. Then, the subdifferential of $Q^s(\cdot)$ at x^k is given by $\partial Q^s(x^k) = (-B^s)^\top \pi^s(x^k)$, where $\pi^s(x^k)$ are the optimal solutions of the dual problem to problem (7.67) at $x = x^k$. Let $\pi^{k,s}$ be one of these optimal dual solutions. Then, the definition of a subgradient implies

$$Q^s(x) \geq Q^s(x^k) - \langle (B^s)^\top \pi^{k,s}, x - x^k \rangle \text{ for all } x \in \mathbb{R}^n,$$

which can be written as follows:

$$Q^s(x) \geq \alpha^{k,s} + (g^{k,s})^\top x, \text{ for all } x \in \mathbb{R}^n, \tag{7.68}$$

where $g^{k,s} = -(B^s)^\top \pi^{k,s}$ and $\alpha^{k,s} = (b^s)^\top \pi^{k,s}$. Equation (7.68) defines an *objective cut.*

If any of the problems (7.66) is infeasible at $x = x^k$, an inequality can be derived, which is satisfied at every $x \in \text{dom } Q^s$. Suppose the infeasible problem occurs at scenario s. We consider the following problem:

$$\begin{aligned} &\min_{x,y} \|u\| \\ &\text{s.t. } A^s y + u = b^s - B^s x, \\ &\quad\;\; y \in \mathcal{Y}^s. \end{aligned}$$

Its optimal value function $U^s(\cdot)$ is convex and $U^s(x) = 0$ if and only if $Q^s(x) < \infty$. Using the subgradient inequality in a similar way as for the objective cuts an inequality of the following form can be constructed:

$$0 = U^s(x) \geq \beta^{k,s} + (r^{k,s})^\top x. \tag{7.69}$$

The inequality (7.69) is called a *feasibility cut.* It is violated at x^k and excludes this point from the feasible set of the first-stage solution.

7.6.1 Decomposition Method with Inverse Event Cuts

We present in detail the method with inverse event cuts, introduced in § 7.2.2, as the most effective for the problem at hand.

First, we adapt several results to the increasing convex order case. We easily obtain a counterpart of Theorem 7.5, which we shall use in our method.

Theorem 7.21. *For two integrable random variables V and W, $V \preceq_{(2)} W$ if and only if*

$$\mathbb{E}[V|\mathcal{A}] \leq \frac{1}{P(\mathcal{A})} \bar{F}_W^{(-2)}(1 - P(\mathcal{A})) \quad \forall\, \mathcal{A} \in \mathcal{F} : 0 < P(\mathcal{A}) < 1. \tag{7.70}$$

Proof. We have shown that $V \preceq_{(2)} W$ if and only if $-V \succeq_{(2)} -W$. Theorem 7.5 establishes that $-V \succeq_{(2)} -W$ holds if and only if

$$\mathbb{E}[-V|\mathcal{A}] \geq \frac{1}{P(\mathcal{A})} F_{-W}^{(-2)}\big(P(\mathcal{A})\big) \quad \forall\, \mathcal{A} \in \mathcal{F} : 0 < P(\mathcal{A}) < 1.$$

We transform the right-hand side of the above inequality as follows:

$$\mathbb{E}[-V|\mathcal{A}] \geq -\frac{1}{P(\mathcal{A})} \int_{1-P(\mathcal{A})}^{1} F_W^{(-1)}(t)\, \mathrm{d}t. \tag{7.71}$$

The last relation is equivalent to (7.70). ◻

Furthermore, it is sufficient to consider only the events of form $\{V > t\}$ for $t \in \mathbb{R}$ in order to verify whether $V \preceq_{(2)} Y$ holds. We obtain the following implication, which is a counterpart of Corollary 7.6.

Theorem 7.22. *Suppose $V, W \in \mathcal{L}_1(\Omega, \mathcal{F}, P)$. Then $V \preceq_{(2)} W$ if and only if*

$$\mathbb{E}[V|V > t] \leq \frac{1}{\bar{F}_V(t)} \bar{F}_W^{(-2)}(F_V(t)) \quad \forall\, t \in \mathbb{R} : \bar{F}_V(t) > 0. \tag{7.72}$$

Proof. Corollary 7.6 and the relations between the second-order dominance and the increasing convex order imply the characterization

$$\mathbb{E}[-V| -V \leq t] \geq \frac{1}{F_{-V}(t)} F_{-W}^{(-2)}\big(F_{-V}(t)\big) \quad \forall\, t \in \mathbb{R} : F_{-V}(t) > 0.$$

This is equivalent to the following inequality for all $t \in \mathbb{R}$ with $F_{-V}(-t) > 0$:

$$\mathbb{E}[-V|V \geq t] \geq \frac{1}{F_{-V}(-t)} F_{-W}^{(-2)}\big(F_{-V}(-t)\big) = \frac{1}{P(V \geq t)} F_{-W}^{(-2)}\big(P(V \geq t)\big).$$

We have shown in the proof of Theorem 7.21, formula (7.71), that

$$\frac{1}{P(V \geq t)} F_{-W}^{(-2)}\big(P(V \geq t)\big) = -\frac{1}{P(V \geq t)} \int_{1-P(V\geq t)}^{1} F_W^{(-1)}(t)\, \mathrm{d}t.$$

Combining the last two relations, we obtain for all $t \in \mathbb{R}$ with $P(V \geq t) > 0$:

$$\mathbb{E}[V|V \geq t] \leq \frac{1}{P(V \geq t)} \bar{F}_W^{(-2)}(P(V < t)). \tag{7.73}$$

Now, we observe that the function $t \mapsto P(V \geq t)$ is left-continuous. Therefore, both the left- and the right-hand side of (7.73) are left-continuous. Thus, for every $t \in \mathbb{R}$ such that $P(V > t) > 0$, using (7.73) with t_k such that $t_k \uparrow t$ and passing to the limit, we obtain (7.72). ◻

We define the set $\mathcal{T}(x) = \{t \in \mathbb{R} : \bar{F}_{Q(x,D)}(t) > 0\}$. The problem (7.64)–(7.65) takes on the form:

$$\begin{aligned}
&\min_{x\in\mathcal{X}} \ f(x) + \mathbb{E}[Q(x,D)]\\
&\text{s.t. } \mathbb{E}\big[Q(x,D)|Q(x,D) > t\big] \le \frac{1}{\bar{F}_{Q(x,D)}(t)} \bar{F}_W^{(-2)}\big(F_{Q(x,D)}(t)\big) \qquad (7.74)\\
&\qquad\qquad \text{for all } t \in \mathcal{T}(x).
\end{aligned}$$

We shall use the multicut method for two-stage problems in which the approximation of the ordering relation by event cuts is embedded.

Let k denote the iteration number. For each scenario $s = 1, \dots, S$, we denote the set of objective cuts (7.68) obtained for the function $Q^s(\cdot)$ until iteration k by $J_{\text{o}}^k(s)$. The corresponding feasibility cuts (7.69) are gathered in $J_{\text{f}}^k(s)$, and J_{e}^k denotes the set of events generated by the sequential approximation of the stochastic order relation. The cuts constructed until iteration k are used in the following *master problem*:

$$\min_{x,v} f(x) + \sum_{s=1}^{S} p_s v^s \qquad (7.75)$$

$$\text{s.t. } \alpha^{j,s} + (g^{j,s})^\top x \le v^s, \quad j \in J_{\text{o}}^k(s), \quad s = 1, \dots, S, \qquad (7.76)$$

$$\beta^{j,s} + (r^{j,s})^\top x \le 0, \ j \in J_{\text{f}}^k(s), \quad s = 1, \dots, S, \qquad (7.77)$$

$$\frac{1}{P(\mathcal{A}^j)} \sum_{s\in\mathcal{A}^j} p_s v^s \le \frac{1}{P(\mathcal{A}^j)} \bar{F}_W^{(-2)}\big(1 - P(\mathcal{A}^j)\big), \quad j \in J_{\text{e}}^k, \qquad (7.78)$$

$$x \in \mathcal{X}. \qquad (7.79)$$

Here $v = (v^1, \dots, v^S)^\top$ are treated as the realizations of a random variable V, attained with the probabilities $p_1, \dots, p_S$. The inequalities (7.78) are the *inverse event cuts* which approximate the order constraint $V \preceq_{(2)} W$. Our intention is to satisfy this order constraint and the inequality $Q(x) \le V$.

The following method solves the two-stage problem with the increasing convex order constraint on the recourse function.

Multi-cut decomposition with inverse event cuts

Step 0. Set $k = 1$, $J_{\text{e}}^0 = \emptyset$, $J_{\text{o}}^0(s) = \emptyset$, $J_{\text{f}}^0(s) = \emptyset$, and $v^{1,s} = -\infty$, for all $s = 1, \dots, S$, $\mathcal{A}^0 = \{1, \dots, S\}$. Choose $x^1 \in \mathcal{X}$.

Step 1. For $s = 1, \dots, S$ solve subproblem (7.67) with $x = x^k$.

(a) If $Q^s(x^k) < \infty$, then set $J_{\text{f}}^k(s) = J_{\text{f}}^{k-1}(s)$, $J_{\text{o}}^k(s) = J_{\text{o}}^{k-1}(s) \cup \{k\}$ and construct an objective cut (7.68), if $Q^s(x^k) > v^{k,s}$.

(b) If $Q^s(x^k) = \infty$, then set $J_{\text{f}}^k(s) = J_{\text{f}}^{k-1}(s) \cup \{k\}$, $J_{\text{o}}^k(s) = J_{\text{o}}^{k-1}(s)$, and construct a feasibility cut (7.69).

Step 2. If $Q^s(x^k) < \infty$ for all $s = 1, \dots, S$, then consider the sets $\mathcal{B}_s = \{Q(x^k) \ge Q^s(x^k)\}$, $s = 1, \dots S$, and calculate

$$\delta_k = \max_{1 \le s \le S} \left\{ \mathbb{E}\left[Q(x^k) \mid \mathcal{B}_s\right] - \frac{1}{P(\mathcal{B}_s)} \overline{F}_W^{(-2)}(1 - P(\mathcal{B}_s)) \right\}. \tag{7.80}$$

If $\delta_k \le 0$, then set $J_e^k = J_e^{k-1}$. Otherwise, let s^* be the maximizer in (7.80). Set $\mathcal{A}^k = \mathcal{B}_{s^*}$ and $J_e^k = J_e^{k-1} \cup \{k\}$.

Step 3. If $\mathbb{E}[Q(x^k)] = \sum_{s=1}^S p_s v^{k,s}$ and $\delta_k \le 0$, then stop; otherwise continue.

Step 4. Solve the master problem (7.75)–(7.79). If it is infeasible, then stop. Otherwise, denote its solution by (x^{k+1}, v^{k+1}), increase k by one, and go to Step 1.

We have the following convergence result.

Theorem 7.23. *Assume that the set $\mathcal{X}$ is a compact convex polyhedron. If the method stops at Step 4 discovering the infeasibility of problem* (7.75)–(7.79), *then problem* (7.66)–(7.67) *is infeasible as well. If problem* (7.66)–(7.67) *has a feasible solution, then the method stops after finitely many iterations k^* with an optimal solution x^{k^*} and y^{s,k^*}, $s = 1, \dots, S$ of problem* (7.66)–(7.67).

Proof. The solution of the master problem (7.75)–(7.79) at iteration k satisfies

$$v^{k,s} = \max_{j \in J_o^{k-1}(s)} \{\alpha^{j,s} + (g^{j,s})^\top x\} = \underline{Q}^s(x^k),$$

where $\underline{Q}^s(x)$ is the piecewise linear approximation of the function $Q^s(x)$ created by the objective cuts. Thus, $\sum_{s=1}^S p_s v^s$ approximates the expected recourse function $\sum_{s=1}^S p_s Q^s(x)$ from below. Furthermore, the feasibility cuts provide an outer approximation of the domain of the functions $Q^s(\cdot)$. Substituting the random variable V in constraint (7.78) instead of $Q(x, D)$ provides a relaxation of constraint (7.74) because V is a lower bound of $Q(x, D)$. The relation $V \preceq_{(2)} W$ is relaxed further because the inequality (7.70) is imposed only on some of the events in problem (7.75)–(7.79). Therefore, the master problem is a relaxation of (7.66)–(7.67). Hence, if the method stops at Step 4 because the master problem (7.75)–(7.79) is infeasible, then the original problem (7.66)–(7.67) is infeasible as well.

Now, we consider the case when problem (7.66)–(7.67) has a feasible solution. The domains of the functions $U^s(\cdot)$, $s = 1, \dots, S$, are polyhedral and will be approximated exactly after finitely many feasibility cuts. Thus, we can assume that a number $\ell_f \in \mathbb{N}$ exists such that for all iterations $k \ge \ell_f$ no new feasibility cuts will be added to the master problem. This implies that $Q(x^k, D) \in \mathcal{L}_\infty(\Omega, \mathcal{F}, P)$ for $k > \ell_f$, which, together with $v^s(x) \le Q^s(x)$, yields $V^k \in \mathcal{L}_\infty(\Omega, \mathcal{F}, P)$ for $k > \ell_f$.

Furthermore, the functions $Q^s(\cdot)$, $s = 1, \ldots, S$, are polyhedral and only finitely many cuts of form (7.68) can be generated. Thus, a number $\ell_o \in N$ exists such that for all iterations $k \geq \ell_o$ no new objective cuts will be added to the master problem. This implies that, in iterations $k \geq \max(\ell_o, \ell_f)$, we have $\underline{Q}^s(x^k) = Q^s(x^k)$ and only event cuts are constructed. Due to Theorem 7.5 and Corollary 7.7 in a finite probability space only finitely many event cuts can be generated and the method will stop at Step 3. Let $k^* \in N$ be the exit iteration. Thus, every feasible solution of problem (7.75)–(7.79) is also feasible for (7.66)–(7.67). Let φ^* denote the optimal value of problem (7.66) and let $\varphi(\cdot)$ be the optimal value of the master problem (7.75)–(7.79). Due to the fact that the master problem (7.75)–(7.79) is a relaxation of (7.66)–(7.67), we have $\varphi^* \geq \varphi(x^{k^*})$. Since x^{k^*} is feasible for (7.66)–(7.67), we obtain

$$\varphi(x^{k^*}) \geq \varphi^* \geq \varphi(x^{k^*}).$$

We conclude that x^{k^*} is an optimal solution of (7.66) and the corresponding points y^{s,k^*}, $s = 1, \ldots, S$, solve (7.67), as stated. □

Remark 7.24.

(r1) We point out that an alternative way to construct an event cut in the multi-cut decomposition method using the upper Lorenz function is to replace constraint (7.78) by the inequality

$$\sum_{s=j}^{S} p_s v_{[s]} - \bar{F}_W^{(-2)}\Big(\sum_{s=j}^{S} p_s\Big) \leq 0. \tag{7.81}$$

This type of inequalities are called *unscaled event cuts* and the method can be modified to make use of the Lorenz function without scaling.

(r2) Note that we do not need a discrete benchmark W for this method. It is sufficient to be able to calculate the values of the function $\bar{F}_W^{(-2)}(\cdot)$.

□

7.6.2 Decomposition Method with Shortfall Event Cuts

In this section, we represent the increasing convex order relation in our first-stage model by the excess function inequality and focus on the problem

$$\begin{aligned} \min_{x \in \mathbb{R}^n} \ & f(x) + \mathbb{E}[Q(x, D)] \\ \text{s.t } & \mathbb{E}[(Q(x, D) - \eta)_+] \leq \mathbb{E}[(W - \eta)_+], \quad \text{for all } \eta \in \mathbb{R}, \\ & x \in \mathcal{X}, \end{aligned} \tag{7.82}$$

where $Q(x, D)$ stands for the optimal value of the second-stage problem (7.65).

Denote $r(\eta) = \mathbb{E}[(W - \eta)_+]$. Due to the finiteness of the probability space, the function $r(\cdot)$ is convex and piecewise linear, with the breakpoints at the realizations w_l, $l = 1, \dots, S$, of W. Assuming, for simplicity, that the realizations w_l are ordered, $w_1 < w_2 < \cdots < w_S$, we have

$$r(w_l) = \sum_{s=l+1}^{S} p_s(w_s - w_l), \quad l = 1, \dots, S.$$

We shall embed a successive approximation of the excess function of the recourse by event cuts within the multicut method. We identify events $\mathcal{A}^k$ associated with a breakpoint η_k and include an event cut of the following form:

$$\sum_{s \in \mathcal{A}^k} p_s(Q^s - \eta_k) \le r(\eta_k), \tag{7.83}$$

into the master problem. The method works as follows.

Multicut decomposition method with shortfall event cuts

Step 0. Set $k = 1$, $J_e^0 = \emptyset$, $J_o^0(s) = \emptyset$, $J_f^0(s) = \emptyset$, and $v^{1,s} = -\infty$, for all $s = 1, \dots, S$, $\mathcal{A}^0 = \{1, \dots, S\}$. Choose $x^1 \in \mathcal{X}$.

Step 1. For $s = 1, \dots, S$ solve subproblem (7.67) with $x = x^k$.

(a) If $Q^s(x^k) < \infty$, then set $J_f^k(s) = J_f^{k-1}(s)$, $J_o^k(s) = J_o^{k-1}(s) \cup \{k\}$ and construct an objective cut (7.68), if $Q^s(x^k) > v^{k,s}$.

(b) If $Q^s(x^k) = \infty$, then set $J_f^k(s) = J_f^{k-1}(s) \cup \{k\}$, $J_o^k(s) = J_o^{k-1}(s)$, and construct a feasibility cut (7.69).

Step 2. If $Q^s(x^k) < \infty$ for all $s = 1, \dots, S$, then consider the sets $\mathcal{B}_i = \{Q(x^k) > w_i\}$, $i = 1, \dots, S$, and let

$$\delta_k = \max_{1 \le i \le S} \Big\{ \sum_{s \in \mathcal{B}_i} p_s(Q^s(x^k) - w_i) - r(w_i) \Big\}. \tag{7.84}$$

If $\delta_k \le 0$, then set $J_e^k = J_e^{k-1}$. Otherwise, let i^* be the maximizer in (7.84). Set $\mathcal{A}^k = \mathcal{B}_{i^*}$, $\eta_k = w_{i^*}$, and $J_e^k = J_e^{k-1} \cup \{k\}$.

Step 3. If $\mathbb{E}[Q(x^k)] = \sum_{s=1}^{S} p_s v^{k,s}$ and $\delta_k \le 0$, then stop; otherwise continue.

Step 4. Solve the master problem

$$
\begin{aligned}
\min_{x,v}\ & f(x) + \sum_{s=1}^{S} p_s v^s \\
\text{s.t. } & \alpha^{j,s} + (g^{j,s})^\top x \le v^s, \quad j \in J_{\text{o}}^k(s), \quad s = 1, \dots, S, \\
& \beta^{j,s} + (r^{j,s})^\top x \le 0, \quad j \in J_{\text{f}}^k(s), \quad s = 1, \dots, S, \\
& \sum_{s \in \mathcal{A}^j} p_s(v^s - \eta_j) \le r(\eta_j), \quad j \in J_{\text{e}}^k, \\
& x \in \mathcal{X}.
\end{aligned}
\tag{7.85}
$$

If it is infeasible, then stop, because problem (7.82) is infeasible. Otherwise, denote the solution by (x^{k+1}, v^{k+1}), increase k by one, and go to Step 1.

The convergence of the method can be proved in the same way as in the preceding section.

Theorem 7.25. *Assume that the set $\mathcal{X}$ is a compact convex polyhedron. If the method stops at Step 4 with infeasibility of problem* (7.85)*, then problem* (7.66)–(7.67) *has no feasible solution. If problem* (7.66)–(7.67) *has a feasible solution, then the method stops after finitely many iterations with an optimal solution of it.*

Proof. We notice that only as many events $\mathcal{A}^j$ may be constructed as the realizations of the benchmark exist. Therefore, only finitely many event cuts can be generated. The remaining arguments about the convergence of this decomposition method are the same as for the convergence of the multicut decomposition method with inverse event cuts, which is presented in § 7.6.1. □

Chapter 8
Risk-Averse Control of Markov Systems

This chapter is devoted to dynamic risk measurement and risk-averse control in discrete-time systems described by controlled Markov kernels. The main idea of our approach is to exploit the structure of the canonical probability space to represent conditional risk measures by simpler structures, called transition risk mappings, which evaluate the risk of a function of the system's state.

8.1 Finite Horizon Models

We introduce fundamental concepts and properties of dynamic risk measures for stochastic processes in discrete time. In subsection 8.1.1, we present the notion of stochastic conditional time consistency, which is a stronger requirement on the dynamic risk measure than the standard time consistency, and which is particularly useful for controlled stochastic processes. Based on this concept, we derive the structure of dynamic risk measures involving transition risk mappings: a family of static risk measures on the space of functions of the state. This allows for the generalization of the theory and methods of dynamic programming.

8.1.1 Risk Measures for Discrete-Time Processes

We consider a Borel subset $(\mathcal{X}, \mathcal{B}(\mathcal{X}))$ of a Polish space, which we call the *state space*, and a discrete-time process $\{X_t\}_{t=1,\dots,T}$ with values in this space. The canonical measurable space $\mathcal{X}^T$ with the product σ-algebra $\mathcal{B}(\mathcal{X})^T$ play the roles of the probability space Ω and its σ-algebra of random events. Thus, $\{X_t\}_{t=1,\dots,T}$ is the process of canonical projections. We also define $\mathcal{H}_t = \mathcal{X}^t$

D. Dentcheva, A. Ruszczyński, *Risk-Averse Optimization and Control*, Springer Series in Operations Research and Financial Engineering,
https://doi.org/10.1007/978-3-031-57988-2_8

to be the space of possible histories up to time t, and we use h_t for a generic element of $\mathcal{H}_t$: a specific history up to time t. The random vector $(X_1, \ldots, X_t)$ will be denoted by H_t.

The probability measure on the canonical space is defined by transition kernels. Specifically, we assume that for all $t = 1, \ldots, T-1$, the conditional distribution of X_{t+1}, given the history h_t, is described by a transition kernel: a measurable function

$$Q_t : \mathcal{X}^t \to \mathcal{P}(\mathcal{X}), \tag{8.1}$$

where $\mathcal{P}(\mathcal{X})$ is the set of probability measures on $(\mathcal{X}, \mathcal{B}(\mathcal{X}))$. These kernels, together with the initial distribution P_1 of X_1, define a unique probability measure $P = P_1 \otimes Q_1 \otimes \cdots \otimes Q_{T-1}$ on the product space $\mathcal{X}^T$ with the product σ-algebra. This specification of the probability space will allow us to further advance the theory of dynamic measures of risk.

For a stochastic system described above, we consider a sequence of random variables $\{Z_t\}_{t=1,\ldots,T}$ taking values in $\mathbb{R}$; we assume that lower values of Z_t are preferred (Z_t represents a "cost" at time t). We require $\{Z_t\}_{t=1,\ldots,T}$ to be bounded and adapted to $\{\mathcal{F}_t\}_{t=1,\ldots,T}$: the natural filtration generated by the process X, and thus each Z_t is a measurable function of $X_1, \ldots, X_t$. We introduce for $t = 1, \ldots, T$ the following spaces:

$$\mathcal{Z}_t = \Big\{ Z : \mathcal{X}^T \to \mathbb{R} \;\Big|\; Z \text{ is } \mathcal{F}_t\text{-measurable and bounded} \Big\}. \tag{8.2}$$

For brevity, we use the notation $\mathcal{Z}_{t:T} \triangleq \mathcal{Z}_t \times \cdots \times \mathcal{Z}_T$, $t = 1, \ldots, T$.

We now refine the concept of time consistency for dynamic risk measures.

Definition 8.1. A dynamic risk measure $\{\varrho_{t,T}\}_{t=1,\ldots,T}$ is *stochastically conditionally time-consistent with respect to* $\{Q_t\}_{t=1,\ldots,T-1}$ if for any $1 \le t \le T-1$, for any $h_t \in \mathcal{X}^t$, and for all $(Z_t, \ldots, Z_T)$ and $(W_t, \ldots, W_T)$ in $\mathcal{Z}_{t:T}$, the conditions

$$\begin{cases} Z_t(h_t) = W_t(h_t), \\ \big(\varrho_{t+1,T}(Z_{t+1}, \ldots, Z_T) \mid H_t = h_t\big) \preceq_{(1)} \big(\varrho_{t+1,T}(W_{t+1}, \ldots, W_T) \mid H_t = h_t\big), \end{cases} \tag{8.3}$$

imply

$$\varrho_{t,T}(Z_t, \ldots, Z_T)(h_t) \le \varrho_{t,T}(W_t, \ldots, W_T)(h_t), \tag{8.4}$$

where the relation $\preceq_{(1)}$ stands for the stochastic dominance understood as follows:

$$\begin{aligned} Q_t(h_t)&\Big(\big\{x \in \mathcal{X} : \varrho_{t+1,T}(Z_{t+1}, \ldots, Z_T)(h_t, x) > \eta\big\}\Big) \\ &\le Q_t(h_t)\Big(\big\{x \in \mathcal{X} : \varrho_{t+1,T}(W_{t+1}, \ldots, W_T)(h_t, x) > \eta\big\}\Big), \quad \forall\, \eta \in \mathbb{R}. \end{aligned}$$

When the choice of the underlying transition kernels is clear from the context, we will simply say that the dynamic risk measure is stochastically conditionally time-consistent. In all considerations in this chapter, we assume that *all risk measures are normalized.*

Proposition 8.2. *The stochastic conditional time consistency implies the time consistency and the local property.*

Proof. If $\{\varrho_{t,T}\}_{t=1,\dots,T}$ satisfies the translation and normalization properties and is stochastically conditionally time-consistent, then it satisfies Definition 4.3 and is time-consistent.

Let us prove by induction on t from T down to 1 that $\varrho_{t,T}$ satisfy the local property. Clearly, $\varrho_{T,T}$ does: if $A \in \mathcal{F}_T$, then by the translation and normalization properties $\mathbb{1}_A \varrho_{T,T}(Z_T) = \mathbb{1}_A Z_T = \varrho_{T,T}(\mathbb{1}_A Z_T)$.

Suppose $\varrho_{t+1,T}$ satisfies the local property for some $1 \le t < T$, and consider any $A \in \mathcal{F}_t$, any $h_t \in \mathcal{X}^t$, and any $(Z_t, \dots, Z_T) \in \mathcal{Z}_{t:T}$. Two cases may occur.

Case 1. If $\mathbb{1}_A(h_t) = 0$, then $[\mathbb{1}_A Z_t](h_t) = 0$ and the local property for $t+1$ yields:

$$\big[\varrho_{t+1,T}(\mathbb{1}_A Z_{t+1}, \dots, \mathbb{1}_A Z_T)\big](h_t, \cdot) = \\ \big[\mathbb{1}_A \varrho_{t+1,T}(Z_{t+1}, \dots, Z_T)\big](h_t, \cdot) = 0.$$

By the stochastic conditional time consistency,

$$\varrho_{t,T}(\mathbb{1}_A Z_t, \mathbb{1}_A Z_{t+1}, \dots, \mathbb{1}_A Z_T)(h_t) = \varrho_{t,T}(0, \dots, 0)(h_t) = 0.$$

Case 2. If $\mathbb{1}_A(h_t) = 1$, then $[\mathbb{1}_A Z_t](h_t) = Z_t(h_t)$ and the local property for $t+1$ implies that

$$\begin{aligned}\big[\varrho_{t+1,T}(\mathbb{1}_A Z_{t+1}, \dots, \mathbb{1}_A Z_T)\big](h_t, \cdot) \\ = \big[\mathbb{1}_A \varrho_{t+1,T}(Z_{t+1}, \dots, Z_T)\big](h_t, \cdot) \\ = \varrho_{t+1,T}(Z_{t+1}, \dots, Z_T)(h_t, \cdot).\end{aligned}$$

By the stochastic conditional time consistency, $\varrho_{t,T}(\mathbb{1}_A Z_t, \dots, \mathbb{1}_A Z_T)(h_t) = \varrho_{t,T}(Z_t, \dots, Z_T)(h_t)$.

In both cases, $\varrho_{t,T}(\mathbb{1}_A Z_t, \dots, \mathbb{1}_A Z_T)(h_t) = \big[\mathbb{1}_A \varrho_{t,T}(Z_t, \dots, Z_T)\big](h_t)$. ∎

The following proposition shows that the stochastic conditional time consistency implies that the one-step conditional risk mappings $\rho_t(\cdot)$ of Corollary 4.6 can be equivalently represented by law-invariant risk measures on $\mathcal{V}$.

Theorem 8.3. *A dynamic risk measure $\{\varrho_{t,T}\}_{t=1,\dots,T}$ is normalized, has the translation property, and is stochastically conditionally time-consistent, if and only if functionals $\sigma_t : \mathcal{X}^t \times \mathcal{P}(\mathcal{X}) \times \mathcal{V} \to \mathbb{R}$, $t = 1, \dots, T-1$, exist, such that*

(i) *for all* $t = 1, \ldots, T-1$ *and all* $h_t \in \mathcal{X}^t$, *the functional* $\sigma_t(h_t, Q_t(h_t), \cdot)$ *is a normalized, monotonic, and law-invariant risk measure on* $\mathcal{V}$, *with the corresponding probability space* $(\mathcal{X}, \mathcal{B}(\mathcal{X}), Q_t(h_t))$;

(ii) *for all* $t = 1, \ldots, T-1$, *for all* $(Z_t, \ldots, Z_T) \in \mathcal{Z}_{t:T}$, *and for all* $h_t \in \mathcal{X}^t$,

$$\varrho_{t,T}(Z_t, \ldots, Z_T)(h_t) = Z_t(h_t) + \sigma_t\big(h_t, Q_t(h_t), \varrho_{t+1,T}(Z_{t+1}, \ldots, Z_T)(h_t, \cdot)\big). \quad (8.5)$$

Moreover, for all $t = 1, \ldots, T-1$, *for every* $h_t \in \mathcal{X}^t$, *and every* $v \in \mathcal{V}$,

$$\sigma_t(h_t, Q_t(h_t), v) = \varrho_{t,T}(0, V, 0, \ldots, 0)(h_t), \quad (8.6)$$

where $V \in \mathcal{Z}_{t+1}$ *satisfies the equation* $V(h_t, \cdot) = v(\cdot)$, *and may be arbitrary elsewhere.*

Proof. Suppose $\{\varrho_{t,T}\}_{t=1,\ldots,T}$ is normalized, has the translation property, and is stochastically conditionally time-consistent. For any $h_t \in \mathcal{X}^t$, the formula (8.6) defines a normalized and monotonic risk measure on the space $\mathcal{V}$. Setting

$$V(h_{t+1}) = \begin{cases} \varrho_{t+1,T}(Z_{t+1}, \ldots, Z_T)(h_t, x), & \text{if } h_{t+1} = (h_t, x), \\ 0, & \text{otherwise}, \end{cases}$$

we obtain, by the translation property and normalization,

$$\varrho_{t+1,T}(V, 0, \ldots, 0)(h_t, \cdot) = V(h_t, \cdot) = \varrho_{t+1,T}(Z_{t+1}, \ldots, Z_T)(h_t, \cdot).$$

Thus, by the translation property and the stochastic conditional time consistency,

$$\begin{aligned} \varrho_{t,T}(Z_t, \ldots, Z_T)(h_t) &= Z_t(h_t) + \varrho_{t,T}(0, Z_{t+1}, \ldots, Z_T)(h_t) \\ &= Z_t(h_t) + \varrho_{t,T}(0, V, 0, \ldots, 0)(h_t) \\ &= Z_t(h_t) + \sigma_t(h_t, Q_t(h_t), v). \end{aligned}$$

This chain of relations proves also the uniqueness of $\sigma_t(h_t, Q_t(h_t), \cdot)$. We need only verify the postulated law invariance of $\sigma_t(h_t, Q_t(h_t), \cdot)$. If $V, V' \in \mathcal{Z}_{t+1}$ have the same conditional distribution, given h_t, then Definition 8.1 implies that $\varrho_{t,T}(0, V, 0, \ldots, 0)(h_t) = \varrho_{t,T}(0, V', 0, \ldots, 0)(h_t)$, and law invariance follows from (8.6).

On the other hand, if such operators $\sigma_t(\cdot, \cdot, \cdot)$ exist, then $\{\varrho_{t,T}\}_{t=1,\ldots,T}$ is stochastically conditionally time-consistent by the monotonicity and law invariance of $\sigma_t(h_t, Q_t(h_t), \cdot)$. We can now use (8.5) to obtain for any $t = 1, \ldots, T-1$, and for all $h_t \in \mathcal{X}^t$ the following identity:

$$\begin{aligned}\varrho_{t,T}(0, Z_{t+1}, \dots, Z_T)(h_t) &= \sigma_t\big(h_t, \varrho_{t+1,T}(Z_{t+1}, \dots, Z_T)(h_t, \cdot)\big)\\ &= \varrho_{t,T}(Z_t, \dots, Z_T)(h_t) - Z_t(h_t),\end{aligned}$$

which is the translation property of $\varrho_{t,T}$. □

The operators $\sigma_t(\cdot, \cdot, \cdot)$ are called *transition risk mappings*.

Remark 8.4. For a fixed history h_t, the measure $Q_t(h_t)$ is known, and there is no direct need to include it as a separate argument of the transition risk mapping. We include it for two reasons: to stress the relevance of the probability measure $Q_t(h_t)$ in the risk model on $\mathcal{V}$, and in anticipation of the application to controlled processes, where the kernel depends on control variables. □

Example 8.5. In the theory of risk-sensitive Markov decision processes, the following family of entropic risk measures is employed:

$$\varrho_{t,T}(Z_t, \dots, Z_T) = \frac{1}{\kappa} \ln\bigg(\mathbb{E}\Big[\exp\Big(\kappa \textstyle\sum_{s=t}^{T} Z_s\Big) \;\Big|\; \mathcal{F}_t\Big]\bigg), \quad \varkappa > 0.$$

It is stochastically conditionally time-consistent and corresponds to the transition risk mapping

$$\sigma_t(h_t, q, v) = \frac{1}{\kappa} \ln\big(\mathbb{E}_q[\mathrm{e}^{\kappa v}]\big) = \frac{1}{\kappa} \ln\bigg(\int_{\mathcal{X}} \mathrm{e}^{\kappa v(x)}\, q(\mathrm{d}x)\bigg). \tag{8.7}$$

We could also make κ in (8.7) dependent on the time t, the current state x_t, or even the entire history h_t, and still obtain a stochastically conditionally time-consistent dynamic risk measure. □

Example 8.6. The following transition risk mapping satisfies the condition of Theorem 8.3 and corresponds to a stochastically conditionally time-consistent dynamic risk measure:

$$\begin{aligned}\sigma_t(h_t, q, v) = &\int_{\mathcal{X}} v(x)\, q(\mathrm{d}x) + \\ &\kappa_t(h_t) \bigg(\int_{\mathcal{X}} \Big[\Big(v(x) - \int_{\mathcal{X}} v(y)\, q(\mathrm{d}y)\Big)_+\Big]^p q(\mathrm{d}x)\bigg)^{1/p},\end{aligned} \tag{8.8}$$

where $\kappa_t : \mathcal{X}^t \to [0, 1]$ is a measurable function, and $p \in [1, +\infty)$. It is an analog of the static mean–semideviation measure of risk of Example 2.4. In the construction of a dynamic risk measure, we use $q = Q_t(h_t)$. If κ_t depends on x_t only, the mapping (8.8) corresponds to a Markov risk measure discussed in § 8.1.3 and § 8.1.4. □

Example 8.7. Our use of the stochastic dominance relation in the definition of stochastic conditional time consistency rules out some candidates for transition risk mappings. Suppose $\sigma_t(h_t, q, v) = v(x_1)$, where $x_1 \in \mathcal{X}$ is a selected state. Such a mapping is a coherent measure of risk, as a function of the last argument, and may be law-invariant. In particular, it is law-invariant with $\mathcal{X} = \{x_1, x_2\}$, $q(x_1) = 1/3$, $q(x_2) = 2/3$, $v(x_1) = 3$, $v(x_2) = 1$, $w(x_1) = 2$, $w(x_2) = 4$. For this mapping, we have $v \preceq_{(1)} w$ under q, but $\sigma_t(h_t, q, v) > \sigma_t(h_t, q, w)$, and thus the condition of stochastic conditional time consistency is violated. This is due to the fact that the probability of reaching x_1, no matter how small, does not affect the value of the risk measure. We exclude such cases because in controlled systems the second argument (q) is the only one that depends on our decisions. It should be included in the definition of our preferences if practically meaningful results are to be obtained. □

8.1.2 Risk Measures for Controlled Stochastic Processes

We now extend the setting of Section 8.1.1 by allowing the kernels (8.1) to depend on control variables.

We still work with the process $\{X_t\}_{t=1,\dots,T}$ on the space $\mathcal{X}^T$ and introduce a control space $(\mathfrak{U}, \mathcal{B}(\mathfrak{U}))$, which is assumed to be a Borel subset of a Polish space. At each time t, we observe the state $x_t \in \mathcal{X}$, and then apply a control $u_t \in \mathfrak{U}$. We assume that the admissible control sets and the transition kernels (conditional distributions of the next state) depend on all currently known state and control values. More precisely, we make the following assumptions.

(i) For all $t = 1, \dots, T$, we require that $u_t \in \mathcal{U}_t(x_1, u_1, \dots, x_{t-1}, u_{t-1}, x_t)$, where $\mathcal{U}_t : \mathcal{G}_t \rightrightarrows \mathfrak{U}$ is a measurable multifunction, and $\mathcal{G}_1, \dots, \mathcal{G}_T$ are the sets of histories of all currently-known state and control values before applying each control:

$$\begin{cases} \mathcal{G}_1 = \mathcal{X}, \\ \mathcal{G}_{t+1} = \operatorname{graph}(\mathcal{U}_t) \times \mathcal{X} \subseteq (\mathcal{X} \times \mathfrak{U})^t \times \mathcal{X}, \quad t = 1, \dots, T-1; \end{cases}$$

(ii) For all $t = 1, \dots, T-1$, the conditional distribution of X_{t+1}, given currently-known states and controls, is described by controlled transition kernels:

$$Q_t : \operatorname{graph}(\mathcal{U}_t) \to \mathcal{P}(\mathcal{X}), \quad t = 1, \dots, T-1. \tag{8.9}$$

For this controlled process, a (deterministic) *history-dependent admissible policy* $\Pi = (\pi_1, \dots, \pi_T)$ is a sequence of measurable selectors, called *decision rules*, $\pi_t : \mathcal{G}_t \to \mathfrak{U}$ such that $\pi_t(g_t) \in \mathcal{U}_t(g_t)$ for all $g_t \in \mathcal{G}_t$. We can easily prove by induction on t that for such an admissible policy Π, each π_t reduces to a measurable function on $\mathcal{X}^t$, as $u_s = \pi_s(h_s)$ for all $s = 1, \dots, t-1$. We are still using π_s to denote the decision rule; it will not lead to any misunderstanding. The set of admissible policies is

$$\varPi = \big\{\Pi = (\pi_1,\dots,\pi_T) : \forall t,\ \pi_t(x_1,\dots,x_t) \in \mathcal{U}_t(x_1,\pi_1(x_1),\dots \\ \dots, x_{t-1},\pi_{t-1}(x_1,\dots,x_{t-1}),x_t)\big\}. \quad (8.10)$$

For any fixed policy $\Pi \in \varPi$, the transition kernels can be rewritten as measurable functions from $\mathcal{X}^t$ to $\mathcal{P}(\mathcal{X})$:

$$Q_t^{\Pi}(x_1,\dots,x_t) \triangleq Q_t\big(x_1,\pi_1(x_1),\dots,x_t,\pi_t(x_1,\dots,x_t)\big), \\ t = 1,\dots,T-1, \quad (8.11)$$

just like the transition kernels of the uncontrolled case given in (8.1), but indexed by Π. Thus, for any policy $\Pi \in \varPi$, we can consider $\{X_t\}_{t=1,\dots,T}$ as a process on the probability space $\big(\mathcal{X}^T, \mathcal{B}(\mathcal{X})^T, P^{\Pi}\big)$ with P^{Π} defined by P_1 and $\{Q_t^{\Pi}\}_{t=1,\dots,T-1}$. The process $\{X_t\}_{t=1,\dots,T}$ is adapted to the policy-independent filtration $\{\mathcal{F}_t\}_{t=1,\dots,T}$. As before, $h_t \in \mathcal{X}^t$ stands for $(x_1,\dots,x_t)$.

We still use the same spaces $\mathcal{Z}_t$, $t = 1,\dots,T$, as defined in (8.2) for the costs incurred at each stage; these spaces also allow us to consider control-dependent costs as collections of policy-indexed costs in $\mathcal{Z}_{1:T}$. Thus, we can define and analyze stochastically conditionally time-consistent dynamic risk measures ϱ^{Π} for each fixed $\Pi \in \varPi$, as in Section 8.1.1. Note that ϱ^{Π} are defined on the same spaces independently of Π, because the filtration and the spaces $\mathcal{Z}_t$, $t = 1,\dots,T$, do not depend on Π; however, we do need to index the measures of risk by the policy Π, because the transition kernels and, consequently, the probability measure on the space $\mathcal{X}^T$, do depend on Π.

We need to compare risk between different policies, so a meaningful concept of time consistency of a family of risk measures $\big\{\varrho_{t,T}^{\Pi}\big\}_{t=1,\dots,T}^{\Pi\in\varPi}$ is needed.

Definition 8.8. A family of dynamic risk measures $\big\{\varrho_{t,T}^{\Pi}\big\}_{t=1,\dots,T}^{\Pi\in\varPi}$ is *stochastically conditionally time-consistent* if for any $\Pi, \Pi' \in \varPi$, for any $1 \le t < T$, for all $h_t \in \mathcal{X}^t$, all $(Z_t,\dots,Z_T)$ and $(W_t,\dots,W_T)$ in $\mathcal{Z}_{t:T}$, the conditions

$$Z_t(h_t) = W_t(h_t), \quad \text{and} \\ \big(\varrho_{t+1,T}^{\Pi}(Z_{t+1},\dots,Z_T) \mid H_t^{\Pi} = h_t\big) \\ \preceq_{(1)} \big(\varrho_{t+1,T}^{\Pi'}(W_{t+1},\dots,W_T) \mid H_t^{\Pi'} = h_t\big),$$

imply

$$\varrho_{t,T}^{\Pi}(Z_t,\dots,Z_T)(h_t) \le \varrho_{t,T}^{\Pi'}(W_t,\dots,W_T)(h_t).$$

Remark 8.9. As in Definition 8.1, the stochastic dominance relation "$\preceq_{(1)}$" should be understood as follows: for all $\eta \in \mathbb{R}$ we have

$$Q_t^{\Pi}(h_t)\Big(\big\{x : \varrho_{t+1,T}^{\Pi}(Z_{t+1},\ldots,Z_T)(h_t,x) > \eta\big\}\Big) \\ \leq Q_t^{\Pi'}(h_t)\Big(\big\{x : \varrho_{t+1,T}^{\Pi'}(W_{t+1},\ldots,W_T)(h_t,x) > \eta\big\}\Big).$$

□

The above definition plays a pivotal role in building an intrinsic connection among the dynamic risk measures $\{\varrho_{t,T}^{\Pi}\}_{t=1,\ldots,T}^{\Pi\in\mathit{\Pi}}$, as we explain it below.

If a family of dynamic risk measures $\{\varrho_{t,T}^{\Pi}\}_{t=1,\ldots,T}^{\Pi\in\mathit{\Pi}}$ is stochastically conditionally time-consistent, then for each fixed $\Pi \in \mathit{\Pi}$ the dynamic risk measure $\{\varrho_{t,T}^{\Pi}\}_{t=1,\ldots,T}$ is stochastically conditionally time-consistent in the sense of Definition 8.1. By virtue of Proposition 8.3, for each $\Pi \in \mathit{\Pi}$, there exist functionals $\sigma_t^{\Pi} : \mathcal{X}^t \times \mathcal{P}(\mathcal{X}) \times \mathcal{V} \to \mathbb{R}$, $t = 1 \ldots T-1$, such that for all $t = 1,\ldots,T-1$, all $h_t \in \mathcal{X}^t$, the functional $\sigma_t^{\Pi}(h_t, Q_t^{\Pi}(h_t), \cdot)$ is a law-invariant risk measure on $\mathcal{V}$ with respect to the distribution $Q_t^{\Pi}(h_t)$, and for all $h_t \in \mathcal{X}^t$,

$$\varrho_{t,T}^{\Pi}(Z_t,\ldots,Z_T)(h_t) = Z_t(h_t) + \\ \sigma_t^{\Pi}\big(h_t, Q_t^{\Pi}(h_t), \varrho_{t+1,T}^{\Pi}(Z_{t+1},\ldots,Z_T)(h_t,\cdot)\big).$$

Consider any $\Pi, \Pi' \in \mathit{\Pi}$, $h_t \in \mathcal{X}^t$, and $(Z_t,\ldots,Z_T)$ and $(W_t,\ldots,W_T)$ in $\mathcal{Z}_{t:T}$, such that

$$\begin{cases} Z_t(h_t) = W_t(h_t), \\ Q_t^{\Pi}(h_t) = Q_t^{\Pi'}(h_t), \\ \varrho_{t+1,T}^{\Pi}(Z_{t+1},\ldots,Z_T)(h_t,\cdot) = \varrho_{t+1,T}^{\Pi'}(W_{t+1},\ldots,W_T)(h_t,\cdot). \end{cases}$$

Then we have

$$\big(\varrho_{t+1,T}^{\Pi}(Z_{t+1},\ldots,Z_T)\big|H_t^{\Pi} = h_t\big) \overset{\mathcal{D}}{\sim} \big(\varrho_{t+1,T}^{\Pi'}(W_{t+1},\ldots,W_T)\big|H_t^{\Pi'} = h_t\big),$$

where the relation $\overset{\mathcal{D}}{\sim}$ means the equality in law. Because of the stochastic conditional time consistency,

$$\varrho_{t,T}^{\Pi}(Z_t,\ldots,Z_T)(h_t) = \varrho_{t,T}^{\Pi'}(W_t,\ldots,W_T)(h_t),$$

whence

$$\sigma_t^{\Pi}\big(h_t, Q_t^{\Pi}(h_t), \varrho_{t+1,T}^{\Pi}(Z_{t+1},\ldots,Z_T)(h_t,\cdot)\big) \\ = \sigma_t^{\Pi'}\big(h_t, Q_t^{\Pi'}(h_t), \varrho_{t+1,T}^{\Pi'}(W_{t+1},\ldots,W_T)(h_t,\cdot)\big).$$

This proves that in fact σ^{Π} does not depend on Π directly, and all dependence on Π is carried by the controlled kernel Q_t^{Π}. This is a highly desirable property, when we apply dynamic risk measures to a control problem. We summarize this important observation in the following theorem, which extends Theorem 8.3 to the case of controlled processes.

Theorem 8.10. *A family of dynamic risk measures* $\{\varrho_{t,T}^{\Pi}\}_{t=1,\dots,T}^{\Pi\in\varPi}$ *is normalized, has the translation property, and is stochastically conditionally time-consistent, if and only if transition risk mappings* $\sigma_t : \mathcal{X}^t \times \mathcal{P}(\mathcal{X}) \times \mathcal{V} \to \mathbb{R}$, $t = 1 \dots T-1$, *exist such that*

(i) *For all* $t = 1, \dots, T-1$ *and all* $h_t \in \mathcal{X}^t$, $\sigma_t(h_t, \cdot, \cdot)$ *is a normalized and law-invariant risk form*[†] *on* $(\mathcal{P}(\mathcal{X}), \mathcal{V})$ *which has the following property of monotonicity with respect to stochastic dominance:*

$$\forall q^1, q^2 \in \{ Q_t^{\Pi}(h_t) : \Pi \in \varPi \},\ \forall v^1, v^2 \in \mathcal{V},$$
$$(v^1; q^1) \preceq_{(1)} (v^2; q^2) \implies \sigma_t(h_t, q^1, v^1) \le \sigma_t(h_t, q^2, v^2),$$

where $(v; q) = q \circ v^{-1}$ *means "the distribution of* v *under* q*;"*

(ii) *For all* $\Pi \in \varPi$, *for all* $t = 1, \dots, T-1$, *for all* $(Z_t, \dots, Z_T) \in \mathcal{Z}_{t:T}$, *and for all* $h_t \in \mathcal{X}^t$,

$$\varrho_{t,T}^{\Pi}(Z_t, \dots, Z_T) = Z_t(h_t) + \sigma_t\big(h_t, Q_t^{\Pi}(h_t), \varrho_{t+1,T}^{\Pi}(Z_{t+1}, \dots, Z_T)(h_t, \cdot)\big). \quad (8.12)$$

Moreover, for all $t = 1, \dots, T-1$, *for every* $h_t \in \mathcal{X}^t$, *for every* $\Pi \in \varPi$, *and for every* $v \in \mathcal{V}$,

$$\sigma_t(h_t, Q_t^{\Pi}(h_t), v) = \varrho_{t,T}^{\Pi}(0, V, 0, \dots, 0)(h_t), \quad (8.13)$$

where $V \in \mathcal{Z}_{t+1}$ *satisfies the equation* $V(h_t, \cdot) = v(\cdot)$, *and can be arbitrary elsewhere.*

Proof. We have shown the existence of $\{\sigma_t\}_{t=1,\dots,T}$ satisfying (8.12) and (8.13) in the discussion preceding the theorem. We can verify the monotonicity with respect to stochastic dominance by (8.13) and Definition 8.8. □

It follows that the transition risk mappings of Examples 8.5 and 8.6 are suitable for controlled processes as well.

[†] In this chapter, for notational convenience, we list the two arguments of a risk form in the reverse order.

8.1.3 Markov Risk Measures

Our results can be refined in the case when $\{X_t\}$ is a controlled Markov system, in which we assume the following conditions:

(i) The admissible control sets are measurable multifunctions of the current state, that is, $\mathcal{U}_t : \mathcal{X} \rightrightarrows \mathfrak{U}, \quad t = 1, \dots, T$;

(ii) The dependence in the transition kernel (8.9) on the history is carried only through the last state and control, that is, $Q_t : \text{graph}(\mathcal{U}_t) \to \mathcal{P}(\mathcal{X})$, $t = 1, \dots, T-1$;

(iii) The stage-wise costs are only dependent on the current state and control: $Z_t = c_t(X_t, U_t)$, $t = 1, \dots, T$, where $c_t : \text{graph}(\mathcal{U}_t) \to \mathbb{R}$, $t = 1, \dots, T$, are measurable bounded functions.[†]

Let $\mathit{\Pi}$ be the set of admissible history-dependent policies:

$$\mathbb{\Pi} = \big\{ \Pi = (\pi_1, \dots, \pi_T) : \pi_t(x_1, \dots, x_t) \in \mathcal{U}_t(x_t),\ t = 1, \dots, T \big\}.$$

To alleviate notation, for all $\Pi \in \mathbb{\Pi}$ and for all measurable $c = (c_1, \dots, c_T)$, we write

$$v_t^{c,\Pi}(h_t) \triangleq \varrho_{t,T}^{\Pi}\big(c_t(X_t, \pi_t(H_t)), \dots, c_T(X_T, \pi_T(H_T))\big)(h_t).$$

The following result is a direct consequence of Theorem 8.10 in the Markovian case.

Corollary 8.11. *For a controlled Markov system, a family of dynamic risk measures* $\big\{\varrho_{t,T}^{\Pi}\big\}_{t=1,\dots,T}^{\Pi \in \mathbb{\Pi}}$ *is normalized, has the translation property, and is stochastically conditionally time-consistent, if and only if functionals* $\sigma_t : \mathcal{X}^t \times \mathcal{P}(\mathcal{X}) \times \mathcal{V} \to \mathbb{R}$, $t = 1 \dots T-1$, *exist, such that*

(i) *For all* $t = 1, \dots, T-1$ *and all* $h_t \in \mathcal{X}^t$, $\sigma_t(h_t, \cdot, \cdot)$ *is a normalized risk form which is strongly monotonic with respect to stochastic dominance on* $\big\{ Q_t(x_t, u) : u \in \mathcal{U}_t(x_t) \big\}$;

(ii) *For all* $\Pi \in \mathbb{\Pi}$, *for all bounded measurable* c, *for all* $t = 1, \dots, T-1$, *and for all* $h_t \in \mathcal{X}^t$,

$$v_t^{c,\Pi}(h_t) = c_t(x_t, \pi_t(h_t)) + \sigma_t\big(h_t, Q_t(x_t, \pi_t(h_t)), v_{t+1}^{c,\Pi}(h_t, \cdot)\big). \tag{8.14}$$

Proof. To verify the "if and only if" statement, we can show that (8.13) is true if σ_t satisfies (8.14) for all measurable bounded c. □

Consider a Markov policy Π composed of state-dependent measurable decision rules $\pi_t : \mathcal{X} \to \mathfrak{U}$, $t = 1, \dots, T$. Because of the Markov property of the transition kernels, for a Markov policy Π, the future evolution of the process

[†] The case of stage-wise costs depending on the next state is discussed in § 8.1.5.

$\{X_\tau\}_{\tau=t,\dots,T}$ is solely dependent on the current state x_t, so is the distribution of the future costs $c_\tau(X_\tau, \pi_\tau(X_\tau))$, $\tau = t, \dots, T$. Therefore, it is reasonable to assume that the dependence of the conditional risk measure on the history is also carried by the current state only.

Definition 8.12. A family of dynamic risk measures $\{\varrho^{\Pi}_{t,T}\}^{\Pi \in \mathit{\Pi}}_{t=1,\dots,T}$ for a controlled Markov system is *Markov* if for all Markov policies $\Pi \in \mathit{\Pi}$, for all measurable $c = (c_1, \dots, c_T)$, and for all $h_t = (x_1, \dots, x_t)$ and $h'_t = (x'_1, \dots, x'_t)$ in $\mathcal{X}^t$ such that $x_t = x'_t$, we have $v^{c,\Pi}_t(h_t) = v^{c,\Pi}_t(h'_t)$.

Proposition 8.13. *Under the translation property and stochastic conditional time consistency,* $\{\varrho^{\Pi}_{t,T}\}^{\Pi \in \mathit{\Pi}}_{t=1,\dots,T}$ *is Markov if and only if the dependence of* σ_t *on* h_t *is carried only by* x_t, *for all* $t = 1, \dots, T-1$.

Proof. Definition 8.12 implies that for all $t = 1, \dots, T-1$, for all $h_t, h'_t \in \mathcal{X}^t$ such that $x_t = x'_t$, for all $u \in \mathcal{U}_t(x_t)$ and for all $v \in \mathcal{V}$, a Markov $\Pi \in \mathit{\Pi}$ exists such that $\pi_t(x_t) = u$. By setting $c = (0, \dots, 0, c_{t+1}, 0, \dots, 0)$ with $c_{t+1} : (x', u') \mapsto v(x')$, we obtain

$$\sigma_t(h_t, Q_t(x_t, u), v) = v^{c,\Pi}_t(h_t) = v^{c,\Pi}_t(h'_t) = \sigma_t(h'_t, Q_t(x_t, u), v).$$

Therefore, σ_t is indeed memoryless, that is, its dependence on h_t is carried by x_t only.

If σ_t, $t = 1, \dots, T-1$, are all memoryless, we can prove by induction backward in time that for all $t = T, \dots, 1$, $v^{c,\Pi}_t(h_t) = v^{c,\Pi}_t(h'_t)$ for all Markov π and all $h_t, h'_t \in \mathcal{X}^t$ such that $x_t = x'_t$. □

Theorem 8.14. *For a controlled Markov system, a family of dynamic risk measures* $\{\varrho^{\Pi}_{t,T}\}^{\Pi \in \mathit{\Pi}}_{t=1,\dots,T}$ *has the translation property, is stochastically conditionally time-consistent, and is Markov, if and only if functionals*

$$\sigma_t : \mathcal{X} \times \mathcal{P}(\mathcal{X}) \times \mathcal{V} \to \mathbb{R}, \quad t = 1 \dots T-1,$$

exist, such that:

(i) *For all* $t = 1, \dots, T-1$ *and all* $x \in \mathcal{X}$, $\sigma_t(x, \cdot, \cdot)$ *is a normalized risk form which is monotonic with respect to stochastic dominance on* $\{ Q_t(x, u) : u \in \mathcal{U}_t(x) \}$;

(ii) *For all* $\Pi \in \mathit{\Pi}$, *for all measurable bounded* c, *for all* $t = 1, \dots, T-1$, *and for all* $h_t \in \mathcal{X}^t$,

$$v^{c,\Pi}_t(h_t) = c_t(x_t, \pi_t(h_t)) + \sigma_t\big(x_t, Q_t(x_t, \pi_t(h_t)), v^{c,\Pi}_{t+1}(h_t, \cdot)\big). \tag{8.15}$$

Theorem 8.14 provides us with a simple recursive formula (8.15) for the evaluation of the risk of a Markov policy Π:

$$v_T^{c,\Pi}(x) = c_T(x, \pi_T(x)), \quad x \in \mathcal{X},$$

and for $t = T-1, \ldots, 1$,

$$v_t^{c,\Pi}(x) = c_t(x, \pi_t(x)) + \sigma_t\big(x, Q_t(x, \pi_t(x)), v_{t+1}^{c,\Pi}\big), \quad x \in \mathcal{X}.$$

It involves the calculation of the policy value functions $v_t^{c,\Pi}(\cdot)$ on the state space $\mathcal{X}$.

8.1.4 Dynamic Programming

In this section, we fix the cost functions $c_1, \ldots, c_T$ and consider a family of dynamic risk measures $\{\varrho_{t,T}^{\Pi}\}_{t=1,\ldots,T}^{\Pi\in\mathit{\Pi}}$ which is normalized, has the translation property, is stochastically conditionally time-consistent, and is Markovian. Our objective is to analyze the risk minimization problem:

$$\min_{\Pi\in\mathit{\Pi}} v_1^{\Pi}(x_1), \quad x_1 \in \mathcal{X}. \tag{8.16}$$

For this purpose, we introduce the family of *value functions*:

$$v_t^*(h_t) = \inf_{\Pi\in\mathit{\Pi}_{t,T}(h_t)} v_t^{\Pi}(h_t), \quad t = 1, \ldots, T, \quad h_t \in \mathcal{X}^t, \tag{8.17}$$

where $\mathit{\Pi}_{t,T}(h_t)$ is the set of feasible deterministic policies $\Pi = \{\pi_t, \ldots, \pi_T\}$. As stated in Theorem 8.14, transition risk mappings $\sigma_t(\cdot,\cdot,\cdot)$ exist, such that

$$v_t^{\Pi}(h_t) = c_t(x_t, \pi_t(h_t)) + \sigma_t\big(x_t, Q_t(x_t, \pi_t(h_t)), v_{t+1}^{\Pi}(h_t, \cdot)\big),$$
$$t = 1, \ldots, T-1,\ \Pi \in \mathit{\Pi},\ h_t \in \mathcal{X}^t. \tag{8.18}$$

Our intention is to prove that the value functions $v_t^*(\cdot)$ are *memoryless*, that is, for all $h_t = (x_1, \ldots, x_t)$ and $h_t' = (x_1', \ldots, x_t')$ such that $x_t = x_t'$, we have $v_t^*(h_t) = v_t^*(h_t')$. In this case, with a slight abuse of notation, we shall simply write $v_t^*(x_t)$.

To formulate the main result of this subsection, we equip the space $\mathcal{P}(\mathcal{X})$ of probability measures on $\mathcal{X}$ with the topology of weak convergence.

Theorem 8.15. *In addition to the general assumptions, we assume the following conditions:*

(i) *The transition kernels $Q_t(\cdot,\cdot)$, $t = 1, \ldots, T$, are continuous;*

(ii) *For every lower semicontinuous $v \in \mathcal{V}$ the transition risk mappings $\sigma_t(\cdot,\cdot,v)$, $t = 1, \ldots, T$, are lower semicontinuous;*

(iii) *The functions $c_t(\cdot,\cdot)$, $t = 1, \ldots, T$, are lower semicontinuous;*

(iv) *The multifunctions* $\mathcal{U}_t(\cdot)$, $t = 1, \ldots, T$, *are measurable, compact-valued, and upper semicontinuous.*

Then the functions $v_t^*(\cdot)$, $t = 1, \ldots, T$, *are measurable, memoryless, lower semicontinuous, and satisfy the following dynamic programming equations:*

$$v_T^*(x) = \min_{u \in \mathcal{U}_T(x)} c_T(x, u), \quad x \in \mathcal{X},$$

and for $t = T - 1, \ldots, 1$,

$$v_t^*(x) = \min_{u \in \mathcal{U}_t(x)} \left\{ c_t(x, u) + \sigma_t\big(x, Q_t(x, u), v_{t+1}^*\big) \right\}, \quad x \in \mathcal{X}.$$

Moreover, an optimal Markov policy $\Pi^* = \{\pi_1^*, \ldots, \pi_T^*\}$ *exists and satisfies the equations:*

$$\begin{aligned} \pi_T^*(x) &\in \operatorname*{arg\,min}_{u \in \mathcal{U}_T(x)} c_T(x, u), \quad x \in \mathcal{X}, \\ \pi_t^*(x) &\in \operatorname*{arg\,min}_{u \in \mathcal{U}_t(x)} \left\{ c_t(x, u) + \sigma_t\big(x, Q_t(x, u), v_{t+1}^*\big) \right\}, \\ & \qquad x \in \mathcal{X}, \quad t = T - 1, \ldots, 1. \end{aligned}$$

Proof. We prove the memoryless property of $v_t^*(\cdot)$ and construct the optimal Markov policy by induction backward in time. For all $h_T \in \mathcal{X}^T$ we have

$$v_T^*(h_T) = \inf_{\Pi \in \boldsymbol{\Pi}} c_T(x_T, \pi_T(h_T)) = \inf_{u \in \mathcal{U}_T(x_T)} c_T(x_T, u). \tag{8.19}$$

Since $c_T(\cdot, \cdot)$ is lower semicontinuous, it is a *normal integrand*, that is, its epigraphical mapping $x \mapsto \{(u, \alpha) \in \mathfrak{U} \times \mathbb{R} : c_T(x, u) \le \alpha\}$ is closed-valued and measurable. Due to assumption (iv), the mapping

$$\bar{c}_T(x, u) = \begin{cases} c_T(x, u) & \text{if } u \in \mathcal{U}_T(x), \\ +\infty & \text{otherwise}, \end{cases}$$

is a normal integrand as well. Thus, the infimum in (8.19) is attained and is a measurable function of x_T. Hence, $v_T^*(\cdot)$ is measurable and memoryless. By assumptions (iii) and (iv) and the Berge theorem, it is also lower semicontinuous. Moreover, the optimal solution mapping $\Psi_T(x) = \big\{u \in \mathcal{U}_T(x) : c_T(x, u) = v_T^*(x)\big\}$ is measurable and has nonempty and closed values. Therefore, a measurable selector π_T^* of Ψ_T exists.

Suppose $v_{t+1}^*(\cdot)$ is measurable, memoryless, and lower semicontinuous, and Markov decision rules $\{\pi_{t+1}^*, \ldots, \pi_T^*\}$ exist such that

$$v_{t+1}^*(x_{t+1}) = v_{t+1}^{\{\pi_{t+1}^*, \ldots, \pi_T^*\}}(x_{t+1}), \quad \forall\, h_{t+1} \in \mathcal{X}^{t+1}.$$

Then for any $h_t \in \mathcal{X}^t$ we have

$$\begin{aligned} v_t^*(h_t) &= \inf_{\Pi \in \Pi} v_t^{\Pi}(h_t) \\ &= \inf_{\Pi \in \Pi} \Big\{ c_t(x_t, \pi_t(h_t)) + \sigma_t\big(x_t, Q_t(x_t, \pi_t(h_t)), v_{t+1}^{\Pi}(h_t, \cdot)\big) \Big\}. \end{aligned}$$

On the one hand, since $v_{t+1}^{\Pi}(h_t, \cdot) \geq v_{t+1}^*(\cdot)$ and σ_t is non-decreasing with respect to the last argument, we obtain

$$\begin{aligned} v_t^*(h_t) &\geq \inf_{\Pi \in \Pi} \big\{ c_t(x_t, \pi_t(h_t)) + \sigma_t\big(x_t, Q_t(x_t, \pi_t(h_t)), v_{t+1}^*\big) \big\} \\ &= \inf_{u \in \mathcal{U}_t(x_t)} \big\{ c_t(x_t, u) + \sigma_t\big(x_t, Q_t(x_t, u), v_{t+1}^*\big) \big\}. \end{aligned} \tag{8.20}$$

By assumptions (i)–(iii), the mapping $(x, u) \mapsto c_t(x, u) + \sigma_t(x, Q_t(x, u), v_{t+1}^*)$ is lower semicontinuous. Invoking assumption (iv) again, exactly as in the case of $t = T$, we conclude that the optimal solution mapping

$$\begin{aligned} \Psi_t(x) = \Big\{ u \in \mathcal{U}_t(x) : c_t(x, u) + \sigma_t\big(x, Q_t(x, u), v_{t+1}^*\big) \\ = \inf_{u \in \mathcal{U}_t(x)} \big\{ c_t(x, u) + \sigma_t\big(x, Q_t(x, u), v_{t+1}^*\big) \big\} \Big\} \end{aligned}$$

is measurable and has nonempty and closed values; hence, a measurable selector π_t^* of Ψ_t exists. Substitution of this selector into (8.20) yields

$$\begin{aligned} v_t^*(h_t) \geq c_t(x_t, \pi_t^*(x_t)) + \sigma_t\big(x_t, Q_t(x_t, \pi_t^*(x_t)), v_{t+1}^{\{\pi_{t+1}^*, \dots, \pi_T^*\}}\big) \\ = v_t^{\{\pi_t^*, \dots, \pi_T^*\}}(x_t). \end{aligned}$$

In the last equation, we used (8.18) and the fact that the decision rules $\pi_t^*, \dots, \pi_T^*$ are Markov.

On the other hand,

$$v_t^*(h_t) = \inf_{\Pi \in \Pi} v_t^{\Pi}(h_t) \leq v_t^{\{\pi_t^*, \dots, \pi_T^*\}}(x_t).$$

Therefore, $v_t^*(h_t) = v_t^{\{\pi_t^*, \dots, \pi_T^*\}}(x_t)$ is measurable, memoryless, and

$$\begin{aligned} v_t^*(x_t) &= \min_{u \in \mathcal{U}_t(x_t)} \big\{ c_t(x_t, u) + \sigma_t\big(x_t, Q_t(x, u), v_{t+1}^*\big) \big\} \\ &= c_t(x_t, \pi_t^*(x_t)) + \sigma_t\big(x_t, Q_t(x_t, \pi_t^*(x_t)), v_{t+1}^*\big). \end{aligned}$$

By assumptions (ii), (iii), (iv), and the Berge theorem, $v_t^*(\cdot)$ is lower semicontinuous. This completes the induction step. □

Let us verify the weak lower semicontinuity assumption (ii) of the mean–semideviation transition risk mapping of Example 8.6. To make the mapping Markovian, we assume that the parameter κ depends on x only, that is,

$$\sigma(x,q,v) = \int_{\mathcal{X}} v(s)\, q(\mathrm{d}s) \\ + \kappa(x)\left(\int_{\mathcal{X}}\left[\left(v(s) - \int_{\mathcal{X}} v(s')\, q(\mathrm{d}s')\right)_+\right]^p q(\mathrm{d}s)\right)^{1/p}. \tag{8.21}$$

As before, $p \in [1,\infty)$. For simplicity, we skip the subscript t of σ and κ.

Lemma 8.16. *Suppose $\kappa(\cdot)$ is continuous. Then for every lower semicontinuous function v, the mapping $(x,q) \mapsto \sigma(x,q,v)$ in (8.21) is lower semicontinuous.*

Proof. Let $q_k \to q$ weakly and $x_k \to x$. For all $s \in \mathcal{X}$ we have the inequality

$$0 \le \left[v(s) - \int_{\mathcal{X}} v(s')\, q(\mathrm{d}s')\right]_+ \\ \le \left[v(s) - \int_{\mathcal{X}} v(s')\, q_k(\mathrm{d}s')\right]_+ + \left[\int_{\mathcal{X}} v(s')\, q_k(\mathrm{d}s') - \int_{\mathcal{X}} v(s')\, q(\mathrm{d}s')\right]_+.$$

By the triangle inequality for the norm in $\mathcal{L}_p(\mathcal{X}, \mathcal{B}(\mathcal{X}), q_k)$,

$$\left(\int_{\mathcal{X}}\left[v(s) - \int_{\mathcal{X}} v(s')\, q(\mathrm{d}s')\right]_+^p q_k(\mathrm{d}s)\right)^{1/p} \\ \le \left(\int_{\mathcal{X}}\left[v(s) - \int_{\mathcal{X}} v(s')\, q_k(\mathrm{d}s')\right]_+^p q_k(\mathrm{d}s)\right)^{1/p} \\ + \left[\int_{\mathcal{X}} v(s')\, q_k(\mathrm{d}s') - \int_{\mathcal{X}} v(s')\, q(\mathrm{d}s')\right]_+.$$

Adding $\int_{\mathcal{X}} v(s)\, q(\mathrm{d}s)$ to both sides, we obtain

$$\int_{\mathcal{X}} v(s)\, q(\mathrm{d}s) + \left(\int_{\mathcal{X}}\left[v(s) - \int_{\mathcal{X}} v(s')\, q(\mathrm{d}s')\right]_+^p q_k(\mathrm{d}s)\right)^{1/p} \\ \le \left(\int_{\mathcal{X}}\left[v(s) - \int_{\mathcal{X}} v(s')\, q_k(\mathrm{d}s')\right]_+^p q_k(\mathrm{d}s)\right)^{1/p} \\ + \max\left[\int_{\mathcal{X}} v(s)\, q_k(\mathrm{d}s), \int_{\mathcal{X}} v(s)\, q(\mathrm{d}s)\right].$$

By the lower semicontinuity of v and weak convergence of q_k to q, we have

$$\int_{\mathcal{X}} v(s)\, q(\mathrm{d}s) \le \liminf_{k\to\infty} \int_{\mathcal{X}} v(s)\, q_k(\mathrm{d}s),$$

that is, for every $\varepsilon > 0$, we can find k_ε such that for all $k \ge k_\varepsilon$

$$\int_{\mathcal{X}} v(s)\, q_k(\mathrm{d}s) \ge \int_{\mathcal{X}} v(s)\, q(\mathrm{d}s) - \varepsilon.$$

Therefore, for these k we obtain

$$\begin{aligned}
&\int_{\mathcal{X}} v(s)\, q(d\mathrm{d}s) + \Big(\int_{\mathcal{X}} \Big[v(s) - \int_{\mathcal{X}} v(s')\, q(\mathrm{d}s')\Big]_+^p q_k(\mathrm{d}s)\Big)^{1/p}\\
&\quad \le \int_{\mathcal{X}} v(s)\, q_k(\mathrm{d}s) + \Big(\int_{\mathcal{X}} \Big[v(s) - \int_{\mathcal{X}} v(s')\, q_k(\mathrm{d}s')\Big]_+^p q_k(\mathrm{d}s)\Big)^{1/p} + \varepsilon.
\end{aligned}$$

Taking the "lim inf" of both sides and using the weak convergence of q_k to q and the lower semicontinuity of the functions integrated, we conclude that

$$\begin{aligned}
&\int_{\mathcal{X}} v(s)\, q(\mathrm{d}s) + \Big(\int_{\mathcal{X}} \Big[v(s) - \int_{\mathcal{X}} v(s')\, q(\mathrm{d}s')\Big]_+^p q(\mathrm{d}s)\Big)^{1/p}\\
&\qquad \le \liminf_{k\to\infty} \Big\{ \int_{\mathcal{X}} v(s)\, q_k(\mathrm{d}s) +\\
&\qquad\quad \Big(\int_{\mathcal{X}} \Big[v(s) - \int_{\mathcal{X}} v(s')\, q_k(\mathrm{d}s')\Big]_+^p q_k(\mathrm{d}s)\Big)^{1/p} \Big\} + \varepsilon.
\end{aligned}$$

As $\varepsilon > 0$ was arbitrary, the last relation proves the lower semicontinuity of σ in the case when $\kappa(x) \equiv 1$. The case of a continuous $\kappa(x) \in [0, 1]$ can be now easily analyzed by noticing that $\sigma(x, q, v)$ is a convex combination of the expected value, and the risk measure of the last displayed relation:

$$\begin{aligned}
\sigma(x,q,v) &= \big(1-\kappa(x)\big) \int_{\mathcal{X}} v(s)\, q(\mathrm{d}s)\\
&+ \kappa(x) \Big\{ \int_{\mathcal{X}} v(s)\, q(\mathrm{d}s) + \Big(\int_{\mathcal{X}} \Big[v(s) - \int_{\mathcal{X}} v(s')\, q(\mathrm{d}s')\Big]_+^p q(\mathrm{d}s)\Big)^{1/p} \Big\}.
\end{aligned}$$

As both components are lower semicontinuous in (x, q), so is their sum. □

8.1.5 Models with Random Costs

In some applications, it is natural to consider stage-wise costs dependent not only on the current state–control pair but also on the next state. Specifically, we assume that we have a sequence of measurable and bounded functions $c_t : \mathcal{X} \times \mathcal{U} \times \mathcal{X} \to \mathbb{R}$, $t = 1, \dots, T-1$, and a function $c_T : \mathcal{X} \times \mathcal{U} \to \mathbb{R}$, so that the stage-wise cost sequence (adapted to $\{\mathcal{F}_t\}$) has the form:

$$\begin{aligned} Z_1 &= 0, \\ Z_t &= c_{t-1}(X_{t-1}, U_{t-1}, X_t), \quad t = 2, \dots, T, \\ Z_{T+1} &= c_T(X_T, U_T). \end{aligned}$$

We introduce the element Z_{T+1}, which is in fact $\mathcal{F}_T$-measurable, for purely formal reasons. In this case, as in § 8.1.3, it is convenient to consider the "value functions"

$$v_t^{c,\Pi}(h_t) \triangleq \varrho_{t,T}^{\Pi}\big(0, Z_{t+1}, \dots, Z_T + Z_{T+1}\big)(h_t), \quad t = 1, \dots, T.$$

Then Corollary 8.11 remains valid, with the recursive equation (8.14) replaced by

$$\begin{aligned} v_t^{c,\Pi}(h_t) &= \sigma_t\big(h_t, Q_t(x_t, \pi_t(h_t)), c_t(x_t, \pi_t(h_t), \cdot) + v_{t+1}^{c,\Pi}(h_t, \cdot)\big), \\ &\qquad\qquad t = 1, \dots, T-1, \qquad (8.22) \\ v_T^{c,\Pi}(h_T) &= c_T(x_T, u_T). \end{aligned}$$

Furthermore, Definition 8.12 and Proposition 8.13 remain unchanged. In Theorem 8.14, the only difference is in formula (8.15), which now reads

$$v_t^{c,\Pi}(h_t) = \sigma_t\big(x_t, Q_t(x_t, \pi_t(h_t)), c_t(x_t, \pi_t(h_t), \cdot) + v_{t+1}^{c,\Pi}(h_t, \cdot)\big). \quad (8.23)$$

The resulting recursive formula for the evaluation of risk of a Markov policy Π takes on the form:

$$v_T^{c,\Pi}(x) = c_T(x, \pi_T(x)), \quad x \in \mathcal{X},$$

and for $t = T-1, \dots, 1$,

$$v_t^{c,\Pi}(x) = \sigma_t\big(x, Q_t(x, \pi_t(x)), c_t(x, \pi_t(x), \cdot) + v_{t+1}^{c,\Pi}(\cdot)\big), \quad x \in \mathcal{X}.$$

As before, it involves the calculation of the values of functions $v_t^{c,\Pi}(\cdot)$ on the state space $\mathcal{X}$. After all these formal adjustments, we can state an analog of Theorem 8.15.

Theorem 8.17. *In addition to the general assumptions, we assume the following conditions:*

(i) *The transition kernels $Q_t(\cdot,\cdot)$, $t = 1,\dots,T$, are continuous;*

(ii) *For every lower semicontinuous $v \in \mathcal{V}$ the transition risk mappings $\sigma_t(\cdot,\cdot,v)$, $t = 1,\dots,T$, are lower semicontinuous;*

(iii) *The functions $c_t(\cdot,\cdot,\cdot)$, $t = 1,\dots,T-1$, and $c_T(\cdot,\cdot)$ are lower semicontinuous;*

(iv) *The multifunctions $\mathcal{U}_t(\cdot)$, $t = 1,\dots,T$, are measurable, compact-valued, and upper semicontinuous.*

Then the functions $v_t^*(\cdot)$, $t = 1,\dots,T$, are measurable, memoryless, lower semicontinuous, and satisfy the following dynamic programming equations:

$$v_T^*(x) = \min_{u\in\mathcal{U}_T(x)} c_T(x,u), \quad x \in \mathcal{X},$$

and for $t = T-1,\dots,1$,

$$v_t^*(x) = \min_{u\in\mathcal{U}_t(x)} \sigma_t\big(x, Q_t(x,u), c_t(x,u,\cdot) + v_{t+1}^*(\cdot)\big), \quad x \in \mathcal{X}.$$

Moreover, an optimal Markov policy $\Pi^* = \{\pi_1^*,\dots,\pi_T^*\}$ exists and satisfies the equations:

$$\begin{aligned}
\pi_T^*(x) &\in \operatorname*{arg\,min}_{u\in\mathcal{U}_T(x)} c_T(x,u), \quad x \in \mathcal{X},\\
\pi_t^*(x) &\in \operatorname*{arg\,min}_{u\in\mathcal{U}_t(x)} \sigma_t\big(x, Q_t(x,u), c_t(x,u,\cdot) + v_{t+1}^*(\cdot)\big),\\
&\qquad x \in \mathcal{X}, \quad t = T-1,\dots,1.
\end{aligned}$$

The proof is identical, with only minor formal adjustments.

Finally, if the stage-wise costs are random and given by kernels $R_t : \mathcal{X}\times\mathfrak{U} \to \mathcal{P}(\mathbb{R})$, we may consider the cost $C_t \sim R_t(x_t,u_t)$ as a part of the next-stage augmented state (x_{t+1}, C_t), augment the transition kernel to $\bar{Q}_t \triangleq (Q_t, R_t)$, and apply the above theory to this case as well.

8.2 Infinite-Horizon Discounted Models

Our next step is to define an infinite horizon risk-averse model. Let $\{\mathcal{F}_t\}_{t\ge1}$ be a filtration on $(\Omega,\mathcal{F})$, with $\mathcal{F}_1 = \{\Omega,\emptyset\}$, and let $\{Z_t\}_{t=1,2\dots}$ be an adapted sequence of random variables. Similar to the construction in § 8.1.3, we consider the spaces $\mathcal{Z}_t = \mathcal{L}_p(\Omega,\mathcal{F}_t,P)$, $t = 1,2,\dots$, with $p \in [1,\infty)$, and we define the space $\mathcal{Z} = \mathcal{Z}_1 \times \mathcal{Z}_2 \times \cdots$. A sequence $Z \in \mathcal{Z}$ is almost surely bounded, if

$$\sup_t \operatorname{ess\,sup} |Z_t(\omega)| < \infty.$$

Our intention is to evaluate the risk of infinite sequences $\{Z_t\}_{t=1,2,\dots}$. Our idea is to consider a discounted sequence $\{Z_1, \alpha Z_2, \dots, \alpha^{T-1} Z_T\}$, with $\alpha \in (0,1)$, evaluate its risk, and pass to the limit with $T \to \infty$. For this program to succeed, we restrict the measures of risk to those that have coherent transition risk mappings.

8.2.1 Discounted Measures of Risk

Suppose we have a collection of finite-horizon risk measures $\{\varrho_{1,T}\}$, $T = 1, 2, \dots$, which satisfy the relations

$$\varrho_{1,T}(Z_1, \dots, Z_T) = \varrho_{1,T+1}(Z_1, \dots, Z_T, 0), \quad T = 1, 2, \dots. \tag{8.24}$$

If they are time-consistent, normalized, and satisfy the translation property, then Corollary 4.6 yields the formula (4.6) for all $1 \le t \le T$, with a sequence of monotonic one-step conditional risk mappings $\rho_t : \mathcal{Z}_{t+1} \to \mathcal{Z}_t$, $t = 1, 2 \dots$. These one-step mappings are common for all dynamic measures of risk $\varrho_{1,T}$, due to (8.24).

Suppose all one-step conditional risk mappings ρ_t, $t = 1, 2, \dots$, are coherent. Fix the discount factor $\alpha \in (0,1)$. For $T = 1, 2, \dots$ we define the functionals $\varrho^{\alpha}_{1,T} : \mathcal{Z}_1 \times \dots \times \mathcal{Z}_T \to \mathbb{R}$ as follows:

$$\begin{aligned}
\varrho^{\alpha}_{1,T}(Z_1, Z_2, \dots, Z_T) &= \varrho_{1,T}(Z_1, \alpha Z_2, \dots, \alpha^{T-1} Z_T) \\
&= Z_1 + \rho_1\Big(\alpha Z_2 + \rho_2(\alpha^2 Z_3 + \dots + \rho_{T-1}(\alpha^{T-1} Z_T)\cdots)\Big) \\
&= Z_1 + \alpha\rho_1\Big(Z_2 + \alpha\rho_2(Z_3 + \dots + \alpha\rho_{T-1}(Z_T)\cdots)\Big).
\end{aligned} \tag{8.25}$$

They are the same as (4.6) for $t = 1$, but with discounting applied to the sequence $\{Z_t\}$. Finally, we define the functional $\varrho^{\alpha}_{1,\infty} : \mathcal{Z} \to \mathbb{R}$ as

$$\varrho^{\alpha}_{1,\infty}(Z) = \lim_{T\to\infty} \varrho^{\alpha}_{1,T}(Z_1, Z_2, \dots, Z_T). \tag{8.26}$$

We call it the *discounted measure of risk*.

Theorem 8.18. *The discounted measure of risk $\varrho^{\alpha}_{1,\infty}$ is well defined on the set of almost surely bounded sequences $Z \in \mathcal{Z}$, and has the following properties:*

(i) *It is convex;*

(ii) *For all $Z, W \in \mathcal{Z}$, if $Z_t \le W_t$ for all $t = 1, 2, \dots$, then*

$$\varrho^{\alpha}_{1,\infty}(Z) \le \varrho^{\alpha}_{1,\infty}(W);$$

(iii) *For all* $Z \in \mathcal{Z}$, *all* $t = 1, 2 \ldots$, *and all* $W_t \in \mathcal{Z}_t$ *we have*

$$\varrho^{\alpha}_{1,\infty}(Z_1, \ldots, Z_t, Z_{t+1} + W_t, \ldots) = \varrho^{\alpha}_{1,\infty}(Z_1, \ldots, Z_t + \alpha W_t, Z_{t+1}, \ldots);$$

(iv) $\varrho^{\alpha}_{1,\infty}(\beta Z) = \beta \varrho^{\alpha}_{1,\infty}(Z)$, $\forall\, Z \in \mathcal{Z},\ \beta \geq 0$.

Proof. Each functional (8.25) can be regarded as defined on the space $\mathcal{Z}$. We first prove that every $\varrho^{\alpha}_{1,T}(\cdot)$ satisfies the conditions (i)–(iv). For $T = 1$ these conditions hold trivially. Supposing they are satisfied for T, we shall prove them for $T + 1$. We have

$$\begin{aligned} &\varrho^{\alpha}_{1,T+1}(Z_1, Z_2, \ldots, Z_{T+1}) \\ &\quad = Z_1 + \rho_1\Big(\alpha Z_2 + \cdots + \rho_{T-1}\big(\alpha^{T-1} Z_T + \rho_T(\alpha^T Z_{T+1})\big)\cdots\Big) \\ &\quad = \varrho^{\alpha}_{1,T}\big(Z_1, Z_2, \ldots, Z_T + \alpha \rho_T(Z_{T+1})\big). \end{aligned} \tag{8.27}$$

Thus $\varrho^{\alpha}_{1,T+1}(Z_1, Z_2, \ldots, Z_{T+1})$ is a composition of $\varrho^{\alpha}_{1,T}(\cdot)$ and the mapping

$$(Z_1, \ldots, Z_T, Z_{T+1}) \mapsto \big(Z_1, \ldots, Z_T + \alpha \rho_T(Z_{T+1})\big).$$

The first function is convex and nondecreasing, by virtue of (i) and (ii) for T. The second function is convex and nondecreasing, owing to the convexity and monotonicity of the conditional risk mapping $\rho_T(\cdot)$. Therefore, their composition is convex and nondecreasing as well. The positive homogeneity property follows in the same way. Let us verify (iii) for $T + 1$. Observe that when $t \leq T$ we can apply property (iii) for T to the right-hand side of (8.27), and obtain it for $T + 1$ on the left-hand side. It remains to consider the case of $t = T + 1$. By (8.27), the translation property of $\rho_T(\cdot)$, and condition (iv) for $t = T$, we have

$$\begin{aligned} &\varrho^{\alpha}_{1,T+1}(Z_1, Z_2, \ldots, Z_{T+1} + W_t) \\ &\quad = \varrho^{\alpha}_{1,T}(Z_1, Z_2, \ldots, Z_T + \alpha \rho_T(Z_{T+1} + W_T)) \\ &\quad = \varrho^{\alpha}_{1,T}(Z_1, Z_2, \ldots, Z_T + \alpha W_T + \alpha \rho_T(Z_{T+1})). \end{aligned}$$

Comparing this to (8.27) we conclude that (iv) holds true for $T + 1$.

We now prove that for every almost surely bounded $Z \in \mathcal{Z}$ the limit in (8.26) exists. Fix T. Let $C = \max_t \operatorname{ess\,sup} |Z_t(\omega)|$. As $-C \leq Z_{T+1} \leq C$, from axiom (A2) we obtain

$$-C \leq \rho_T(Z_{T+1}) \leq C. \tag{8.28}$$

Using this in (8.27), we get

$$\begin{aligned} \varrho^{\alpha}_{1,T}(Z_1, \ldots, Z_T - \alpha C) \leq \varrho^{\alpha}_{1,T+1}(Z_1, \ldots, Z_T, Z_{T+1}) \\ \leq \varrho^{\alpha}_{1,T}(Z_1, \ldots, Z_T + \alpha C). \end{aligned}$$

We can now apply property (iii), which we have just proved for $\varrho^{\alpha}_{1,T}$, to both sides of the last inequality. We do it $T-1$ times moving the constants $\pm\alpha C$ to the first argument, and we conclude that

$$\begin{aligned}\varrho^{\alpha}_{1,T}(Z_1,\dots,Z_T)-\alpha^T C \le \varrho^{\alpha}_{1,T+1}(Z_1,\dots,Z_T,Z_{T+1})\\ \le \varrho^{\alpha}_{1,T}(Z_1,\dots,Z_T)+\alpha^T C.\end{aligned}$$

This implies that the sequence in (8.26) is a Cauchy sequence. As the functions $\varrho^{\alpha}_{1,T}(\cdot)$, $T=1,2,\dots$, satisfy conditions (i)–(iv), the limit function $\varrho^{\alpha}_{1,\infty}(\cdot)$ satisfies these conditions as well. □

Remark 8.19. The derivations in this section are the main reason for focusing on coherent conditional risk mappings. Absent coherence, the only relevant conditional risk measures that allow for a meaningful infinite-horizon composition for discounted costs are the entropic measures of Example 8.5. The corresponding discounted measure of risk takes on the form:

$$\begin{aligned}\varrho^{\alpha}_{1,\infty}(Z_1,Z_2,\dots) &= \lim_{T\to\infty}\frac{1}{\kappa}\ln\Big(\mathbb{E}\Big[\exp\Big(\kappa\textstyle\sum_{t=1}^{T}\alpha^{t-1}Z_t\Big]\Big)\\ &= \frac{1}{\kappa}\ln\Big(\mathbb{E}\Big[\exp\Big(\kappa\textstyle\sum_{t=1}^{\infty}\alpha^{t-1}Z_t\Big]\Big), \quad \varkappa>0.\end{aligned}$$

For bounded costs $\{Z_t\}$, the limit is well-defined. This approach is studied in the theory of *risk-sensitive* Markov Decision Processes, with the use of highly specialized techniques that do not extend to other measures of risk.

For general convex conditional risk mappings $\rho_t(\cdot)$, one could directly define, instead of the derivation in (8.25), the functional

$$\tilde{\varrho}^{\alpha}_{1,T}(Z_1,Z_2,\dots,Z_T) = Z_1+\alpha\rho_1\Big(Z_2+\alpha\rho_2\big(Z_3+\cdots+\alpha\rho_{T-1}(Z_T)\cdots\big)\Big).$$

Then our considerations would go through, formally, with minor adjustments. We decided not to follow this path because it diverges from the common understanding of discounting. □

8.2.2 Dynamic Programming Equations. Value Iteration

Consider now the application of the discounted risk measure (8.26) to a controlled Markov process $\{x_t\}$, $t=1,2,\dots$.

We assume that the controlled Markov model is *stationary*, that is, there exist a control set $\mathcal{U}:\mathcal{X}\rightrightarrows\mathfrak{U}$, a controlled transition kernel $Q:\text{graph}(\mathcal{U})\to\mathcal{P}(\mathcal{X})$, and a cost function $c : \text{graph}(\mathcal{U}) \to \mathbb{R}$, such that $\mathcal{U}_t = \mathcal{U}$, $Q_t = Q$, and $c_t = c$, for $t = 1, 2, \ldots$.

We also make the following assumption about the family of dynamic risk measures $\left\{\varrho_{t,T}^{\Pi}\right\}_{1\le t\le T}^{\Pi\in\varPi}$, which we use to construct the discounted risk measures.

Assumption 8.20. *The dynamic risk measures* $\left\{\varrho_{t,T}^{\Pi}\right\}_{1\le t\le T}^{\Pi\in\varPi}$ *are normalized, have the translation property, are stochastically conditionally time-consistent, and Markov, for all* $T = 1, 2, \ldots$. *Furthermore, they share the same transition risk mapping* $\sigma(\cdot,\cdot,\cdot)$ *such that for all* $x \in \mathcal{X}$ *and all* $\mu \in \mathcal{P}(\mathcal{X})$ *the function* $\sigma(x, \mu, \cdot)$ *is a coherent measure of risk on* $\mathcal{V}$, *with the corresponding probability space* $(\mathcal{X}, \mathcal{B}(\mathcal{X}), \mu)$.

Each policy $\Pi = \left\{\pi_t\right\}_{t=1}^{\infty}$ results in a cost sequence $Z_t = c(X_t, U_t)$, $t = 1, 2, \ldots$. We denote, for every $t = 1, 2, \ldots$, for every history $h_t \in H_t$, and for every $T \ge t$,

$$v_{t,T}^{\Pi}(h_t) = \varrho_{t,T}^{\Pi}\big(\alpha^{t-1}c(x_t, u_t), \alpha^t c(X_{t+1}, U_{t+1}), \ldots, \alpha^{T-1}c(X_T, U_T)\big),$$

and define

$$v_{t,\infty}^{\Pi}(h_t) = \lim_{T\to\infty} v_{t,T}^{\Pi}(h_t). \tag{8.29}$$

As discussed in the previous section, the limit in (8.29) exists if the costs are uniformly bounded, and Assumption 8.20 is satisfied. Moreover, for a transition risk mapping $\sigma : \mathcal{X} \times \mathcal{P}(\mathcal{X}) \times \mathcal{V} \to \mathbb{R}$, $t = 1, 2, \ldots$, where $\mathcal{V}$ is the set of bounded measurable functions on $\mathcal{X}$, we have:

$$v_{t,T}^{\Pi}(h_t) = \alpha^{t-1}c(x_t, \pi_t(h_t)) + \sigma\big(x_t, Q(x_t, \pi_t(h_t)), v_{t+1,T}^{\Pi}(h_t, \cdot)\big). \tag{8.30}$$

We are interested in the infinite horizon problem

$$\min_{\Pi\in\varPi} v_{1,\infty}^{\Pi}(x_1). \tag{8.31}$$

In order to analyze this problem we need to make several assumptions. It is most convenient to state these assumptions within the formulation of our main result. Its proof will be deferred until after several technical results.

As in § 8.1.4, we equip the space $\mathcal{P}(\mathcal{X})$ of probability measures on $(\mathcal{X}, \mathcal{B}(\mathcal{X}))$ with the topology of weak convergence.

Theorem 8.21. *Suppose Assumption* 8.20 *and the following conditions are satisfied:*

(i) *The transition kernel* $Q(\cdot,\cdot)$ *is continuous;*

(ii) *For every lower semicontinuous* $v \in \mathcal{V}$, *the transition risk mapping* $\sigma(\cdot,\cdot,v)$ *is lower semicontinuous;*

(iii) *The function* $c(\cdot,\cdot)$ *is bounded and lower semicontinuous;*

(iv) *The multifunction* $\mathcal{U}(\cdot)$ *is measurable, compact-valued, and upper semicontinuous.*

Then problem (8.31) *has an optimal solution and its optimal value* $v^*(x)$*, as a function of the initial state* $x_1 = x$*, is lower semicontinuous and satisfies the following dynamic programming equation:*

$$v(x) = \min_{u \in \mathcal{U}(x)} \big\{c(x,u) + \alpha\sigma\big(x, Q(x,u), v\big)\big\}, \quad x \in \mathcal{X}. \tag{8.32}$$

Moreover, an optimal stationary Markov policy $\Pi^* = \{\pi^*, \pi^*, \ldots\}$ *exists and satisfies the equation:*

$$\pi^*(x) \in \operatorname*{arg\,min}_{u \in \mathcal{U}(x)} \big\{c(x,u) + \alpha\sigma\big(x, Q(x,u), v^*\big)\big\}, \quad x \in \mathcal{X}. \tag{8.33}$$

Conversely, every solution of the equation (8.32) *is the optimal value of problem* (8.31)*, and every measurable solution of* (8.33) *defines an optimal stationary Markov policy.*

If C is the uniform upper bound on $|c(\cdot,\cdot)|$, then from Theorem 8.18(ii)-(iii) we obtain the following uniform bound on $\hat{v}(x)$:

$$\hat{v}(x) = \varrho^{\alpha}_{1,\infty}(Z_1, Z_2, Z_3, \ldots) \le \varrho^{\alpha}_{1,\infty}(C, C, C, \ldots) = \frac{C}{1-\alpha}.$$

Thus the optimal value $v^*(\cdot)$ is well-defined and is uniformly bounded. By the same token, all limits in (8.29) exist.

In order to prove Theorem 8.21, we need to establish several auxiliary results. For a measurable selector $\pi : \mathcal{X} \to \mathfrak{U}$ of $\mathcal{U}(\cdot)$ we define the operator $\mathfrak{D}_\pi : \mathcal{V} \to \mathcal{V}$ as follows:

$$[\mathfrak{D}_\pi v](x) = c(x, \pi(x)) + \alpha\sigma\big(x, Q(x, \pi(x), v)\big), \quad x \in \mathcal{X}.$$

We also define the operator $\mathfrak{D} : \mathcal{V} \to \mathcal{V}$ by the following relation:

$$[\mathfrak{D}v](x) = \inf_{u \in \mathcal{U}(x)} \big\{c(x,u) + \alpha\sigma\big(x, Q(x,u), v\big)\big\}, \quad x \in \mathcal{X}.$$

With this notation, equation (8.32) reads

$$v = \mathfrak{D}v, \tag{8.34}$$

and thus the properties of the operator $\mathfrak{D}$ are germane to our problem.

Lemma 8.22. *The operators* $\mathfrak{D}_\pi$ *and* $\mathfrak{D}$ *are nondecreasing in the sense that* $v \ge w$ *implies* $\mathfrak{D}_\pi v \ge \mathfrak{D}_\pi w$ *and* $\mathfrak{D}v \ge \mathfrak{D}w$.

Proof. Both results follow from the fact that $\sigma(x, Q(x,u), \cdot)$ is nondecreasing. □

Recall that $\mathcal{V}$ is the space of bounded measurable functions on $(\mathcal{X}, \mathcal{B}(\mathcal{X}))$. For $v \in \mathcal{V}$, we write $\|v\|_\infty = \sup_{x\in\mathcal{X}} |v(x)|$.

Lemma 8.23. *The operators $\mathfrak{D}_\pi$ and $\mathfrak{D}$ are contraction mappings with modulus α on the space $\mathcal{V}$, that is, for all $v, w \in \mathcal{V}$ we have*

$$\|\mathfrak{D}_\pi v - \mathfrak{D}_\pi w\|_\infty \le \alpha\|v - w\|_\infty, \tag{8.35}$$

$$\|\mathfrak{D} v - \mathfrak{D} w\|_\infty \le \alpha\|v - w\|_\infty. \tag{8.36}$$

Proof. Since $\sigma(x, Q(x,u), \cdot)$ is coherent, the monotonicity and translation properties yield:

$$\begin{aligned}\sigma(x, Q(x,u), v) &\le \sigma\big(x, Q(x,u), w + \|v - w\|_\infty\big)\\ &= \sigma(x, Q(x,u), w) + \|v - w\|_\infty.\end{aligned} \tag{8.37}$$

Thus, for all $\pi(x) \in U(x)$ we have

$$\mathfrak{D}_\pi v(x) \le \mathfrak{D}_\pi w(x) + \alpha\|v - w\|_\infty, \quad x \in \mathcal{X}. \tag{8.38}$$

Reversing the roles of v and w we obtain (8.35). Inequality (8.36) follows from taking for every $x \in \mathcal{X}$ the infimum of both sides of (8.38) with respect to $\pi(x)$, and then reversing v and w again. □

This allows us to evaluate any stationary Markov policy.

Theorem 8.24. *The value $v^\Pi_{1,\infty}$ of a stationary Markov policy $\Pi = \{\pi, \pi, \dots\}$ is the unique solution of the equation*

$$w = \mathfrak{D}_\pi w. \tag{8.39}$$

Proof. For a deterministic Markov rule π, the resulting system is Markovian, and thus $v^\Pi_{t,T}(\cdot)$ is in fact a function of x_t only, for all $t = 1, 2, \dots, T$. Equation (8.30) yields:

$$v^\Pi_{t,T}(x_t) = \alpha^{t-1}c_t(x_t, \pi(x_t)) + \sigma\big(x_t, Q(x_t, \pi(x_t)), v^\Pi_{t+1,T}(\cdot)\big).$$

Observe that

$$v^\Pi_{t,T}(\cdot) = \alpha^{t-1}v^\Pi_{1,T-t+1}(\cdot), \quad t = 1, 2, \dots, T.$$

These equations imply (with the convention that $v^\Pi_{1,0} \equiv 0$)

$$v^\Pi_{1,T-t+1} = \mathfrak{D}_\pi v^\Pi_{1,T-t}, \quad t = 1, 2, \dots, T,$$

and thus

$$v^\Pi_{1,T} = [\mathfrak{D}_\pi]^T v^\Pi_{1,0}, \quad T = 1, 2, \dots. \tag{8.40}$$

In view of (8.35), by virtue of Banach's contraction mapping theorem, the limit of (8.40) exists and is the unique fixed point of the operator $\mathfrak{D}_\pi$. This proves that v^Π satisfies equation (8.39); it is a *fixed point* of $\mathfrak{D}_\pi$. □

As a by-product of the proof, we obtained an algorithm (8.40) for finding a fixed point of $\mathfrak{D}_\pi$. Evidently, we may start from an arbitrary initial function, instead of $v^\Pi_{1,0} \equiv 0$.

We can now pass to the analysis of the optimal value function.

Lemma 8.25. *If $w \in \mathcal{V}$ is lower semicontinuous, then $\mathfrak{D}w$ is lower semicontinuous, and*

(i) *If $w \geq \mathfrak{D}w$, then $w \geq v^*$;*

(ii) *If $w \leq \mathfrak{D}w$, then $w \leq v^*$.*

Proof. The lower semicontinuity of $\mathfrak{D}w$ can be established exactly as in the proof of Theorem 8.15. We verify the two implications.
(i) Let π be the measurable selector of U for which

$$w \geq \mathfrak{D}_\pi w. \tag{8.41}$$

It exists, due to assumptions (i)–(iv) of Theorem 8.21, exactly as in the proof of Theorem 8.15. Applying the operator $\mathfrak{D}_\pi$ to both sides of inequality (8.41) and using Lemma 8.22, we obtain

$$w \geq [\mathfrak{D}_\pi]^T w, \quad T = 1, 2, \ldots.$$

By virtue of Theorem 8.24, the limit of the right-hand side is the value of the policy Π:

$$w \geq \lim_{T\to\infty} [\mathfrak{D}_\pi]^T w = v^\Pi_{1,\infty}.$$

This proves (i).

(ii) Consider an arbitrary (not necessarily Markov) policy $\Pi = \{\pi_1, \pi_2, \ldots\}$ and the resulting cost sequence $Z_t = c(X_t, U_t)$, $t = 1, \ldots, T$, in a finite horizon problem with the terminal cost $w(X_{T+1})$, We write $u_t = \pi_t(h_t)$, where h_t is the known history of the process up to time t. We have

$$\begin{aligned}
&\varrho^\alpha_{1,T+1}(Z_1, \ldots, Z_T, w(X_{T+1})) \\
&\quad = Z_1 + \rho_1\Big(\alpha Z_2 + \cdots + \rho_{T-1}\big(\alpha^{T-1} Z_T + \rho_T(\alpha^T w(X_{T+1}))\big)\cdots\Big) \\
&\quad = c_1 + \rho_1\Big(\alpha Z_2 + \cdots + \rho_{T-1}\big(\alpha^{T-1}[Z_T + \alpha\rho_T(w(X_{T+1}))]\big)\cdots\Big).
\end{aligned}$$

By assumption (ii), the expression in brackets can be estimated as follows:

$$Z_T + \alpha\rho_T(w(X_{T+1})) \geq [\mathfrak{D}w](X_T) \geq w(X_T).$$

Therefore, $\varrho^{\alpha}_{1,T+1}(Z_1,\ldots,Z_T,w(X_{T+1})) \geq \varrho^{\alpha}_{1,T}(Z_1,\ldots,Z_{T-1},w(X_T))$. Proceeding in this way, we conclude that

$$\varrho^{\alpha}_{1,T+1}(Z_1,\ldots,Z_T,w(X_{T+1})) \geq w(x_1).$$

Let C be an upper bound on $|w(x)|$, $x \in \mathcal{X}$. By property (ii) and (iii) of Theorem 8.18, the cost of policy Π can be estimated as follows:

$$\varrho^{\alpha}_{1,T}(Z_1,\ldots,Z_T) \geq \varrho^{\alpha}_{1,T+1}(Z_1,\ldots,Z_T,w(X_{T+1})) - C\alpha^T.$$

Combining the last two inequalities and passing to the limit with $T \to \infty$ we conclude that for every policy Π

$$v^{\Pi}_{1,\infty}(x_1) \geq w(x_1).$$

Therefore the infimum over all $\Pi \in \mathbb{\Pi}$ is bounded from below by $w(x_1)$, as claimed. □

We are now ready to provide the proof of the main theorem.

Proof of Theorem 8.21. According to Lemma 8.23, the operator $\mathfrak{D}$ is a contraction in $\mathcal{V}$. By Banach's contraction mapping principle, equation (8.34), which is a compact form of (8.32), has a unique solution v in $\mathcal{V}$. This solution can be obtained as the limit of the iterative procedure

$$v^{k+1} = \mathfrak{D}v^k, \quad k = 1,2,\ldots, \tag{8.42}$$

In view of Lemma 8.25, if v^1 is lower semicontinuous, all functions v^k, $k = 1,2,\ldots$ are lower semicontinuous. As the convergence is uniform, the limit v is lower semicontinuous as well. Owing to points (i) and (ii) of Lemma 8.25, $v = v^*$.

Consider the problem on the right-hand side of (8.33). Owing to the assumptions of Theorem 8.21, an optimal solution to this problem exists, exactly in the same way as in the proof of Theorem 8.15. Applying the operator $\mathfrak{D}_{\pi^*}$ to both sides of the equation $v^* = \mathfrak{D}_{\pi^*}v^*$, we obtain

$$v^* = [\mathfrak{D}_{\pi^*}]^T\, v^*, \quad T = 1,2,\ldots.$$

Passing to the limit with $T \to \infty$ and using Theorem 8.24, we conclude that v^* is the cost of the stationary policy $\Pi^* = \{\pi^*,\pi^*,\ldots\}$ in the infinite horizon problem.

Suppose another optimal Markov policy $\Pi = \{\pi,\pi,\ldots\}$ exists, but $v^* \neq \mathfrak{D}_\pi v^*$. By Theorem 8.24, the value $v^{\Pi}_{1,\infty}$ of the policy Π is the unique fixed point of $\mathfrak{D}_\pi$ Thus $v \neq v^*$, a contradiction. □

Our analysis of the dynamic programming equation (8.32) suggests the iterative method (8.42) corresponding to the classical *value iteration* method in risk-neutral dynamic programming. If conditions of Theorem 8.21 are satisfied, then the sequence $\{v^k\}$ generated by the value iteration method is convergent linearly in $\mathcal{V}$ to the optimal value function v^*, with the quotient α. This follows from the contraction property of $\mathfrak{D}$ in $\mathcal{V}$, established in Lemma 8.23, by virtue of Banach's contraction mapping principle.

8.2.3 Policy Iteration

Consider now another approach to solving the dynamic programming equations, extending the risk-neutral *policy iteration method.* In the new method, for $k = 1, 2, \ldots$, given a stationary Markov policy $\Pi^k = \{\pi^k, \pi^k, \ldots\}$, we find the corresponding value function $v^k \in \mathcal{V}$ by solving the equation

$$v(x) = c(x, \pi^k(x)) + \alpha\sigma\big(x, Q(x, \pi^k(x)), v\big), \quad x \in \mathcal{X}. \tag{8.43}$$

Then we find the next decision rule $\pi^{k+1}(\cdot)$ as a measurable function satisfying the relation

$$\pi^{k+1}(x) \in \operatorname*{arg\,min}_{u \in \mathcal{U}(x)} \big\{c(x, u) + \alpha\sigma\big(x, Q(x, u), v^k\big)\big\}, \quad x \in \mathcal{X}. \tag{8.44}$$

It defines the next stationary policy $\Pi^{k+1} = \{\pi^{k+1}, \pi^{k+1}, \ldots\}$. After that, k is increased by 1 and the iteration continues.

To guarantee the existence of the minimizer in (8.44), we select $v^0 \in \mathcal{V}^{\mathrm{lc}}$ and choose π^1 by (8.44) for $k = 0$.

Clearly, equations (8.43)–(8.44) can be compactly written as follows:

$$v^k = \mathfrak{D}_{\pi^k} v^k, \tag{8.45}$$

$$\mathfrak{D}_{\pi^{k+1}} v^k = \mathfrak{D} v^k. \tag{8.46}$$

Theorem 8.26. *Suppose the assumptions of Theorem* 8.21 *are satisfied. Then the sequence of functions* v^k, $k = 1, 2, \ldots$, *is nonincreasing and convergent to the unique bounded solution* $v^*(\cdot)$ *of the dynamic programming equation* (8.32).

Proof. By the same argument as in the proof of Theorem 8.15, all functions $v^k \in \mathcal{V}^{\mathrm{lc}}$ and the minimizers in (8.44) are well defined. It follows from (8.46) and (8.45) that

$$\mathfrak{D}_{\pi^{k+1}} v^k = \mathfrak{D} v^k \le \mathfrak{D}_{\pi^k} v^k = v^k.$$

Using the monotonicity property of the operator $\mathfrak{D}_{\pi^{k+1}}$ (Lemma 8.22), we obtain that

$$\left[\mathfrak{D}_{\pi^{k+1}}\right]^T v^k \le \mathfrak{D}v^k \le v^k, \quad T = 1, 2, \ldots.$$

By Lemma 8.23 and Banach's contraction mapping principle, the left-hand side of this inequality converges to v^{k+1}, as $T \to \infty$. Thus

$$v^{k+1} \le \mathfrak{D}v^k \le v^k. \tag{8.47}$$

Therefore, the sequence $\{v^k\}$ is nonincreasing. Each v^k, as the value of not necessarily optimal policy Π^k, satisfies the inequality $v^k \ge v^*$. From the first part of (8.47) we deduce that

$$0 \le v^{k+1} - v^* \le \mathfrak{D}v^k - v^* = \mathfrak{D}v^k - \mathfrak{D}v^*.$$

Owing to the contraction property of the operator $\mathfrak{D}$ established in Lemma 8.23, we conclude that

$$\|v^{k+1} - v^*\|_\infty \le \alpha \|v^k - v^*\|_\infty.$$

Hence, the sequence $\{v^k\}$ is convergent to v^*. □

If $\mathfrak{D}_{\pi^{k+1}} v^k = v^k$, then it follows from (8.46) that $v^k = \mathfrak{D}v^k$. By virtue of Theorem 8.21, Π^k and Π^{k+1} are optimal policies, and v^k is the optimal value function.

Consider the special case of finitely many possible states and controls. This implies that the number of possible policies is finite as well. Only finitely many different value functions v^k can be generated by the policy iteration method, and they form a nonincreasing sequence. Thus, a value function can be repeated only if it occurs in two consecutive steps, but then we know that it is optimal. Consequently, the policy iteration method is convergent in finitely many steps.

At every step of the policy iteration method, to evaluate the current policy Π^k, we have to solve the nonlinear equation (8.43). Observe that the mapping $v \mapsto \sigma\big(x, Q(x, \pi^k(x)), v\big)$ is nonsmooth, in general. Because of that, equation (8.43) is much harder than the corresponding *linear* equation in the risk-neutral case.

We aim to propose a version of the nonsmooth Newton method for solving this equation. In our case, owing to the special form of equation (8.43), we can use the simplest algorithm with linear auxiliary problems. We can also provide a direct proof of its global and monotonic convergence.

To simplify notation, we suppress the index k of the current iteration of the policy method, and we use ℓ to denote iterations of the Newton method.

Consider equation (8.43), which we write with the use of the dual representation of the transition risk mapping as follows

$$v(x) = \bar{c}(x) + \alpha \sup_{\mu \in \bar{\mathcal{A}}(x)} \langle v, \mu \rangle, \quad x \in \mathfrak{X}, \tag{8.48}$$

with $\bar{c}(x) = c(x, \pi^k(x))$ and $\bar{\mathcal{A}}(x) = \mathcal{A}(x, Q(x, \pi^k(x)))$. Suppose at iteration ℓ of the Newton method we have a certain approximation v_ℓ of the solution of (8.48). We calculate a kernel μ_ℓ by

$$\mu_\ell(x) \in \operatorname*{arg\,max}_{\mu \in \bar{\mathcal{A}}(x)} \langle v_\ell, \mu \rangle, \quad x \in \mathcal{X}.$$

Under the assumptions of Theorem 8.26, the set $\bar{\mathcal{A}}(x)$ is weakly* compact, and thus an optimal kernel μ_ℓ exists. It may not be unique, but this will not play any role in our method. Then we solve the *linear* equation

$$v(x) = \bar{c}(x) + \alpha\langle v, \mu_\ell(x)\rangle, \quad x \in \mathcal{X}. \tag{8.49}$$

We denote its solution by $v_{\ell+1}$, increase ℓ by one, and continue.

Observe that the linear equation (8.49) is the evaluation of the expected cost in the process with transition kernel μ_ℓ, which can be done in the same way as in the policy iteration method for a risk-neutral model. The essence of our method in this case is to construct a sequence of risk-neutral models, for $\ell = 1, 2, \ldots$, to evaluate discounted risk.

Theorem 8.27. *Assume the conditions of Theorem* 8.21. *Then for every initial function v_1 the sequence $\{v_\ell\}$ generated by the Newton method is convergent to the unique solution $\bar{v}$ of equation* (8.48). *Moreover, the sequence is monotonic: $v_{\ell+1} \ge v_\ell$, $\ell = 2, 3, \ldots$.*

Proof. Let us denote by $\mathfrak{M}_\ell$ the affine operator appearing on the right-hand side of equation (8.49):

$$[\mathfrak{M}_\ell v](x) = \bar{c}(x) + \alpha\langle \mu_\ell(x), v\rangle.$$

The operator $\mathfrak{M}_\ell$ is a contraction in $\mathcal{V}$. By virtue of Banach's contraction mapping principle, equation (8.49) has a unique solution $v_{\ell+1} \in \mathcal{V}$.

By construction, for every $v \in \mathcal{V}$, we have

$$\mathfrak{M}_\ell v \le \mathfrak{D}_{\pi^k} v. \tag{8.50}$$

As $\mu_\ell(x)$ is a probability measure for every x, the operator $\mathfrak{M}_\ell$ is nondecreasing. Applying $\mathfrak{M}_\ell$ to both sides of (8.50) we get

$$[\mathfrak{M}_\ell]^2 v \le \mathfrak{M}_\ell \mathfrak{D}_{\pi^k} v \le [\mathfrak{D}_{\pi^k}]^2 v.$$

Continuing in this way, we conclude that for every $T = 1, 2, \ldots$

$$[\mathfrak{M}_\ell]^T v \le [\mathfrak{D}_{\pi^k}]^T v.$$

Let us pass to the limit on both sides of this inequality with $T \to \infty$. Owing to the contraction properties of $\mathfrak{M}_\ell$ and $\mathfrak{D}_{\pi^k}$, the left hand side converges to $v_{\ell+1}$, the unique solution of (8.49), while the right hand side converges to the value function $\bar{v}$, the unique solution of the dynamic programming equation (8.48). Consequently, $v_{\ell+1} \leq \bar{v}$.

In view of the argument in the preceding paragraph, we have $v_\ell \leq \bar{v}$ for all $\ell \geq 2$. Then

$$v_\ell \leq \mathfrak{D}_{\pi^k} v_\ell = \mathfrak{M}_\ell v_\ell.$$

Applying the monotone operator $\mathfrak{M}_\ell$ to both sides we conclude that

$$v_\ell \leq \mathfrak{D}_{\pi^k} v_\ell \leq [\mathfrak{M}_\ell]^T v_\ell \overset{T\to\infty}{\longrightarrow} v_{\ell+1}. \tag{8.51}$$

It follows that the sequence $\{v_\ell\}$ is nondecreasing and bounded from above. Consequently, it has a limit v_∞. Passing to the limit with $\ell \to \infty$ in the relations (8.51), we conclude that

$$v_\infty \leq \mathfrak{D}_{\pi^k} v_\infty \leq v_\infty.$$

This means that the limit v_∞ is the unique solution $\bar{v}$ of equation (8.48). □

8.2.4 Models with Transition-Dependent Costs

Similar to § 8.1.5, we briefly consider the case when the stage-wise costs dependent not only on the current state–control pair, but also on the next state. We assume that a bounded and lower semicontinuous function $c : \mathcal{X} \times \mathcal{U} \times \mathcal{X} \to \mathbb{R}$ exists so that the stage-wise cost sequence (adapted to $\{\mathcal{F}_t\}$) has the form:

$$Z_t = c_{t-1}(X_{t-1}, U_{t-1}, X_t), \quad t = 2, 3, \ldots$$

We define

$$v^\Pi_{t,T}(h_t) = \varrho^\Pi_{t,T}\big(0, \alpha^{t-1} c(x_t, u_t, X_{t+1}), \alpha^t c(X_{t+1}, U_{t+1}, X_{t+2}), \ldots, \\ \alpha^{T-1} c(X_T, U_T, X_{T+1})\big),$$

and

$$v^\Pi_{t,\infty}(h_t) = \lim_{T\to\infty} v^\Pi_{t,T}(h_t).$$

Then Theorem 8.21 remains valid, with the dynamic programming equation (8.32) replaced by

$$v(x) = \min_{u\in\mathcal{U}(x)} \sigma\big(x, Q(x,u), c(x,u,\cdot) + \alpha v(\cdot)\big), \quad x \in \mathcal{X}. \tag{8.52}$$

Moreover, an optimal stationary Markov policy $\Pi^* = \{\pi^*, \pi^*, \ldots\}$ exists and satisfies the equation:

$$\pi^*(x) \in \operatorname*{arg\,min}_{u \in \mathcal{U}(x)} \sigma\big(x, Q(x,u), c(x,u,\cdot) + \alpha v^*(\cdot)\big), \quad x \in \mathcal{X}. \tag{8.53}$$

The proof uses the operators $\mathfrak{D}_\pi : \mathcal{V} \to \mathcal{V}$ and $\mathfrak{D} : \mathcal{V} \to \mathcal{V}$ defined as follows:

$$[\mathfrak{D}_\pi v](x) = \sigma\big(x, Q(x,\pi(x)), c(x,\pi(x),\cdot) + \alpha v(\cdot)\big), \quad x \in \mathcal{X},$$
$$[\mathfrak{D} v](x) = \inf_{u \in \mathcal{U}(x)} \sigma\big(x, Q(x,u), c(x,u,\cdot) + \alpha v(\cdot)\big), \quad x \in \mathcal{X}.$$

Furthermore, the value iteration methods and the policy iteration methods are convergent as well. All these results follow identically, as in the previous case.

8.3 Total Risk Problems for Transient Models

In this section, we are interested in infinite-horizon problems for a controlled Markov process, in which the future costs *are not discounted.* We assume that the controlled Markov model is stationary, that is, there exist a control set $U : \mathcal{X} \rightrightarrows \mathfrak{U}$, a controlled transition kernel Q : graph$(U) \to \mathcal{P}(\mathcal{X})$, and a cost function c : graph$(U) \to \mathbb{R}$, such that $\mathcal{U}_t = \mathcal{U}$, $Q_t = Q$, and $c_t = c$, for $t = 1, 2, \ldots$.

For an arbitrary policy $\Pi = \{\pi_1, \pi_2, \ldots\}$, we consider the finite-horizon risk

$$v_{1,T}^{\Pi}(x_1) = \varrho_{1,T}^{\Pi}\big(c(X_1,U_1), c(X_2,U_2), \ldots, c(X_T,U_T)\big), \tag{8.54}$$

where $X_1 = x_1$ and $U_t = \pi_t(H_t)$, $t = 1, 2, \ldots$. We are interested in establishing the existence of the limit

$$v_{1,\infty}^{\Pi}(x_1) = \lim_{T\to\infty} v_{1,T}^{\Pi}(x_1), \tag{8.55}$$

and in finding a policy that minimizes this limit, which we call the *total risk* of the sequence $\{c(x_t,u_t)\}_{t=1,2,\ldots}$.

As in the expected-value case, the limit (8.55) exists in some special cases only. We focus here on the most important case of transient models. We assume that the state space contains a special *absorbing state* x_A, such that $\mathcal{U}(x_A) = \{0\}$, $Q(x_A, 0) = \delta_{x_A}$ (the Dirac measure concentrated at x_A), and $c(x_A, 0) = 0$. We also assume that policies exist such that the system reaches the absorbing state with probability one and we focus on such policies exclusively. We denote the set of transient states by $\widetilde{\mathcal{X}} = \mathcal{X} \setminus \{x_A\}$.

In order to leverage the results of § 8.1.3, we make Assumption 8.20 about the family of dynamic risk measures which we use in (8.54).

8.3.1 Evaluation of a Stationary Markov Policy

Consider a stationary policy $\Pi = \{\pi, \pi, \dots\}$ and define the cost until absorption by (8.55), where each $v^{\Pi}_{1,T}(x_1)$ is defined by the formula (8.54):

$$\begin{aligned} v^{\Pi}_{1,T}(x_1) &= \varrho^{\Pi}_{1,T}\big(c(x_1,\pi(x_1)), c(X_2,\pi(X_2)),\dots, c(X_T,\pi(X_T))\big) \\ &= c(x_1,\pi(x_1)) + \rho^{\Pi}_1\Big(c(X_2,\pi(X_2)) + \dots + \rho^{\Pi}_{T-1}\big(c(X_T,\pi(X_T))\big)\cdots\Big). \end{aligned} \tag{8.56}$$

The first question to answer is when the total risk (8.55) is finite. This question is nontrivial, because even for uniformly bounded costs $Z_t = c(X_t, \pi(X_t))$, $t = 1, 2, \dots$, and for a transient finite-state Markov chain, the limit in (8.55) may be infinite, as the following example demonstrates.

Example 8.28. Consider a transient Markov chain with two states and with the following transition probabilities: $Q_{11} = Q_{12} = \frac{1}{2}$, $Q_{22} = 1$. Only one control is possible in each state, the cost of each transition from state 1 is equal to 1, and the cost of the transition from 2 to 2 is 0. Clearly, the time until absorption is a geometric random variable with parameter $\frac{1}{2}$. Let $x_1 = 1$. If the limit (8.55) is finite, then (skipping the dependence on π) we have

$$v_{1,\infty}(1) = \lim_{T\to\infty} v_{1,T}(1) = \lim_{T\to\infty} \rho_1\big(1 + v_{1,T-1}(x_2)\big) = \rho_1\big(1 + v_{1,\infty}(x_2)\big).$$

In the last equation, we used the continuity of $\rho_1(\cdot)$. Clearly, $v_{1,\infty}(2) = 0$.

Suppose we are using the Average Value at Risk with $0 < \alpha \le \frac{1}{2}$, to define $\rho_1(\cdot)$. Using Theorem 2.16, we obtain

$$\begin{aligned} v_{1,\infty}(1) &= \inf_{\eta\in\mathbb{R}} \Big\{\eta + \frac{1}{\alpha}\mathbb{E}\big[\big(1 + v_{1,\infty}(x_2) - \eta\big)_+\big]\Big\} \\ &= 1 + \inf_{\eta\in\mathbb{R}} \Big\{\eta + \frac{1}{\alpha}\mathbb{E}\big[\big(v_{1,\infty}(x_2) - \eta\big)_+\big]\Big\} = 1 + \frac{1}{\alpha}\int_{1-\alpha}^{1} F^{-1}(\beta)\, d\beta, \end{aligned} \tag{8.57}$$

where $F(\cdot)$ is the distribution function of $v_{1,\infty}(x_2)$. As all β-quantiles of $v_{1,\infty}(x_2)$ for $\beta \ge \frac{1}{2}$ are equal to $v_{1,\infty}(1)$, the last equation yields $v_{1,\infty}(1) = 1 + v_{1,\infty}(1)$, a contradiction. It follows that a composition of Average Values at Risk has no finite limit, if $0 < \alpha \le \frac{1}{2}$.

On the other hand, if $\frac{1}{2} < \alpha < 1$, then

$$F^{-1}(\beta) = \begin{cases} v_{1,\infty}(2) = 0 & \text{if } 1-\alpha \le \beta < \frac{1}{2}, \\ v_{1,\infty}(1) & \text{if } \frac{1}{2} \le \beta \le 1. \end{cases}$$

Formula (8.57) then yields $v_{1,\infty}(1) = 1 + \frac{1}{2\alpha} v_{1,\infty}(1)$. This equation has a solution $v_{1,\infty}(1) = 2\alpha/(2\alpha - 1)$.

If we use the mean-semideviation model of Example 8.6, we obtain

$$\begin{aligned} v_{1,\infty}(1) &= \mathbb{E}\big[1 + v_{1,\infty}(x_2)\big] + \kappa \mathbb{E}\Big[\Big(1 + v_{1,\infty}(x_2) - \mathbb{E}\big[1 + v_{1,\infty}(x_2)\big]\Big)_+\Big] \\ &= 1 + \frac{1}{2} v_{1,\infty}(1) + \frac{\kappa}{2}\Big(v_{1,\infty}(1) - \frac{1}{2} v_{1,\infty}(1)\Big) = 1 + \frac{2+\kappa}{4} v_{1,\infty}(1). \end{aligned}$$

Thus $v_{1,\infty}(1) = 4/(2-\kappa)$, which is finite for all $\kappa \in [0, 1]$, that is, for all values of κ for which the model defines a coherent measure of risk. □

It follows that deeper properties of the measures of risk and their interplay with the transition kernel need to be investigated to answer the question about the finiteness of the dynamic measure of risk in this case.

Definition 8.29. We call the Markov model with a stationary Markov policy $\Pi = \{\pi, \pi, \dots\}$ *risk-transient* if

$$\lim_{T_1\to\infty} \sup_{x_1\in\mathcal{X}} \sup_{T_2\ge T_1} \varrho^{\Pi}_{1,T_2}\big(0, \dots, 0, \mathbb{1}_{\widetilde{\mathcal{X}}}(X_{T_1+1}), \dots, \mathbb{1}_{\widetilde{\mathcal{X}}}(X_{T_2})\big) = 0.$$

If the limit above is uniform for all stationary Markov policies, the model is called *uniformly risk-transient*.

We can now provide sufficient conditions for the finiteness of the limit (8.55).

Theorem 8.30. *Suppose Assumption* 8.20 *is satisfied. If the cost function* $c(\cdot,\cdot)$ *is uniformly bounded and the model is risk-transient for the policy* Π*, then the limit*

$$v^{\Pi}_{1,\infty}(\cdot) = \lim_{T\to\infty} v^{\Pi}_{1,T}(\cdot) \tag{8.58}$$

exists in $\mathcal{V}$.

Proof. By the coherence of the one-step conditional risk mappings, each conditional risk measure $\varrho_{1,T}(\cdot)$ is convex and positively homogeneous, and thus subadditive. For any $1 < T_1 < T_2$ we obtain the following estimate of (8.56):

$$\begin{aligned} v^{\Pi}_{1,T_2}(x_1) &= \varrho^{\Pi}_{1,T_2}(Z_1, \dots, Z_{T_2}) \\ &\le \varrho^{\Pi}_{1,T_2}(Z_1, \dots, Z_{T_1}, 0, \dots, 0) + \varrho^{\Pi}_{1,T_2}(0, \dots, 0, Z_{T_1+1}, \dots, Z_{T_2}) \\ &= v^{\Pi}_{1,T_1}(x_1) + \varrho^{\Pi}_{1,T_2}(0, \dots, 0, Z_{T_1+1}, \dots, Z_{T_2}). \end{aligned}$$

In a similar way, from subadditivity we also deduce that

$$\begin{aligned} v^{\Pi}_{1,T_1}(x_1) &= \varrho^{\Pi}_{1,T_2}(Z_1, \dots, Z_{T_1}, 0 \dots, 0) \\ &\le v^{\Pi}_{1,T_2}(x_1) + \varrho^{\Pi}_{1,T_2}(0, \dots, 0, -Z_{T_1+1}, \dots, -Z_{T_2}). \end{aligned}$$

As all $|Z_t| \le C \mathbb{1}_{\widetilde{\mathcal{X}}}(X_t)$ with some constant C,

$$\left| v^{\Pi}_{1,T_1}(x_1) - v^{\Pi}_{1,T_2}(x_1) \right| \le C \varrho^{\Pi}_{1,T_2}\big(0, \dots, 0, \mathbb{1}_{\widetilde{\mathcal{X}}}(X_{T_1+1}), \dots, \mathbb{1}_{\widetilde{\mathcal{X}}}(X_{T_2})\big). \tag{8.59}$$

Thus, for every $x_1 \in \mathcal{X}$, the sequence $\left\{\varrho^{\Pi}_{1,T}(Z_1, \dots, Z_T)\right\}$ is a Cauchy sequence. Furthermore, the bound (8.59) is uniform over all $x_1 \in \mathcal{X}$ and thus convergence of the sequence $\left\{v^{\Pi}_{1,T}\right\}$ in $\mathcal{V}$ follows. □

This analysis allows us to derive policy evaluation equations for the infinite horizon problem, in the case of a fixed Markov policy.

Theorem 8.31. *Suppose Assumption* 8.20 *is satisfied and the cost function* $c(\cdot, \cdot)$ *is uniformly bounded. If the model is risk-transient for the stationary Markov policy* $\Pi = \{\pi, \pi, \dots\}$ *then a function* $v \in \mathcal{V}$ *satisfies the equations*

$$v(x) = c(x, \pi(x)) + \sigma\big(x, Q(x, \pi(x)), v(\cdot)\big), \quad x \in \widetilde{\mathcal{X}}, \tag{8.60}$$

$$v(x_{\mathrm{A}}) = 0, \tag{8.61}$$

if and only if $v(x) = v^{\Pi}_{1,\infty}(x)$ *for all* $x \in \mathcal{X}$.

Proof. Suppose a function $v \in \mathcal{V}$ satisfies equations (8.60)–(8.61). Iterating (8.60), we obtain for all $x_1 \in \mathcal{X}$ the following equation:

$$v(x_1) = c(x_1, \pi(x_1)) + \rho^{\Pi}_1\Big(c(X_2, \pi(X_2)) + \cdots \\ \cdots + \rho^{\Pi}_{T-1}\big(c(X_T, \pi(X_T)) + \rho^{\Pi}_T(v(X_{T+1}))\big) \cdots\Big).$$

Denote $Z_t = c(X_t, \pi(X_t))$. The last equation, subadditivity, and monotonicity of the conditional risk measures yield:

$$\begin{aligned} v(x_1) &= \varrho^{\Pi}_{1,T+1}\big(Z_1, \dots, Z_T, v(X_{T+1})\big) \\ &\le \varrho^{\Pi}_{1,T}\big(Z_1, \dots, Z_T\big) + \varrho^{\Pi}_{1,T+1}\big(0, \dots, 0, v(X_{T+1})\big) \\ &\le \varrho^{\Pi}_{1,T}\big(Z_1, \dots, Z_T\big) + \varrho^{\Pi}_{1,T+1}\big(0, \dots, 0, |v(X_{T+1})|\big). \end{aligned}$$

In a similar way,

$$\begin{aligned} &\varrho^{\Pi}_{1,T}\big(Z_1, \dots, Z_T\big) \\ &\quad \le \varrho^{\Pi}_{1,T+1}\big(Z_1, \dots, Z_T, v(X_{T+1})\big) + \varrho^{\Pi}_{1,T+1}\big(0, 0, \dots, -v(X_{T+1})\big) \\ &\quad \le v(x_1) + \varrho^{\Pi}_{1,T+1}\big(0, \dots, 0, |v(X_{T+1}|)\big). \end{aligned}$$

We conclude that

$$\left| v(x_1) - \varrho^{\Pi}_{1,T}\big(Z_1, \dots, Z_T\big) \right| \le \varrho^{\Pi}_{1,T+1}\big(0, \dots, 0, |v(X_{T+1})|\big).$$

Passing to the limit with $T \to \infty$ and using Definition 8.29, we conclude that $v \equiv \varrho^{\Pi}_{1,\infty}$.

To prove the converse implication we use the fact that all conditional risk measures $\rho^{\Pi}_t(\cdot)$ share the same transition risk mapping $\sigma(\cdot,\cdot,\cdot)$ to rewrite (8.56) as follows:

$$v^{\Pi}_{1,T}(x_1) = c(x_1, \pi(x_1)) + \sigma\big(x_1, Q(x_1, \pi(x_1)), v^{\Pi}_{1,T-1}\big). \tag{8.62}$$

The function $\sigma(x_1, \mu, \cdot)$, as a finite-valued coherent measure of risk on a Banach lattice $\mathcal{V}$, is continuous (see Theorem 2.32). By Theorem 8.30, the sequence $\big\{v^{\Pi}_{1,T}(\cdot)\big\}$ is convergent to $v^{\Pi}_{1,\infty}(\cdot)$ in the space $\mathcal{V}$. Therefore, we can pass to the limit in (8.62) to obtain equation (8.60). Equation (8.61) is obvious. □

8.3.2 Dynamic Programming Equations. Value Iteration

We shall now focus on the optimal value function

$$v^*(x) = \inf_{\Pi \in \mathbb{\Pi}} v^{\Pi}_{1,\infty}(x), \quad x \in \mathcal{X}, \tag{8.63}$$

where $\mathbb{\Pi}$ is a set of deterministic policies.

It is convenient to state our assumptions within the formulation of the main theorem of this section. The proof will be deferred to after several auxiliary results. We denote by $\mathcal{V}^{\text{lc}}$ the set of bounded lower semicontinuous functions on $\mathcal{X}$.

Theorem 8.32. *Suppose Assumption* 8.20 *and the following conditions are satisfied:*

(i) *The transition kernel $Q(\cdot,\cdot)$ is continuous;*

(ii) *For every lower semicontinuous $v \in \mathcal{V}$, the transition risk mapping $\sigma(\cdot,\cdot,v)$ is lower semicontinuous;*

(iii) *For all $x \in \mathcal{X}$ and $u \in \mathcal{U}(x)$, the function $\sigma(x,u,\cdot)$ is a coherent measure of risk on the space $\mathcal{L}_p(\mathcal{X}, \mathcal{B}(\mathcal{X}), Q(x,u))$, with some $p \in [1,\infty)$;*

(iv) *The function $c(\cdot,\cdot)$ is nonnegative, bounded, and lower semicontinuous;*

(v) *The multifunction $\mathcal{U}(\cdot)$ is measurable, compact-valued, and upper semicontinuous;*

(vi) *The model is risk-transient for every stationary Markov policy $\Pi \in \mathbb{\Pi}$.*

Then a function $v \in \mathcal{V}^{\text{lc}}$ satisfies the equations

$$v(x) = \inf_{u \in \mathcal{U}(x)} \big\{c(x,u) + \sigma\big(x, Q(x,u), v\big)\big\}, \quad x \in \mathcal{X}, \tag{8.64}$$

$$v(x_{\text{A}}) = 0, \tag{8.65}$$

if and only if $v(\cdot) \equiv v^*(\cdot)$. *Moreover, any measurable minimizer* $u^* = \pi^*(x)$, $x \in \mathcal{X}$, *on the right hand side of* (8.64) *defines an optimal deterministic Markov policy* $\Pi^* = \{\pi^*, \pi^*, \dots\}$.

Before proceeding to the proof, we derive a supporting result.

To simplify notation, we define the operators $\mathfrak{D} : \mathcal{V} \to \mathcal{V}$ and $\mathfrak{D}_\pi : \mathcal{V} \to \mathcal{V}$ as follows:

$$[\mathfrak{D}v](x) = \inf_{u \in \mathcal{U}(x)} \big\{ c(x,u) + \sigma\big(x, Q(x,u), v\big) \big\}, \quad x \in \mathcal{X}, \tag{8.66}$$

$$[\mathfrak{D}_\pi v](x) = c(x, \pi(x)) + \sigma\big(x, Q(x, \pi(x)), v\big), \quad x \in \mathcal{X}. \tag{8.67}$$

where $\pi \lessdot \mathcal{U}$. By an argument identical to that used in the proof of Theorem 8.15, if $v \in \mathcal{V}^{\rm lc}$, then the minimum in (8.66) exists and $\mathfrak{D}v \in \mathcal{V}^{\rm lc}$ as well.

Due to the monotonicity of $\sigma(x, \mu, \cdot)$, both operators are nondecreasing. By construction, $\mathfrak{D}v \le \mathfrak{D}_\pi v$ for all $v \in \mathcal{V}$ and all π.

First, we analyze the *value iteration method*, in which we construct a sequence of functions $\{v^k\}$ in $\mathcal{V}^{\rm lc}$ by the following procedure:

$$v^{k+1} = \mathfrak{D}v^k, \quad k = 1, 2, \dots \tag{8.68}$$

with $v^0 \equiv 0$. By construction, each v^k is the optimal value in a finite-horizon problem:

$$v^k(x_1) = \inf_{\Pi \in \mathit{\Pi}} \varrho_{1,k+1}\big(c(x_1, \pi_1(x_1)), \dots, c(X_k, \pi_k(X_k)), v^0\big).$$

Theorem 8.33. *If the assumptions of Theorem* 8.32 *are satisfied then the sequence of functions* $\{v^k\}$ *generated by the value iteration method is convergent a function* $v^\infty \in \mathcal{V}^{\rm lc}$, *which satisfies the equations* (8.64)–(8.65). *Moreover, a stationary Markov policy* Π^* *exists, such that* $v^\infty = v^{\Pi^*}$.

Proof. Since $c(\cdot, \cdot) \ge 0$, we have the following inequalities

$$0 \le v^k \le v^{k+1} \le v^*, \quad k = 0, 1, 2, \dots \tag{8.69}$$

The lower semicontinuity of $v^k(\cdot)$ has been established in Theorem 8.15. Therefore, the sequence $\{v^k\}$ is monotonically convergent to some function $v^\infty \in \mathcal{V}^{\rm lc}$ (the lower semicontinuity can be verified by a standard analysis argument). By the Lebesgue theorem, $\{v^k\}$ is convergent to v^∞ in the space $\mathcal{L}_p(\mathcal{X}, \mathcal{B}(\mathcal{X}), Q(x,u))$. As $\sigma(x, Q(x,u), \cdot)$ is a coherent measure of risk on that space, it is continuous (Theorem 2.32), and thus for all $x \in \mathcal{X}$ and all $u \in \mathcal{U}(x)$

$$\lim_{k \to \infty} \sigma(x, Q(x,u), v^k) = \sigma(x, Q(x,u), v).$$

Therefore, for every x and every decision rule π

$$\lim_{k\to\infty}\big[\mathfrak{D}_\pi v^k\big](x) = \big[\mathfrak{D}_\pi v^\infty\big](x).$$

Moreover,

$$\lim_{k\to\infty}\big[\mathfrak{D} v^k\big](x) = \big[\mathfrak{D} v^\infty\big](x),$$

for all $x \in \mathcal{X}$.

The sequence $\{\mathfrak{D} v^k\} = \{v^{k+1}\}$ has the same limit v^∞. It follows that

$$v^\infty = \mathfrak{D} v^\infty, \tag{8.70}$$

which is (8.64). Equation (8.65) is true by construction, because $v^k(x_A) = 0$ for all k.

Suppose π^* is the minimizer on the right-hand side of (8.66) for $v = v^\infty$. It is well defined because v^∞ is lower semicontinuous. By Theorem 8.31, $v^\infty = v^{\Pi^*}$. In view of (8.69), $v^\infty = v^*$, and the stationary Markov policy $\Pi^* = \{\pi^*, \pi^*, \dots\}$ is optimal. □

We are now ready to prove the main theorem.

Proof of Theorem 8.32. The equations (8.64)–(8.65) for the optimal value function v^* follow from Theorem 8.33.

To prove the converse implication, suppose $v \in \mathcal{V}^{\text{lc}}$ satisfies (8.64)–(8.65). As in the proof of Theorem 8.15, a measurable minimizer $\pi^*(x)$ on the right hand side of (8.64) exists. We obtain the equation

$$v(x) = c(x, \pi^*(x)) + \sigma\big(x, Q(x, \pi^*(x)), v\big), \quad x \in \mathcal{X}.$$

Due to Theorem 8.31 and assumption (vi), for the stationary Markov policy $\Pi^* = \{\pi^*, \pi^*, \dots\}$, we have

$$v(x) = v^{\Pi^*}_{1,\infty}(x) \geq v^*(x), \quad x \in \mathcal{X}. \tag{8.71}$$

Theorem 8.33 implies that an optimal stationary Markov policy exists. It remains to prove that no stationary Markov policy is better than Π^*.

It follows from (8.64) that for any stationary Markov policy $\Pi = \{\pi, \pi, \dots\}$, we have

$$v(x) \leq c(x, \pi(x)) + \sigma\big(x, Q(x, \pi(x)), v\big), \quad x \in \mathcal{X}, \quad t = 1, 2, \dots \tag{8.72}$$

The transition risk mapping σ is nondecreasing with respect to the last argument. Therefore, iterating inequality (8.72) for $t = 1, 2, \dots$ we obtain:

$$v(x_1) \leq \varrho^{\Pi}_{1,T+1}\big(c(x_1, \pi(x_1)), \dots, c(X_T, \pi(X_T)), v(X_{T+1})\big).$$

By subadditivity,

$$v(x_1) \le v^{\Pi}_{1,T}(x_1) + \varrho^{\Pi}_{1,T+1}\big(0,\dots,0,|v(X_{T+1})|\big).$$

Passing to the limit with $T \to \infty$ and using the fact that the system is risk transient for the policy Π, we obtain the inequality

$$v(\cdot) \le v^{\Pi}_{1,\infty}(\cdot).$$

The last estimate together with (8.71) implies that $v(\cdot) \equiv v^*(\cdot)$ and that the stationary Markov policy $\Pi^* = \{\pi^*, \pi^*, \dots\}$ is optimal. □

Example 8.34. Let us consider the classical example of asset selling. Offers S_t arriving in time periods $t = 1, 2, \dots$ are independent random variables with values in the interval $[0, M]$, distributed according to a measure P. At each time we may accept the highest offer received so far, or we may wait, in which case a waiting cost $c_0 > 0$ is incurred. Denoting the random stopping time by τ we see that the total "cost" equals $Z = c_0(\tau - 1) - \max_{1\le t\le\tau} S_t$. The problem is an example of an *optimal stopping problem*, a structure of considerable theoretical and practical relevance.

Formally, we introduce the state space $\mathcal{X} = \{x_{\rm A}\} \cup [0, M]$, where $x_{\rm A}$ is the absorbing state reached after the transaction, and the other states represent the highest offer received so far. The control space is $\mathcal{U} = \{0, 1\}$, with 0 representing "wait" and 1 representing "sell." The state evolves according to the equation

$$x_{t+1} = \begin{cases} \max(x_t, S_{t+1}) & \text{if } u_t = 0,\\ x_{\rm A} & \text{if } u_t = 1. \end{cases}$$

Denoting by $F_S(\cdot)$ the distribution function of a generic offer S, we can write the controlled transition kernel Q as follows:

$$Q(\mathrm{d}y|x,u) = \begin{cases} F_S(x)\,\delta_x(\mathrm{d}y) & \text{if } u = 0,\ y \in [0, x],\\ P(\mathrm{d}y) & \text{if } u = 0,\ y \in (x, M],\\ \delta_{x_{\rm A}}(\mathrm{d}y) & \text{if } u = 1. \end{cases}$$

The cost function is

$$c(x,u) = \begin{cases} c_0 & \text{if } u = 0,\\ -x & \text{if } u = 1. \end{cases}$$

The expected value version of this problem has a known solution: accept the first offer greater than or equal to the solution $\hat{x}$ of the equation

$$c_0 = \int_0^M (s - \hat{x})_+ \, P(\mathrm{d}s). \tag{8.73}$$

We shall solve the risk-averse version of the problem.

Suppose $\sigma : \mathcal{X} \times \mathcal{P}(\mathcal{X}) \times \mathcal{V} \to \mathbb{R}$ is a coherent law-invariant transition risk mapping. Then

$$c(x,u) + \sigma\big(x, Q(x,u), v(\cdot)\big) = \begin{cases} c_0 + \sigma\big(x, Q(x,0), v(\cdot)\big) & \text{if } u = 0, \\ -x & \text{if } u = 1. \end{cases}$$

Equation (8.64) takes on the form:

$$v(x) = \min\Big\{ -x, c_0 + \sigma\big(x, Q(x,0), v(\cdot)\big)\Big\}, \quad x \in [0, M]. \tag{8.74}$$

Although our problem does not exactly fit the assumptions of Theorem 8.32, because the costs are not non-negative, we can prove the convergence of the value iteration method (8.68) starting from $v^0 \equiv -M$ exactly as in the proof of Theorem 8.33. This is due to the fact that the resulting sequence $\{v^k\}$ is nondecreasing and $v^k \le v^* \le 0$. Therefore the optimal value function is indeed the solution of (8.74).

Since σ is law-invariant and the distribution of $v(\cdot)$ with respect to the measure $Q(x,0)$ on $[0, M]$ is the same as the distribution of $v\big(\max(x, S)\big)$ under the measure P of S, we obtain

$$\sigma\big(x, Q(x,0), v(\cdot)\big) = \sigma\big(x, P, v\big(\max(x,\cdot)\big)\big).$$

Suppose our attitude to risk does not depend on the current state, that is, $\sigma(\cdot,\cdot,\cdot)$ does not depend on its first argument. Using the dual representation (2.28), we may rewrite the last equation as follows:

$$\sigma\big(x, Q(x,0), v(\cdot)\big) = \max_{\zeta \in \mathcal{A}} \int_0^M \zeta(s)\, v\big(\max(x,s)\big)\; P(ds).$$

The convex closed set of probability density functions $\mathcal{A}$ is fixed, because $\sigma(\cdot,\cdot,\cdot)$ does not depend on its first argument, and the second argument, P, does not change. Equation (8.74) takes on the form

$$v(x) = \min\Big\{ -x, c_0 + \max_{\zeta \in \mathcal{A}} \int_0^M \zeta(s)\, v\big(\max(x,s)\big)\; P(ds)\Big\}, \quad x \in [0, M]. \tag{8.75}$$

Observe that $v(x) \le -x$ and thus $v\big(\max(x,s)\big) \le -\max(x,s)$. The last displayed inequality implies that

$$\begin{aligned} v(x) &\le \min\Big\{ -x, c_0 + \max_{\zeta \in \mathcal{A}} \int_0^M \zeta(s)\Big[-\max(x,s)\Big]\; P(ds)\Big\} \\ &= \min\Big\{ -x, c_0 - \min_{\zeta \in \mathcal{A}} \int_0^M \zeta(s)\, \max(x,s)\; P(ds)\Big\}, \quad x \in [0, M]. \end{aligned}$$

If the offer at level x is accepted, then $v(x) = -x$. After simplifications, we obtain the inequality:

$$\min_{\zeta\in\mathcal{A}} \int_0^M (s-x)_+\zeta(s)\ P(\mathrm{d}s) \le c_0.$$

This suggests the solution: *accept any offer that is greater than or equal to the solution x^* of the equation*

$$\min_{\zeta\in\mathcal{A}} \int_0^M (s-x)_+\zeta(s)\ P(\mathrm{d}s) = c_0; \tag{8.76}$$

if $x < x^*$, then wait. The corresponding value function is $v^*(x) = -\max(x, x^*)$. Equation (8.75) can be verified by direct substitution.

Let us verify directly that the solution of equation (8.76) exists and is unique. To guarantee that a solution of (8.76) exists, it is sufficient to assume that

$$\mathbb{E}_\zeta[S] = \int_0^M s\zeta(s)\ P(\mathrm{d}s) > c_0.$$

for all $\zeta \in \mathcal{A}$. Since the left-hand side of (8.76), as a function of x, is continuous, nonincreasing, and equals 0, at $x = M$, it must attain the value c_0 at some point x^*. Define ζ^* as the density in $\mathcal{A}$ for which the minimum in (8.76) with $x = x^*$ is attained. It exists because $\mathcal{A}$ is weakly* compact. We thus have

$$\int_0^M (s-x^*)_+\zeta^*(s)\ P(\mathrm{d}s) = c_0.$$

It follows that

$$\int_{x^*}^M \zeta^*(s)\ P(\mathrm{d}s) > 0.$$

Therefore, for every $x > x^*$ we obtain

$$\begin{aligned}\min_{\zeta\in\mathcal{A}} \int_0^M (s-x)_+\zeta(s)\ P(\mathrm{d}s) &\le \int_0^M (s-x)_+\zeta^*(s)\ P(\mathrm{d}s)\\ &= \int_{x^*}^M (s-x)_+\zeta^*(s)\ P(\mathrm{d}s) < \int_{x^*}^M (s-x^*)\,\zeta^*(s)\ P(\mathrm{d}s) = c_0.\end{aligned}$$

Consequently, x^* is the largest solution of (8.76). No other solution may exist, because they cannot both be the largest.

The solution (8.76) of the risk-averse problem is closely related to the solution (8.73) of the expected value problem. The only difference is that we have to account for the least favorable distribution of the offers. If $\mathbb{1} \in \mathcal{A}$ (which is the typical situation), then the critical level $x^* \le \hat{x}$. □

8.3.3 Policy Iteration

Consider a sequence of Markov deterministic policies $\Pi^k = \{\pi^k, \pi^k, \ldots\}$, $k = 1, 2, \ldots$ constructed in the following way. We choose any function $v^0 \in \mathcal{V}^{\text{lc}}$. For $k = 0, 1, \ldots$ we determine $\pi^{k+1}(\cdot)$ as the measurable solution of the problem

$$\min_{u \in \mathcal{U}(x)} \big\{c(x, u) + \sigma\big(x, Q(x, u), v^k\big)\big\}, \quad x \in \widetilde{\mathcal{X}}, \tag{8.77}$$

which exists by the lower semicontinuity of $v^k(\cdot)$. The corresponding value of the Markov policy $\Pi^{k+1} = \{\pi^{k+1}, \pi^{k+1}, \ldots\}$ is the function $v^{k+1} = v_{1,\infty}^{\Pi^{k+1}} \in \mathcal{V}^{\text{lc}}$, and the iteration continues.

Theorem 8.35. *Assume all conditions of Theorem* 8.32. *Then the policy iteration method generates a sequence of stationary Markov policies* $\{\pi^k, \pi^k, \ldots\}$, $k = 1, 2, \ldots$, *such that the sequence* $\big\{v_{1,\infty}^{\Pi^k}\big\}$ *is nonincreasing and convergent to the optimal value function* v^*.

Proof. By construction, the sequences $\{\pi^k\}$ and $\{v^k\}$ satisfy the relations:

$$\mathfrak{D}_{\pi^{k+1}} v^k = \mathfrak{D} v^k \le \mathfrak{D}_{\pi^k} v^k = v^k. \tag{8.78}$$

Applying the operator $\mathfrak{D}_{\pi^{k+1}}$ to this relation, we deduce from its monotonicity that

$$[\mathfrak{D}_{\pi^{k+1}}]^T v^k \le \mathfrak{D}_{\pi^{k+1}} v^k = \mathfrak{D} v^k \le v^k, \quad T = 2, 3, \ldots. \tag{8.79}$$

Relation (8.79) can be equivalently written as

$$\varrho_{1,T+1}^{\Pi^{k+1}}\big(Z_1, \ldots, Z_T, v^k(X_T)\big) \le [\mathfrak{D} v^k](x_1) \le v^k(x_1),$$

where $Z_t = c(X_t, U_t)$, $t = 1, 2, \ldots, T$, is the cost sequence resulting from the policy Π^{k+1}. Passing to the limit with $T \to \infty$, we conclude that the sequence $\{v^k\}$ is nonincreasing:

$$v^{k+1}(x) = v_{1,\infty}^{\Pi^{k+1}}(x) \le [\mathfrak{D} v^k](x) \le v^k(x), \quad x \in \mathcal{X}, \quad k = 0, 1, 2, \ldots. \tag{8.80}$$

Since $v^k \ge v^*$, the sequence $\{v^k\}$ is monotonically convergent to some limit $v^\infty \ge v^*$. By Lebesgue Theorem, it is convergent in the space $\mathcal{L}_p(\mathcal{X}, \mathcal{B}(\mathcal{X}), \mu)$. As the function $\sigma\big(x, \mu, \cdot\big)$ is a coherent measure of risk on that space, it follows from Theorem 2.32 that it is continuous and thus

$$\sigma\big(x, Q(x, u), v^k\big) \downarrow \sigma\big(x, Q(x, u), v^\infty\big), \quad \text{as} \quad k \to \infty, \quad \forall\, u \in \mathcal{U}(x). \tag{8.81}$$

The left inequality in (8.80) also implies that

$$v^{k+1}(x) \le c(x,u) + \sigma\big(x, Q(x,u), v^k\big), \quad \forall\, u \in \mathcal{U}(x). \tag{8.82}$$

Passing to the limit with $k \to \infty$ on both sides of (8.82) and using (8.81), we conclude that

$$v^{\infty}(x) \le c(x,u) + \sigma\big(x, Q(x,u), v^{\infty}\big), \quad \forall\, u \in \mathcal{U}(x).$$

Since this is true for all $x \in \mathcal{X}$ and all $u \in \mathcal{U}(x)$, it follows that

$$v^{\infty}(x) \le [\mathfrak{D}v^{\infty}](x) = \inf_{u \in \mathcal{U}(x)} \big\{c(x,u) + \sigma\big(x, Q(x,u), v^{\infty}\big)\big\}, \quad x \in \mathcal{X}. \tag{8.83}$$

Iterating this inequality, we conclude that for every feasible decision rule π and every $T = 1, 2, \ldots$

$$v^{\infty} \le [\mathfrak{D}]^T v^{\infty} \le [\mathfrak{D}_{\pi}]^T v^{\infty}.$$

The right-hand side of this inequality is convergent to $v_{1,\infty}^{\Pi}(\cdot)$, when $T \to \infty$. Since π was arbitrary, $v^{\infty} \le v^*$. Consequently, $v^{\infty} = v^*$. □

As in subsection 8.2.3, if the state and control spaces are finite, then the policy iteration method is convergent in finitely many steps.

8.4 Partially Observable Systems

8.4.1 Partially Observable Markov Decision Processes

We consider a partially observable Markov process $\{X_t, Y_t\}_{t=1,\ldots,T}$, in which $\{X_t\}$ is observable and $\{Y_t\}$ is not[†]. The state space of the model is defined as $\mathcal{X} \times \mathcal{Y}$ where $(\mathcal{X}, \mathcal{B}(\mathcal{X}))$ and $(\mathcal{Y}, \mathcal{B}(\mathcal{Y}))$ are two Borel spaces (Borel subsets of Polish spaces). The measurable space that we will work with is then given by $\Omega = (\mathcal{X} \times \mathcal{Y})^T$ endowed with the product σ-field $\mathcal{F}$, and we use X_t and Y_t to denote the canonical projections at time t.

Let $\{\mathcal{F}_t\}_{t=1,\ldots,T}$ denote the natural filtration generated by the process (X, Y) and $\{\mathcal{F}_t^X\}_{t=1,\ldots,T}$ be the filtration generated by the process X.

The control space is given by a Borel space $\mathfrak{U}$, and since only X is observable, the set of admissible controls at step t is given by a measurable multifunction $\mathcal{U}_t : \mathcal{X} \rightrightarrows \mathfrak{U}$ with nonempty closed values.The transition kernel at time t is $K_t : \mathcal{X} \times \mathcal{Y} \times \mathfrak{U} \to \mathcal{P}(\mathcal{X} \times \mathcal{Y})$, where $\mathcal{P}(\mathcal{X} \times \mathcal{Y})$ is the space of probability measures on $\mathcal{X} \times \mathcal{Y}$. In other words, if the state at time t is (x, y) and we apply control u, the distribution of the next state is $K_t(x, y, u)$.

[†] In the control literature, the unobservable part of the state is frequently denoted by X and the observation by Y, but our notation allows us to directly use the results of sections 8.1.1 and 8.1.2.

At time t, the observation history is $h_t = (x_1, x_2, \ldots, x_t)$. We use $\mathcal{H}_t = \mathcal{X}^t$ to denote the spaces of possible histories h_t. We also write H_t for $(X_1, \ldots, X_t)$.

For this controlled process, a (deterministic) history-dependent admissible policy $\Pi = (\pi_1, \ldots, \pi_T)$ is a sequence of measurable functions $\pi_t : \mathcal{H}_t \to \mathfrak{U}$ such that $\pi_t(h_t) \in \mathcal{U}_t(x_t)$ for all possible h_t (such a policy exists, due to the measurable selector theorem).† The set of admissible policies is

$$\varPi = \big\{ \Pi = (\pi_1, \ldots, \pi_T) \mid \pi_t(x_1, \ldots, x_t) \in \mathcal{U}_t(x_t),\ t = 1, \ldots, T \big\}.$$

For a random Y_1, any policy $\Pi \in \varPi$ defines a process $\{X_t, Y_t, U_t\}_{t=1,\ldots,T}$ on the probability space $(\Omega, \mathcal{F}, P^{\Pi})$, with $U_t = \pi_t(X_1, \ldots, X_t)$.

We assume that the cost process Z_t^{Π}, $t = 1, \ldots, T$, is bounded and adapted to $\{\mathcal{F}_t^X\}$, *i.e.*, $Z_t^{\Pi} \in \mathcal{Z}_t$ for all Π and t, where

$$\mathcal{Z}_t = \Big\{ Z : \Omega \to \mathbb{R} \,\Big|\, Z \text{ is } \mathcal{F}_t^X\text{-measurable and bounded} \Big\}, \qquad t = 1, \ldots, T.$$

We allow the cost process to depend on the policy Π in order to cover the case of control-dependent costs, such as $Z_t^{\Pi} = c_t(X_t, \pi_t(X_t))$, or more general cases discussed below. Any $Z \in \mathcal{Z}_t$ can be viewed as a measurable and bounded function of $(X_1, \ldots, X_t)$.

In the setting of partially observable systems, the theory of measures of risk for processes with history-dependent transition kernels, presented in sections 8.1.1 and 8.1.2, applies directly. We only need to describe the conditional distribution of X_{t+1}, given h_t, under a policy Π:

$$Q_t^{\Pi}(h_t)(C) = \mathbb{P}^{\Pi}[X_{t+1} \in C \mid H_t = h_t], \quad \forall C \in \mathcal{B}(\mathcal{X}). \tag{8.84}$$

The fundamental concept in this context is the conditional distribution of the current unobservable state Y_t, given h_t. We call it the *belief state* and denote by $\Xi_t(h_t)$. It is an element of $\mathcal{P}(\mathcal{Y})$ given by the following formula:

$$\Xi_t(h_t)(B) = \mathbb{P}^{\Pi}[Y_t \in B \mid H_t = h_t], \quad \forall B \in \mathcal{B}(\mathcal{Y}). \tag{8.85}$$

If we knew the belief state $\xi_t = \Xi_t(h_t)$, we could apply the total probability formula:

$$Q_t^{\Pi}(h_t) = \int_{\mathcal{Y}} K_t^X(x_t, y, \pi_t(h_t))\, \xi_t(\mathrm{d}y), \tag{8.86}$$

where $K_t^X(x_t, y, \pi_t(h_t))$ is the marginal distribution of $K_t(x_t, y, \pi_t(h_t))$ on $\mathcal{X}$.

The belief states can be updated recursively with the use of the Bayes kernels. Suppose the current observable part of the state is x, the control is u, and the conditional distribution of the unobservable state (the current belief state), given

† As in § 8.1.2, considering decision rules involving state and control histories $(x_1, u_1, x_2, u_2, \ldots, x_t)$ can be reduced to this case.

the history of the process, is ξ. After a transition of the system to a new state, with x' being its observable part, we can find the conditional distribution of the unobservable component y' of the state.

Let us start with a general construction of the Bayes kernel. Assuming the above setup, for given $(x, \xi, u) \in \mathcal{X} \times \mathcal{P}(\mathcal{Y}) \times \mathcal{U}$, we define a new measure $m_t(x, \xi, u)$ on $\mathfrak{X} \times \mathfrak{Y}$, initially on all measurable rectangles $A \times B$, as

$$m_t(x, \xi, u)(A \times B) = \int_{\mathfrak{Y}} K_t(A \times B \,|\, x, y, u)\, \xi(\mathrm{d}y).$$

This uniquely defines a probability measure on $\mathfrak{X} \times \mathfrak{Y}$. If the measurable space $(\mathfrak{Y}, \mathcal{B}(\mathfrak{Y}))$ is standard Borel, *i.e.*, isomorphic to a Borel subset of $\mathbb{R}$, we can disintegrate $m_t(x, \xi, u)$ into its marginal $\lambda_t(x, \xi, u) \in \mathcal{P}(\mathfrak{X})$ and a transition kernel $\Phi_t(x, \xi, u) : \mathfrak{X} \to \mathcal{P}(\mathfrak{Y})$:

$$m_t(x, \xi, u) = \lambda_t(x, \xi, u) \otimes \Phi_t(x, \xi, u).$$

We call Φ_t the *Bayes kernel* and write $\Phi_t(x, \xi, u, x')$ instead of $\Phi_t(x, \xi, u)(x')$. In the following, we always assume that the space $\mathfrak{Y}$ is standard Borel.

Example 8.36. In models with *parametric uncertainty*, the unobserved state y does not move; it is an unknown parameter of the transition kernel K_t^X in the space $\mathfrak{X}$. If $\mathfrak{X}$ and $\mathfrak{Y}$ are finite sets, then

$$\Phi_t(x, \xi, u, x')(y) \;\propto\; \xi(y)\, K_t^X(x, y, u)(x'), \quad y \in \mathfrak{Y},$$

with the symbol $\propto$ meaning "proportional to," subject to normalization. □

Example 8.37. In a more general case, but still with finite spaces $\mathfrak{X}$ and $\mathfrak{Y}$, we have

$$\Phi_t(x, \xi, u, x')(y') \;\propto\; \sum_{y \in \mathfrak{Y}} \xi(y)\, K_t(x, y, u)(x', y'), \quad y' \in \mathfrak{Y}.$$

□

Example 8.38. Assume that each transition kernel $K_t(x, y, u)$ has a density $q_t(\cdot, \cdot \,|\, x, y, u)$ with respect to a product probability measure $\mu_X \times \mu_Y$ on $\mathfrak{X} \times \mathfrak{Y}$. Then the Bayes kernel has the form

$$\big[\Phi_t(x, \xi, u, x')\big](B) \;\propto\; \int_B \int_{\mathfrak{Y}} q_t(x', y' \,|\, x, y, u)\, \xi(\mathrm{d}y)\, \mu_Y(\mathrm{d}y'), \quad B \in \mathcal{B}(\mathfrak{Y}).$$

If the denominator above is zero for some (x, ξ, u, x'), we can formally define $\Phi_t(x, \xi, u, x')$ to be an arbitrarily selected distribution on $\mathfrak{Y}$. □

It follows that we can calculate Q_t^{Π} and Ξ_t (defined in (8.84) and (8.85)) recursively with the help of the Bayes kernels:

- The initial belief $\Xi_1(x_1)$ is the conditional distribution of Y_1 given $X_1 = x_1$;
- The Bayes kernel provides us the following formula to update the belief states:
$$\Xi_{t+1}(h_{t+1}) = \Phi_t\big(x_t, \Xi_t(h_t), \pi_t(h_t), x_{t+1}\big);$$
- The conditional distribution of X_{t+1} can be calculated by (8.86).

8.4.2 Markov Risk Measures

We make the following additional assumption:

Assumption 8.39. *The costs $Z_1, \dots, Z_T$ are only dependent on the current observable state, the current belief state, and the current control, that is,*

$$Z_t^{\Pi}(h_t) = r_t\big(x_t, \Xi_t(h_t), \pi_t(h_t)\big), \quad t = 1, \dots, T. \tag{8.87}$$

For example, in the expected value models, we have

$$r_t(x, \xi, u) = \int_{\mathcal{Y}} c_t(x, y, u)\, \xi(\mathrm{d}y),$$

where $c_t : \mathcal{X} \times \mathcal{Y} \times \mathcal{U} \to \mathbb{R}$ is the running cost function, but more general risk forms $\varrho\big[\xi_t, c_t(x, \cdot, u)\big]$ may be used here.[†]

Summing up, with the use of the *extended state* $(X, \Xi) \in \mathcal{X} \times \mathcal{P}(\mathcal{Y})$, the partially observable system can be viewed as an observable system on a larger space, which is still a Borel space. Therefore, all our considerations of § 8.1.2 can be directly translated to the partially observable case. Nonetheless, it is a very special case in which additional research reveals important properties.

We adapt the definition of a Markov policy in the following way.

Definition 8.40. A policy $\Pi \in \mathbb{\Pi}$ is *Markov* if $\pi_t(h_t) = \pi_t(h_t')$ for all $t = 1, \dots, T$ and all $h_t, h_t' \in \mathcal{X}^t$ such that $x_t = x_t'$ and $\Xi_t(h_t) = \Xi_t(h_t')$.

With the extended state, Definition 8.12 of a Markov risk measure applies as well.

For a fixed Markov policy Π, the future of the process $\{(X_\tau, \Xi_\tau)\}_{\tau=t,\dots,T}$ is solely dependent on the current extended state $(x_t, \Xi_t(h_t))$, and so is the distribution of the future risk functions $r_\tau(X_\tau, \Xi_\tau, \pi_\tau(X_\tau, \Xi_\tau))$, $\tau = t, \dots, T$. Therefore, we can still define the Markov property of the risk measures, as in Definition 8.12.

To alleviate notation, for all $\Pi \in \mathbb{\Pi}$ and for a measurable and bounded $r = (r_1, \dots, r_T)$, we write

---

[†] The theory of risk forms is discussed in Section 2.7.1.

$$v_t^{\Pi}(h_t) \triangleq \rho_{t,T}^{\Pi}\big(r_t(X_t, \Xi_t, \pi_t(H_t)), \ldots, r_T(X_T, \Xi_T, \pi_T(H_T))\big)(h_t). \quad (8.88)$$

The following theorem summarizes our observations.

Theorem 8.41. *A family of dynamic risk measures* $\{\rho_{t,T}^{\Pi}\}_{t=1,\ldots,T}^{\Pi \in \varPi}$ *for a partially observable system is normalized, translation-invariant, stochastically conditionally time-consistent, and Markov if and only if transition risk mappings*

$$\sigma_t : \mathcal{X} \times \mathcal{P}(\mathcal{Y}) \times \mathcal{P}(\mathcal{X}) \times \mathcal{V} \to \mathbb{R}, \quad t = 1 \ldots T-1,$$

exist, such that

(i) *for all* $t = 1, \ldots, T-1$ *and all* $(x, \xi) \in \mathcal{X} \times \mathcal{P}(\mathcal{Y})$, *the risk form* $\sigma_t(x, \xi, \cdot, \cdot)$ *is normalized and monotonic with respect to stochastic dominance on* $\big(\mathcal{P}(\mathcal{X}), \mathcal{V}\big)$;

(ii) *for all* $\Pi \in \varPi$, *for all measurable bounded* r, *for all* $t = 1, \ldots, T-1$, *and for all* $h_t \in \mathcal{X}^t$,

$$v_t^{\Pi}(h_t) = r_t\big(x_t, \Xi_t(h_t), \pi_t(h_t)\big) + \sigma_t\big(x_t, \Xi_t(h_t), Q_t^{\Pi}(h_t), v_{t+1}^{\Pi}(h_t, \cdot)\big). \quad (8.89)$$

This allows us to evaluate the risk of Markov policies recursively.

Corollary 8.42. *Under the conditions of Theorem* 8.41, *for any Markov policy* Π, *the function* (8.88) *depends on* r, $\pi_t, \ldots, \pi_T$, *and* (x_t, ξ_t) *only, and the following relation is true:*

$$v_t^{\pi_t,\ldots,\pi_T}(x_t, \xi_t) = r_t(x_t, \xi_t, \pi_t(x_t, \xi_t)) + \\ \sigma_t\Big(x_t, \xi_t, \int_{\mathcal{Y}} K_t^X(x_t, y, \pi_t(x_t, \xi_t))\, \xi_t(\mathrm{d}y), V_{t+1}^{\Pi}\big(\cdot; x_t, \xi_t, \pi_t(x_t, \xi_t)\big)\Big). \quad (8.90)$$

where

$$V_{t+1}^{\Pi}\big(x'; x, \xi, u\big) = v_{t+1}^{\pi_{t+1},\ldots,\pi_T}(x', \Phi_t(x, \xi, u, x')), \quad x' \in \mathcal{X}.$$

8.4.3 Dynamic Programming

We consider a family of dynamic risk measures $\{\rho_{t,T}^{\Pi}\}_{t=1,\ldots,T}^{\Pi \in \varPi}$ which is normalized, translation-invariant, stochastically conditionally time-consistent, and Markov. Our objective is to analyze the risk minimization problem:

$$\min_{\Pi \in \varPi} v_1^{\Pi}(x_1, \Xi_1(x_1)), \quad x_1 \in \mathcal{X}.$$

For this purpose, we introduce the family of *value functions*:

$$v_t^*(h_t) = \inf_{\Pi \in \Pi_{t,T}} v_t^{\Pi}(h_t), \quad t = 1, \dots, T, \quad h_t \in \mathcal{X}^t, \tag{8.91}$$

where $\Pi_{t,T}$ is the set of feasible deterministic policies $\Pi = \{\pi_t, \dots, \pi_T\}$. By Theorem 8.41, transition risk mappings $\{\sigma_t\}_{t=1,\dots,T-1}$ exist, such that equations (8.89) hold.

The derivation of the dynamic programming equations mirrors the theory presented in § 8.1.4. The only difficulty is that the belief state is an element of the space $\mathcal{P}(\mathcal{Y})$ of probability measures. We assume that the spaces $\mathcal{P}(\mathcal{X})$ and $\mathcal{P}(\mathcal{Y})$ are equipped with the topology of weak convergence, and the space $\mathcal{V}$ is equipped with the topology of pointwise convergence. All continuity statements are made with respect to the said topologies.

We also assume that the kernels $K_t(x, y, u)$ have densities $q_t(\cdot, \cdot \mid x, y, u)$ with respect to a product probability measure $\mu_X \otimes \mu_Y$ on $\mathcal{X} \times \mathcal{Y}$, as in Example 8.38. In this case,

$$\begin{aligned}&\left[\int_{\mathcal{Y}} K_t^X(x, y, u)\, \xi(\mathrm{d}y)\right](\mathrm{d}x') \\ &\qquad = \left[\int_{\mathcal{Y}} \int_{\mathcal{Y}} q_t(x', y' \mid x, y, u)\, \xi(\mathrm{d}y)\, \mu_Y(\mathrm{d}y')\right] \mu_X(\mathrm{d}x'). \end{aligned} \tag{8.92}$$

Our main result is that the value functions (8.91) are *memoryless*, that is, they depend on (x_t, ξ_t) only, and that they satisfy a generalized form of a dynamic programming equation. The equation also allows us to identify the optimal policy.

Theorem 8.43. *We assume the following conditions:*

(i) *The functions* $(x, u) \mapsto q_t(x', y'|x, y, u)$ *are continuous, uniformly over* (x', y', y)*;*

(ii) *The transition risk mappings* $\sigma_t(\cdot, \cdot, \cdot, \cdot)$, $t = 1, \dots, T$, *are lower semicontinuous;*

(iii) *The functions* $r_t(\cdot, \cdot, \cdot)$, $t = 1, \dots, T$, *are lower semicontinuous;*

(iv) *The multifunctions* $\mathcal{U}_t(\cdot)$, $t = 1, \dots, T$, *are compact-valued and upper-semicontinuous.*

Then the functions v_t^*, $t = 1, \dots, T$ *are memoryless, lower semicontinuous, and satisfy the following dynamic programming equations:*

$$\begin{aligned}
v_T^*(x,\xi) &= \min_{u\in\mathcal{U}_T(x)} r_T(x,\xi,u), \quad x\in\mathcal{X}, \quad \xi\in\mathcal{P}(\mathcal{Y}),\\
v_t^*(x,\xi) &= \min_{u\in\mathcal{U}_t(x)} \Big\{ r_t(x,\xi,u) + \\
&\quad \sigma_t\Big(x,\xi,\int_{\mathcal{Y}} K_t^X(x,y,u)\,\xi(\mathrm{d}y), V_{t+1}^*(\cdot;x,\xi,u)\Big)\Big\},\\
&\quad x\in\mathcal{X}, \quad \xi\in\mathcal{P}(\mathcal{Y}), \quad t=T-1,\dots,1.
\end{aligned}$$

with

$$V_{t+1}^*(x';x,\xi,u) \triangleq v_{t+1}^*\big(x',\Phi_t(x,\xi,u,x')\big).$$

Moreover, an optimal Markov policy Π^* *exists and satisfies the equations:*

$$\begin{aligned}
\pi_T^*(x,\xi) &\in \operatorname*{arg\,min}_{u\in\mathcal{U}_T(x)} r_T(x,\xi,u), \quad x\in\mathcal{X}, \quad \xi\in\mathcal{P}(\mathcal{Y}),\\
\pi_t^*(x,\xi) &\in \operatorname*{arg\,min}_{u\in\mathcal{U}_t(x)} \Big\{ r_t(x,\xi,u) + \\
&\quad \sigma_t\Big(x,\xi,\int_{\mathcal{Y}} K_t^X(x,y,u)\,\xi(\mathrm{d}y), V_{t+1}^*(\cdot;x,\xi,u)\Big)\Big\},\\
&\quad x\in\mathcal{X}, \quad \xi\in\mathcal{P}(\mathcal{Y}), \quad t=T-1,\dots,1.
\end{aligned}$$

Proof. For all $h_T \in \mathcal{X}^T$ we have

$$v_T^*(h_T) = \inf_{\pi_T} r_T(x_T,\xi_T,\pi_T(h_T)) = \inf_{u\in\mathcal{U}_T(x_T)} r_T(x_T,\xi_T,u). \tag{8.93}$$

By assumptions (iii) and (iv), owing to the Berge theorem, the infimum in (8.93) is attained and is a lower semicontinuous function of (x_T,ξ_T). Hence, v_T^* is memoryless. Moreover, the optimal solution mapping $\Psi_T(x,\xi) = \big\{u \in \mathcal{U}_T(x) : r_T(x,\xi,u) = v_T^*(x,\xi)\big\}$ has nonempty and closed values and is measurable. Therefore, a measurable selector π_T^* of Ψ_T exists, and

$$v_T^*(h_T) = v_T^*(x_T,\xi_T) = v_T^{\pi_T^*}(x_T,\xi_T).$$

We prove the theorem by induction backward in time. Suppose $v_{t+1}^*(\cdot)$ is memoryless, lower semicontinuous, and Markov decision rules $\{\pi_{t+1}^*,\dots,\pi_T^*\}$ exist such that

$$v_{t+1}^*(h_{t+1}) = v_{t+1}^*(x_{t+1},\xi_{t+1}) = v_{t+1}^{\{\pi_{t+1}^*,\dots,\pi_T^*\}}(x_{t+1},\xi_{t+1}), \ \forall h_{t+1}\in\mathcal{X}^{t+1}.$$

Then for any $h_t \in \mathcal{X}^t$ formula (8.89), after substituting (8.86), yields

$$v_t^*(h_t) = \inf_{\Pi \in \Pi_{t,T}} \Big\{ r_t(x_t, \xi_t, \pi_t(h_t)) + \sigma_t\Big(x_t, \xi_t, \int_{\mathcal{Y}} K_t^X(x_t, y, \pi_t(h_t))\, \xi_t(\mathrm{d}y), v_{t+1}^{\Pi}(h_t, \cdot)\Big)\Big\}.$$

Since $v_{t+1}^{\Pi}(h_t, \cdot) \geq V_{t+1}^*(\cdot; x_t, \xi_t, \pi_t(h_t))$ and σ_t is non-decreasing with respect to the last argument, we obtain

$$\begin{aligned} v_t^*(h_t) &\geq \inf_{\Pi \in \Pi_{t,T}} \Big\{ r_t(x_t, \xi_t, \pi_t(h_t)) \\ &\quad + \sigma_t\Big(x_t, \xi_t, \int_{\mathcal{Y}} K_t^X(x_t, y, \pi_t(h_t))\, \xi_t(\mathrm{d}y), V_{t+1}^*(\cdot; x, \xi, \pi_t(h_t))\Big)\Big\} \\ &= \inf_{u \in \mathcal{U}_t(x_t)} \Big\{ r_t(x_t, \xi_t, u) \\ &\quad + \sigma_t\Big(x_t, \xi_t, \int_{\mathcal{Y}} K_t^X(x_t, y, u)\, \xi_t(\mathrm{d}y), V_{t+1}^*(\cdot; x, \xi, u)\Big)\Big\}. \end{aligned} \tag{8.94}$$

In order to complete the induction step, we need to establish the lower semicontinuity of the mapping

$$(x, \xi, u) \mapsto \sigma_t\Big(x, \xi, \int_{\mathcal{Y}} K_t^X(x, y, u)\, \xi(\mathrm{d}y), V_{t+1}^*(\cdot; x, \xi, u)\Big). \tag{8.95}$$

To this end, suppose $x^{(k)} \to x$, $\xi^{(k)} \to \xi$ (weakly), $u^{(k)} \to u$, as $k \to \infty$.

First, we verify that the mapping $(x, \xi, u) \mapsto \int_{\mathcal{Y}} K_t^X(x, y, u)\, \xi(\mathrm{d}y)$ appearing in the third argument of σ_t is weakly continuous. By formula (8.92), for any bounded continuous function $f : \mathcal{X} \to \mathbb{R}$ we have

$$\begin{aligned} &\int_{\mathcal{X}} f(x') \Big[\int_{\mathcal{Y}} K_t^X(x^{(k)}, y, u^{(k)})\, \xi^{(k)}(\mathrm{d}y) \Big](\mathrm{d}x') \\ &= \int_{\mathcal{X}} f(x') \Big[\int_{\mathcal{Y}} \int_{\mathcal{Y}} q_t(x', y' \mid x^{(k)}, y, u^{(k)})\, \xi^{(k)}(\mathrm{d}y)\, \mu_Y(\mathrm{d}y') \Big]\, \mu_X(\mathrm{d}x'). \end{aligned} \tag{8.96}$$

By assumption (i),

$$\lim_{k \to \infty} \int_{\mathcal{Y}} [q_t(x', y' \mid x^{(k)}, y, u^{(k)}) - q_t(x', y' \mid x, y, u)] \xi^{(k)}(\mathrm{d}y) = 0, \tag{8.97}$$

uniformly over x', y'. Moreover, by the Lebesgue theorem, the function

$$y \mapsto \int_{\mathcal{X}} f(x') \int_{\mathcal{Y}} q_t(x', y' \mid x, y, u)\ \mu_Y(\mathrm{d}y')\ \mu_X(\mathrm{d}x') \tag{8.98}$$

is continuous. Therefore, combining (8.96) and (8.97), we obtain the chain of equations:

$$\begin{aligned}
&\lim_{k\to\infty} \int_{\mathcal{X}} f(x') \left[\int_{\mathcal{Y}} K_t^X(x^{(k)}, y, u^{(k)})\ \xi^{(k)}(\mathrm{d}y) \right](\mathrm{d}x') \\
&= \lim_{k\to\infty} \int_{\mathcal{X}} f(x') \left[\int_{\mathcal{Y}} \int_{\mathcal{Y}} q_t(x', y' \mid x, y, u)\, \xi^{(k)}(\mathrm{d}y)\, \mu_Y(\mathrm{d}y') \right] \mu_X(\mathrm{d}x') \\
&= \lim_{k\to\infty} \int_{\mathcal{Y}} \left[\int_{\mathcal{X}} f(x') \int_{\mathcal{Y}} q_t(x', y' \mid x, y, u)\ \mu_Y(\mathrm{d}y')\ \mu_X(\mathrm{d}x') \right] \xi^{(k)}(\mathrm{d}y) \\
&= \int_{\mathcal{Y}} \left[\int_{\mathcal{X}} f(x') \int_{\mathcal{Y}} q_t(x', y' \mid x, y, u)\ \mu_Y(\mathrm{d}y')\ \mu_X(\mathrm{d}x') \right] \xi(\mathrm{d}y) \\
&= \int_{\mathcal{X}} f(x') \left[\int_{\mathcal{Y}} K_t^X(x, y, u)\ \xi(\mathrm{d}y) \right](\mathrm{d}x').
\end{aligned}$$

The last by one equation follows from the weak convergence of $\xi^{(k)}$ to ξ and from the continuity of the function (8.98). Thus, the third argument of σ_t in (8.95) is continuous with respect to (x, ξ, u).

Let us examine the last argument of σ_t in (8.95). By (8.97), for every continuous bounded function $f(\cdot)$ on $\mathcal{Y}$, and for each fixed $x' \in \mathcal{X}$,

$$\begin{aligned}
&\lim_{k\to\infty} \int_{\mathcal{Y}} f(y')\ \Phi_t(x^{(k)}, \xi^{(k)}, u^{(k)}, x')(\mathrm{d}y') \\
&\qquad = \lim_{k\to\infty} \frac{\int_{\mathcal{Y}} f(y') \int_{\mathcal{Y}} q_t(x', y' \mid x^{(k)}, y, u^{(k)})\ \xi^{(k)}(\mathrm{d}y)\ \mu_Y(\mathrm{d}y')}{\int_{\mathcal{Y}} \int_{\mathcal{Y}} q_t(x', y' \mid x^{(k)}, y, u^{(k)})\ \xi^{(k)}(\mathrm{d}y)\ \mu_Y(\mathrm{d}y')} \\
&\qquad\qquad = \frac{\int_{\mathcal{Y}} f(y') \int_{\mathcal{Y}} q_t(x', y' \mid x, y, u)\ \xi(\mathrm{d}y)\ \mu_Y(\mathrm{d}y')}{\int_{\mathcal{Y}} \int_{\mathcal{Y}} q_t(x', y' \mid x, y, u)\ \xi(\mathrm{d}y)\ \mu_Y(\mathrm{d}y')},
\end{aligned}$$

provided that (x, ξ, u, x') is such that

$$\int_{\mathcal{Y}} \int_{\mathcal{Y}} q_t(x', y' \mid x, y, u)\ \xi(\mathrm{d}y)\ \mu_Y(\mathrm{d}y') > 0. \tag{8.99}$$

Therefore, the operator $\Phi_t(\cdot, \cdot, \cdot, x')$ is weakly continuous at these points. Let x, ξ, u be fixed. Consider the sequence of functions $V_{t+1}^{(k)} : \mathcal{X} \to \mathbb{R}$, $k = 1, 2, \ldots$ defined as follows:

$$V_{t+1}^{(k)}(x') = v_{t+1}^*\big(x', \Phi_t(x^{(k)}, \xi^{(k)}, u^{(k)}, x')\big),$$

Since $v_{t+1}^*(\cdot,\cdot)$ is lower-semicontinuous and $\Phi_t(\cdot,\cdot,\cdot,x')$ is continuous, whenever condition (8.99) is satisfied, we infer that $V_{t+1}^*(x') \le \liminf_{k\to\infty} V_{t+1}^{(k)}(x')$, at all $x' \in \mathcal{X}$ at which (8.99) holds. As v_{t+1}^* and Φ_t are measurable, both $V_{t+1}^*(\cdot)$ and $\liminf_{k\to\infty} V_{t+1}^{(k)}(\cdot)$ are measurable as well.

According to Theorem 8.41, the mapping σ_t is monotonic with respect to the stochastic dominance $\preceq_{(1)}$ of the last argument with respect to the probability measure $\int_{\mathcal{Y}} K_t^X(x,y,u)\;\xi(\mathrm{d}y)$. Since the probability od transition from (x,ξ) to a point x' such that

$$\int_{\mathcal{Y}}\int_{\mathcal{Y}} q_t(x',y' \mid x,y,u)\;\xi(\mathrm{d}y)\;\mu_Y(\mathrm{d}y') = 0$$

is zero, then the value of $\liminf_{k\to\infty} V_{t+1}^{(k)}(x')$ at the set of x' at which (8.99) is violated, is irrelevant. Consequently, by assumption (ii), with the view at the already established continuity of the third argument, we obtain the following chain of relations:

$$\begin{aligned}
&\sigma_t\Big(x,\xi,\int_{\mathcal{Y}} K_t^X(x,y,u)\;\xi(\mathrm{d}y), V_{t+1}^*(\cdot;x,\xi,u)\Big)\\
&\qquad \le \sigma_t\Big(x,\xi,\int_{\mathcal{Y}} K_t^X(x,y,u)\;\xi(\mathrm{d}y), \liminf_{k\to\infty} V^{(k)}\Big)\\
&\qquad = \sigma_t\Big(x,\xi,\lim_{k\to\infty}\int_{\mathcal{Y}} K_t^X(x^{(k)},y,u^{(k)})\;\xi^{(k)}(\mathrm{d}y), \liminf_{k\to\infty} V_{t+1}^{(k)}\Big)\\
&\qquad \le \liminf_{k\to\infty} \sigma_t\Big(x^{(k)},\xi^{(k)},\int_{\mathcal{Y}} K_t^X(x^{(k)},y,u^{(k)})\;\xi^{(k)}(\mathrm{d}y), V_{t+1}^{(k)}\Big).
\end{aligned}$$

Consequently, the mapping (8.95) is lower semicontinuous.

Using assumptions (ii) and (iv) and invoking the Berge theorem again, we deduce that the infimum in (8.94) is attained and is a lower semicontinuous function of (x_t,ξ_t). Moreover, the optimal solution mapping, that is, the set of $u \in \mathcal{U}_T(x)$ at which the infimum in (8.94) is attained, is nonempty, closed-valued, and measurable. Therefore, a minimizer π_t^* in (8.94) exists and is a measurable function of (x_t,ξ_t). Substituting this minimizer into (8.94), we obtain

$$\begin{aligned}
v_t^*(h_t) \ge\;& r_t\big(x_t,\xi_t,\pi_t^*(x_t,\xi_t)\big)\\
&+\sigma_t\Big(x_t,\xi_t,\int_{\mathcal{Y}} K_t^X\big(x_t,y,\pi_t^*(x_t,\xi_t)\big)\;\xi_t(\mathrm{d}y), V_{t+1}^*\big(x,\xi,\pi_t^*(x_t,\xi_t)\big)\Big)\\
&\qquad = v_t^{\{\pi_t^*,\dots,\pi_T^*\}}(x_t,\xi_t).
\end{aligned}$$

In the last equation, we used Corollary 8.42. On the other hand, we have

$$v_t^*(h_t) = \inf_{\Pi \in \Pi_{t,T}} v_t^{\Pi}(h_t) \le v_t^{\{\pi_t^*,\dots,\pi_T^*\}}(x_t, \xi_t).$$

Therefore $v_t^*(h_t) = v_t^{\{\pi_t^*,\dots,\pi_T^*\}}(x_t, \xi_t)$ is memoryless, lower semicontinuous, and

$$\begin{aligned} v_t^*(x_t, \xi_t) &= \min_{u \in \mathcal{U}_t(x_t)} \Big\{ r_t(x_t, \xi_t, u) \\ &+ \sigma_t\Big(x_t, \xi_t, \int_{\mathcal{Y}} K_t^X(x_t, y, u)\, \xi_t(\mathrm{d}y), V_{t+1}^*(x_t, \xi_t, u)\Big)\Big\} \\ &= r_t\big(x_t, \xi_t, \pi_t^*(x_t, \xi_t)\big) \\ &\quad + \sigma_t\Big(x_t, \xi_t, \int_{\mathcal{Y}} K_t^X\big(x_t, y, \pi_t^*(x_t, \xi_t)\big)\, \xi_t(\mathrm{d}y), V_{t+1}^*\big(x_t, \xi_t, \pi_t^*(x_t, \xi_t)\big)\Big). \end{aligned}$$

This completes the induction step. □

The most essential assumption of Theorem 8.43 is the assumption (ii) of the lower semicontinuity of the transition risk mappings $\sigma_t(\cdot, \cdot, \cdot, \cdot)$. If these mappings are derived from convex or coherent risk measures, their lower semicontinuity with respect to the last argument follows from the corresponding property of the risk measure. In particular, Theorem 2.32 derives the continuity from the monotonicity on Banach lattices. The semicontinuity with respect to the third argument, the probability measure, is a more complex issue. Lemma 8.16 verifies this condition for the mean-semideviation risk measure. Similar remarks apply to assumption (iii) about the state–control risk functions.

We illustrate our ideas with the following example.

Example 8.44. In clinical trials, the potency of a drug is characterized by an unknown parameter $y \in \mathcal{Y}$, where $\mathcal{Y}$ is a finite set. The purpose of the trials is to estimate y and to determine the optimal dose. Let us assume for simplicity that y *is* the optimal dose and that $\mathcal{U} = \mathcal{Y}$. At time $t = 1$ we have a prior distribution $\xi_1 \in \mathcal{P}(\mathcal{Y})$.

If a dose $u_t \in \mathcal{Y}$ is administered to a patient at time t, a response X_{t+1} is observed (the subscript $t + 1$ indicates that the response is not known when u_t is determined). In many cases, X_{t+1} is a Bernoulli random variable, with $X_{t+1} = 1$ representing toxic response, and $X_{t+1} = 0$ nontoxic. The probability of a toxic response is a function of y and u_t, that is, $P[X_{t+1} = 1] = \Psi(y, u_t)$. We assume that we know the family of functions $\Psi(\cdot, \cdot)$. The "cost" is $c_t(y, u_t)$; it depends on both the applied and best doses. For example, it may have the form $c_t(y, u) = |u - y|$ to penalize for the over- and under-dosage. It is not observed.

The process continues for T stages, with u_T being the final dose recommendation, whose cost is equal to $c_T(y, u_T)$ (for example, $M|u_T - y|$, with large M).

The problem can be cast to our setting. The state space $\mathcal{X}$ is $\{0, 1\}$. The transition kernel does not depend on X at all; the distribution of the next observation X_{t+1} depends on y and u_t only. The unobserved state y does not move, so we have the parametric uncertainty model of Example 8.36. Thus, for all $y \in \mathcal{Y}$

$$\Phi_t(\xi_t, u_t, x_{t+1})(y) \propto \begin{cases} \xi_t(y)\, \Psi(y, u_t) & \text{if } x_{t+1} = 1, \\ \xi_t(y)\,\big(1 - \Psi(y, u_t)\big) & \text{if } x_{t+1} = 0. \end{cases}$$

The patients participating in the trials are in dire need of treatment. The consequences of under- or over-dosage can be severe, including death. Ethical issues motivate risk-averse trial designs.

Define, as in the expected-value model,

$$r_t(\xi, u) = \sum_{y \in \mathcal{Y}} c_t(y, u)\, \xi(y), \quad t = 1, \dots, T.$$

Observe that the only component of the extended state (x, ξ) that carries the information from the past is the belief state ξ. With an additional assumption that the transition risk mapping does not directly depend on its first argument ξ, the dynamic programming equations of Theorem 8.43 simplify:

$$v_T^*(\xi) = \min_{u \in \mathcal{Y}} r_T(\xi, u), \quad \xi \in \mathcal{P}(\mathcal{Y}),$$

$$v_t^*(\xi) = \min_{u \in \mathcal{U}_t(x)} \Big\{ r_t(\xi, u) + \sigma_t\Big(\textstyle\sum_{y \in \mathcal{Y}} \xi(y)\, \Psi(y, u), v_{t+1}^*\big(\Phi_t(\xi, u, \cdot)\big)\Big)\Big\},$$
$$\xi \in \mathcal{P}(\mathcal{Y}), \quad t = T-1, \dots, 1.$$

In the expected-value model, we would have

$$\begin{aligned} \sigma_t\Big(\textstyle\sum_{y \in \mathcal{Y}} \Psi(y, u)\, \xi(y), v_{t+1}^*\big(\Phi_t(\xi, u, \cdot)\big)\Big) = &\\ \sum_{y \in \mathcal{Y}} \xi(y)\, \Psi(y, u)\, v_{t+1}^*\big(\Phi_t(\xi, u, 1)\big) &\\ + \Big(1 - \sum_{y \in \mathcal{Y}} \xi(y)\, \Psi(y, u)\Big)\, v_{t+1}^*\big(\Phi_t(\xi, u, 0)\big). \end{aligned}$$

In the risk-averse case, we may use the worst-case risk form as σ_t:

$$\sigma_t\Big(\sum_{y\in\mathcal{Y}}\Psi(y,u)\,\xi(y),v^*_{t+1}\big(\Phi_t(\xi,u,\cdot)\big)\Big) = \max\Big\{v^*_{t+1}\big(\Phi_t(\xi,u,1)\big),v^*_{t+1}\big(\Phi_t(\xi,u,0)\big)\Big\}.$$

In this application, it has the advantage of not depending directly on the probability of a toxic response.

As only two possible outcomes may occur, convex combinations of the two versions of a transition risk mapping cover the entire spectrum of operators that may be used here. □

8.5 Relation to Min–Max Markov Control Models

Close relations exist between our risk-averse control models and min–max Markov control problems.

We start from the finite horizon model. If the transition risk mappings $\sigma_t(x, Q(x,u),\cdot)$ are coherent measures of risk on the space $\mathcal{L}_p(\mathcal{X},\mathcal{B}(\mathcal{X}),Q(x,u))$, for some $p\in[1,\infty)$, they have the dual representation (2.28):

$$\sigma_t(x,Q(x,u),v) = \max_{\mu\in\mathcal{A}_t(x,Q_t(x,u))}\langle\mu,v\rangle,\quad v\in\mathcal{V}.$$

The risk-averse preferences may be interpreted as distributionally robust preferences, with ambiguity in the transition kernel: the distribution of the next state is chosen from the set $\mathcal{S}_t(x_t,u_t)=\mathcal{A}_t(x_t,Q_t(x_t,u_t))$. We can, therefore, envisage a *min–max control model*, in which two players are active. Player 1 chooses at each time t and at each state x_t a control $u_t\in\mathcal{U}_t(x_t)$. Player 2, the opponent, given x_t and u_t, selects a probability measure $\mu_t\in\mathcal{S}_t(x_t,u_t)$ which describes the distribution of the next state x_{t+1}.

Let us start from the finite horizon problem. We denote, as before, by $\Pi=(\pi_1,\dots,\pi_T)$ a policy of Player 1, where $\pi_t:\mathcal{H}_t\to\mathfrak{U}$, with $\pi_t(x_1,\dots,x_t)\in\mathcal{U}_t(x_t)$. Similar, $\Gamma=(\gamma_1,\dots,\gamma_T)$ is a policy of Player 2, where $\gamma_t:\mathcal{H}_t\times\mathfrak{U}\to\mathcal{P}(\mathcal{X})$. The feasible set of Player 2 depends on the policy of Player 1: $\mathbb{G}(\Pi)=\{\Gamma:\gamma_t(x_1,\dots,x_t,u_t)\in\mathcal{S}_t(x_t,u_t),\ t=1,2,\dots\}$. The min–max control problem is defined as follows:

$$\min_{\Pi\in\mathit{\Pi}}\ \max_{\Gamma\in\mathbb{G}(\Pi)}\ \mathbb{E}\left[\sum_{t=1}^{T}c_t(x_t,u_t)+c_{T+1}(x_{T+1})\right],\tag{8.100}$$

where for each t we have $u_t=\pi_t(x_1,\dots,x_t)$, and the conditional distribution of x_{t+1} given $\{x_1,\dots,x_t,u_t\}$ is $\mu_t=\gamma_t(x_1,\dots,x_t,u_t)$.

Corollary 8.45. *Suppose all assumptions of Theorem* 8.15 *are satisfied and the transition risk mappings* $\sigma_t(x,Q(x,u),\cdot)$ *are coherent measures of risk on the* *space* $\mathcal{L}_p(\mathfrak{X}, \mathcal{B}(\mathfrak{X}), Q(x,u))$, *for some* $p \in [1,\infty)$. *Then every optimal solution of problem* (8.16) *is also an optimal solution to problem* (8.100). *Furthermore, for every Markov optimal policy* $\Pi^* = \{\pi_1^*, \dots, \pi_T^*\}$ *of Player 1, the sequence of transition kernels* $\{\mu_1^*, \dots, \mu_T^*\}$ *such that*

$$\mu_t^*(x) \in \underset{\mu \in \mathcal{S}_t(x,\pi_t^*(x))}{\arg\max} \langle v_{t+1}^*, \mu \rangle, \quad x \in \mathfrak{X}, \quad t = 1, \dots, T,$$

is an optimal Markov policy of Player 2.

A similar observation holds for the *stationary infinite horizon discounted problem.* A policy of Player 1 is a sequence $\Pi = (\pi_1, \pi_2, \dots)$ with each $\pi_t : \mathcal{H}_t \to \mathfrak{U}$ such that $\pi_t(x_1, \dots, x_t) \in U(x_t)$. A policy of Player 2 is a sequence $\Gamma = (\gamma_1, \gamma_2, \dots)$ with each $\gamma_t : \mathcal{H}_t \times \mathfrak{U} \to \mathcal{P}(\mathfrak{X})$. The feasible set of Player 2, $\mathbb{G}(\Pi)$, is defined by the conditions $\gamma_t(x_1, \dots, x_t, u_t) \in \mathcal{S}(x_t, u_t) = \mathcal{A}(x_t, Q(x_t, u_t))$, $t = 1, 2, \dots$, and depends on the policy of Player 1. For a given $\alpha \in (0,1)$, our aim is to find a policy $\Pi = \{\pi_t\}_{t=1}^{\infty}$ so as to minimize the worst expected discounted cost:

$$\min_{\Pi} \max_{\Gamma \in \mathbb{G}(\Pi)} \mathbb{E}\left[\sum_{t=1}^{\infty} \alpha^{t-1} c(x_t, u_t)\right]. \tag{8.101}$$

In the problem above, for each t the control of Player 1 is $u_t = \pi_t(x_1, \dots, x_t)$, and the conditional distribution of x_{t+1} given $x_1, \dots, x_t, u_t$ is the control of Player 2: $\mu_t = \gamma_t(x_1, \dots, x_t, u_t)$.

Corollary 8.46. *Suppose all assumptions of Theorem* 8.21 *are satisfied and the transition risk mapping* $\sigma(x, Q(x,u), \cdot)$ *is a coherent measures of risk on the space* $\mathcal{L}_p(\mathfrak{X}, \mathcal{B}(\mathfrak{X}), Q(x,u))$, *for some* $p \in [1,\infty)$. *Then every optimal solution of problem* (8.31) *is also an optimal solution to problem* (8.101). *Furthermore, for every stationary Markov optimal policy* $\Pi^* = \{\pi^*, \pi^*, \dots\}$ *of Player 1, the sequence of transition kernels* $\{\mu^*, \mu^*, \dots\}$ *such that*

$$\mu^*(x) \in \underset{\mu \in \mathcal{S}(x,\pi^*(x))}{\arg\max} \langle v^*, \mu \rangle, \quad x \in \mathfrak{X},$$

is an optimal stationary Markov policy of Player 2.

It may be worth stressing that the min-max problems (8.100) and (8.101) are specific in the sense that the feasible sets of the actions of Player 2 depend on the actions of Player 1. Therefore the "min" and the "max" operators cannot be, in general, interchanged, even if we allow mixed strategies of the players.

Bibliographical Remarks

Chapter 1

The expected utility theory of von Neumann and Morgernstern [275], and the dual utility theory of Quiggin [219] and Yaari [283], are classical topics in mathematical economics; see, e.g., [117]. Our objective is to relate them to our main interest in risk and optimization. In our presentation, we follow [74].

The existence of a continuous numerical representation of a continuous total preorder on a separable and connected topological space is a classical result in topology [97, §6]. The generalization attempts in [57, Thm. II] and [223, Thm. 1] suffered from errors, which were corrected in [58]. A short and clear proof was eventually provided by [143]. For extensions and further discussion, see [35, 38].

The axioms of §1.2 are exactly the conditions assumed in the pioneering work of [275]; see also [113, §8.2–3], [116, §2.2],[134]. The property of quasi-concavity is called uncertainty aversion in [121, 254]. The main theorem of this section is due to [275]. Its complicated constructive proof has been reproduced in many sources or emulated in the setting of mixture sets (see, e.g., [113, Thm. 8.4], [116, Thm. 2, Ch. 2], [134], and the references therein). For the separation theorem for sets with a nonempty algebraic interior, see [88, 96, 151].

The method of §1.2.3 is standard; see [113, §10], [116, Ch. 3, Thm. 1–4]. The conditional expectations in infinitely-dimensional spaces are discussed in [163, §2.1].

In [283], the dual utility theory considers the space of uniformly bounded random variables on an implicitly assumed atomless probability space. The reader may consult [93, Ch. III] for information about integration with respect to a finitely additive measure and spaces of bounded functions.

D. Dentcheva, A. Ruszczyński, *Risk-Averse Optimization and Control*, Springer Series in Operations Research and Financial Engineering,
https://doi.org/10.1007/978-3-031-57988-2

Quiggin [219] considers discrete distributions and derives, from a different system of axioms, the existence of an "anticipated utility functional"; see also [220, Ch. 11] and [221, Thm. 3.2]. The term "rank-dependent utility function" is borrowed from this theory.

The assertion of Theorem 1.18, in the case of distributions supported on [0, 1], is similar to the assertion of [283, Thm. 1]. However, our assumptions are weaker, and we could not resort to the expected utility theory applied to the quantile functions, as in the proof of [283, Thm. 1]. Formula (1.14) is a special case of the Choquet integral of the function $F^{-1}(\cdot)$; see [49].

The extreme points of the set defined by the convex order in the proof of Theorem 1.33 are derived in [252]. The Krein–Milman theorem is due to [155].

Chapter 2

The first attempts to construct an axiomatic theory of risk measures are due to [148] in the context of mean–risk models, albeit without the monotonicity axiom. This was supplemented in [200] by monotonicity with respect to stochastic dominance. The complete sets of axioms, as formulated in §2.1, are due to [8] for measures defined on finite probability spaces. We may remark here that the translation axiom was used in the expected utility theory as "delta property"; see [137].

Mean–semideviation measures were introduced in [200, 201], building on an earlier work [152]. The mean–deviation point of view was further advanced in [231, 235]. The inverse measures of Example 2.5 were analyzed in [20, 83, 157]. The entropic risk measure has been introduced in [144, 145] to model risk in dynamic programming models. The entropic value at risk is due to [2]; see also [3, 215]. The optimized certainty equivalent is due to [22, 23]; see also [273].

The Value at Risk is a classical concept in finance, see, e.g., [60, 92]. The Average (Conditional) Value at Risk was introduced in [8] as a Tail Conditional Expectation. The extremal representation of Theorem 2.16 was derived in [230] in the smooth case; see also [1, 208]. The relation to the Lorenz curve [171], the duality Theorem 2.15, and Corollary 2.17 are due to [202]. The buffered probability of exceedance is due to [175]. Generalized quantile measures were discussed in [21]. The Wang transform was proposed in [276]. Polyhedral risk measures were introduced in [95].

Duality theory for risk measures was developed in [8] (for finite probability spaces), and then in [59, 119, 165] (for the space of essentially bounded random variables). Theorems 2.25 and 2.32 are due to [250]; see also [46, 47, 110, 277]. For the background of convex analysis in infinite-dimensional spaces and Fenchel–Moreau duality see [213, 229, 284]. The Namioka–Klee theorem is due to [192]; see also Theorem 4.15.

The Kusuoka representation of law-invariant coherent risk measures was derived in [161] for spaces of essentially bounded random variables. For an extension to integrable variables, see [212].

Risk measures on the space of the quantile functions were considered in [75].

Systemic risk measures with aggregation functions were discussed in [43] in finite probability spaces and further extended in [158], where the authors analyzed convex risk measures defined on a general measurable space and proposed examples of aggregation functions suitable for financial systems. Different versions of the translation property were discussed in [36, 99, 242]. Our presentation follows [5].

The theory of risk forms is due to [77, 79]. For background information about set-valued analysis and integration, see [12] and [204].

Chapter 3

The subdifferential of the expected value of a family of convex functions was first derived in [264] for positively homogeneous functions (we recall it as Theorem 3.13). For generalizations to convex functions and convex integrands, see [142, 227, 228] and the books [39, 166]. Convex lattice-valued operators and their subgradients were considered by [270]. The first results on the differentiation of a composition of a risk measure and a convex operator are due to [250, 258]. In §3.1.2 and §3.1.3 we improve on these results by exploiting the local property of the operator. The Hahn–Banach theorem on lattices can be found in [147, Thm. X.5.7].

For the subdifferentiation of Lipschitz continuous functions, see [51, 53], and the Banach space extension of [52, 162]. Generalized differentiable functions were introduced in [195]; see also [185]. The content of §3.1.4–3.1.7 is original. §3.1.8 is based on [33].

Duality associated with risk measures is specific to the risk-averse formulation. It was first analyzed in [250] and then exploited in various contexts in a multitude of applications, mainly in the multistage setting. Duality associated with the nonanticipativity constraints is one of the fundamental ideas in stochastic optimization, due to [98, 105, 278]. It is heavily exploited in multistage optimization; we discuss it later. In the risk-averse context, the topic was first analyzed in [250].

Early applications of risk measures in portfolio optimization were discussed, inter alia, by [152, 181, 182, 186, 251].

Stochastic optimization models in machine learning with robustness considerations were discussed, *inter alia*, in [126, 174, 272] and the references therein. Stochastic subgradients of mini-batch risk functions were analyzed in [79]. The stochastic subgradient method for convex problems is due to [101, 102], deriving from the classical idea of stochastic approximation [225], and the deterministic subgradient method [100, 216, 260]. Finite iteration error bounds are due to [120, 217, 218]. Our analysis for generalized differentiable functions uses the differential inclusion method proposed in [19, 274], further developed in [24, 25], and specialized to nonsmooth stochastic optimization in [91, 177]. For background information on differential inclusions, see [11], and on the differential equation method for stochastic approximation, see [34, 160, 170]. The nonsmooth and nonconvex version of the method with dual averaging is due to [244, 249]; for related research in the weakly convex case, see [176].

Chapter 4

The question of consistency of preferences over time was already considered in the utility theory by [153] and [156]. In the context of dynamic measures of risk, time consistency was considered by [9] and [48]. Our formulation is based on [248].

Conditional risk measures and their dual representations were considered in [45, 86, 118, 224, 253]. Our presentation is to a large extent original; it combines some ideas of [111], with the material of §3.1.1, and refines the classical duality theory for convex lattice-valued operators of [285].

The composition of multikernels discussed in §4.2.4 corresponds to the concept of a product-type (rectangular) collection of probability measures of [237].

Construction of scenario trees originated in [138], with the use of conditional moments, and was further developed in [131, 209–211], with the use of probability metrics.

Optimality conditions and duality theory for convex multistage stochastic optimization problems, with objective functionals defined as expected values of convex integrands, were developed in [98, 105, 203, 232–234]. Modern contributions are [205, 206]. Our approach, based on [78], overcomes the difficulties associated with the use of the spaces of essentially bounded functions. The concepts of metric regularity and subregularity of multifunctions employed there are at the core of modern variational analysis. For the early contributions, see [149, 150, 189, 226]. A comprehensive exposition can be found in [89, 141, 159]. References [37, 140] discuss the use in the derivation of the optimality conditions. For the theory of causal operators, see [56]. The superpositional measurability is covered in [6].

Decomposition methods are standard approaches to two- and multistage stochastic programming problems. For two-stage linear expected-value problems, cutting plane methods were proposed in [271] and improved in [30, 243]. The extension to multistage problems (frequently referred to as Stochastic Dual Dynamic Programming) is due to [29, 194, 207]; for other advances, see

[42, 245] and the review [247]. Primal decomposition of mean-risk two-stage problems was developed in [4]. General coherent conditional risk measures in two-stage problems were considered in [187]. Specialized mean–AVaR models were considered in [196]. Multistage problems with mean-AVaR criteria were discussed in [124, 136, 154, 256, 257]. Polyhedral risk measures in and multistage problems were treated in [125]. A method for general coherent measures of risk was proposed in [214]. Our analysis of the primal methods for convex problems in §4.5 is original.

Dual (scenario) decomposition methods for expected value problems were introduced in [191, 246]. A risk-averse version was developed in [54].

Chapter 5

To the best of our knowledge the first general notion of a stochastic order is introduced in [172]. It is related to the theory of weak majorization as discussed in [7]. A comparison of random variables with respect to the first-order stochastic dominance is used in the context of statistical tests in the early works [31, 164, 179]. The stochastic dominance relations have received considerable attention after the works [18, 112, 127, 128, 222, 238]. The characterization of the first and second-order dominance via families of (differentiable) utility functions was stated in [127] and rigorously proved in [18], where also the third-order dominance is characterized. Stochastic dominance of fractional order $p \geq 1$ was introduced for non-negative bounded random variables in [114], where also the monotonicity of the relation was proved. The form of the generators of the stochastic dominance relations of general order p given here is new. A thorough survey and analysis of stochastic order relations is contained in the monographs [190] and [255].

The properties of the integrated distribution functions in Lemma 5.4 are shown for integer p in [201]. The moment inequalities in Theorem 5.5 were shown first in [115]; our presentation follows the exposition in [199]. The characterization of the stochastic dominance relation of first and second order by distortion functions (or rank-dependent utility functions) in Theorem 5.14 was derived in [67]. The conjugate duality relation for the upper Lorenz function was shown in [61]. The consistency of the stochastic dominance relation of second order with law-invariant measures of risk was first shown for specific measures of risk in [200–202] and in a more general setting in [165]. The consistency of the mean-semideviations with the corresponding stochastic dominance relation in Theorem 5.23 is shown for orders with integer $p \geq 2$ in [201, Theorem 1]. We provide a unifying proof for all $p \geq 2$. Theorem 5.28 is a generalization of the result in [188] where it is assumed additionally that $\mathbb{E}[X] = \mathbb{E}[Y]$. Theorem 5.30 is a generalization of the result in [252] where random variables on the

Lebesgue space on $(0, 1)$ are considered such that $X \succeq_{(2)} Y$ and $\mathbb{E}[X] = \mathbb{E}[Y]$. Theorems 5.28 and 5.30 were first shown in [64].

Optimization problems with stochastic dominance constraints were introduced in [63] for integer orders and further analyzed in [65]. The presentation for fractional orders is new. The optimality conditions in Lagrangian form, which show the role of utility functions of the generator as Lagrange multipliers, are established first in [63] in a simpler problem formulation. The optimality conditions and the duality result for $p = 2$ under convexity assumptions are further generalized to the form of Theorems 5.36 and 5.37 in [65]. Ref. [94] refines these results to obtain the second welfare theorem in persuasion economics.

The optimality conditions for non-convex problems with second-order stochastic dominance constraints are established in [69]; Theorems 5.45 and 5.47 are generalizations of the results there. Problems with stochastic dominance constraint of second order using the Lorenz function (5.91)–(5.93) are introduced in [67] where the optimality conditions in Theorem 5.49 are proved as well as a duality to rank-dependent utility functions (distortions) in Theorem 5.50. The duality relations to risk functionals and optimality conditions using coherent measures of risk for problem (5.91)–(5.93) are proved in [70]. Optimization with stochastic dominance constraints relates also to chance constraints or Average (Conditional) Value-at-Risk constraints, see [68]. The optimality conditions in subgradient form presented in Theorem 5.41 are new. Stochastic optimization problems with first-order stochastic dominance constraints are formulated and analyzed in [66].

Another aspect not covered here is the stability of the dominance-constrained problems with respect to the probability measure. We refer the readers to [62, 82, 169] and the references therein.

Chapter 6

The definition of stochastic orders for random vectors, which is adopted here, is a modification of the notion in [190, Definition 3.5.1] called increasing positive linear convex order, where the requirement $c^\top X \preceq_{(2)} c^\top Y$ is imposed for all $c \in \mathbb{S}_+^m$. Similar notions with scalarization in more general subsets of $\mathbb{R}^m$ are adopted in [135, 139]. The set-valued mapping was introduced in [81]. Optimization problems with multivariate stochastic dominance constraints of the second order were first analyzed in [72], where also the form of the generator was established. The optimality conditions in the subgradient form given in Theorem 6.12 are new. The problem formulation based on the Lorenz function and the optimality conditions for that formulation are proved in [84]. The relation to measures of risk in Corollary 6.17 is new.

The comparison of stochastic processes has a long history. We mention [31, 236, 240, 262, 263, 267, 282] for some of the early publications; for an overview see [190, 255]. Relevant conditional orders on processes are discussed in [241, 280, 281].

Stochastic ordering constraints in dynamic stochastic optimization are discussed in [71, 85, 103, 104, 129]. The definition in [71] leads to comparisons and policies that are not time-consistent, and it is not included here. The proposal in [129] addresses the limiting distribution in an infinite-time horizon average-cost Markov decision problem. In [103, 104], time-consistent constraints are imposed in a manner akin to Average (Conditional) Value-at-Risk constraints for a multistage stochastic optimization problem. The representation of the Average Value-at-Risk as the expected value of a nonlinear function allows for the problem to be treated by the theory and methods of risk-neutral optimization. Two-stage stochastic programming problems with stochastic dominance constraint on the recourse function were introduced in [122, 123].

The consistency of preferences in the stochastic comparisons of sequences was first raised in [85]. The general definition of time-consistent comparison of sequences presented here is new and so are the optimality conditions and the analysis of the multistage problem in this chapter.

Chapter 7

The reformulation in Theorem 7.2 was proposed in [173]. The statement in Lemma 7.1 is due to [32]. The shortfall event method for finite probability space was proposed in [239]; the convergence for general probability spaces is new. The inverse event methods based on the Lorenz function, which are presented in §7.2.2, were proposed in [73], where also their convergence was proved. The dual method was proposed for the first time to address problems with multivariate dominance constraints in [80], together with the method for verification of the multivariate stochastic dominance using DC optimization techniques of [268]. The augmented Lagrangian method was proposed for solving two-stage optimization problems with multivariate stochastic dominance constraints in [84]. The numerical solution of two-stage problems with multivariate stochastic dominance constraints was first discussed in [81], where also a new combinatorial method for testing the multivariate dominance relation is proposed. Similar problem formulations are discussed in [197]. These types of problems and the numerical methods for their solution are not included in our presentation. The idea to use the exact penalty method for problems with stochastic dominance constraints was proposed in [265], where several penalty functions were formulated. Our presentation combines the ideas from [265] with the approximations discussed before. A version of the exact penalty method with additional smoothing was developed for sample-based problems with second-order stochastic dominance constraints in [266]. The exact penalty method was also proposed for problems with multivariate second-order stochastic dominance constraints on a finite probability space in [184]. Next to the dual method for problems with multivariate dominance constraints proposed in [80], another method was proposed in [135]. The methods for two-stage problems that are included in our exposition are developed in [61]. For the background in the risk-neutral setting, see [247].

We do not include methods for solving problems with the first-order constraints. Such methods can be found in [90, 123, 173, 198].

Chapter 8

For general information about Markov Decision Processes, we refer the Reader to [15, 26, 132, 133].

In the extant literature, three basic approaches to risk aversion in MDPs have been employed.

In [144] lexicographic moment optimality is analyzed, with the use of the moment-generating function of the discounted reward (exponential utility). The general utility is discussed in [145]. Ref. [16] evaluates a certainty equivalent utility on the total discounted cost. In [146], the recursive nonlinear utility model of [153] is applied. In general, major theoretical and computational difficulties are associated with this approach, unless the utility functional is decomposable by stage.

Extensive efforts have been devoted to the risk-sensitive models, which use the dynamic entropic risk measure. This line of research was initiated in [137]. For various aspects of the theory, see [10, 17, 27, 50, 55, 87, 167, 183].

Mean–variance models were considered already in [108, 109, 261]; see the review [279]. Specialized methods were developed in [180].

Specific measures of risk, such as AVaR or the optimized certainty equivalent, allow for the reduction of risk-averse dynamic models to risk-neutral ones, with the addition of an additional control variable; such a line of research has been pursued in many works; see [14, 269].

Our approach uses Markov dynamic risk measures introduced in [248], and further developed and applied in [13, 40, 41, 107, 168, 178, 259]. An alternative approach, using occupation measures for finite state and control models was developed in [130]. For partially observable risk-averse models, see [28, 106].

Continuous-time Markov risk measures were discussed in [76, 193]. For application to investment problems, see [44].

References

1. C. Acerbi, D. Tasche, On the coherence of expected shortfall. J. Bank. Finance **26**, 1491–1507 (2002)
2. A. Ahmadi-Javid, Entropic value-at-risk: a new coherent risk measure. J. Optim. Theory Appl. **155**, 1105–1123 (2012)
3. A. Ahmadi-Javid, A. Pichler, An analytical study of norms and Banach spaces induced by the entropic value-at-risk. Math. Financ. Econ. **11**(4), 527–550 (2017)
4. S. Ahmed, Convexity and decomposition of mean-risk stochastic programs. Math. Program. **106**, 433–446 (2006)
5. A. Almen, D. Dentcheva, On risk evaluation and control of distributed multi-agent systems (2023). arXiv preprint arXiv:2311.10287
6. J. Appell, P.P. Zabrejko, *Nonlinear Superposition Operators* (Cambridge University Press, Cambridge, 1990)
7. B.C. Arnold, *Majorization and the Lorenz Order: A Brief Introduction*, vol. 43 (Springer Science & Business Media, Berlin, 2012)
8. P. Artzner, F. Delbaen, J.-M. Eber, D. Heath, Coherent measures of risk. Math. Finance **9**, 203–228 (1999)
9. P. Artzner, F. Delbaen, J.-M. Eber, D. Heath, H. Ku, Coherent multiperiod risk adjusted values and Bellman's principle. Ann. Oper. Res. **152**, 5–22 (2007)
10. H. Asienkiewicz, A. Jaśkiewicz, A note on a new class of recursive utilities in Markov decision processes. Appl. Math. **44**, 149–161 (2017)
11. J.-P. Aubin, A. Cellina, *Differential Inclusions: Set-Valued Maps and Viability Theory*, vol. 264 (Springer Science & Business Media, Berlin, 2012)
12. J.-P. Aubin, H. Frankowska, *Set-Valued Analysis* (Birkhäuser, Boston, 2009)

D. Dentcheva, A. Ruszczyński, *Risk-Averse Optimization and Control*, Springer Series in Operations Research and Financial Engineering,
https://doi.org/10.1007/978-3-031-57988-2

13. N. Bäuerle, A. Glauner, Markov decision processes with recursive risk measures. Eur. J. Oper. Res. **296**(3), 953–966 (2022)
14. N. Bäuerle, J. Ott, Markov decision processes with average-value-at-risk criteria. Math. Methods Oper. Res. **74**(3), 361–379 (2011)
15. N. Bäuerle, U. Rieder, *Markov Decision Processes with Applications to Finance* (Springer Science & Business Media, Berlin, 2011)
16. N. Bäuerle, U. Rieder, More risk-sensitive Markov decision processes. Math. Oper. Res. **39**(1), 105–120 (2013)
17. N. Bäuerle, U. Rieder, Partially observable risk-sensitive Markov decision processes. Math. Oper. Res. **42**(4), 1180–1196 (2017)
18. V.S. Bawa, Optimal rules for ordering uncertain prospects. J. Financ. Econ. **2**, 95–121 (1975)
19. V.Z. Belenkii, V.A. Volkonskii, S.A. Ivankov, A.B. Pomaranski, A.D. Shapiro, *Iterative Methods in Game Theory and Programming* (Nauka, Moscow, 1974)
20. F. Bellini, E.R. Gianin, On Haezendonck risk measures. J. Bank. Finance **32**(6), 986–994 (2008)
21. F. Bellini, B. Klar, A. Müller, E.R. Gianin, Generalized quantiles as risk measures. Insur. Math. Econ. **54**, 41–48 (2014)
22. A. Ben-Tal, M. Teboulle, Expected utility, penalty functions, and duality in stochastic nonlinear programming. Manag. Sci. **32**(11), 1445–1466 (1986)
23. A. Ben-Tal, M. Teboulle, An old-new concept of convex risk measures: the optimized certainty equivalent. Math. Finance **17**(3), 449–476 (2007)
24. M. Benaïm, J. Hofbauer, S. Sorin, Stochastic approximations and differential inclusions. SIAM J. Control Optim. **44**(1), 328–348 (2005)
25. M. Benaïm, J. Hofbauer, S. Sorin, Stochastic approximations and differential inclusions, part ii: applications. Math. Oper. Res. **31**(4), 673–695 (2006)
26. D. Bertsekas, *Dynamic Programming and Optimal Control: Volume I*, vol. 4 (Athena Scientific, Belmont, 2012)
27. T. Bielecki, D. Hernández-Hernández, S.R. Pliska, Risk sensitive control of finite state Markov chains in discrete time, with applications to portfolio management. Math. Methods Oper. Res. **50**(2), 167–188 (1999)
28. T.R. Bielecki, I. Cialenco, A. Ruszczyński, Risk filtering and risk-averse control of Markovian systems subject to model uncertainty. Math. Methods Oper. Res. **98**(2), 231–268 (2023)
29. J.R. Birge, Decomposition and partitioning methods for multistage stochastic linear programs. Oper. Res. **33**(5), 989–1007 (1985)
30. J.R. Birge, F.V. Louveaux, A multicut algorithm for two-stage stochastic linear programs. Eur. J. Oper. Res. **34**(3), 384–392 (1988)
31. D. Blackwell, Comparison of experiments, in *Proceedings of the Second Berkeley Symposium on Mathematical Statistics and Probability*, vol. 2 (University of California Press, Berkeley, 1951), pp. 93–103

32. D. Blackwell, Equivalent comparisons of experiments. Ann. Math. Stat. **24**, 265–272 (1953)
33. J. Bolte, T. Le, E. Pauwels, Subgradient sampling for nonsmooth nonconvex minimization. SIAM J. Optim. **33**(4), 2542–2569 (2023)
34. V.S. Borkar, *Stochastic Approximation: A Dynamical Systems Viewpoint* (Springer, New York, 2009)
35. D.S. Bridges, G.B. Mehta, *Representations of Preferences Orderings*, vol. 422. Lecture Notes in Economics and Mathematical Systems (Springer, Berlin, 1995)
36. C. Burgert, L. Rüschendorf, Consistent risk measures for portfolio vectors. Insur. Math. Econ. **38**(2), 289–297 (2006)
37. J.V. Burke, Calmness and exact penalization. SIAM J. Control Optim. **29**(2), 493–497 (1991)
38. J.C. Candeal, E. Induráin, G.B. Mehta, Utility functions on locally connected spaces. J. Math. Econ. **40**(6), 701–711 (2004)
39. C. Castaing, M. Valadier, *Convex Analysis and Measurable Multifunctions*. Lecture Notes in Mathematics, vol. 580 (Springer, Berlin, 1977)
40. O. Çavus, A. Ruszczyński, Computational methods for risk-averse undiscounted transient Markov models. Oper. Res. **62**(2), 401–417 (2014)
41. O. Çavus, A. Ruszczyński, Risk-averse control of undiscounted transient Markov models. SIAM J. Control Optim. **52**(6), 3935–3966 (2014)
42. Z.-L. Chen, W.B. Powell, Convergent cutting-plane and partial-sampling algorithm for multistage stochastic linear programs with recourse. J. Optim. Theory Appl. **102**, 497–524 (1999)
43. C. Chen, G. Iyengar, C.C. Moallemi, An axiomatic approach to systemic risk. Manag. Sci. **59**(6), 1373–1388 (2013)
44. Z. Chen, G. Li, Y. Zhao, Time-consistent investment policies in Markovian markets: a case of mean-variance analysis. J. Econ. Dynam. Control **40**, 293–316 (2014)
45. P. Cheridito, M. Kupper, Composition of time-consistent dynamic monetary risk measures in discrete time. Int. J. Theor. Appl. Finance **14**(1), 137–162 (2011)
46. P. Cheridito, T. Li, Dual characterization of properties of risk measures on Orlicz hearts. Math. Financ. Econ. **2**, 29–55 (2008)
47. P. Cheridito, T. Li, Risk measures on Orlicz hearts. Math. Finance: Int. J. Math. Stat. Financ. Econ. **19**(2), 189–214 (2009)
48. P. Cheridito, F. Delbaen, M. Kupper, Dynamic monetary risk measures for bounded discrete-time processes. Electron. J. Probab. **11**, 57–106 (2006)
49. G. Choquet, Theory of capacities. Ann. Inst. Fourier Grenoble **5**(1953–1954), 131–295 (1955)
50. K.J. Chung, M.J. Sobel, Discounted MDPs: distribution functions and exponential utility maximization. SIAM **25**, 49–62 (1987)
51. F.H. Clarke, Generalized gradients and applications. Trans. Am. Math. Soc. **205**, 247–262 (1975)

52. F.H. Clarke, Generalized gradients of Lipschitz functionals. Adv. Math. **40**(1), 52–67 (1981)
53. F.H. Clarke, *Optimization and Nonsmooth Analysis*. Canadian Mathematical Society Series of Monographs and Advanced Texts. A Wiley-Interscience Publication (Wiley, New York, 1983)
54. R.A. Collado, D. Papp, A. Ruszczyński, Scenario decomposition of risk-averse multistage stochastic programming problems. Ann. Oper. Res. **200**, 147–170 (2012)
55. S.P. Coraluppi, S.I. Marcus, Risk-sensitive and minimax control of discrete-time, finite-state Markov decision processes. Automatica **35**(2), 301–309 (1999)
56. C. Corduneanu, *Functional Equations with Causal Operators* (CRC Press, Boca Raton, 2005)
57. G. Debreu, Representation of a preference ordering by a numerical function, in *Decision Processes*, ed. by C.D.A. Thrall, B. Combs (Wiley, New York, 1954), pp. 159–165
58. G. Debreu, Continuity properties of Paretian utility. Int. Econ. Rev. **5**, 285–293 (1964)
59. P. Delbaen, Coherent risk measures on general probability spaces, in *Essays in Honour of Dieter Sondermann* (Springer, Berlin, 2002)
60. M.A.H. Dempster, *Risk Management: Value at Risk and Beyond* (Cambridge University Press, Cambridge, 2002)
61. D. Dentcheva, M.G. Martinez, Two-stage stochastic optimization problems with stochastic ordering constraints on the recourse. Eur. J. Oper. Res. **219**, 1–8 (2012)
62. D. Dentcheva, W. Römisch, Stability and sensitivity of stochastic dominance constrained optimization models. SIAM J. Optim. **23**(3), 1672–1688 (2013)
63. D. Dentcheva, A. Ruszczyński, Optimization with stochastic dominance constraints. SIAM J. Optim. **14**, 548–566 (2003)
64. D. Dentcheva, A. Ruszczyński, Convexification of stochastic ordering. C. R. Acad. Bulg. Sci. **57**, 11–16 (2004)
65. D. Dentcheva, A. Ruszczyński, Optimality and duality theory for stochastic optimization problems with nonlinear dominance constraints. Math. Program. **99**, 329–350 (2004)
66. D. Dentcheva, A. Ruszczyński, Semi-infinite probabilistic optimization: first-order stochastic dominance constrain. Optimization **53**(5–6), 583–601 (2004)
67. D. Dentcheva, A. Ruszczyński, Inverse stochastic dominance constraints and rank dependent expected utility theory. Math. Program. **108**, 297–311 (2006)
68. D. Dentcheva, A. Ruszczyński, Portfolio optimization under stochastic dominance constraints. J. Bank. Finance **30**, 433–451 (2006)

69. D. Dentcheva, A. Ruszczyński, Composite semi-infinite optimization. Control Cybern. **36**(3), 633–646 (2007)
70. D. Dentcheva, A. Ruszczyński, Duality between coherent risk measures and stochastic dominance constraints in risk-averse optimization. Pac. J. Optim. **4**(3), 433–446 (2008)
71. D. Dentcheva, A. Ruszczyński, Stochastic dynamic optimization with discounted stochastic dominance constraints. SIAM J. Control Optim. **47**(5), 2540–2556 (2008)
72. D. Dentcheva, A. Ruszczyński, Optimization with multivariate stochastic dominance constraints. Math. Program. **117**(1–2), 111–127 (2009)
73. D. Dentcheva, A. Ruszczyński, Inverse cutting plane methods for optimization problems with second-order stochastic dominance constraints. Optimization **59**(3), 323–338 (2010)
74. D. Dentcheva, A. Ruszczyński, Common mathematical foundations of expected utility and dual utility theories. SIAM J. Optim. **23**(1), 381–405 (2013)
75. D. Dentcheva, A. Ruszczyński, Risk preferences on the space of quantile functions. Math. Program. **148**(1–2), 181–200 (2014)
76. D. Dentcheva, A. Ruszczyński, Time-coherent risk measures for continuous-time Markov chains. SIAM J. Financ. Math. **9**(2), 690–715 (2018)
77. D. Dentcheva, A. Ruszczyński, Risk forms: representation, disintegration, and application to partially observable two-stage systems. Math. Program. **181**, 297–317 (2020)
78. D. Dentcheva, A. Ruszczyński, Subregular recourse in nonlinear multistage stochastic optimization. Math. Program. **189**(1–2), 249–270 (2021)
79. D. Dentcheva, A. Ruszczyński, Mini-batch risk forms. SIAM J. Optim. **33**(2), 615–637 (2023)
80. D. Dentcheva, E. Wolfhagen, Optimization with multivariate stochastic dominance constraints. SIAM J. Optim. **25**(1), 564–588 (2015)
81. D. Dentcheva, E. Wolfhagen, Two-stage optimization problems with multivariate stochastic order constraints. Math. Oper. Res. **41**(1), 1–22 (2016)
82. D. Dentcheva, R. Henrion, A. Ruszczyński, Stability and sensitivity of optimization problems with first order stochastic dominance constraints. SIAM J. Optim. **18**(1), 322–337 (2007)
83. D. Dentcheva, S. Penev, A. Ruszczyński, Kusuoka representation of higher order dual risk measures. Ann. Oper. Res. **181**, 325–335 (2010)
84. D. Dentcheva, G. Martinez, E. Wolfhagen, Augmented Lagrangian methods for solving optimization problems with stochastic-order constraints. Oper. Res. **64**(6), 1451–1465 (2016)
85. D. Dentcheva, M. Ye, Y. Yi, Risk-averse sequential decision problems with time-consistent stochastic dominance constraints, in *2022 IEEE 61st Conference on Decision and Control (CDC)* (IEEE, 2022), pp. 3605–3610

86. K. Detlefsen, G. Scandolo, Conditional and dynamic convex risk measures. Finance Stoch. **9**(4), 539–561 (2005)
87. G.B. Di Masi, L. Stettner, Risk-sensitive control of discrete-time Markov processes with infinite horizon. SIAM J. Control Optim. **38**(1), 61–78 (1999)
88. J. Dieudonné, Sur le théorème de Hahn-Banach. Revue Sci. (Rev. Rose Illus.) **79**, 642–643 (1941)
89. A.L. Dontchev, R.T. Rockafellar, *Implicit Functions and Solution Mappings* (Springer, Berlin, 2009)
90. D. Drapkin, R. Schultz, An algorithm for stochastic programs with first-order dominance constraints induced by linear recourse. Discrete Appl. Math. **158**(4), 291–297 (2010)
91. J.C. Duchi, F. Ruan, Stochastic methods for composite and weakly convex optimization problems. SIAM J. Optim. **28**(4), 3229–3259 (2018)
92. D. Duffie, J. Pan, An overview of value at risk. J. Deriv. **4**(3), 7–49 (1997)
93. N. Dunford, J.T. Schwartz, *Linear Operators. Part I* (Wiley, New York, 1958)
94. P. Dworczak, G. Martini, The simple economics of optimal persuasion. J. Polit. Econ. **127**(5), 1993–2048 (2019)
95. A. Eichhorn, W. Römisch, Polyhedral risk measures in stochastic programming. SIAM J. Optim. **16**, 69–95 (2005)
96. M. Eidelheit, Zur Theorie der konvexen Mengen in linearen normierten Räumen. Stud. Math. **6**, 104–111 (1936)
97. S. Eilenberg, Ordered topological spaces. Am. J. Math. **63**, 39–45 (1941)
98. M.J. Eisner, P. Olsen, Duality for stochastic programming interpreted as LP in L_p-space. SIAM J. Appl. Math. **28**(4), 779–792 (1975)
99. I. Ekeland, W. Schachermayer, Law-invariant risk measures on $L_\infty(\mathbb{R}^d)$). Stat. Risk Model. **28**(3), 195–225 (2011)
100. Y.M. Ermoliev, Methods for solving nonlinear extremal problems. Kibernetika **2**(4), 1–17 (1966)
101. Y.M. Ermoliev, Z.V. Nekrylova, Some methods of stochastic optimization. Kibernetika **2**, 77–78 (1966)
102. Y.M. Ermoliev, Z.V. Nekrylova, The stochastic gradient method and its applications. Teoriya Optim. Reshenii (1), 24–47 (1967)
103. L.F. Escudero, M.A. Garín, M. Merino, G. Pérez, On time-consistent stochastic dominance induced by mixed integer-linear recourse in multistage stochastic programs. Eur. J. Oper. Res. **249**(1), 164–176 (2016)
104. L.F. Escudero, J.F. Monge, D.R. Morales, On the time-consistent stochastic dominance risk averse measure for tactical supply chain planning under uncertainty. Comput. Oper. Res. **100**, 270–286 (2018)
105. I.V. Evstigneev, Lagrange multipliers for the problems of stochastic programming, in *Warsaw Fall Seminars in Mathematical Economics 1975*, ed. by M. Łoś, J. Łoś, A. Wieczorek (Springer, Berlin, 1976), pp. 34–48

106. J. Fan, A. Ruszczyński, Risk measurement and risk-averse control of partially observable discrete-time Markov systems. Math. Methods Oper. Res. **88**, 161–184 (2018)
107. J. Fan, A. Ruszczyński, Process-based risk measures and risk-averse control of discrete-time systems. Math. Program. **191**, 113–140 (2022)
108. J.A. Filar, H.-M. Lee, Gain/variability tradeoffs in undiscounted Markov decision processes, in *1985 24th IEEE Conference on Decision and Control* (IEEE, 1985), pp. 1106–1112
109. J.A. Filar, L.C.M. Kallenberg, H.-M. Lee, Variance-penalized Markov decision processes. Math. Oper. Res. **14**(1), 147–161 (1989)
110. D. Filipović, G. Svindland, The canonical model space for law-invariant convex risk measures is l1. Math. Finance: Int. J. Math. Stat. Financ. Econ. **22**(3), 585–589 (2012)
111. D. Filipovic, M. Kupper, N. Vogelpoth, Approaches to conditional risk. SIAM J. Financ. Math. **3**(1), 402–432 (2012)
112. P.C. Fishburn, *Decision and Value Theory*, vol. 43. Operations Research Society of America. Publications in Operations Research (Wiley, New York, 1964)
113. P.C. Fishburn, *Utility Theory for Decision Making*. Publications in Operations Research, No. 18 (Wiley, New York, 1970)
114. P.C. Fishburn, Continua of stochastic dominance relations for bounded probability distributions. J. Math. Econ. **3**, 295–311 (1976)
115. P.C. Fishburn, Stochastic dominance and moments of distributions. Math. Oper. Res. **5**, 94–100 (1980)
116. P.C. Fishburn, *The Foundations of Expected Utility*, vol. 31. Theory and Decision Library (D. Reidel Publishing Co., Dordrecht, 1982)
117. P.C. Fishburn, *Nonlinear Preference and Utility Theory*, vol. 5. Johns Hopkins Series in the Mathematical Sciences (Johns Hopkins University Press, Baltimore, 1988)
118. H. Föllmer, I. Penner, Convex risk measures and the dynamics of their penalty functions. Stat. Decis. **24**(1), 61–96 (2006)
119. H. Föllmer, A. Schied, Convex measures of risk and trading constraints. Finance Stoch. **6**, 429–447 (2002)
120. S. Ghadimi, G. Lan, Stochastic first- and zeroth-order methods for nonconvex stochastic programming. SIAM J. Optim. **23**(4), 2341–2368 (2013)
121. I. Gilboa, D. Schmeidler, Maxmin expected utility with nonunique prior. J. Math. Econ. **18**(2), 141–153 (1989)
122. R. Gollmer, U. Gotzes, F. Neise, R. Schultz, Risk modeling via stochastic dominance in power systems with dispersed generation. Manuscript, Department of Mathematics, University of Duisburg-Essen, Duisburg (2007)
123. R. Gollmer, F. Neise, R. Schultz, Stochastic programs with first-order dominance constraints induced by mixed-integer linear recourse. SIAM J. Optim. **19**(2), 552–571 (2008)

124. V. Guigues, SDDP for some interstage dependent risk-averse problems and application to hydro-thermal planning. Comput. Optim. Appl. **57**, 167–203 (2014)
125. V. Guigues, W. Römisch, SDDP for multistage stochastic linear programs based on spectral risk measures. Oper. Res. Lett. **40**(5), 313–318 (2012)
126. M. Gürbüzbalaban, A. Ruszczyński, L. Zhu, A stochastic subgradient method for distributionally robust non-convex and non-smooth learning. J. Optim. Theory Appl. **194**(3), 1014–1041 (2022)
127. J. Hadar, W.R. Russell, Rules for ordering uncertain prospects. Am. Econ. Rev. **59**(1), 25–34 (1969)
128. G. Hanoch, H. Levy, The efficiency analysis of choices involving risk. Rev. Econ. Stud. **36**(3), 335–346 (1969)
129. W.B. Haskell, R. Jain, Stochastic dominance-constrained Markov decision processes. SIAM J. Control Optim. **51**(1), 273–303 (2013)
130. W.B. Haskell, R. Jain, A convex analytic approach to risk-aware Markov decision processes. SIAM J. Control Optim. **53**(3), 1569–1598 (2015)
131. H. Heitsch, W. Römisch, Scenario tree modeling for multistage stochastic programs. Math. Program. **118**(2), 371–406 (2009)
132. O. Hernández-Lerma, J.B. Lasserre, *Discrete-Time Markov Control Processes: Basic Optimality Criteria*, vol. 30 (Springer Science & Business Media, Berlin, 2012)
133. O. Hernández-Lerma, J.B. Lasserre, *Further Topics on Discrete-Time Markov Control Processes*, vol. 42 (Springer Science & Business Media, Berlin, 2012)
134. I.N. Herstein, J. Milnor, An axiomatic approach to measurable utility. Econometrica **21**, 291–297 (1953)
135. T. Homem-de Mello, S. Mehrotra, A cutting-surface method for uncertain linear programs with polyhedral stochastic dominance constraints. SIAM J. Optim. **20**(3), 1250–1273 (2010)
136. T. Homem-de Mello, B.K. Pagnoncelli, Risk aversion in multistage stochastic programming: a modeling and algorithmic perspective. Eur. J. Oper. Res. **249**(1), 188–199 (2016)
137. R.A. Howard, J.E. Matheson, Risk-sensitive Markov decision processes. Manag. Sci. **18**(7), 356–369 (1972)
138. K. Høyland, S.W. Wallace, Generating scenario trees for multistage decision problems. Manag. Sci. **47**(2), 295–307 (2001)
139. J. Hu, T. Homem-de Mello, S. Mehrotra, Sample average approximation of stochastic dominance constrained programs. Math. Program. **133**(1–2), 171–201 (2012)
140. A.D. Ioffe, Necessary and sufficient conditions for a local minimum. 1: a reduction theorem and first order conditions. SIAM J. Control Optim. **17**(2), 245–250 (1979)

141. A.D. Ioffe, *Variational Analysis of Regular Mappings* (Springer, Berlin, 2017)
142. A.D. Ioffe, V.L. Levin, Subdifferentials of convex functions. Trudy Moskovskogo Matematicheskogo Obshchestva **26**, 3–73 (1972)
143. J.-Y. Jaffray, Existence of a continuous utility function: an elementary proof. Econometrica **43**(5–6), 981–983 (1975)
144. S.C. Jaquette, Markov decision processes with a new optimality criterion: discrete time. Ann. Stat. **1**(3), 496–505 (1973)
145. S.C. Jaquette, A utility criterion for Markov decision processes. Manag. Sci. **23**(1), 43–49 (1976)
146. A. Jaśkiewicz, J. Matkowski, A.S. Nowak, Persistently optimal policies in stochastic dynamic programming with generalized discounting. Math. Oper. Res. **38**(1), 108–121 (2013)
147. L.V. Kantorovich, G.P. Akilov, *Functional Analysis*, 3rd edn. (Nauka, Moscow, 1984)
148. M. Kijima, M. Ohnishi, Mean-risk analysis of risk aversion and wealth effects on optimal portfolios with multiple investment opportunities. Ann. Oper. Res. **45**, 147–163 (1993)
149. D. Klatte, B. Kummer, Constrained minima and Lipschitzian penalties in metric spaces. SIAM J. Optim. **13**(2), 619–633 (2002)
150. D. Klatte, B. Kummer, *Nonsmooth Equations in Optimization: Regularity, Calculus, Methods and Applications*, vol. 60 (Springer Science & Business Media, Berlin, 2006)
151. V.L. Klee, Convex sets in linear spaces. Duke Math. J. **18**, 443–466, 875–883 (1951)
152. H. Konno, H. Yamazaki, Mean-absolute deviation portfolio optimization model and its application to Tokyo stock market. Manag. Sci. **37**, 519–531 (1991)
153. T.C. Koopmans, Stationary ordinal utility and impatience. Econometrica **28**, 287–309 (1960)
154. V. Kozmík, D.P. Morton, Evaluating policies in risk-averse multistage stochastic programming. Math. Program. **152**, 275–300 (2015)
155. M. Krein, D. Milman, On extreme points of regular convex sets. Stud. Math. (IM PAN) **9**, 133–138 (1940)
156. D.M. Kreps, E.L. Porteus, Temporal resolution of uncertainty and dynamic choice theory. Econometrica **46**(1), 185–200 (1978)
157. P.A. Krokhmal, Higher moment coherent risk measures. Quant. Finance **7**(4), 373–387 (2007)
158. E. Kromer, L. Overbeck, K. Zilch, Systemic risk measures on general measurable spaces. Math. Methods Oper. Res. **84**, 323–357 (2016)
159. A.Y. Kruger, D.R. Luke, N.H. Thao, Set regularities and feasibility problems. Math. Program. **168**(1–2), 279–311 (2018)

160. H.J. Kushner, G. Yin, *Stochastic Approximation and Recursive Algorithm and Applications*, vol. 35. Application of Mathematics (Springer, Berlin, 2003)
161. S. Kusuoka, On law-invariant coherent risk measures, in *Advances in Mathematical Economics*, vol. 3, ed. by S. Kusuoka, T. Maruyama (Springer, Tokyo, 2001), pp. 83–95
162. G. Lebourg, Generic differentiability of Lipschitzian functions. Trans. Am. Math. Soc. **256**, 125–144 (1979)
163. M. Ledoux, M. Talagrand, *Probability in Banach Spaces*, vol. 23. Ergebnisse der Mathematik und ihrer Grenzgebiete (3) [Results in Mathematics and Related Areas (3)] (Springer, Berlin, 1991). Isoperimetry and processes
164. E. Lehmann, Testing multiparameter hypotheses. Ann. Math. Stat. **23**, 541–552 (1952)
165. J. Leitner, A short note on second-order stochastic dominance preserving coherent risk measures. Math. Finance **15**, 649–651 (2005)
166. V.L. Levin, *Convex Analysis in Spaces of Measurable Functions and Its Applications in Economics* (Nauka, Moskva, 1985) (Russian)
167. S. Levitt, A. Ben-Israel, On modeling risk in Markov decision processes, in *Optimization and Related Topics (Ballarat/Melbourne, 1999)*, vol. 47. Appl. Optim. (Kluwer Acad. Publ., Dordrecht, 2001), pp. 27–40
168. K. Lin, S.I. Marcus, Dynamic programming with non-convex risk-sensitive measures, in *American Control Conference (ACC), 2013* (IEEE, 2013), pp. 6778–6783
169. Y. Liu, H. Xu, Stability analysis of stochastic programs with second order dominance constraints. Math. Program. **142**, 435–460 (2013)
170. L. Ljung, Analysis of recursive stochastic algorithms. IEEE Trans. Automat. Control **22**(4), 551–575 (1977)
171. M.O. Lorenz, Methods of measuring the concentration of wealth. Publ. Am. Stat. Assoc. **9**(70), 209–219 (1905)
172. O. Lorenz, Methods of measuring concentration of wealth. J. Am. Stat. Assoc. **9**, 209–219 (1905)
173. J. Luedtke, New formulations for optimization under stochastic dominance constraints. SIAM J. Optim. **19**(3), 1433–1450 (2008)
174. A. Mądry, A. Makelov, L. Schmidt, D. Tsipras, A. Vladu, Towards deep learning models resistant to adversarial attacks. Stat **1050**, 9 (2017)
175. A. Mafusalov, S. Uryasev, Buffered probability of exceedance: mathematical properties and optimization. SIAM J. Optim. **28**(2), 1077–1103 (2018)
176. V. Mai, M. Johansson, Convergence of a stochastic gradient method with momentum for non-smooth non-convex optimization, in *International Conference on Machine Learning* (PMLR, 2020), pp. 6630–6639

177. S. Majewski, B. Miasojedow, E. Moulines, Analysis of nonsmooth stochastic approximation: the differential inclusion approach (2018). arXiv preprint. arXiv:1805.01916
178. M. Malekipirbazari, Ö. Çavuş, Risk-averse allocation indices for multi-armed bandit problem. IEEE Trans. Automat. Control **66**(11), 5522–5529 (2021)
179. H. Mann, D. Whitney, On a test of whether one of two random variables is stochastically larger than the other. Ann. Math. Stat. **18**, 50–60 (1947)
180. S. Mannor, J.N. Tsitsiklis, Algorithmic aspects of mean-variance optimization in Markov decision processes. Eur. J. Oper. Res. **231**(3), 645–653 (2013)
181. R. Mansini, W. Ogryczak, M.G. Speranza, LP solvable models for portfolio optimization: a classification and computational comparison. IMA J. Manag. Math. **14**(3), 187–220 (2003)
182. R. Mansini, W. Ogryczak, M.G. Speranza, Conditional value at risk and related linear programming models for portfolio optimization. Ann. Oper. Res. **152**, 227–256 (2007)
183. S.I. Marcus, E. Fernández-Gaucherand, D. Hernández-Hernández, S. Coraluppi, P. Fard, Risk sensitive Markov decision processes, in *Systems and Control in the Twenty-First Century (St. Louis, MO, 1996)*, vol. 22. Progr. Systems Control Theory (Birkhäuser, Boston, 1997), pp. 263–279
184. R. Meskarian, J. Fliege, H. Xu, Stochastic programming with multivariate second order stochastic dominance constraints with applications in portfolio optimization. Appl. Math. Optim. **70**, 111–140 (2014)
185. V.S. Mikhalevich, A.M. Gupal, V.I. Norkin, *Nonconvex Optimization Methods* (Nauka, Moscow, 1987)
186. N. Miller, A. Ruszczyński, Risk-adjusted probability measures in portfolio optimization with coherent measures of risk. Eur. J. Oper. Res. **191**, 193–206 (2008)
187. N. Miller, A. Ruszczyński, Risk-averse two-stage stochastic linear programming: modeling and decomposition. Oper. Res. **59**(1), 125–132 (2011)
188. L. Mirsky, On a convex set of matrices. Arch. Math. **10**, 88–92 (1959)
189. B.S. Mordukhovich, Complete characterization of openness, metric regularity, and Lipschitzian properties of multifunctions. Trans. Am. Math. Soc. **340**(1), 1–35 (1993)
190. A. Müller, D. Stoyan, *Comparison Methods for Stochastic Models and Risks* (Wiley, Chichester, 2002)
191. J.M. Mulvey, A. Ruszczyński, A new scenario decomposition method for large-scale stochastic optimization. Oper. Res. **43**(3), 477–490 (1995)
192. I. Namioka, *Partially Ordered Linear Topological Spaces*, number 24 (American Mathematical Society, Providence, 1957)

193. M. Nendel, Markov chains under nonlinear expectation. Math. Finance **31**(1), 474–507 (2021)
194. M.-C. Noël, Y. Smeers, Nested decomposition of multistage nonlinear programs with recourse. Math. Program. **37**, 131–152 (1987)
195. V.I. Norkin, Generalized-differentiable functions. Cybern. Syst. Anal. **16**(1), 10–12 (1980)
196. N. Noyan, Risk-averse two-stage stochastic programming with an application to disaster management. Comput. Oper. Res. **39**(3), 541–559 (2012)
197. N. Noyan, G. Rudolf, Optimization with multivariate conditional value-at-risk constraints. Oper. Res. **61**(4), 990–1013 (2013)
198. N. Noyan, A. Ruszczyński, Valid inequalities and restrictions for stochastic programming problems with first order stochastic dominance constraints. Math. Program. **114**(2), 249–275 (2008)
199. G.L. O'Brien, Stochastic dominance and moment inequalities. Math. Oper. Res. **9**(3), 475–477 (1984)
200. W. Ogryczak, A. Ruszczyński, From stochastic dominance to mean–risk models: semideviations as risk measures. Eur. J. Oper. Res. **116**, 33–50 (1999) (IIASA report IR-97-027, 1997)
201. W. Ogryczak, A. Ruszczyński, On consistency of stochastic dominance and mean-semideviation models. Math. Program. **89**, 217–232 (2001)
202. W. Ogryczak, A. Ruszczyński, Dual stochastic dominance and related mean risk models. SIAM J. Optim. **13**, 60–78 (2002)
203. J.V. Outrata, W. Römisch, On optimality conditions for some nonsmooth optimization problems over L_p spaces. J. Optim. Theory Appl. **126**(2), 411–438 (2005)
204. K.R. Parthasarathy, *Probability Measures on Metric Spaces*, vol. 352 (American Mathematical Society, Providence, 2005)
205. T. Pennanen, Convex duality in stochastic optimization and mathematical finance. Math. Oper. Res. **36**(2), 340–362 (2011)
206. T. Pennanen, A.-P. Perkkiö, *Convex Stochastic Optimization* (Springer, New York, 2024)
207. M.V. Pereira, L.M. Pinto, Multi-stage stochastic optimization applied to energy planning. Math. Program. **52**, 359–375 (1991)
208. G. Pflug, Some remarks on the value-at-risk and the conditional value-at-risk, in *Probabilistic Constrained Optimization - Methodology and Applications* (Kluwer Academic Publishers, Dordrecht, 2000), pp. 272–281
209. G.C. Pflug, Version-independence and nested distributions in multistage stochastic optimization. SIAM J. Optim. **20**(3), 1406–1420 (2010)
210. G.C. Pflug, A. Pichler, A distance for multistage stochastic optimization models. SIAM J. Optim. **22**(1), 1–23 (2012)
211. G.C. Pflug, A. Pichler, Dynamic generation of scenario trees. Comput. Optim. Appl. **62**(3), 641–668 (2015)

212. G. Pflug, W. Römisch, *Modeling, Measuring and Managing Risk* (World Scientific, Singapore, 2007)
213. R.R. Phelps, *Convex Functions, Monotone Operators and Differentiability*, vol. 1364 (Springer, Berlin, 2009)
214. A. Philpott, V. de Matos, E. Finardi, On solving multistage stochastic programs with coherent risk measures. Oper. Res. **61**(4), 957–970 (2013)
215. A. Pichler, R. Schlotter, Entropy based risk measures. Eur. J. Oper. Res. **285**(1), 223–236 (2020)
216. B.T. Polyak, A general method for solving extremal problems. Dokl. Akad. Nauk **174**(1), 33–36 (1967)
217. B.T. Polyak, New stochastic approximation type procedures. Automat. i Telemekh **7**(98–107), 2 (1990)
218. B.T. Polyak, A.B. Juditsky, Acceleration of stochastic approximation by averaging. SIAM J. Control Optim. **30**(4), 838–855 (1992)
219. J. Quiggin, A theory of anticipated utility. J. Econ. Behav. Organ. **3**, 323–343 (1982)
220. J. Quiggin, *Generalized Expected Utility Theory: The Rank-Dependent Model* (Kluwer, Norwell, 1993)
221. J. Quiggin, P. Wakker, The axiomatic basis of anticipated utility: a clarification. J. Econ. Theory **64**(2), 486–499 (1994)
222. J. Quirk, R. Saposnik, Admissibility and measurable utility functions. Rev. Econ. Stud. **29**, 140–146 (1962)
223. T. Rader, The existence of a utility function to represent preferences. Rev. Econ. Stud. **30**, 229–232 (1963)
224. F. Riedel, Dynamic coherent risk measures. Stoch. Process. Appl. **112**, 185–200 (2004)
225. H. Robbins, S. Monro, A stochastic approximation method. Ann. Math. Stat. **22**, 400–407 (1951)
226. S.M. Robinson, Generalized equations and their solutions, Part I: basic theory, in *Point-to-Set Maps and Mathematical Programming* (Springer, Berlin, 1979), pp. 128–141
227. R.T. Rockafellar, Integrals which are convex functionals. Pac. J. Math. **24**(3), 525–539 (1968)
228. R.T. Rockafellar, Integrals which are convex functionals. II. Pac. J. Math. **39**(2), 439–469 (1971)
229. R.T. Rockafellar, *Conjugate Duality and Optimization* (SIAM, Philadelphia, 1974)
230. R.T. Rockafellar, S. Uryasev, Optimization of conditional value at risk. J. Risk **2**, 21–41 (2000)
231. R.T. Rockafellar, S. Uryasev, The fundamental risk quadrangle in risk management, optimization and statistical estimation. Surv. Oper. Res. Manag. Sci. **18**(1–2), 33–53 (2013)

232. R.T. Rockafellar, R.J.-B. Wets, Stochastic convex programming: basic duality. Pac. J. Math. **62**(1), 173–195 (1976)
233. R.T. Rockafellar, R.J.-B. Wets, Stochastic convex programming: relatively complete recourse and induced feasibility. SIAM J. Control Optim. **14**(3), 574–589 (1976)
234. R.T. Rockafellar, R.J.-B. Wets, Stochastic convex programming: singular multipliers and extended duality singular multipliers and duality. Pac. J. Math. **62**(2), 507–522 (1976)
235. R.T. Rockafellar, S. Uryasev, M. Zabarankin, Generalized deviations in risk analysis. Finance Stoch. **10**, 51–74 (2006)
236. T. Rolski, D. Stoyan, On the comparison of waiting times in gi/g/1 queues. Oper. Res. **24**(1), 197–200 (1976)
237. B. Roorda, J.M. Schumacher, J. Engwerda, Coherent acceptability measures in multiperiod models. Math. Finance: Int. J. Math. Stat. Financ. Econ. **15**(4), 589–612 (2005)
238. M. Rothschild, J.E. Stiglitz, Increasing risk: I. A definition, in *Uncertainty in Economics*, pp. 99–121 (Elsevier, Amsterdam, 1978)
239. G. Rudolf, A. Ruszczyński, Optimization problems with second order stochastic dominance constraints: duality, compact formulations, and cut generation methods. SIAM J. Optim. **19**(3), 1326–1343 (2008)
240. L. Rüschendorf, Asymptotic distributions of multivariate rank order statistics. Ann. Stat. **4**, 912–923 (1976)
241. L. Rüschendorf, On conditional stochastic ordering of distributions. Adv. Appl. Probab. **23**(1), 46–63 (1991)
242. L. Rüschendorf, *Mathematical Risk Analysis* (Springer, Berlin, 2013)
243. A. Ruszczyński, A regularized decomposition method for minimizing a sum of polyhedral functions. Math. Program. **35**, 309–333 (1986)
244. A. Ruszczyński, A linearization method for nonsmooth stochastic programming problems. Math. Oper. Res. **12**(1), 32–49 (1987)
245. A. Ruszczyński, Parallel decomposition of multistage stochastic programming problems. Math. Program. **58**, 201–228 (1993)
246. A. Ruszczyński, On convergence of an augmented Lagrangian decomposition method for sparse convex optimization. Math. Oper. Res. **20**(3), 634–656 (1995)
247. A. Ruszczyński, Decomposition methods, in *Stochastic Programming*, ed. by A. Ruszczyński, A. Shapiro. Handbooks in Operations Research and Management Science (Elsevier, Amsterdam, 2003), pp. 141–211
248. A. Ruszczyński, Risk-averse dynamic programming for Markov decision processes. Math. Program. **125**(2, Ser. B), 235–261 (2010)
249. A. Ruszczyński, Convergence of a stochastic subgradient method with averaging for nonsmooth nonconvex constrained optimization. Optim. Lett. **14**(7), 1615–1625 (2020)

250. A. Ruszczyński, A. Shapiro, Optimization of convex risk functions. Math. Oper. Res. **31**, 433–452 (2006)
251. A. Ruszczyński, R. Vanderbei, Frontiers of stochastically nondominated portfolios. Econometrica **71**, 1287–1297 (2003)
252. J.V. Ryff, Orbits of L_1-functions under doubly stochastic transformation. Trans. Am. Math. Soc. **117**, 92 (1965)
253. G. Scandolo, Risk Measures in a Dynamic Setting. PhD thesis, Università degli Studi di Milano, Milan (2003)
254. D. Schmeidler, Subjective probability and expected utility without additivity. Econometrica **57**(3), 571–587 (1989)
255. M. Shaked, J.G. Shanthikumar, *Stochastic Orders and Their Applications* (Academic Press, Boston, 1994)
256. A. Shapiro, Analysis of stochastic dual dynamic programming method. Eur. J. Oper. Res. **209**(1), 63–72 (2011)
257. A. Shapiro, W. Tekaya, J.P. da Costa, M.P. Soares, Risk neutral and risk averse stochastic dual dynamic programming method. Eur. J. Oper. Res. **224**(2), 375–391 (2013)
258. A. Shapiro, D. Dentcheva, A. Ruszczyński, *Lectures on Stochastic Programming: Modeling and Theory* (SIAM, Philadelphia, 2021)
259. Y. Shen, W. Stannat, K. Obermayer, Risk-sensitive Markov control processes. SIAM J. Control Optim. **51**(5), 3652–3672 (2013)
260. N.Z. Shor, An application of the method of gradient descent to the solution of the network transportation problem. Materialy Nauchnovo Seminara po Teoret. i Priklad. Voprosam Kibernetiki i Issled. Operacii, Nauchnyi Sov. po Kibernetike, Akad. Nauk Ukrain. SSSR **1**, 9–17 (1962)
261. M.J. Sobel, The variance of discounted Markov decision processes. J. Appl. Probab. **19**(4), 794–802 (1982)
262. H. Stoyan, Monotonieeigenschaften der Kundenwartezeiten im Modell GI/G/1. ZAMM - Zeitschrift Fur Angewandte Mathematik Und Mechanik **49**, 729–734 (1969)
263. D. Stoyan, *Comparison Methods for Queues and Other Stochastic Models* (Wiley, New York, 1983)
264. V. Strassen, The existence of probability measures with given marginals. Ann. Math. Stat. **38**, 423–439 (1965)
265. H. Sun, H. Xu, R. Meskarian, Y. Wang, Exact penalization, level function method, and modified cutting-plane method for stochastic programs with second order stochastic dominance constraints. SIAM J. Optim. **23**(1), 602–631 (2013)
266. H. Sun, H. Xu, Y. Wang, A smoothing penalized sample average approximation method for stochastic programs with second-order stochastic dominance constraints. Asia-Pac. J. Oper. Res. **30**(03), 1340002 (2013)
267. R. Szekli, On the concavity of the waiting-time distribution in some GI/G/1 queues. J. Appl. Probab. **23**, 555–561 (1986)

268. P.D. Tao, L.H. An, Convex analysis approach to DC programming: theory, algorithms and applications. Acta Math. Vietnamica **22**(1), 289–355 (1997)
269. K. Uğurlu, Controlled Markov decision processes with AVaR criteria for unbounded costs. J. Comput. Appl. Math. **319**, 24–37 (2017)
270. M. Valadier, Sous-différentiabilité de fonctions convexes à valeurs dans un espace vectoriel ordonné. Math. Scand. **30**(1), 65–74 (1972)
271. R.M. Van Slyke, R. Wets, L-shaped linear programs with applications to optimal control and stochastic programming. SIAM J. Appl. Math. **17**(4), 638–663 (1969)
272. V. Vapnik, *The Nature of Statistical Learning Theory* (Springer Science & Business Media, Berlin, 2013)
273. A. Vinel, P.A. Krokhmal, Certainty equivalent measures of risk. Ann. Oper. Res. **249**(1–2), 75–95 (2017)
274. V.A. Volkonskii, S.A. Ivankov, Convergence of iterative processes of finding the equilibrium point in the presence of random errors, and the estimation of its rate. Sib. Math. J. **11**(4), 582–591 (1970)
275. J. von Neumann, O. Morgenstern, *Theory of Games and Economic Behavior* (Princeton University Press, Princeton, 1944)
276. S.S. Wang, A class of distortion operators for pricing financial and insurance risks. J. Risk Insur. **67**, 15–36 (2000)
277. S. Weber, Distribution-invariant risk measures, information, and dynamic consistency. Math. Finance: Int. J. Math. Stat. Financ. Econ. **16**(2), 419–441 (2006)
278. R.J.-B. Wets, On the relation between stochastic and deterministic optimization, in *Control Theory, Numerical Methods and Computer Systems Modelling* (Springer, Berlin, 1975), pp. 350–361
279. D.J. White, Mean, variance, and probabilistic criteria in finite Markov decision processes: a review. J. Optim. Theory Appl. **56**(1), 1–29 (1988)
280. W. Whitt, Uniform conditional stochastic order. J. Appl. Probab. **17**(1), 112–123 (1980)
281. W. Whitt, Uniform conditional variability ordering of probability distributions. J. Appl. Probab. **22**(3), 619–633 (1985)
282. W. Whitt, D. Stoyan, D.J. Daley, Comparison methods for queues and other stochastic models. J. Am. Stat. Assoc. **81**, 266 (1986)
283. M.E. Yaari, The dual theory of choice under risk. Econometrica **55**, 95–115 (1987)
284. C. Žalinescu, *Convex Analysis in General Vector Spaces* (World Scientific, River Edge, 2002)
285. J. Zowe, A duality theorem for a convex programming problem in order complete vector lattices. J. Math. Anal. Appl. **50**(2), 273–287 (1975)

Index

D. Dentcheva, A. Ruszczyński, *Risk-Averse Optimization and Control*, Springer Series in Operations Research and Financial Engineering,
https://doi.org/10.1007/978-3-031-57988-2

Printed in the USA
CPSIA information can be obtained
at www.ICGtesting.com
CBHW060252170724
11581CB00014BA/209

9 783031 579875